AF900338

Alain Dufresne
Nanocellulose

Alain Dufresne

Nanocellulose

From Nature to High Performance Tailored Materials

DE GRUYTER

Author

Prof. Dr. Alain Dufresne
Grenople INP Pagora
461 Rue de la Papeterie
38402 Saint Martin d'Hères cedex
France

The book has 136 figures and 41 tables.

The cover image shows a transmission electron micrograph of cellulose nanocrystals extracted by sulfuric acid hydrolysis of ramie fibers.

ISBN 978-3-11-025456-3
e-ISBN 978-3-11-025460-0

Library of Congress Cataloging-in-Publication Data

A CIP catalog record for this book has been applied for at the Library of Congress.

Bibliographic information published by the Deutsche Nationalbibliothek

The Deutsche Nationalbibliothek lists this publication in the Deutsche Nationalbibliografie; detailed bibliographic data are available in the internet at http://dnb.d-nb.de.

© 2012 by Walter de Gruyter GmbH, Berlin/Boston.
The publisher, together with the authors and editors, has taken great pains to ensure that all information presented in this work (programs, applications, amounts, dosages, etc.) reflects the standard of knowledge at the time of publication. Despite careful manuscript preparation and proof correction, errors can nevertheless occur. Authors, editors and publisher disclaim all responsibility and for any errors or omissions or liability for the results obtained from use of the information, or parts thereof, contained in this work.

Typesetting: PTP-Berlin Protago-TEX-Production GmbH, Berlin
Printing: Hubert & Co., Göttingen

Printed in Germany.
∞ Printed on acid-free paper.
www.degruyter.com

To LiNoLeï

Preface

Cellulosic nanoparticles or nanocellulose provide a unique renewable building block on which materials with improved performance and new functionality can be prepared. Basically divided into two main families (cellulose nanocrystals and microfibrillated cellulose) depending on the way they are extracted from trees, plants, or other cellulose-containing species, they have the potential to play a major role in the 21st century in the development of advanced materials.

In all terrestrial and aquatic plant species, the primary cell wall is a dynamic structure and its constituting material must be synthesized in a form that is competent to undergo extension. In this system, the function of cellulose is to provide the mechanical stiffness. The specific elastic modulus of crystalline cellulose is potentially stronger than steel and similar to Kevlar as reported in Chapter 1. Natural cellulose-based materials have been used by our society as engineering materials for thousands of years and this continues today given the huge world-wide markets of industries in forest products, paper, textiles... However, in nature cellulose is generally closely associated with other materials and the release of cellulose nanoparticles involves a chemical purification step followed by high mechanical shearing treatment leading to microfibrillated cellulose or hydrolyzing treatment leading to cellulose nanocrystals. The preparation of these cellulosic nanoparticles is described in Chapters 2 and 3, respectively. Cellulose can also be synthesized in pure and highly crystalline microfibrillar form by bacteria. This specific form of cellulose, called bacterial cellulose, is discussed in Chapter 4. Cellulosic nanoparticles have high aspect ratio, low density, and a highly reactive surface as a result of the high density of hydroxyl groups borne by the cellulose molecule. This latter characteristic facilitates the surface chemical modification of nanocellulose and grafting of chemical species which is addressed in Chapter 5. This surface functionalization provides new functionality and allows tailoring of cellulose nanoparticle surface chemistry to facilitate self-assembly, regulated dispersion in a wide range of polymer matrices, and control of both the particle-particle and particle-matrix bond strength. Because of their high specific surface area and reactive surface, cellulose nanoparticle suspensions show particular rheological behavior. Besides, the rod-like morphology of hydrolyzed cellulose nanocrystals induces liquid crystalline behavior of the suspensions that can be tuned and captured in solid films. These aspects are detailed in Chapter 6. The function of cellulose in nature is to confer its mechanical properties to higher plant cells. This property can be potentially transferred to a polymeric matrix if a proper processing technique is used. The way of processing strongly impacts the morphology of the material and therefore the end-use properties of the obtained heterogeneous material. Chapter 7 reviews the different processing methods that have been described in the literature to prepare polymer nanocomposite materials reinforced with nanocellulose. The thermal, mechanical and barrier properties of ensuing nanocomposites are discussed in the following three chapters. Besides, cellulose is not the only polysaccharide that can be used to produce renewable nanoparticles. Starch and chitin are also suitable substrates to provide highly crystalline nanoscale particles with different morphologies and potential different properties as described in Chapter 11.

Nanocellulose has long been regarded as a laboratory curiosity. In recent years an incredible enthusiasm has arisen for these renewable nanoparticles and their extraordinary possibilities. The field has generated an exceptional appeal throughout the world and the literature has literally exploded. This made writing this book difficult but exciting. This book includes both general and advanced chapters and should be used for numerous applications, i.e. teaching, research and industrial applications. It has been intended to be pedagogic to allow even novices to discover the basic knowledge on nanocellulose.

I would like to deeply thank the Master and PhD students, as well as post-doc researchers who have worked with me during the last 15 years. Their intensive and dedicated work has contributed greatly, enriching the knowledge of this old rediscovered material and this book would not have been possible without their highly appreciated help. I do not name them but they will recognize themselves.

November, 2012 — Alain Dufresne

Contents

1 Cellulose and potential reinforcement

There are numerous examples where animals or plants synthesize extracellular high-performance skeletal biocomposites consisting of a matrix reinforced by fibrous biopolymers. Cellulose, the most abundant polymer on Earth, is a classic example of these reinforcing elements. It is a ubiquitous structural polymer that confers its mechanical properties to higher plant cells. Natural cellulose-based materials have been used by our society as engineering materials for thousands of years and this continues today given the worldwide huge markets and industries for forest products, paper, textiles ... A closer look to this material reveals a hierarchical structure design which is the source of its functionality, flexibility and high strength/weight performance.

1.1 Polysaccharides

Polysaccharides form part of the group of molecules known as carbohydrates and have been proposed as the first biopolymers to have formed on Earth (Tolstoguzov, 2004). This term was applied originally to compounds with the general formula $C_x(H_2O)_y$ but now it is also used to describe a variety of derivatives including nitrogen- and sulfur-containing compounds. Carbohydrates were once thought to represent "hydrated carbon". However, the arrangement of atoms in carbohydrates has little to do with water molecules. They are classified on the basis of their main monosaccharide components and the sequences and linkages between them, as well as the anomeric configuration of linkages, the ring size (furanose or pyranose), the absolute configuration (D- or L-) and any other substituents present.

Three common "single" sugars or monosaccharides, viz. glucose, galactose and fructose, share the same molecular formula $C_6H_{12}O_6$, and because of their six carbon atoms, each is a hexose (Figure 1.1). Although all three share the same molecular formula, the arrangement of atoms differs in each case and these substances, which have different structural formulas, are known as structural isomers.

Two monosaccharides can be linked together to form a "double" sugar or disaccharide. Three common disaccharides can be found: sucrose (common table sugar,

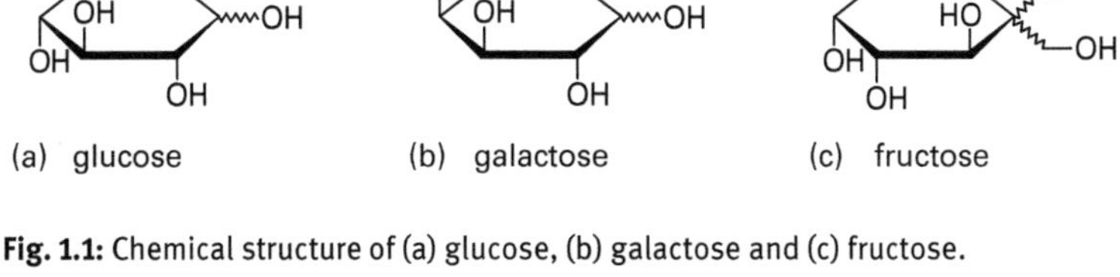

Fig. 1.1: Chemical structure of (a) glucose, (b) galactose and (c) fructose.

consisting of glucose + fructose), lactose (major sugar in milk, consisting of glucose + galactose) and maltose (product of starch digestion, consisting of two glucose units). Although the process of linking of two monomers is rather complex, the end result in each case is the loss of a hydrogen atom from one of the monosaccharides and a hydroxyl group from the other. The resulting linkage between the sugars is called a glycosidic bond. All sugars are highly soluble in water because of their many hydroxyl groups and although not as concentrated a fuel as fats, they are the most important source of energy for many cells. Further linkages of disaccharides lead to polysaccharides.

Certain structural characteristics, such as chain conformation and intermolecular associations, will influence the physico-chemical properties of polysaccharides. The most stable arrangement of atoms in a polysaccharide will be that which satisfies both the intra- and inter-molecular forces. Regular ordered polysaccharides are in general capable of assuming only a limited number of conformations due to severe steric restrictions on the freedom of rotation of sugar units about the interunit glycosidic bonds. There is also a clear correlation between allowed conformations and linkage structure. The structural non-starch polysaccharides, such as cellulose and xylan, have preferred orientations that automatically support extended conformations. Storage polysaccharides, such as the chains in amylopectin, tend to adopt wide helical conformations. The degree of stiffness and regularity of polysaccharide chains is likely to affect the rate and extent of their fermentation. Pentose sugars, such as arabinose and xylose, can adopt one of two specific conformations, furanose rings (often formed by arabinose) that can oscillate and are more flexible, and pyranose rings (usually formed by xylose and glucose) which are less flexible.

Carbohydrates, especially those containing large numbers of hydroxyl groups, are often thought of as being hydrophilic but they are also capable of generating apolar surfaces depending on the monomer ring conformation, the epimeric structure, and the stereochemistry of the glycosidic linkages. Apolarity has been demonstrated for dextrin, α-(1→4)-linked glucans, while dextran α-(1→6)-glucans, and cellulose, β-(1→4)-glucans, are much less hydrophobic (in solution) and unable to project an apolar surface. Hydrophobicity will also be affected by the degree of polysaccharide hydration, particularly the amount of intra-molecular hydrogen bonding. Hydrophobicity will affect their availability for fermentation in the gut, and their binding to bile acids.

Polysaccharides are more hydrophobic if they have a greater number of internal hydrogen bonds, and as their hydrophobicity increases there is less direct interaction with water. Carbohydrates contain hydroxyl (alcohol) groups that preferentially interact with two water molecules each if they are not in interaction with other hydroxyl groups on the molecule. Interaction with hydroxyl groups on the same or neighboring residues will necessarily reduce the polysaccharide's hydration status. β-linkages to the 3- and 4- positions in mannose or glucose homopolymers allow strong inflexible inter-residue hydrogen bonding, so reducing polymer hydration, and giving rise to rigid inflexible structural polysaccharides, whereas α-linkages to the 2-, 3- and 4-positions in mannose or glucose homopolymers give rise to greater aqueous hydration and more flexible linkages (Almond, 2005).

1.2 Chemical structure of the cellulose macromolecule

Though cellulose has been used for centuries in highly diverse applications, its chemical composition, structure and morphology remained very long ignored. Advances in the state of knowledge on the molecular structure of cellulose is intimately linked to the evolution of characterization techniques such as X-ray diffraction, electron microscopy, ^{13}C solid state nuclear magnetic resonance (NMR) spectroscopy, neutron scattering ...

The early work of Braconnot concerning the acid hydrolysis of the substance constituting plant cell walls goes back to the early XIX century (Braconnot, 1819). However, it was Anselme Payen who established that the fibrous component of all plant cells has a unique chemical structure (Payen, 1838) and first used the term "cellulose" in 1838. He discovered that when plant tissue, cotton linters, root tips, pit and ovules from the flowers of trees are purified with an acid-ammonia treatment, followed by an extraction in water, a constant fibrous material was formed. It required 75 more years for the basic cellulose formula to be established by Willstätter and Zechmeister (1913).

This fibrous, tough, and water-insoluble substance is found in the protective cell walls of plants, particularly in stalks, stems, trunks and all woody portions of plant tissues. More generally, lower (fresh-water and marine algae) and higher (bushes and trees) plants, bacteria, fungi, and animals (for example, tunicates-cellulose from the mantle of tunicates is named tunicin) as well as some amoebas are well-known natural sources of cellulose. Cellulose is often said to be the most abundant polymer on Earth. It is certainly one of the most important structural elements in plants and other living species serving to maintain their structure. Each of these living species, from tree to bacteria, produces cellulose day-by-day, e.g. a tree produces about 10 g of cellulose per day and the total production of cellulose all over the world is estimated to be $1.3 \cdot 10^{10}$ tons per year (Sandermann, 1973). Other sources indicate that the global annual production of cellulose is estimated at $1.5 \cdot 10^{12}$ tons (Klemm et al., 2005).

The paper and cardboard industries are the largest consumers of cellulose. Only 2% of the cellulose used (3.2 million tons in 2003) is used to produce fibers, films of regenerated cellulose and for the synthesis of cellulose esters or ethers. Technological development, especially in the field of molecular biology, in these areas offers new opportunities. Some animals, particularly ruminants and termites, can digest cellulose with the help of symbiotic microorganisms. Cellulose is not digestible by humans, and is often referred to as "dietary fiber" or "roughage". To date, several reviews have been published on cellulose research, structure and applications (Gardner and

Blackwell, 1974; Preston, 1975; Sarko, 1987; Okamura, 1991; Hon, 1994; O'Sullivan, 1997; Zugenmaier, 2001; Kovalenko, 2010).

Cellulose structure has been under investigation since the early days of polymer science. It was shown (Irvine and Hirst, 1923; Freudenberg and Braun, 1928) that 2,3,6 trimethyl glucose was the sole quantitative product resulting from methylation and hydrolysis of cellulose. This work evidenced that in cellulose the carbon atoms 2, 3 and 6 carried free hydroxyls available for reaction. The basic chemical structure of cellulose is presented in Figure 1.2. It is composed of β-1,4-linked D-glucopyranose rings. The adjacent monomer units are arranged so that glucosidic oxygens point in opposite directions and the repeating unit of the polymer chain of cellulose is composed of two β-D-glucopyranose rings rotated with respect to each other to form a so-called cellobiose unit (Figure 1.2). The numbering system for carbon atoms in anhydroglucose unit of cellulose is also indicated in Figure 1.2. The C—O—C bond angle between two β-D-glucopyranose rings is ~ 116° (Tarchevsky and Marchenko, 1991). Nowadays, the conformation of the glucopyranose ring is well established because numerous crystallographic investigations of D-glucose and cellobiose (Chu and Jeffrey, 1968) and other physico-chemical studies (Marszalek et al., 1998) provided evidence that the ring adopts a chair conformation designated $^{4}C1$ (in cellulose esters or ethers, the ring retains this conformation (Kovalenko, 2010)). As a result of its equatorial-equatorial glycosidic linkage, the cellulose chains have their units positioned so that their adjacent rings can form hydrogen bonds between the ring oxygen atom of one glycosil unit and the hydrogen atom of the C-3 hydroxyl group of the preceding ring. These hydrogen bonds hinder the free rotation of the rings along their linking glycoside bonds resulting in the stiffening of the chain.

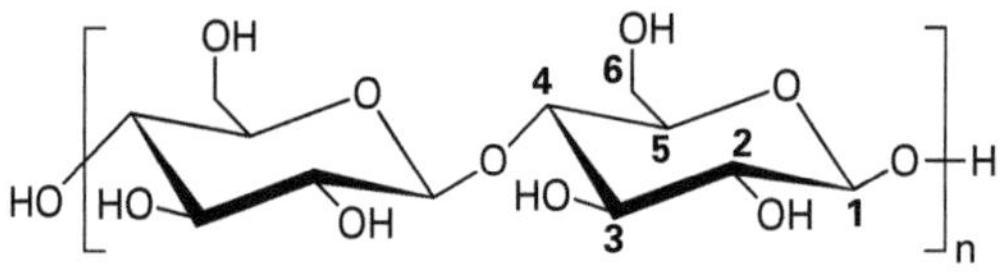

Fig. 1.2: Basic chemical structure of cellulose and numbering system for carbon atoms in anhydroglucose unit of cellulose.

These cellobiose units are covalently linked to form an extended, insoluble, straight chain of linear homopolymer consisting of between 2,000–27,000 residues. The degree of polymerization (DP) of native cellulose depends on the source and cellulose chains are supposed to consist of approximately 10,000 glucopyranose units in wood cellulose and 15,000 in native cotton cellulose (Sjoström, 1981). There is some evidence for a lower DP in primary cell walls compared with secondary cell walls. *Valonia* presents a DP around 26,500. Given a glucose unit as 0.515 nm (5.15 Å) long, for DP values ranging from 2,000 to 27,000, the cellulose molecules may have average lengths ranging between 1 and 14 μm by considering stretched chains.

However, chain lengths of such large, insoluble molecules are rather difficult to measure and the DP of native cellulose is not well established. The combination of procedures required to isolate, purify and solubilize cellulose generally causes enzymatic and mechanical degradation during analysis resulting in chains scission. The values of DP obtained are therefore minimal and depend on the method used to determine it. For the same reasons the distribution of chain lengths of cellulose is not well established. Nonetheless, some authors suggest that the molecular mass distribution should be homogeneous for a given source of cellulose (Marx-Figini, 1964).

One of the most specific characteristic of cellulose is that each of its monomers bears three hydroxyl groups. These hydroxyl groups and their hydrogen bonding ability play a major role in directing crystalline packing and in governing important physical properties of these highly cohesive materials. The two chain ends are chemically different. One end has a D-glucopyranose unit in which the anomeric carbon atom is involved in a glycosidic linkage and has a free secondary alcohol function on the C4. The other end has a D-glucopyranose unit in which the anomeric carbon atom is free. This cyclic hemiacetal function is in an equilibrium in which a small proportion is an aldehyde which gives rise to reducing properties at this end of the chain so that the cellulose chain has a chemical polarity. This end is called reductive because it has the ability to reduce Cu^{2+} ions into Cu^{+} ions in a Fehling's solution. This gives native cellulose a certain chemical polarity. Determination of the relative orientation of cellulose chains in the three-dimensional structure has been and remains one of the major problems in the study of cellulose.

1.3 Biosynthesis of cellulose

The biosynthesis of cellulose is a very complex phenomenon that reflects two linked processes. The first one is the formation of β-1,4-D-glucopyranose chains by the polymerization of glucose, and the other one is the organization of fibrillar supramolecular architecture which leads to the formation of elongated crystalline structures. The latter, called microfibrils, correspond to a collection of highly oriented cellulose chains. In vascular plants, cellulose is the constituent that ensures the protection and support of plant cells, and it is directly synthesized in the cell wall at the plasma membrane. The polymerization of glucose is provided by an enzyme system whose main family is named *cellulose synthase* (CS). This enzyme family cannot function without the presence of another class of enzymes called SUSY (*sucrose synthase*). The SUSY ensures a continuous supply of UDP-glucose (uridine diphosphate-glucose) required for the operation.

In the presence of CS, the UDP-glucose unit initiates the polymerization process by loss of the UDP unit and dimerization of glucose. Many studies have been undertaken in order to create *in vitro* polymerization of glucose. The biosynthesis of cellulose using bacteria, including *Acetobacter xylinium* was conducted in 1985 and led to the synthesis of fibrils *in vitro* by CS solubilized in the presence of *Acetobacter xylinium* (Lin et al., 1985). Within the plant cell, the model of Delmer and Amor proposed in 1995 (Delmer and Amor, 1995) represents the protein complex through the cytoplasmic membrane (Figure 1.3). This model describes the growth of glycosidic chains and the catalytic role of the main enzymes.

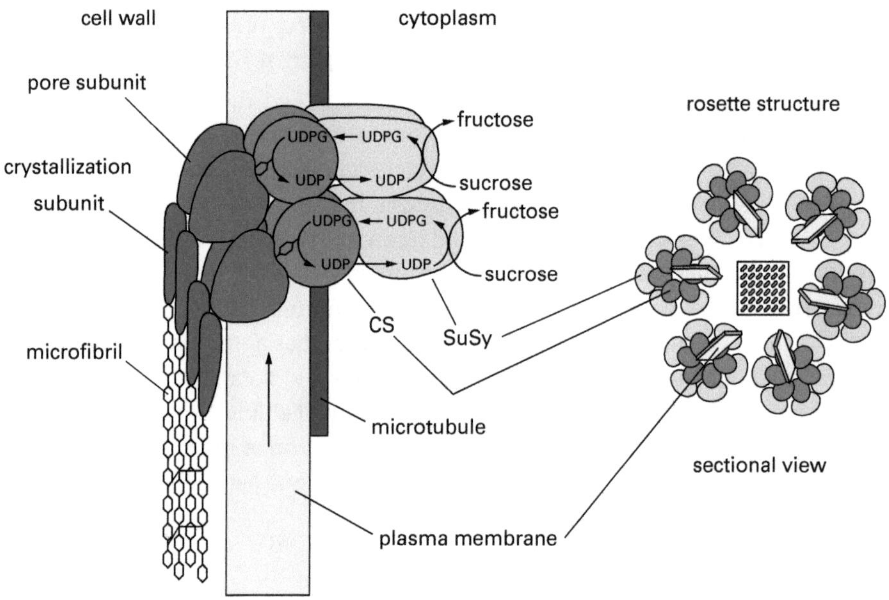

Fig. 1.3: Enzymatic system of polymerization of glucose across the plasma membrane: hypothetical model of a cellulose synthase complex in the plasma membrane (adapted from Delmer and Amor, 1995).

SUSY hydrolyzes sucrose to fructose by creating a UDP-glucose unit and it is within the CS that polymerization occurs. This observation shows that microfibril bundles grow in a single plane of symmetry. A sectional view shows that these planes are organized as rosettes (Saxena and Brown, 2000a). This simplified view does not report the entire phenomenon. The CS is not a single enzyme but contains several that are classified into major families of cellulose synthase (Saxena and Brown, 2000b). Indeed, there is probably no biochemical reaction in plants that is both so important and so poorly understood at the molecular level.

At the biopolymer chains supramolecular architecture level, electron microscopy studies conducted in 1976 showed that cellulose as a biopolymer displays an arrangement known as microfibrils (Frey-Wyssling, 1976). Glucose chains aggregate together by hydrogen bonds to form the metastable form of cellulose usually called cellulose I (see Section 1.4) (Cousins and Brown, 1995). The process of formation of the microfibrils may be provided by four steps from the polymerization to the supramolecular arrangement of the chains. The first step consists in the enzymatic polymerization of glucose monomers (Figure 1.4, step 0). The chains are subsequently linked by van der Waals forces to form micro-sheets (Figure 1.4, step 1). The micro-sheets join together by hydrogen bonds to form microcrystals (Figure 1.4, step 2). Finally, several microcrystals combine to give the microfibrils (Figure 1.4, step 3) (Cousins and Brown, 1995).

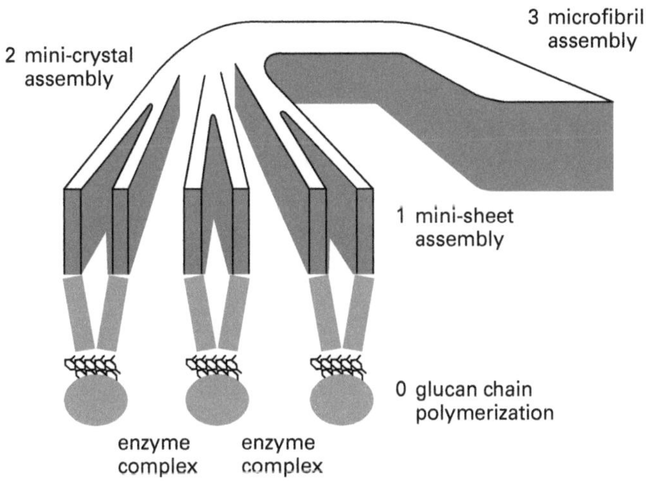

Fig. 1.4: A proposed model for the stages of microfibril formation: (0) glucose monomers are polymerized enzymatically from catalytic sites in the enzyme complex subunits to form glucan chains, (1) the glucan chains associate via van der Waals forces to form mini-sheets, (2) mini-sheets associate and hydrogen bond to form mini- or microcrystals, (3) several microcrystals then associate to form a crystalline microfibril (Cousins and Brown, 1995).

The polymerization process is achieved by successive addition of two glucose units, which allows to say that the polymerization unit is not glucose but cellobiose. At the supramolecular level, the formation of microfibrils explains the parallel growth of polymeric chains. This metastable architecture is the one found in the majority of lignocellulosic plants, called cellulose I. However, during *in vitro* synthesis, the cellulose chains arrange themselves in the crystalline form of cellulose II thermodynamically more stable (see Section 1.4). The designation cellulose I is related to the fact that the crystalline structure of cellulose is not unique, which leads to raise the issue of polymorphism of cellulose.

1.4 Polymorphism of cellulose

The ribbon-like character observed for cellulosic macromolecules allows adjacent cellulose chains to fit closely together in an ordered crystalline region. The free hydroxyl groups present in the cellulose macromolecule are likely to be involved in a number of *intra* and *inter* molecular hydrogen bonds, which may give rise to various ordered crystalline arrangements. In the cellulose biosynthesis, the polymerization of glucopyranose residues and the formation of crystalline domains are interrelated.

Cellulose has several polymorphs. The polymorphism is most typical of crystals of organic compounds whose molecules contain groups capable of hydrogen bonding (Bernstein, 2002). The repeating unit of cellulose, or cellobiose, includes six hydroxyl groups and three oxygen atoms. Therefore, many possibilities of various hydrogen bonding systems result from the presence of six hydrogen bond donors and nine hydrogen bond acceptors. Because of different mutual arrangements of the glucopyranose rings and possibility of conformational changes of the hydroxymethyl groups, cellulose chains can exhibit different crystal packings (Kovalenko, 2010). As a result, cellulose exists in several crystal modifications, differing in unit cell dimensions and, possibly, in chain polarity. The possible transitions between the different cellulose polymorphs are presented schematically in Figure 1.5.

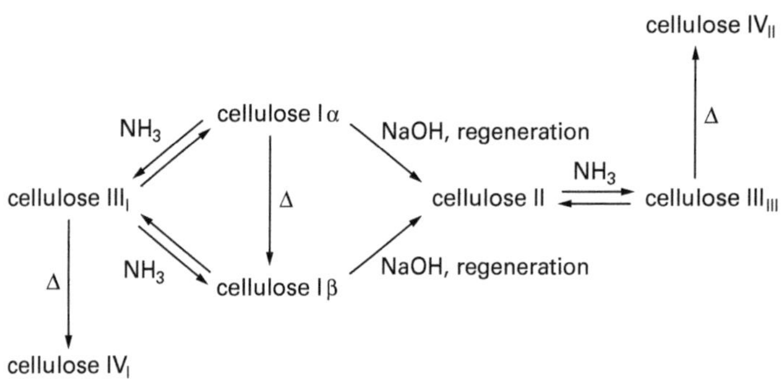

Fig. 1.5: Interconversions of cellulose polymorphs.

1.4.1 Cellulose I

In nature, cellulose is found in the crystalline form of cellulose I (native cellulose). As a first approximation, the structure of native cellulose determined by X-ray diffraction was described as a monoclinic cell containing two cellulose chains (Gardner and Blackwell, 1974). The first assignment of the peaks obtained by ^{13}C cross polarized magic angle spinning (CP/MAS) NMR was performed by Earl and VanderHart (1981). Atalla and VanderHart (1984) have shown by NMR spectroscopic studies that native cellulose is composed of more than one crystalline form and is a mixture consisting of two polymorphs, viz. cellulose Iα and Iβ. Differences were reported in the resonances of the C1 atoms. A singlet and a doublet appeared around 106 ppm for cellulose Iα and Iβ, respectively. Initial ambiguity in the interpretation of crystallographic data for native cellulose may be largely attributed to the co-existence of these two polymorphs. These two crystalline forms have the same conformation of the heavy atom skeleton, but differ in their hydrogen bonding patterns. The Iα form represents a triclinic phase with one-chain-per-unit cell, while the Iβ form represents a monoclinic phase with two-chains-per-unit cell. This description has been numerically simulated (Vietor et al., 2000). Nishiyama et al. (2003a) have confirmed and refined the crystal structures of both Iα and Iβ phases by the determination of the various network systems of hydrogen bonds. The experiments were conducted using jointly X-ray and neutron diffraction on hydrogenated and deuterated oriented fibers. The neutron diffraction experiments have allowed, among other things, replacing the hydroxyl hydrogen atoms by deuterium atoms, to determine the geometry of intra- and interchain hydrogen bonding for both phases.

The ratio of the two allomorphs Iα and Iβ differs greatly depending on the species. The Iα phase is mainly found in celluloses produced by primitive organisms, such as algae or bacteria, while cellulose Iβ lies mainly in the cellulose produced by higher plants (cotton, wood, ...) and animals, such as in the envelope of marine animals (Belton et al., 1989). In wood pulp, for example, the Iβ phase prevails with a proportion of about 64%. Almost pure Iα cellulose can be obtained from bacteria (e.g., *Acetobacter xylinum*) or from the cell walls of fresh-water algae *Glaucosystis nostochinearum* (Nishiyama et al., 2003b; Saxena and Brown, 2005; Witter et al., 2006). Polymorph Iβ is the major part of cellulose found in cotton, wood, and ramie and of tunicin from *Halocynthia roretzi* (Saxena and Brown, 2005; Nishiyama et al., 2002; Nishiyama et al., 2008). Tunicin possesses only the Iβ form but no pure sample of Iα has been found in nature. Bacterial cellulose has the highest content (70%) of polymorph Iα.

The content of polymorphs Iα and Iβ in cellulose can be changed under external actions. Indeed, the polymorph Iα can be converted (not completely, however) into the more stable Iβ phase by annealing at 260°C to 280°C in various media, such as organic solvents or helium (Debzi et al., 1991), hydrolysis (Atalla et al., 1985) or passage through cellulose III. For example, annealing at 260°C in a 0.1 sodium hydroxide solution converts most of the Iα to the Iβ form. The dynamics of the transformation of polymorph Iα into Iβ under heating of cellulose Iα in an inert atmosphere was investigated. Powder X-ray diffraction and two-dimensional Fourier transform infrared correlation spectroscopic studies showed that a high-temperature intermediate was formed around 200°C (Wada et al., 2003; Watanabe et al., 2007). The existence of both polymorphs in cellulose may affect the reactivity of native cellulose as Iα is metastable and thus more reactive than Iβ.

1.4.2 Cellulose II

Native cellulose can be converted irreversibly into the thermodynamically more favorable cellulose II polymorph by swelling native cellulose I (metastable form) in concentrated sodium hydroxide aqueous solutions (17 to 20% wt/vol) and removal of the swelling agent (this alkali treatment is named mercerization process after its inventor Mercer in 1844). Mercerization is used to activate the polymer prior to the production of technical cellulose ether. The mercerization of cellulose leads only to its swelling, but not to dissolution. The insertion of chemical species induces the structural change and the passage from a structure with parallel cellulosic chains to a configuration with anti-parallel chains.

Cellulose II can also be prepared by regeneration, which is the solubilization of cellulose I in a solvent followed by precipitation by dilution into an aqueous medium. This is the typical process for the technical spinning of man-made cellulose fibers. There are several industrial processes for the regeneration of cellulose, viz. fortisan (not used nowadays), viscose, copper ammonium, and N-methylmorpholine N-oxide (NMMO) processes. All these processes involve the dissolution of cellulose followed by the formation of regenerated cellulose fibers.

It was found that regeneration gives a higher level of conversion of cellulose I to cellulose II (Kolpak and Blackwell, 1976). The crystalline structure of cellulose II, determined by Kolpak et al. (1978) and Stipanovic and Sarko (1976), was studied by neutron diffraction to confirm that contrary to cellulose I, the arrangement is antiparallel chains (Lagan et al., 1999) which allows the establishment of a larger number of intermolecular hydrogen bonds than the native form. The transition from cellulose I to cellulose II is irreversible, which suggest that cellulose II is thermodynamically more stable than cellulose I. Cellulose II, like cellulose Iβ, has the monoclinic unit cell. The different arrangement of the chains (parallel in cellulose Iβ and antiparallel in cellulose II) is the most substantial difference between these two polymorphs. Even though the unit cells of cellulose II obtained by the two routes (mercerized and regenerated cellulose) resemble each other closely, small differences have been reported (Wellard, 1954).

Cellulose II is formed naturally by a mutant strain of *Gluconacetobacter xylinum* (Kuga et al., 1993) and occurs in the alga *Halicystis* (Sisson, 1938). They were both very useful to provide an insight into the crystalline structure of cellulose II.

1.4.3 Cellulose III

Treatment with liquid ammonia or with certain organic amines, such as ethylene diamine (EDA), followed by washing with alcohol allows the preparation of cellulose III either from cellulose I (which leads to the form cellulose III_I) or from cellulose II (which leads to the form III_{II}). A hexagonal unit cell is reported for cellulose III and it has been found that the polarity of the resultant cellulose chains resembles that of the starting material. The lattice dimensions of cellulose III_I and cellulose III_{II} are similar.

These transformations are reversible, suggesting that the chain orientation is similar to the one of the starting material. However, at the crystalline level, an extensive decrystallization and fragmentation of the cellulose crystals were observed during the conversion of cellulose I to cellulose III_I (Sarko et al., 1976; Roche and Chanzy, 1981; Reis et al., 1991). During the reverse transition, i.e. conversion back to cellulose I, partial recrystallization takes place but the distortion and fragmentation of the crystals are irreversible and restoration of the damage done to the morphological surface is incomplete. It was shown by ^{13}C NMR that the transformation of cellulose I to cellulose III_I polymorph induces a significant reduction of the lateral dimensions of the crystallites (Sarko et al., 1976). At the same time, the cellulose chains show conformational changes arising from the primary hydroxyl groups.

1.4.4 Cellulose IV

Cellulose III treated at high temperature in glycerol is transformed into cellulose IV. Again two types exist, viz. cellulose IV_I and IV_{II} obtained from cellulose III_I and III_{II}, respectively. The conversions are never totally complete, which explains the difficulties in the production of good quality X-ray diffraction patterns (Buléon and Chanzy, 1980). It is generally accepted that cellulose IV_I is a disordered form of cellulose I. This could explain the reported occurrence of this form in the native state in some plants (primary walls of cotton and some fungi) as determined by X-ray diffraction (Chanzy et al., 1978; Chanzy et al., 1979). It has been confirmed by studies based on X-ray diffraction and ^{13}C solid state CP-MAS NMR experiments (Wada et al., 2004).

1.5 Cellulose microfibrils

As shown previously, the presence of many hydroxyl groups along the cellulose chain results in the formation of a network of intra- and intermolecular hydrogen bonds. In addition, a network of van der Waals connections is established between the chains layers (French et al., 1993). These two link networks allow the establishment of ordered crystalline structures. Intramolecular hydrogen bonds occur primarily between the hydrogen borne by the OH group of the C3 carbon cycle and oxygen from the adjacent ring (O5) (see Figure 1.6). There may also be an interaction between the hydrogen borne by the primary OH group of the C6 carbon and oxygen from the carbon 2 hydroxyl of the adjacent ring. The intermolecular bonds occur between the hydrogen of the HO-6 primary hydroxyl and oxygen in position O3 in a cycle of a neighboring unit.

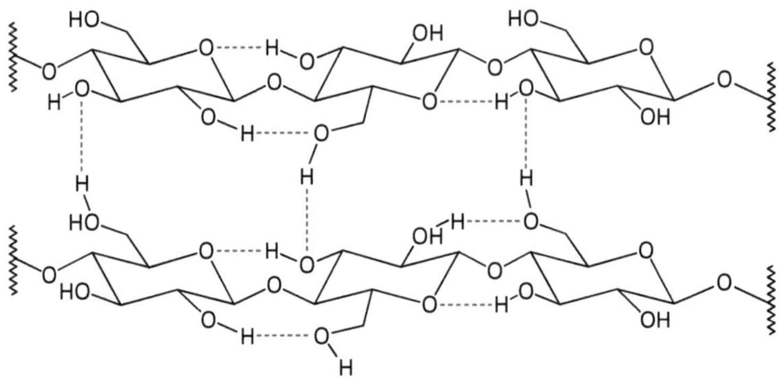

Fig. 1.6: Schematic representation of intra- and intermolecular hydrogen bonds in cellulose.

This ordered molecular arrangement of cellulosic chains parallel to each other is the basis of a crystal structure called microfibrils. The hierarchy of structure and supramolecular organization of cellulose are schematized in Figure 1.7.

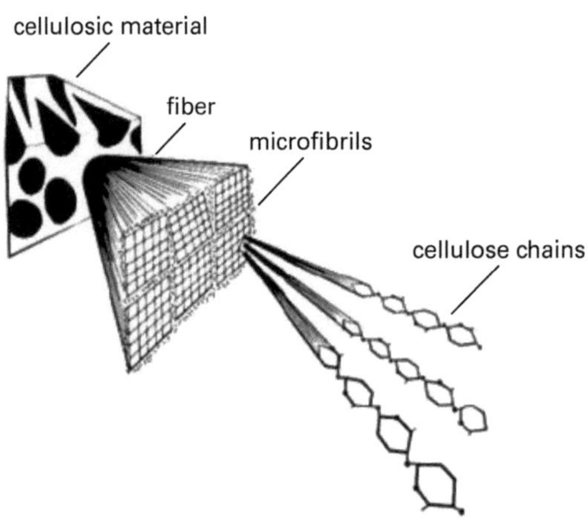

Fig. 1.7: Schematic representation of the hierarchical structure of a lignocellulosic fiber (after Marchessault and Sundararajan, 1983).

The microfibrillar nature of cellulose was established from electron microscopy observations (Frey-Wyssling et al., 1948, Preston and Cronshaw, 1958). To describe the arrangement of the chains within the microfibrils, several models have been proposed. They can be grouped into two categories: the models with stretched chains (Frey-Wyssling, 1954, Hess et al., 1957) and the models with folded chains (Dolmetsch and Dolmetsch, 1962, St John Manley, 1964; Marx-Figini and Schulz, 1966). The latter have been abandoned in favor of a crystal with stretched chains thanks to DP measurements performed on microtomed ramie fibers (Muggli et al., 1969; Mühlethaler, 1969) and small angle X-ray diffraction experiments conducted by Bonard in 1966 (Bonard, 1966). The stretched chains model was validated by studying the mechanical properties of cellulose fibrils and the DP as a function of the length of the sample (Mark et al., 1969). It clearly indicated that the cellulose chains exist in an extended form in the crystal. More recent works on the chain polarity of cellulose microfibrils showed that the chains are parallel to the main axis of the microfibrils. This orientation has been evidenced by experiments highlighting the unidirectional nature of the degradation of microcrystalline cellulose from *Valonia* under the action of exocellulases (Chanzy and Henrissat, 1985).

Microfibrils have widths, lengths, shapes and crystallinities that may vary depending on the origin of cellulose as shown by transmission electron microscopy observations (Chanzy, 1990). However, they are always significantly longer than wide. The width of the microfibrils can vary from 2–3 nm for the cell walls of primary tissues of some plants to 60 nm for certain algae. The section of microfibrils from *Valonia* is assumed to be square (Sassi and Chanzy, 1995) while that of tunicates is rather lozenge-shaped and becomes edged off after hydrolysis (Revol et al., 1992, Van Daele et al., 1992; Sugiyama et al., 1992). The schematic morphological characteristics of microfibrils of different origins are shown in Figure 1.8.

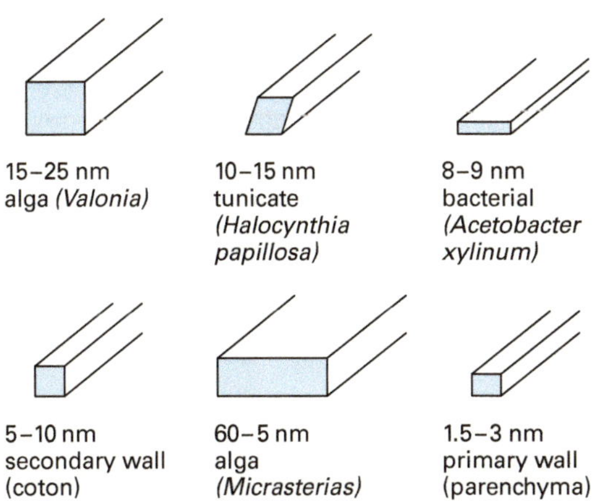

Fig. 1.8: Microfibril morphology depending on the origin of cellulose and order of magnitude of the widths.

For a long time, studies carried out at the sub-microscopic scale emphasized the discontinuous character of microfibrils. Experimental evidence was provided by wide angle (Fink et al., 1987) and small angle X-ray diffraction (Grigoriew and Chmielewski, 1998), ^{13}C CP/MAS solid state NMR experiments (Earl and VanderHart, 1981), and tensile tests performed on cellulose fibers (Ishikawa et al., 1997). Confinement of microfibrils during their biosynthesis generates twists distributed along the chain. In these zones, the crystalline arrangement is destroyed. The periodic nature of this distribution of disordered or amorphous regions has been shown for ramie microfibrils by small angle neutron diffraction (Nishiyama et al., 2003a) with a periodicity of about 150 nm corresponding to the sizes estimated by diffraction of hydrolyzed microfibrils.

The exact organization of crystallites in the microfibril has not been up to now fully elucidated (number, spatial arrangement) but the microfibrillar model of cellulose considers a highly crystalline core surrounded by less organized surface chains (Preston and Cronshaw, 1958). The proportion of surface chains whose solid state NMR signal is different from the one of the crystalline core is directly dependent on the dimensions of the microfibril (Newman, 1999). The fraction of non-crystalline cellulose chains corresponding to surface chains and amorphous regions is higher the finer the microfibril is. The amount of amorphous phase in the alga *Valonia*, which has large diameter microfibrils (15 to 25 nm), is of the order of a few per cent, whereas it is 30 to 35% for cotton linters with a section around 7 nm, and 65 to 70% for primary walls that have very fine microfibrils (1.5–3 nm). Increasing the number of molecules at the surface would result in a corresponding increase of reactivity, since the surface molecules are accessible for chemical or physical modification, while the cellulose molecules hidden inside the microfibril structure are not. The accessibility of amorphous cellulose surface chains to chemical modification may also be useful for determining the size of crystallites.

Therefore, in nature, cellulose occurs as a slender rod-like or threadlike entity, which arises from the linear association of crystallites. This entity is called the microfibril (collection of cellulose chains) and it forms the basic structural unit of the plant cell wall. Each microfibril can be considered as a string of cellulose crystallites, linked along the chain axis by amorphous domains. They are biosynthesized by enzymes and deposited in a continuous fashion. Their structure consists of a predominantly crystalline cellulose core. This is covered with a sheath of paracrystalline polyglucosan material surrounded by hemicelluloses.

Different models have been proposed in the literature for the fibrillar structure of cellulose. The generally most accepted is the one suggested by Fengel (1971) for the ultrastructure organization of the cell wall components in wood which had several layers of hemicellulose molecules between the fibrils (dimensions 12.0 nm) and a monomolecular layer of hemicellulose between the elementary fibrils. Lignin was envisaged as surrounding the total microfibrillar system as shown in Figure 1.9 (Fengel, 1971). Fengel and Wegener (1984) present an accurate model that considers the intimate links between hemicellulose and cellulose on the one hand, and hemicellulose and lignin on the other. In this model, the microfibrils of cellulose, the less ordered cellulose chains and hemicelluloses associate together through many hydrogen bonds. On the other hand, hemicellulose is more strongly linked to lignin by covalent bonds. The Fengel model is relatively comprehensive, but does not fully take into

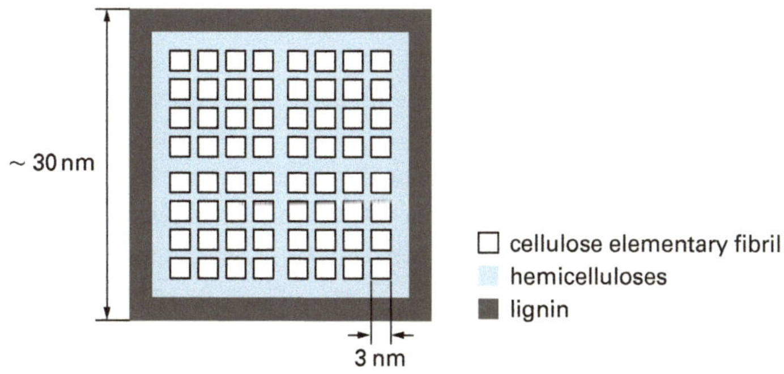

Fig. 1.9: Model of the wood cellulose microfibril structure, consisting of elementary nanofibrils. Adapted from Fengel and Wegener (1984).

account the composite nature of the cellulose microfibrils (crystalline and amorphous parts). Indeed, if one considers a single macromolecule of cellulose, some parts of it are found in the crystalline parts, while others integrate less ordered areas.

Taylor and Wallace (1989) discussed the effect of the hemicellulose xyloglucan binding to the fibrils. The extent of the association between cellulose and xyloglucan is dependent on the source of cellulose (Hayashi and Maclachlan, 1984). Binding of xyloglucan has been suggested as a regulator of cellulose fibrillar size (Sasaki and Taylor, 1984).

1.6 Hierarchical structure of plants and natural fibers

The term "natural fibers" covers a broad range of vegetable, animal, and mineral fibers. However, in the composites industry, it usually refers to wood fibers and agro-based bast, leaf, seed, and stem fibers. Natural fibers are not limited to a macroscopic view of cellulose. A more intimate insight accounts for hierarchical assemblies of microfibers called microfibrils, which are themselves the product of the supramolecular architecture of the basic polymer, namely cellulose as shown in Section 1.5.

Wood and plants are cellular hierarchical biocomposites designed by nature and they are basically semicrystalline cellulose microfibril-reinforced amorphous matrices made of hemicellulose, lignin, waxes, extractives and trace elements. Lignocellulosic fibers consist therefore of microfibrils aggregate. The primary cell wall is then essentially a composite material consisting of a framework of cellulose microfibrils embedded in a cementing matrix of other, mostly hemicelluloses and lignin, polymers. Hemicellulose is not a form of cellulose but falls into a group of polysaccharides (with the exception of pectin) attached to the cellulose after the lignin has been removed. However, their structure contains many different sugar units, apposed to the D-anhydroglucose units in cellulose, and is a highly branched polymer compared to the linearity of cellulose. Lignin is generally considered as a little understood hydrocarbon polymer with a highly complex structure consisting of aliphatic and aromatic constituents and forms the matrix sheath around the fibers that holds the natural structure (e.g. trees) together.

The structure of plants spans many length scales, like many other biological tissues including bones (this basic structure of all vertebrates is made of collagen fibrils embedded in an inorganic apatite matrix) and teeth, in order to provide maximum strength with a minimum of material. Wood, which is approximately 40–50 wt% cellulose with about half in nanocrystalline form and half in amorphous form, is a well-known example (Figure 1.10). Meters describe the whole tree, centimeters describe structures within the cross-section, millimeters describe growth rings, tens of micrometers the cellular anatomy, micrometers describe the layer structure within cell walls, tens of nanometers describe the configuration of cellulose fibrils in a matrix mainly composed of hemicellulose and lignin, and nanometers describe the molecular structures of cellulose, hemicellulose, and lignin and their chemical interactions (Moon, 2008).

From a structural point of view, natural fibers are multicellular in nature and consist of bundles of elongated mostly cylindrical honeycomb cells which have differ-

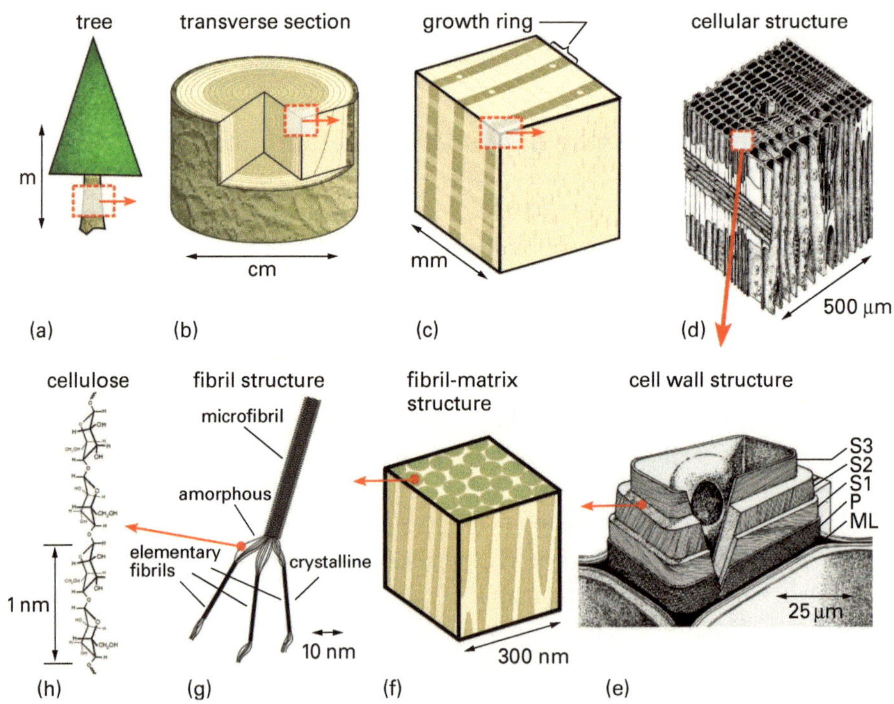

Fig. 1.10: Wood hierarchical structure: from tree to cellulose (Moon, 2008).

ent sizes, shapes and arrangements depending on the source of plant fiber. These cells are cemented together by an intercellular substance which is isotropic, non-cellulosic and ligneous in nature. They are like microscopic tubes, i.e. cell walls surrounding the center lumen that contribute to the water uptake behavior of plant fibers (Figure 1.11). Therefore, natural fibers present a multi-level organization and consist of several cells formed out of semicrystalline oriented cellulose microfibrils connected to a complete layer by lignin, hemicelluloses and in some cases pectins. Table 1.1 reports the mean chemical composition of some natural fibers. These values are obviously only indicative since climatic conditions, age and digestion process influence not only the structure of fibers but also their chemical composition. With the exception of some species, like cotton, nettle and others, the components of natural fibers are cellulose, hemicelluloses and lignin which are the basic components with regard to the physical properties of the fibers.

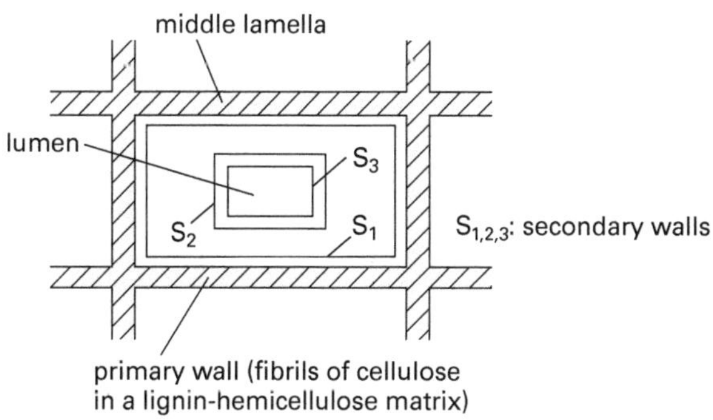

Fig. 1.11: Schematic structure of a natural fiber cell (Bismarck et al., 2002).

Cell wall architectures are the nanodimensional structures composed of multiple elementary nanofibril arrangements. Cellulose fibrils self-assemble in a manner similar to liquid crystals leading to nanodimensional structures seen in typical plant cell walls. The cell walls differ in both their composition and orientation of cellulose microfibrils. In most plant fibers, the cellulose microfibrils in the central walls, the major part of the cell wall representing up to 86% of the cell wall (Fengel and Stoll, 1973) is labeled S2, form a constant angle to the normal axis called the microfibrillar angle (Figure 1.12). The normal axis corresponds to the longitudinal direction of the tracheid (i.e. of the stem or of the branch). In the second largest layer of the cell wall, labeled S1, the fibrils run at a gentle helical slope roughly perpendicular to the ones in the S2 (Fengel and Wegener, 1984; Reiterer et al., 1998).

The characteristic value for this structural parameter varies from one plant fiber to another and the crystallites are therefore arranged in a spiral form, the pitch of which is specific of a given source. The spiral angle in the S2 (as in the other different cell wall layers) can be measured using polarized light microscopy (Preston, 1934;

Fiber	Cellulose (wt%)	Hemicellulose (wt%)	Lignin (wt%)	Waxes (wt%)
Abaca	56–63	20–25	7–9	3
Alfa	45.4	38.5	14.9	2
Bagasse	55.2	16.8	25.3	–
Bamboo	26–43	30	21–31	–
Banana	63–64	19	5	–
Coir	32–43	0.15–0.25	40–45	–
Cotton	85–90	5.7	–	0.6
Curaua	73.6	9.9	7.5	–
Flax	71	18.6–20.6	2.2	1.5
Hemp	68	15	10	0.8
Henequen	60	28	8	0.5
Isora	74	–	23	1.09
Jute	61–71	14–20	12–13	0.5
Kenaf	72	20.3	9	
Kudzu	33	11.6	14	–
Nettle	86	10	–	4
Oil Palm	65	–	29	–
Piassava	28.6	25.8	45	–
Pineapple	81	–	12.7	–
Ramie	68.6–76.2	13–16	0.6–0.7	0.3
Sisal	65	12	9.9	2
Sponge Gourd	63	19.4	11.2	3
Sun Hemp	41–48	8.3–13	22.7	–
Wheat Straw	38–45	15–31	12–20	–

Table 1.1: Chemical composition of various natural fibers (Valadez-Gonzalez et al., 1999; Hattallia et al., 2002; Hoareau et al., 2004).

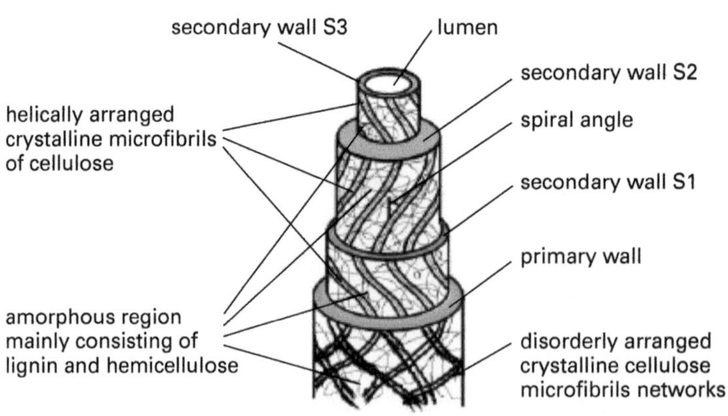

Fig. 1.12: Schematic structure of an elementary plant fiber (cell). The secondary cell wall, S2, makes up about 80 per cent of the total thickness (Rong et al., 2001).

Boyd and Foster, 1974), staining methods for microscopic investigations (Hiller, 1964; El-Osta et al., 1972), X-ray diffraction methods (Kantola and Seitsonen, 1961; Kantola and Seitsonen, 1969; El-Osta et al., 1972; El-Osta, 1973; Paakkari and Serimaa, 1984) and small-angle X-ray scattering (Jakob et al., 1994; Reiterer et al., 1998). Hence the properties of the single fibers depend on the crystallite content, their sizes, shape, orientation, length/diameter (L/D) or aspect ratio of cells, thickness of cell walls, and finally, lumen. In general, the fiber strength increases with increasing cellulose content and decreasing spiral angle with respect to fiber axis. The most important factor controlling the different types of natural fibers is their species because the properties of fibers are different between different species. In addition, the properties of fibers within a species vary depending on area of growth, climate and age of the plant. Lastly, the properties of natural fibers vary greatly depending on the processing method used to break down the lignocellulosic substrate to the fiber level.

The outer cell wall is porous and contains almost all of the non-cellulose compounds, except proteins, inorganic salts and coloring matter and it is this outer cell wall that creates poor absorbency, poor wettability and other undesirable textile properties. In most applications, fiber bundles or strands are used rather than individual fibers. Within each bundle the fiber cells overlap and are bonded together by pectins that give strength to the bundle as a whole. However, the strength of the bundle structure is significantly lower than that of the individual fiber cell. Then the potential of individual fibers is not fully exploited.

Multiple such cellulose-lignin/hemicellulose layers in one primary and three secondary cell walls stick together to a multiple-layer-composite. Such microfibrils have typically a diameter of about 2–20 nm, are made up of 30 to 100 cellulose molecules in extended chain conformation and provide the mechanical strength to the fiber. The degree of crystallinity and typical dimensions of cellulose microfibrils depend on their origin, although the biosynthetic mechanism is the same in all organisms (Sarko and Muggli, 1974; Woodcock and Sarko, 1980).

1.7 Potential reinforcement of cellulose

To be potentially mechanically effective, reinforcement must have a modulus and strength that greatly exceed those of the continuous medium in which it is dispersed. This stiffening is generally obtained at the expense of the ductility or plasticity of the material that becomes more brittle. The modulus of glassy and rigid crystalline polymers is of the order of a few 10^9 Pa, i.e. a few GPa. Therefore, the modulus of cellulosic particles must be significantly higher than this value to be exploitable and potentially usable as a load-bearing element for glassy polymers. In addition, it must be homogeneously dispersed and distributed, and the level of adhesion between both phases should be sufficient to allow proper stress transfer from the matrix to the reinforcing phase across the interface upon loading.

In nature, cellulose is a ubiquitous structural polymer that confers its mechanical properties to higher plant cells. The hierarchical structure of natural fibers, based on their elementary nanofibrillar components, leads to the unique strength and high performance properties of different species of plants. Indeed, the most important attribute of wood and other lignocellulosic materials is their mechanical properties, in particular their unusual ability to provide high mechanical strength and high strength-to-weight ratio while allowing for flexibility to counter large dimensional changes due to swelling and shrinking. In all terrestrial and aquatic plant species, the primary cell wall is a dynamic structure and its constituting material must be synthesized in a form that is competent to undergo extension.

The mechanical properties of cellulose can be characterized by its properties in both the ordered (so-called crystalline) and disordered (so-called amorphous) regions of the molecule. The chain molecules in the disordered regions contribute to the flexibility and the plasticity of the bulk material, while those in the ordered regions contribute to the stiffness and elasticity of the material. As they are almost defect-free, the modulus of cellulosic nanocrytals is close to the theoretical limit for cellulose. The promise behind cellulose-derived composites lies in the fact that the axial specific Young's modulus (modulus-to-density ratio) of the basic cellulose crystal derived from theoretical chemistry is potentially stronger than steel and similar to Kevlar.

1.7.1 Mechanical properties of natural fibers

The properties of natural fibers are strongly influenced by many factors, particularly chemical composition and location in plants. In most natural fibers the microfibrils orient themselves at an angle to the fiber axis called the microfibril angle. A weak correlation between strength and cellulose content and microfibril or spiral angle is found for different plant fibers (Lee and Rowell, 1991). In general, fiber strength increases with increasing cellulose content and decreasing spiral angle with respect to fiber axis. This means that the most efficient cellulose fibers are those with high cellulose content and low microfibril angle. Other factors that may affect the fiber properties are maturity, separating process, microscopic and molecular defects such as pits and nodes, type of soil and weather conditions under which they were grown.

The mechanical properties reported in the literature for some plant fibers are collected in Table 1.2. They are generally determined from tensile tests performed on more or less individual fibers (bundles of fiber) despite a great variability of results. The experimental conditions (temperature and relative humidity) should be strictly controlled because of the great variability of the properties of natural fibers with respect to these parameters. A circular cross section is generally assumed to calculate the cross-sectional area of the sample and thus convert the applied load into stress. For statistical significance a large number of tests are required (Eichhorn et al., 2000). These mechanical properties are much lower when compared to those of the most widely-used competing reinforcing glass fibers. However, because of their lower density, the specific properties (property-to-density ratio), strength, and stiffness of plant fibers are comparable to the values of glass fibers (Bismarck et al., 2005).

Fiber	Density ($g \cdot cm^{-3}$)	Diameter (μm)	Tensile Strength (MPa)	Young's Modulus (GPa)	Elongation at break (%)
Flax	1.5	40–600	345–1,500	27.6	2.7–3.2
Hemp	1.47	25–500	690	70	1.6
Jute	1.3–1.49	25–200	393–800	13–26.5	1.16–1.5
Kenaf			930	53	1.6
Ramie	1.55		400–938	61.4–128	1.2–3.8
Nettle			650	38	1.7
Sisal	1.45	50–200	468–700	9.4–22	3–7
Henequen					
PALF		20–80	413–1,627	34.5–82.5	1.6
Abaca			430–760		
Oil Palm EFB (empty fruit bunch)	0.7–1.55	150–500	248	3.2	25
Oil Palm Mesocarp			80	0.5	17
Cotton	1.5–1.6	12–38	287–800	5.5–12.8	7–8
Coir	1.15–1.46	100–460	131–220	4–6	15–40
E-glass	2.55	< 17	3,400	73	2.5
Kevlar	1.44		3,000	60	2.5–3.7
Carbon	1.78	5–7	3,400–4,800	240–425	1.4–1.8

Table 1.2: Characteristic values for the density, diameter and mechanical properties of vegetable and synthetic fibers (Bismarck et al., 2005).

Raman spectroscopy is an invaluable method for evaluating changes that occur in a fiber structure subjected to a mechanical solicitation, i.e. stress and strain (Eichhorn et al., 2001a). With this method, the molecular deformation mechanisms of the polymeric chains can be examined through the large stress-induced Raman band shifts that can occur during deformation. Therefore, the technique relies on the accurate measurement of the position of a structurally characteristic Raman band as a function of external deformation of the fiber, whether in air or when incorporated into a composite material. It has been successfully applied to study deformation processes for a wide variety of aromatic high-modulus polymeric fibers (Yeh and Young, 1999). The Raman peak shifts towards lower wavenumbers are thought to correspond to the direct deformation of bonds within the cellulose chain structure as first demonstrated and predicted theoretically for polydiacetylene single crystals (Batchelder and Bloor, 1979).

For cellulose, the positions of characteristic bands located at 1414, 1095 and 895 cm^{-1} have been reported to shift (Hamad and Eichhorn, 1997; Eichhorn et al., 2000; Eichhorn and Young, 2003; Eichhorn and Young, 2004; Gierlinger et al., 2006; Peetla et al., 2006; Tze et al., 2006; Tze et al., 2007). For a number of cellulosic materials (regenerated and natural fibers, wood, and paper), it was reported that during tensile deformation the highest intensity Raman band located at 1095 cm^{-1}, corresponding to the stretching mode of C–O within the ring structure of cellulose (Blackwell et al., 1970; Attala, 1976), shifted towards lower wavenumbers and was most indicative of the molecular deformation (Hamad and Eichhorn, 1997; Eichhorn et al., 2001a). An example of the effect of deformation on the position of the Raman band initially located at 1095 cm^{-1} for coir fibers is shown in Figure 1.13(a). This effect is indicative of the stress level in the fibers. An example of a typical shift Raman peak in the 1095 cm^{-1} peak position with strain for a fibrous fragment of microcrystalline cellulose in an epoxy resin is shown in Figure 1.13(b). Moreover, the rate of Raman band shift was shown to be invariant with stress, which is consistent with a fiber structure based on a modified series aggregate model. Since the Raman band located at 1095 cm^{-1} is associated with the backbone of cellulose, its intensity is sensitive to the orientation of these chains with respect to the polarization direction of the laser (Bakri and Eichhorn, 2010). After the fiber failed, the band was found on its original position again, proving the elastic nature of the deformation (Gierlinger et al., 2006).

Fourier transform near-infrared (FT-NIR) Raman microspectroscopy was used to investigate the micromechanical tensile deformation behavior of hemp fibers (Peetla et al., 2006). Mechanical properties were accessed for different dew-retting durations and alkali chemical treatments with aqueous sodium hydroxide solutions (mercerization treatment) of different concentrations. The macroscopic results (tensile tests) were found to be in accordance with the microscopic results. The tensile modulus and

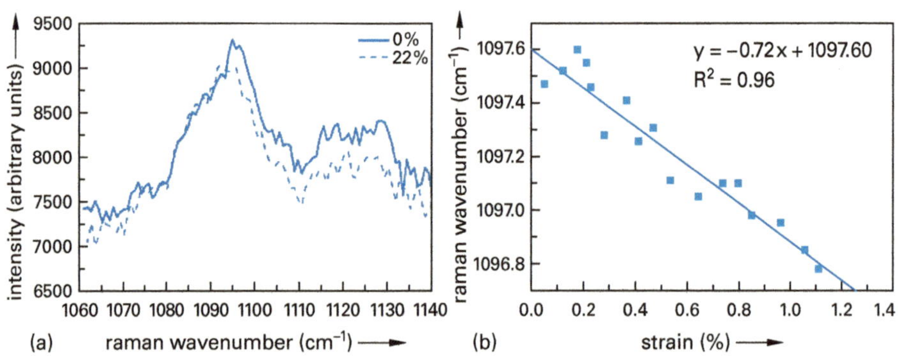

Fig. 1.13: (a) Example of a typical shift in the position of the Raman band initially located at 1095 cm^{-1} for a coir fiber deformed in tension (Bakri and Eichhorn, 2010). The strain rate is indicated in the Figure. (b) Typical shift of the 1095 cm^{-1} Raman peak for a fibrous fragment of microcrystalline cellulose (Eichhorn and Young, 2001).

tensile strength ranged between 7.8 and 12.8 GPa, and 93 and 250 MPa, respectively, depending on the retting level, and decreased down to 4.7 GPa and 74 MPa, respectively, for hemp fibers treated under strong mercerization conditions. The Young's modulus of a particulate form of cellulose, namely microcrystalline cellulose, was estimated from the values of the shift rate of the 1095 cm^{-1} Raman band with strain (Eichhorn and Young, 2001). This shift was monitored and compared to the deformation of natural fibers (flax and hemp). A value of 25 ± 4 GPa was reported. It has been shown that this value is consistent with the degree of crystallinity of microcrystalline cellulose measured from X-ray diffraction experiments.

Another important Raman band at 1414 cm^{-1} is associated with 3-atom band vibrations (HCC, HCO and HOC bending) that ought to be influenced by transverse forces through side-chain hydrogen bonding. The significant shift of this peak upon deformation shows the important role of hydrogen bonding within the structure in stress-transfer between adjacent cellulose chains (Eichhorn et al., 2003). Raman spectroscopy has been shown to be also a useful tool for characterizing the orientation of the fibrillar structure in cellulosic fibers and for following their micromechanical deformation (Bakri and Eichhorn, 2010).

1.7.2 Mechanical properties of cellulose microfibrils

The mechanical properties of constitutive microfibrils released from lignocellulosic fibers should be less dispersed because of a more homogeneous nature. Moreover, the modulus of individual microfibrils must be higher than that of lignocellulosic fibers. However, the analysis of plant primary cell walls tissues is difficult using standard techniques. The properties of the nanoscopic fibrous components cannot be physically measured without extracting them from the tissue, which may result in significant chemical alteration and mechanical damage.

A structural approach has been used to develop a hierarchical description of plant tissue mechanical properties down to the level of cell wall components (Hepworth and Bruce, 2000). This model was used to back calculate cell wall microfibril properties. Force deflection data from the compression of cubes of potato tissues (loading rate 10 $mm \cdot min^{-1}$) were fed in a model containing two structural levels, the cell structure and the cell wall structure. Materials properties were assigned at the level of cell wall microfibrils. The modulus was found to vary with the strain and displayed a maximum value of 130 GPa. The maximum microfibril strain was chosen as the value after a tissue deformation of 22% that corresponds to the maximum tissue deformation at which the constant volume assumption was applicable at this particular rate of tissue deformation. At 8% wall strain, i.e. the value at which failures were suspected to begin, the stress was predicted to be 7.5 GPa. This value is also close to theoretical chemistry predictions of 7–8 GPa for the strength of cellulose microfibrils and failure by chain scission. At larger strains the modulus decreased significantly, showing the influence of non-cellulosic polysaccharides on the microfibril properties.

Atomic force microscopy (AFM) allows the direct and accurate mechanical characterization of nanomaterials. The AFM tip can be used to measure the elastic modulus of suspended single filaments such as cellulose microfibrils by performing a nanoscale three-point bending test, in which the center of the filament is deflected by a known force (Figure 1.14). In this experiment, the AFM cantilever applies the known force on a filament bridging a gap. The stiffness of bacterial cellulose consisting of filaments with diameters ranging from 35 to 90 nm has been determined by this technique (Guhados et al., 2005). The sample was imaged in contact mode to locate isolated bacterial cellulose filaments that spanned the gap of a silicon-nitride-coated silicon grating with a pitch of 3.0 μm and nominal step height of 1000 nm. Once identified, the filaments were imaged at higher resolution to determine their dimensions. The cantilever deflection was recorded as a function of vertical sample displacement. In principle, the deflection of the filament is due to tensile/compressive and shear deformations. However, when the ratio of the length of filament that bridges the gap to the diameter is higher than 16, which was the case in this study, shear can be neglected. No dependence on diameter was observed, indicating that shear forces can be effectively neglected and that the filaments behave mechanically like a homogeneous material. A Young's modulus value of 78 ± 17 GPa was reported. This value is lower than the one reported for single filaments of bacterial cellulose (114 GPa) and obtained by following the shift of the 1095 cm^{-1} Raman band towards lower wavenumbers upon the application of tensile deformation (Hsieh et al., 2008). The authors explained the discrepancy with the value obtained by AFM tip bending of cellulosic filaments (Guhados et al., 2005) by the different solicitation applied to the sample, modulus values obtained in bending being expected to be different to those obtained in tension.

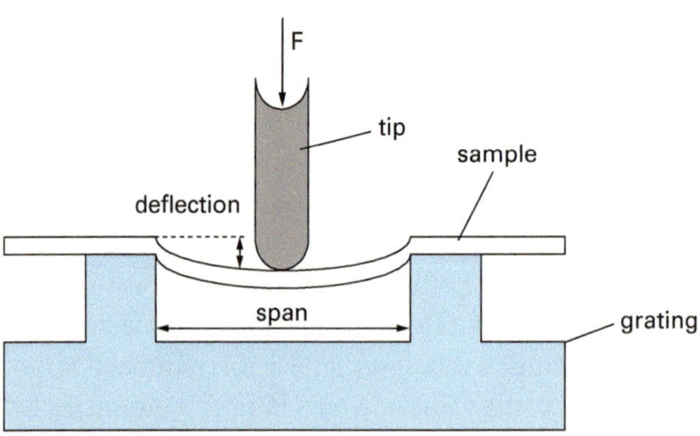

Fig. 1.14: Schematic of the measurement of the elastic modulus of single filaments using AFM by performing a nanoscale three-point bending test.

Similar AFM tip bending experiments were performed on bundles of cellulose microfibrils and the effects of both the isolation process and the cellulose source on the elastic modulus were investigated (Cheng et al., 2009). Regenerated cellulose fibers (Lyocell), pure cellulose flours and pulp fibers were used. Defibrillation was performed by mechanical methods with high shear force, viz. ultrasonic treatment and high-pressure homogenization. Commercial microfibrillated cellulose (MFC) (Daicel Chemical Industries Ltd., Japan) was also used as reference. A broad range of diameters was obtained for cellulose microfibril bundles and only filaments with diameters in the range 150–300 nm were investigated. The cellulosic filaments were suspended over the edged groove in a silicon wafer. The wafer had grooves with 5 µm in width and 1360 nm in depth. The elastic modulus of lyocell microfibril bundles with diameters ranging between 150 and 180 nm was evaluated to be 98 ± 6 GPa. Values of 81 ± 12 GPa and 84 ± 23 were reported for pulp and MFC, respectively. These values decreased sharply for diameters above 180 nm. The different values reported for the stiffness of cellulose microfibrils (or bundles) are collected in Table 1.3.

Material	Method	E_L (GPa)	Reference
Potato Tuber Tissue	Calculation	130	(Hepworth and Bruce, 2000)
Bacterial Cellulose	AFM	78 ± 17	(Guhados et al., 2005)
Bacterial Cellulose	Raman	114	(Hsieh et al., 2008)
Lyocel Microfibrils	AFM	98 ± 6	(Cheng et al., 2009)
Pulp Microfibrils		81 ± 12	
Commercial MFC		84 ± 23	

Table 1.3: Longitudinal (E_L) modulus of cellulose microfibrils.

1.7.3 Mechanical properties of cellulose crystal

The modulus of cellulose microfibrils is expected to result from a mixing rule of the modulus of the crystals, the amorphous fraction and defects/air in the sample. As for any semicrystalline polymer, the crystalline regions of cellulose act as physical cross-links for the material. In this physically cross-linked system, the crystalline regions would also act as filler particles due to their finite size, which would increase the modulus substantially.

The elastic modulus of the crystalline region of cellulose is an important property of this material, especially with respect to the ultimate aim of exploiting its full potential in composite materials. The elastic properties of cellulose crystalline regions have been investigated since the mid-1930s either by theoretical evaluations or by experimental measurements (wave propagation, X-ray diffraction, Raman spectroscopy, and AFM). It was shown in 1936 (Meyer and Lotmar, 1936) that the modulus of elasticity corresponding to the principal chain direction of a polymer crystal of specified nature may be calculated from the force constants of the chemical bonds of the chain derived from vibration frequencies of molecules. Appling the method to the cellulose crystal, the authors obtained for two different estimates of force constants longitudinal modulus values of $7.7 \cdot 10^{11}$ and $12.1 \cdot 10^{11}$ dyn $\cdot$ cm^{-2}, i.e. 77 and 121 GPa. Another calculation of the elastic modulus of polymer crystal has been made following a method applied to other polymers by treating an isolated molecule and considering the changes in bond lengths and bond angles caused by the application of a stress (Treloar, 1960). The changes in these quantities were calculated from the appropriate force constants derived from spectroscopic data. A value of $5.65 \cdot 10^{11}$ dyn $\cdot$ cm^{-2}, i.e. 56 GPa was derived for cellulose. This quite low value was assigned to the neglect of secondary forces derived from spectroscopic data.

The cellulose crystal modulus was first studied experimentally in 1962 for cellulose I (Sakurada et al., 1962) and cellulose II (Mann and Roldan-Gonzales, 1962). For cellulose I, the modulus value was determined from the crystal deformation of highly oriented fibers of bleached ramie. In this study, the experimental determination of the elastic moduli of crystalline regions of other polymers, such as polyethylene, polyvinyl alcohol, polyvinylidene chloride, polypropylene, and polyoxymethylene, was also investigated using highly oriented filaments or fibers. The lattice extension was measured by X-ray diffraction under a constant stress, so that the relaxation had no influence on the result. The fiber specimen around 35 mm long was mounted horizontally in stretching clamps and a constant weight was applied to the fiber by use of a pulley in order to induce a given extension. The stress in the crystalline regions was assumed to be equal to the stress applied to the sample, and this assumption of a homogeneous stress distribution was proven experimentally. At the zero position of the mount, the angle between the X-ray beam and fiber axis was 90°. The fiber axis was tilted by an angle θ to meet the Bragg conditions and obtain the most intense diffraction rays. The lattice extension was measured under this constant stress and similar experiments were carried out by varying the thickness of the fiber and the applied weight. Then, the stress-strain curves were plotted. The calculation of the elastic modulus was based on the assumption of the series model in which crystalline and amorphous regions alternate along the length of the fiber. A value of $134 \cdot 10^{4}$ kg $\cdot$ cm^{-2}, i.e. 134 GPa, was reported for cellulose I. The dimensions of the unit cell were a = 8.35 Å, b = 10.3 Å, c = 7.9 Å (corresponding to the FIP – fiber identity period – or length of 2 glucose units), and β = 84°. Measurements were also performed under various relative humidities (Sakurada et al., 1964), for which the modulus of the macroscopic specimen changed from 12 to 27 GPa. This change was assumed to be due to the strong variability of the properties of amorphous regions upon water vapor adsorption since the modulus of crystalline domains remains unchanged when varying the atmosphere.

X-ray measurements of the elastic modulus of cellulose II crystals were performed using Fortisan H fibers (Mann and Roldan-Gonzales, 1962). The position of the 040 reflection was measured with and without load on the fibers. The crystallographic 040 planes are perpendicular to the chain axes of the cellulosic molecules and the 040 spacing gave a measure of the length of the repeating unit of the chain. An apparent modulus value ranging between 7 and $9 \cdot 10^{11}$ dyn · cm^{-2}, i.e. 70 and 90 GPa, was calculated for crystalline regions on the basis of a series model.

The crystallite modulus of native cellulose along the chain has been calculated based on the X-ray analyzed molecular conformation and the force constants used in the vibrational analysis, i.e. which can well reproduce the actual infrared and Raman spectral data (Tashiro and Kobayashi, 1985). The molecular model was based on the result of the X-ray analysis reported by Gardner and Blackwell (1974). The dimensions of the unit cell were a = 16.34 Å, b = 15.72 Å, c = 10.38 Å, and β = 97°. The intramolecular force constants of the valence-force-field type were adapted from Cael et al. (1975) with some modification. The calculated values were 172.9 and 70.8 GPa when intramolecular hydrogen bonding was taken or was not taken into account, respectively. It evidenced the important role of intramolecular bonding on the determination of the crystallite modulus and chain deformation mechanism.

Molecular mechanics calculations have been performed on cellobiose to predict the modulus of elasticity of the cellulose chain (Kroon-batenburg et al., 1986). Either one or two intramolecular hydrogen bonds in cellobiose, parallel to the glycosidic linkage, were considered. The values derived from the model were 136 and 89 GPa, respectively. A good agreement was observed between these predicted data and experimental values for native (cellulose I) and regenerated (cellulose II) fibers. It was therefore concluded that the essential distinction between the conformations of the cellulose chain in the native and regenerated fibers was the number of intramolecular hydrogen bonds in the monomeric unit. This difference was supposed to be responsible for the respective values of the chain modulus in cellulose I and II.

The crystalline lattice moduli of cellulose I and II were measured by X-ray diffraction using ramie and mercerized ramie fibers (Matsuo et al., 1990). Values in the range 122–135 GPa and 106–112 GPa for cellulose I and II, respectively, were reported. It was shown that the crystal lattice moduli of cellulose I and II measured by X-ray diffraction depend upon morphological properties of the bulk specimen. Effects of the orientation factors of crystal and amorphous chains and crystallinity were considered. Numerical calculation indicated that the crystal lattice modulus measured by X-ray diffraction differs from the intrinsic lattice modulus when a parallel coupling between amorphous and crystalline phases is predominant, while both moduli were almost equal when a series coupling is predominant. The morphological dependence was found less pronounced when increasing the degree of molecular orientation and crystallinity. It was concluded that specimens with a high degree of molecular orientation and crystallinity should be used for measuring crystal lattice moduli by X-ray diffraction methods.

Theoretical evaluation of the three-dimensional elastic constants for the cellulose crystal forms I and II based on lattice dynamical treatment was reported by

Tashiro and Kobayashi (1991). The calculated Young's modulus along the chain axis was 167.5 GPa for form I and 162.1 GPa for form II. The lower value observed for form II was ascribed to the lower force constant value of intramolecular hydrogen bonds, showing again the importance of intramolecular hydrogen bonds, whereas the intermolecular hydrogen bonds were found to play a minor role. Anisotropy of the Young's modulus and linear compressibility in the planes perpendicular to the chain axis were also calculated. The two transverse moduli were 11 GPa and 50 GPa.

X-ray diffraction measurements were also used to determine the elastic modulus of the crystalline regions of cellulose polymorphs in the direction parallel to the chain axis (Nishino et al., 1995). Starting materials for cellulose I and II were purified ramie and polynosic fibers. Values of 138, 88, 87, 58 and 75 GPa were reported for cellulose I, II, III_I, III_{II} and IV_I, respectively. This indicated that the skeleton conformation of the different polymorphs changed upon crystal transitions and that each was completely different from a mechanical point of view. The crystal transition induced a skeletal contraction accompanied by a change in intramolecular hydrogen bonds, which is considered to result in a drastic change in the modulus value of the cellulose polymorphs. Indeed, cellulose I that showed the highest modulus value displayed the longest fiber identity period (FIP). The elastic modulus of cellulose was also calculated for geometries obtained from numerical studies of the structure using a force field (Reiling and Brickmann, 1995). Values of 134 and 135 GPa were reported for cellulose I, using the Reuss and Voigt limits, respectively, and a value of 83 GPa was calculated for cellulose II.

Molecular dynamics modeling was used to investigate the structure and mechanical properties of regenerated cellulose fibers (Ganster and Blackwell, 1996). The longitudinal modulus at room temperature was determined to be 155 GPa, whereas the value in the perpendicular direction varied between 24 and 51 GPa.

Measurements of the elastic modulus of tunicin, the cellulose extracted from tunicate – a sea animal – using Raman spectroscopic technique has been reported (Šturcova et al., 2005). Epoxy/tunicin nanocomposites were deformed using a four-point bending test, and the shift in the characteristic Raman band located at 1095 cm^{-1} was used as an indication of the stress in the material. Furthermore, since no broadening of the Raman band upon deformation was observed, it was shown that this shift was related to direct chain stretching of cellulose and that relatively little amorphous/crystalline effects seen with semicrystalline cellulosic fibers occur. This analysis yielded a value of 143 GPa for the elastic modulus of the cellulose nanocrystal.

The elastic modulus of the cellulose Iβ crystal was also calculated by the molecular mechanics simulation technique (Tanaka and Iwata, 2006). The derived values varied from 124 to 155 GPa. The molecular mechanics modeling of the deformation of a number of proposed structures for the crystalline regions of cellulose Iα, Iβ and II has been reported (Eichhorn and Davies, 2006). Chain stiffness values in the range 136–155 GPa and 116–149 GPa have been reported for cellulose Iα and Iβ, respectively.

For cellulose II the values were in the range 109–166 GPa. By removal of the hydrogen bonding in the structure, the stiffness of the chain decreased to 114–117 GPa, 124–127 GPa and 101–106 GPa for cellulose Iα, Iβ and II, respectively, showing the effect of this important parameter.

The elastic modulus of cellulosic nanocrystals (inappropriately called single microfibrils in the study) prepared from tunicate was also measured by AFM using a three-point bending test (Iwamoto et al., 2009). The tunicin nanocrystals were prepared by two chemical methods, namely by oxidation of cellulose with 2,2,6,6-tetramethylpiperidine-1-oxyl radical (TEMPO) as a catalyst followed by a subsequent mechanical disintegration in water, and by sulfuric acid hydrolysis. The nanocellulosic materials were deposited on a specially designed silicon wafer with grooves 227 nm in width. The three-point bending test was applied using an AFM cantilever in which the AFM tip was used as the third loading point and measured the applied force and the displacement of nanocrystals that bridged the nanoscale grooves fabricated on the substrate. Values of 145 and 150 GPa were reported for nanocrystals prepared by TEMPO-oxidation and acid hydrolysis, respectively.

A procedure was recently developed to calculate the transverse elastic modulus of cellulose nanocrystals by comparing the experimentally measured force-distance curves with 3D finite element calculations of tip indentation on the cellulose nanocrystals (Lahiji et al., 2010). The influence of relative humidity (RH) on the stiffness of cellulose nanocrystals was measured by comparing AFM measurements on the same nanoparticle under different humidity conditions (0.1 and 30% RH). The transverse modulus of an isolated cellulose nanocrystal was estimated to be between 18 and 50 GPa at 0.1% RH (flowing N_2 gas). A minimal effect of RH was reported, confirming the resistance of the cellulose crystals to water penetration. The flexibility of the nanocrystals was also investigated by using the AFM tip as a nanomanipulator. It showed nanocrystal bending, but it was unclear if this resulted from single-crystal bending or multiple cellulose nanocrystals pivoting to their contact point.

The different values reported for the stiffness of cellulose nanocrystals are collected in Table 1.4. These values are comparable to those reported for aromatic ring polymers such as poly-p-phenylene terephtalamide (153–200 GPa) and poly-m-phenylene isophtalamide (88 GPa) (Tashiro et al., 1977). However, it is much lower than that (235 GPa) of polyethylene, which possesses the maximum elastic modulus of the crystalline regions in the direction parallel to the chain axis (Nakamae et al., 1991). However, if the cross-sectional area of each individual molecule is considered, it is found that the modulus value is similar for cellulose and polyethylene. The different values reported for the crystal of cellulose are comparable to those reported for pure crystalline β-chitin produced by the marine diatom *Thalassiosira fluviatilis* (Xu et al., 1994).

Material	Method	E_L (GPa)	E_T (GPa)	Reference
Cellulose I	Calculation	77–121		(Meyer and Lotmar, 1936)
		56		(Treloar, 1960)
Bleached ramie fibers (cellulose I)	X-ray diffraction	134		(Sakurada et al., 1962)
Fortisan H fibers (cellulose II)	X-ray diffraction	70–90		(Mann and Roldan-Gonzales, 1962)
Cellulose I	Calculation	172.9*		(Tashiro and Kobayashi, 1985)
		70.8**		
		76	51–57	(Jaswon et al., 1968
Cellobiose (two hydrogen bonds – cellulose I)	Calculation	136 ± 6		(Kroon-Batenburg et al., 1986)
Cellobiose (one hydrogen bond – cellulose II)		89 ± 4		
Ramie fibers (cellulose I)	X-ray diffraction	122–135		(Matsuo et al., 1990)
Mercerized ramie fibers (cellulose II)		106–112		
Cellulose I	Calculation	167.5	11	(Tashiro and Kobayashi, 1991)
Cellulose II		162.1	50	
Purified ramie fibers (cellulose I)	X-ray diffraction	138		(Nishino et al., 1995)
Polynosics (cellulose II)		88		
Cellulose III_I		87		
Cellulose III_{II}		58		
Cellulose IV_I		75		
Cellulose I	Calculation	134–135		(Reiling and Brickmann, 1995)
Cellulose II		83		
Cellulose II	Calculation	155	24–51	(Ganster and Blackwell, 1996)
Cellulose I_α	Calculation	127.8		(Neyertz et al., 2000)
Cellulose I_β		115.2		
Cellulose I_β	Raman	143		(Šturcova et al., 2005)
Cellulose I_β	Calculation	124–155		(Tanaka and Iwata, 2006)
Cellulose I_α	Calculation	136–155*		(Eichhorn and Davies, 2006)
		114–117**		
Cellulose I_β		116–149*		
		124–127**		
Cellulose II		109–166*		
		101–106**		

Material	Method	E_L (GPa)	E_T (GPa)	Reference
Cellulose I_β	Calculation	156 at 300 K 117 at 500 K		(Bergenstråhle et al., 2007)
Cellulose I	Raman	57–105		(Rusli and Eichhorn, 2008)
Ramie fibers (cellulose I)	Inelastic X-ray scattering	220	15	(Diddens et al., 2008)
TEMPO-oxidized Cellulose I_β	AFM	145		(Iwamoto et al., 2009)
Acid hydrolyzed Cellulose I_β		150		
Wood	AFM		18–50	(Lahiji et al., 2010)
Disaccharide cellulose I_β	Calculation	85.2*/37.6**		(Cintrón et al., 2011)
Disaccharide cellulose I_β		99.7*/33.0**		
Extended cellulose I_β chains (10–40 glucoses)		126.0*/63.3**		

* with intramolecular hydrogen bondings
** without intramolecular hydrogen bondings

Table 1.4: Longitudinal (E_L) and transverse (E_T) moduli of crystalline cellulose.

These impressive mechanical properties make cellulose nanoparticles ideal candidates for the processing of reinforced polymer composites. Incorporating these nanoparticles in a synthetic or natural polymeric matrix consists therefore in biomimeting nature. All what scientists need to do is to try to mimic nature or to exploit natural biocomposites in order to develop novel materials that can be suitable to our needs without being harmful to the environment.

1.8 Cellulose-based materials

There is a growing interest in the utilization of biological materials, such as wood, not only as construction materials or as raw material for pulp and paper production, but also as a new feedstock for the development of advanced materials with tailor-made properties. Indeed, many commonly-used polymeric materials pose problems at the end of their intended life and are derived from petroleum. The fast-paced consumption of petroleum, roughly 100,000 times faster than nature can replenish it, and the general disposal possibilities, incineration and land filling, contribute to the unsustainability of the current situation (Netravali and Chabba, 2003). General solutions to this problem can focus either on the supply side, the life-end side, or on both at the same time. A vast amount of publications are available with work focusing on the development of polymers from renewable materials and on biodegradable polymers and composites (Mohanty et al., 2000; Rouilly and Rigal, 2002; Flieger et al., 2003; Bastioli, 2005; Wool and Sun, 2005; Yu et al., 2006).

Replacement of conventional plastics by degradable polymers, particularly for short-lived applications such as packaging, catering, surgery or hygiene, is of major interest for different actors in socio-economic life (from the plastics industry to the citizen). The potential of biodegradable polymers and more specifically of polymers derived from agro-resources, such as polysaccharides, has long been recognized. However, to date, these agro-polymers largely used in some applications (e.g. the food industry) have not found extensive applications in non-food industries, although they could be an interesting way to overcome the limitation of the petrochemical resources in the future. Material valorization implies some limitations linked to difficulties in achieving accurate and economically viable outlets.

Cellulose and more generally polysaccharides present some well-known advantages, namely low cost, lightweight, renewable character, high specific strength and modulus, availability in a variety of forms throughout the world, reactive surface and the possibility to generate energy, without residue, after burning at the end of their life cycle.

Two main groups of cellulose-based materials can be basically distinguished, viz. thermoplastically processable cellulose derivatives, such as esters, which can be used for extrusion and molding, and cellulose composites suitable only for treatment in conventional processes. Cellulose as a material is used by the natural world in the construction of plants and trees, and by man to make sails, ropes and clothes to name but a few examples. It is also the major constituent of paper and further processing can be performed to make cellophane and rayon.

1.8.1 Thermoplastically processable cellulose derivatives

Cellulose was used to produce the first successful thermoplastic polymer, celluloid, by the Hyatt Manufacturing Company in 1870. The compound was first chemically synthesized (without the use of any biologically derived enzymes) in 1992, by Kobayashi and Shoda (Klemm et al., 2005). As a carbohydrate, the chemistry of cellulose is primarily the chemistry of alcohols and it forms many of the common derivatives of alcohols, such as esters, ethers, etc. The hydroxyl groups of cellulose can be partially or fully reacted with various chemicals to provide derivatives with useful properties. These derivatives form the basis for much of the industrial technology of cellulose in use today. Because of the strong hydrogen bonds that occur between cellulose chains, cellulose does not melt or dissolve in common solvents. Thus, it is difficult to convert the short fibers from wood pulp into the continuous filaments needed for artificial silk, an early goal of cellulose chemistry. Several different cellulose derivatives were examined as early routes to artificial silk, but only two, the acetate and xanthate esters, are of commercial importance for fibers today.

Natural fibers can be used as raw materials for cellulose production. It can be modified into cellulose esters, such as cellulose acetate, cellulose acetate propionate, and butyrate, which are currently used as major components of thermoplastics. Among the esters, cellulose acetate is soluble in organic solvents such as acetone and can be spun into fiber or formed into other shapes. Xanthate esters are formed when cellulose is first treated with strong alkali and then exposed to carbon disulfide. Cellulose xanthate is soluble in aqueous alkali and the resulting solution can be extruded as filaments or films. This is the basis for the viscose process for rayon manufacture. More recently, technology has been developed to form textile fibers (Lyocell) directly from wood pulp without using a derivative to facilitate dissolution. This technology is based on the ability of amine oxides, particularly N-methylmorpholine N-oxide to dissolve unsubstituted cellulose. They are called man-made cellulosic fibers. The inorganic ester nitrocellulose was initially used as an explosive and was an early film forming material.

Ether derivatives include ethylcellulose, a water-insoluble commercial thermoplastic used in coatings, inks, binders, and controlled-release drug tablets. Other ethers are hydroxypropyl cellulose, carboxymethyl cellulose, hydroxypropyl methyl cellulose, used as a viscosity modifier, gelling agent, foaming agent and binding agent, and hydroxyethyl methyl cellulose, used in the production of cellulose films.

1.8.2 Cellulose fiber reinforced composites

Natural fibers such as flax, hemp, straw, kenaf and jute consist mainly of cellulose, hemicellulose and lignin, but they are usually listed as a material when used as fibers in composites. Agricultural fibers include crop residual, such as straw, stems, hulls, and milling by-products (e.g. brans) from wheat, corn, soybean, sorghum, oat, barley, rice, sugarcane, pineapple, banana, coconut and other crops. Large quantities of agricultural fibers are available and these lignocellulosic agricultural byproducts are a cheap source of cellulose fibers. The major composition of these fibers is similar to wood fibers and includes cellulose, lignin and pentosan, and makes them suitable for uses such as composite, textile, pulp and paper manufacture. In addition, biofibers can also be used to produce fuel, chemicals, enzymes and food. Wheat straw is usually used for fuel, manure, cattle feed, mulch and bedding materials for animals (Sampathrajan et al., 1992). Particleboard can be prepared using wheat straw, sunflower stalks, rice straw, cotton stalks, sugarcane bagasse, flax, maize husks and maize cobs. The production processes, structure, properties and suitability of these biofibers for various industrial applications has been analyzed (Reddy and Yang, 2005).

Natural fibers can also be used for composites as harvested. Over the last few years a number of researchers have been involved in investigating the exploitation of cellulosic fibers as load-bearing constituents in composite materials. This considerable interest in both the literature and industry for the possibility of replacing conventional fibers such as glass is due to some well-known advantages of lignocellulosics fibers. The specific properties of this natural product, namely low cost, lightweight, renewable character, high specific strength and modulus, availability in a variety of forms throughout the world, reactive surface, non-abrasive nature and the possibility to generate energy without residue after burning at the end of their life cycle, motivate their association with organic polymers to elaborate composite materials. However, it is well known that different surface properties between the fiber and the matrix, i.e. the former is highly polar and hydrophilic while the latter is, generally, non-polar and relatively hydrophobic, impose the surface modification of the fibers' surface, in order to improve the fiber/polymer compatibility and their interfacial adhesion. Without such a treatment the stress applied to the fiber/polymer composite is not efficiently transferred from the matrix to the fiber and the beneficial reinforcement effect of the fiber remains underexploited. Likewise, the poor ability of the polymer to wet the fiber hinders the homogenous dispersion of short fibers within the polymeric matrix. The potential and applications of lignocellulosic fibers reinforced polymers have been reviewed during the last decade (Eichhorn et al., 2001b; Bledzki and Gassan, 1999; Mohanty et al., 2005).

1.9 Conclusions

There is a considerable interest in the possibility of replacing conventional fibers, such as glass, for polymer composite applications. It is ascribed to some well-known advantages of natural fibers. The promise behind cellulose-derived composites lies in the fact that the longitudinal modulus of the basic cellulose crystal displays a high value, around 150 GPa as determined by theoretical evaluations or experimental measurements. The axial specific Young's modulus (modulus-to-density ratio) of the basic cellulose crystal is therefore potentially stronger than steel and similar to Kevlar. However, the full exploitation of this reinforcing capability requires the release of these nanocrystals from lignocellulosic fibers. In addition, the use of constituting cellulose microfibrils or nanocrystals instead of fibers allows overcoming the big variation properties inherent to natural products. Indeed, it is well known that the fiber properties depend on factors such as maturity, separating process, microscopic and molecular defects, type of soil and weather conditions under which they were grown.

1.10 References

Almond, A. (2005). Towards understanding the interaction between oligosaccharides and water molecules. Carbohydr. Res. *340*, 907–920.

Attala, R.H. (1976). Raman spectral studies of polymorphy in cellulose. Part I: celluloses I and II. Appl. Polym. Symp. *28*, 659–669.

Atalla, R.H. and VanderHart, D.L. (1984). Native cellulose: a composite of two distinct crystalline forms. Science *223*, 283–285.

Atalla, R.H., Whitmore, R.E. and Vanderhart, D.L. (1985). A highly crystalline cellulose from *Rhizoclonium hieroglyphicum*. Biopolymers, *24*, 421–423.

Bakri, B. and Eichhorn, S.J. (2010). Elastic coils: deformation micromechanisms of coir and celery fibres. Cellulose *17*, 1–11.

Bastioli, C. (2005). Handbook of biodegradable polymers (Rapra Technology Limited, Shawbury, Shrewsbury, Shropshire, UK).

Batchelder, D.N. and Bloor, D. (1979). Strain dependence of the vibrational-modes of a diacetylene crystal. J. Polym. Sci. Part B: Polym. Phys. *17*, 569–581.

Belton, P.S., Tanner, S.F., Cartier, N. and Chanzy, H. (1989). High-resolution solid-state ^{13}C nuclear magnetic resonance spectroscopy of tunicin, an animal cellulose. Macromolecules *22*, 1615–1617.

Bergenstråhle, M., Berglund, L.A. and Mazeau, K. (2007). Thermal response in crystalline Iβ cellulose: A molecular dynamics study. J. Phys. Chem. B *111*, 9138–9145.

Bernstein J. (2002). Polymorphism in molecular crystals (Clarendon Press, Oxford, UK).

Bismarck, A., Aranberri-Askargorta, I., Springer, J., Lampke, T., Wielage, B., Stamboulis, A., Shenderovich, I. and Limbach, H.-H. (2002). Surface characterization of flax, hemp and cellulose fibers; surface properties and the water uptake behaviour. Polym. Compos. *23*, 872–894.

Bismarck, A., Mishra, S. and Lampke T. (2005). Plant fibers as reinforcement for green composites. In: Natural fibers, biopolymers and biocomposites, A.K. Mohanty, M. Misra and L.T. Drzal, eds. (CRC Press, Taylor & Francis Group, Bota Raton, FL), pp. 37–108.

Blackwell, J., Vasko, P.D. and Koenig, J.L. (1970). Infrared and Raman spectra of the cellulose from the cell wall of *Valonia ventricosa*. J. Appl. Phys. *41*, 4375–4379.

Bledzki, A.K. and Gassan, J. (1999). Composites reinforced with cellulose based fibres. Prog. Polym. Sci. *24*, 221–274.

Bonard, R. (1966). Kolloidstrukturen in verstreckten Hochpolymeren. Kolloid-Z *211*, 14–33.

Boyd, J.D. and Foster, R.C. (1974). Tracheid anatomy changes as response to changing structural requirements of the tree. Wood Sci. Technol. *8*, 91–105.

Braconnot, H. (1819). Sur la conversion du corps ligneux en gomme, en sucre, et en un acide d'une nature particulière, par le moyen de l'acide sulfurique; conversion de la même substance ligneuse en ulmine par la potasse. Ann. Chim. *12*, 172–195.

Buléon, A. and Chanzy, H. (1980). Single crystals of cellulose IV_{II}: Preparation and properties. J. Polymer Sci, *18*, 1209–1217.

Cael, J.J., Gardner, K.H., Koenig, J.L. and Blackwell, J. (1975). Infrared and Raman spectroscopy of carbohydrates. Paper. V. normal coordinate analysis of cellulose I. J. Chem. Phys. *62*, 1145–1153.

Chanzy, H., Imada, K. and Vuong, R. (1978). Electron diffraction from the primary wall of cotton fibers. Protoplasma *94*, 299–306.

Chanzy, H., Imada, K., Mollard, A., Vuong, R. and Barnoud, F. (1979). Crystallographic aspects of sub-elementary cellulose fibrils occurring in the wall of rose cells cultured in vitro. Protoplasma *100*, 303–316.

Chanzy, H. and Henrissat, B. (1985). Unidirectional degradation of *Valonia* cellulose microcrystals subjected to cellulose action. FEBS Lett. *184*, 285–288.

Chanzy, H. (1990). Aspects of cellulose structure. In: Cellulose sources and exploitation, J.F. Kennedy, G.O. Phillips and P.A. Williams, eds. (Ellis Horwood Ltd, New York), pp. 3–12.

Cheng, Q., Wang, S. and Harper, D.P. (2009). Effects of process and source on elastic modulus of single cellulose fibrils evaluated by atomic force microscopy. Composites Part A *40*, 583–588.

Chu, S.S.C. and Jeffrey, G.A. (1968). The refinement of the crystal structures of β-D-glucose and cellobiose. Acta Cryst. B *24*, 830–838.

Cintrón, M.S., Johnson, G.P. and French, A.D. (2011). Young's modulus calculations for cellulose I_β by MM3 and quantum mechanics. Cellulose *18*, 505–516.

Cousins, S.K. and Brown, R.M.Jr. (1995). Cellulose I microfibril assembly: computational molecular mechanics energy analysis favours bonding by van der Waals forces as the initial step in crystallization. Polymer *36*, 3885–3888.

Debzi, E.M., Chanzy, H., Sugiyama, J., Tekely, P. and Excoffier, G. (1991). The Iα →Iβ transformation of highly crystalline cellulose by annealing in various mediums. Macromolecules, *24*, 6816–6822.

Delmer, D.P. and Amor, Y. (1995). Cellulose biosynthesis. Plant Cell *7*, 987–1000.

Diddens, I., Murphy, B., Krisch, M. and Müller, M. (2008). Anisotropic elastic properties of cellulose measured using inelastic X-ray scattering. Macromolecules *41*, 9755–9759.

Dolmetsch, H. and Dolmetsch, H. (1962). Evidence for the folding of the chains within the cellulose molecule. Kolloid-Z *185*, 106–119.

Earl, W.L. and VanderHart, D.L. (1981). Observations by high-resolution C-13 NMR of cellulose-I related to morphology and crystal-structure. Macromolecules *14*, 570–574.

Eichhorn, S.J., Hughes, M., Snell, R. and Mott, L. (2000). Strain induced shifts in the Raman spectra of natural cellulose fibers. J. Mat. Sci. Lett. *19*, 721–723.

Eichhorn, S.J., Sirichaisit, J. and Young, R.J. (2001a). Deformation mechanisms in cellulose fibres, paper and wood. J. Mater. Sci. *36*, 3129–3135.

Eichhorn, S.J., Baillie, C.A., Zafeiropoulos, N., Mwaikambo, L.Y., Ansell, M.P., Dufresne, A., Entwistle, K.M., Herrera-Franco, P.J., Escamilla, G.C., Groom, L., Hugues, M., Hill, C., Rials, T.G. and Wild, P.M. (2001b). Current international research into cellulosic fibres and composites. J. Mater. Sci. *36*, 2107–2131.

Eichhorn, S.J. and Young, R.J. (2001). The Young's modulus of a microcrystalline cellulose. Cellulose *8*, 197–207.

Eichhorn, S.J. and Young, R.J. (2003). Deformation micromechanics of natural cellulose fibre networks and composites. Compos. Sci. Technol. *63*, 1225–1230.

Eichhorn, S.J., Young, R.J., Davies, R.J. and Riekel, C. (2003). Characterisation of the microstructure and deformation of high modulus cellulose fibres. Polymer *44*, 5901–5908.

Eichhorn, S.J. and Young, R.J. (2004). Deformation micromechanics of hemp fibres and epoxy resin microdroplets. Compos. Sci. Technol. *64*, 767–772.

Eichhorn, S.J. and Davies, G.R. (2006). Modelling the crystalline deformation of native and regenerated cellulose. Cellulose *13*, 291–307.

El-Osta, M.L.M., Wellwood, R.W. and Butters, R.G. (1972). An improved X-ray technique for measuring microfibril angle of coniferous wood. Wood Sci. *5*, 113–117.

El-Osta, M.L.M. (1973). A direct X-ray technique for measuring microfibril angle. Wood and Fiber *5*, 118–129.

Fengel, D. (1971). Ideas on ultrastructural organization of cell-wall components. J. Polym. Sci. Part C *36*, 383–392.

Fengel, D. and Stoll, M. (1973). Über die Veränderungen des Zellquerschnittes, der Dicke der Zellwand und der Wandschichten von Fichtentracheiden innerhalb eines Jahrringes. Holzforschung *27*, 1–7.

Fengel, D. and Wegener, G. (1984). Wood: chemistry, ultrastructure, reactions. Walter de Gruyter, Berlin & New York.

Fink, H.P., Philipp, B., Paul, D., Serimaa, R. and Paakkari, T. (1987). The structure of amorphous cellulose as revealed by wide-angle X-ray scattering. Polymer *28*, 1265–1270.

Flieger, M., Kantorová, M., Prell, A., Řezanka, T. and Votruba, J. (2003). Biodegradable plastics from renewable sources. Folia Microbiol. *48*, 27–44.

French, A.D., Miller, D.P. and Aabloo, A. (1993). Miniature crystal models of cellulose polymorphs and other carbohydrates. Int. J. Biol. Macromol. *15*, 30–36.

Freudenberg, K. and Braun, E. (1928). Methylcellulose 5. Mitteilung über Lignin und Cellulose Ann. *460*, 288–304.

Frey-Wyssling, A., Mühlethaler, K. and Wyckoff, R.W.G. (1948). Mikro-fibrillenbau der pflanzlichen Zellwande. Experientia *4*, 475–476.

Frey-Wyssling, A. (1954). The fine structure of cellulose microfibrils. Science *119*, 80–82

Frey-Wyssling, A. (1976). The plant cell wall (Gebruder Bornträger, Berlin).

Ganster, J. and Blackwell, J. (1996). NpH-MD-simulations of the elastic moduli of cellulose II at room temperature. J. Mol. Model. *2*, 278–285.

Gardner, K.H. and Blackwell, J. (1974). The structure of native cellulose. Biopolymers *13*, 1975–2001.

Gierlinger, N., Schwanninger, M., Reinecke, A. and Burgert, I. (2006). Molecular changes during tensile deformation of single wood fibres followed by Raman microscopy. Biomacromolecules *7*, 2077–2081.

Grigoriew, H. and Chmielewski, A.G. (1998). Capabilities of X-ray methods in studies of processes of permeation through dense membranes. J. Membr. Sci. *142*, 87–95.

Guhados, G., Wan, W. and Hutter, J.L. (2005). Measurement of the elastic modulus of single bacterial cellulose fibers using atomic force microscopy. Langmuir *21*, 6642–6646.

Hamad, W.Y. and Eichhorn, S.J. (1997). Deformation micromechanics of regenerated cellulose fibers using Raman spectroscopy. J. Eng. Mater. Technol. *119*, 309–313.

Hattallia, S., Benaboura, A., Ham-Pichavant, F., Nourmamode, A. and Castellan, A. (2002). Adding value to alfa grass (Stipa tenacissima L.), soda lignin as phenolic resins. 1. Lignin characterization. Polym. Degrad. Stab. *75*, 259–264.

Hayashi, J. and Maclachlan, G. (1984). Pea xyloglucan and cellulose. 2. Macromolecular organization. Cell Physiology *75*, 596–604.

Hepworth, D.G. and Bruce, D.M. (2000). A method of calculating the mechanical properties of nanoscopic plant cell wall components from tissue properties. J. Mat. Sci. *35*, 5861–5865.

Hess, K., Mahl, H. and Gütter, E. (1957). Electron microscopic representation of long periodic intervals in cellulose fibers and comparison with the periods of other kinds of fibers. Kolloid-Z *155*, 1–19.

Hiller, C.H. (1964). Correlation of fibril angle with wall thickness of tracheids in summerwood of slash and loblolly pine. Tappi *47*, 125–128.

Hoareau, W., Trindada, W.G., Siegmund, B., Castellan, A. and Frollini, E. (2004). Sugar-cane bagasse and curaua lignins oxidatized by chlorine dioxide and reacted with furfuryl alcohol: characterization and stability. Polym. Degrad. Stab. *86*, 567–576.

Hon, D.N.-S. (1994). Cellulose: a random walk along its historical path. Cellulose *1*, 1–25.

Hsieh, Y.-C., Yano, H., Nogi, M. and Eichhorn, S.J. (2008). An estimation of the Young's modulus of bacterial cellulose filaments. Cellulose *15*, 507–513.

Irvine, J.C. and Hirst, E.L. (1923). The constitution of polysaccharides. Part VI. The molecular structure of cotton cellulose. J. Chem. Soc. *123*, 518–532.

Ishikawa, A., Okano, T. and Sugiyama, J. (1997). Fine structure and tensile properties of ramie crystalline form of cellulose I, II, III_I and IV_I, Polymer *38*, 463–468.

Iwamoto, S., Kai, W., Isogai, A. and Iwata, T. (2009). Elastic modulus of single cellulose microfibrils from tunicate measured by atomic force microscopy. Biomacromolecules *10*, 2571–2576.

Jakob, H.F., Fratzl, P. and Tschegg, S.E. (1994). Size and arrangement of elementary cellulose fibrils in wood cells: a small-angle X-ray scattering study of *Picea abies*. J. Struct. Biol. *113*, 13–22.

Jaswon, M.A., Gillis, P.P. and Mark, R.E. (1968). The elastic constants of crystalline native cellulose. Proc. Roy. Soc. *306*, 389–412.

Kantola, M. and Seitsonen, S. (1961). X-ray orientation investigations on Finnish Conifers. Ann. Acad. Sci. Fenn. A VI Phys. *80*, 1–15.

Kantola, M. and Seitsonen, S. (1969). On the relation between tracheid length and microfibrillar orientation measured by X-ray diffraction in conifer wood. Ann. Acad. Sci. Fenn. A VI Phys. *300*, 2–10.

Klemm, D., Heublein, B., Fink, H-P. and Bohn, A. (2005). Cellulose: fascinating biopolymer and sustainable raw material. Angew. Chem. Int. Ed., *36*, 3358–3393.

Kolpak, F.J. and Blackwell, J. (1976). Determination of the structure of cellulose II. Macromolecules *9*, 273–278.

Kolpak, F.J., Weih, M. and Blackwell, J. (1978). Mercerization of cellulose: Paper 1. Determination of the structure of mercerized cotton. Polymer *19*, 123–131

Kovalenko, V.I. (2010). Crystalline cellulose: structure and hydrogen bonds. Russ. Chem. Rev. *79*, 231–241.

Kroon-Batenburg, L.M.J., Kroon, J. and Northolt, M.G. (1986). Chain modulus and intermolecular hydrogen bonding in native and regenerated cellulose fibers. Polym. Commun. *27*, 290–292.

Kuga, S., Takagi, S. and Brown, R.M.J. (1993). Native folded-chain cellulose II. Polymer *34*, 3293–3297.

Lahiji, R.R., Xu, X., Reifenberger, R., Raman, A., Rudie, A. and Moon, R.J. (2010). Atomic force microscopy characterization of cellulose nanocrystals. Langmuir *26*, 4480–4488.

Langan, P., Nishiyama, Y. and Chanzy, H. (1999). A revised structure and hydrogen-bonding system in cellulose II from a neutron fiber diffraction analysis. J. Am. Chem. Soc. *121*, 9940–9946.

Lee, S.M. and Rowell, R.M. (1991). International encyclopedia of composites, VCH-Publishers, New York.

Lin, F.C., Brown, R.M.Jr., Cooper, J.B. and Delmer, D.P. (1985). Synthesis of fibrils in vitro by a solubilized cellulose synthase from *Acetobacter xylinum*. Science *230*, 822–825.

Mann, J. and Roldan-Gonzales, L. (1962). X-ray measurements of the elastic modulus of cellulose crystals. Polymer *3*, 549–553.

Mark, R.E., Kaloni, P.N., Tang, R.C. and Gillis, P.P. (1969). Cellulose: Refutation of a Folded Chain Structure. Science *164*, 72–73.

Marszalek, P.E., Oberhauser, A.F., Pang, Y-P. and Fernandez, J.M. (1998). Polysaccharide elasticity governed by chair-boat transitions of the glucopyranose ring. Nature *396*, 661–664.

Marx-Figini, M. (1964). Über die Kinetik der Biosynthese der Cellulose in der Baumwolle. Papier *18*, 546–549.

Marx-Figini, M. and Schultz, G.V. (1966). Zur Biosynthese der Cellulose. Naturwissenschaften *53*, 466–474.

Matsuo, M., Sawatari, C., Iwai, Y. and Ozaki, F. (1990). Effect of orientation distribution and crystallinity on the measurement by X-ray diffraction of the crystal lattice moduli of cellulose I and II. Macromolecules *23*, 3266–3275.

Meyer, K. and Lotmar, W. (1936). Sur l'élasticité de la cellulose. Helv. Chim. Acta *19*, 68–86.

Mohanty, A.K., Misra, M. and Hinrichsen, G. (2000). Biofibres, biodegradable polymers and biocomposites: an overview. Macromol. Mat. Eng. *276/277*, 1–24.

Mohanty, A.K., Misra, M. and Drzal, L.T. (2005). Natural fibers, biopolymers, and biocomposites (CRC Press, Taylor & Francis Group, Boca Raton, FL).

Moon, R.J. (2008). Nanomaterials in the forest products industry (McGraw-Hill Yearbook in Science & Technology, Chicago, IL), pp. 225–228.

Muggli, R., Elias, H.G. and Mühlethaler, K. (1969). Zum Feinbau der Elementarfibrillen der Cellulose. Makromol. Chem. *121*, 290–294.

Mühlethaler, K. (1969). Fine structure of natural polysaccharide systems. J. Polym. Sci. Part C *28*, 305–316.

Nakamae, K., Nishino, T. and Okubo, M. (1991). Elastic modulus of crystalline regions of polyethylene with different microstructures: experimental proof of homogeneous stress distribution. J. Macromol. Sci. Phys. *B30*, 1–7.

Netravali, A.N. and Chabba, S. (2003). Composites get greener. Materials Today *6*, 22–29.

Newman, R.H. (1999). Estimation of the lateral dimensions of cellulose crystallites using ^{13}C NMR signal strengths. Solid state NMR *15*, 21–29.

Neyertz, S., Pizzi, A., Merlin, A., Maigret, B., Borwn, D. and Deglise, X. (2000). A new all-atom force field for crystalline cellulose. I. J. Appl. Polym. Sci. *78*, 1939–1946.

Nishino, T., Takano, K. and Nakamae, K. (1995). Elastic modulus of the crystalline regions of cellulose polymorphs. J. Polym. Sci. Part B: Polym. Phys. *33*, 1647–1651.

Nishiyama, Y., Langan, P., and Chanzy, H. (2002). Crystal structure and hydrogen-bonding system in cellulose 1β from synchrotron X-ray and neutron fiber diffraction. J. Am. Chem. Soc. *124*, 9074–9082.

Nishiyama, Y., Kim, U.J., Kim, D.Y., Katsumata, K.S., May, R.P. and Langan, P. (2003a). Disorder along ramie cellulose microfibrils. Biomacromolecules *4*, 1013–1017.

Nishiyama, Y., Sugiyama, J., Chanzy, H. and Langan, P. (2003b). Crystal structure and hydrogen bonding system in cellulose 1(alpha), from synchrotron X-ray and neutron fiber diffraction. J. Am. Chem. Soc. *125*, 14300–14306.

Nishiyama, Y., Johnson, J.P., French, A.D., Forsyth, V.T. and Langan, P. (2008). Neutron crystallography, molecular dynamics, and quantum mechanics studies of the nature of hydrogen bonding in cellulose I-β. Biomacromolecules *9*, 3133–3140.

Okamura, K. (1991). Structure of cellulose. In: Wood and Cellulosic Chemistry, D.N.-S. Hon and N. Shiraishi, ed. (Marcel Dekker, New York), pp. 89–111.

O'Sullivan, A.C. (1997). Cellulose: the structure slowly unravels. Cellulose *4*, 173–207.

Paakkari, T. and Serimaa, R. (1984). A study of the structure of wood cells by X-ray diffraction. Wood Sci. Technol. *18*, 79–85.

Payen, A. (1838). Mémoire sur la composition du tissu propre des plantes et du ligneux. C.R. Hebd. Seances Acad. Sci. *7*, 1052–1056.

Peetla, P., Schenzel, K.C. and Diepenbrock, W. (2006). Determination of mechanical strength properties of hemp fibers using near-infrared Fourier transform Raman microspectroscopy. Appl. Spectrosc. *60*, 682–691.

Preston, R.D. (1934). The organisation of the walls of conifer tracheids. Phil. Trans. *B 224*, 131–174.

Preston, R.D. and Cronshaw, J. (1958). Constitution of the fibrillar and non-fibrillar components of the walls of *Valonia ventri- cosa*. Nature *181*, 248–250.

Preston, R.D. 1975. X-ray analysis and the structure of the components of plant cell walls. Physics Reports *21*, 183–226.

Reddy, N. and Yang, Y. (2005). Biofibers from agricultural byproducts for industrial applications. Trends in Biotechnology *23*, 22–27.

Reiling, S. and Brickmann, J. (1995). Theoretical investigations on the structure and physical properties of cellulose. Macromol. Theory Simul. *4*, 725–743.

Reis, D., Vian, B., Chanzy, H. and Roland, J.-C. (1991). Liquid crystal-type assembly of native cellulose-glucuronoxylans extracted from plant cell wall. Biology of the Cell *73*, 173–178.

Reiterer, A., Jakob, H.F., Stanzl-Tschegg, S.E. and Fratzl, P. (1998). Spiral angle of elementary cellulose fibrils in cell walls of *Picea abies* determined by small-angle X-ray scattering. Wood Sci. Technol. *32*, 335–345.

Revol, J.F., Bradford, H., Giasson, J., Marchessault, R.H. and Gray, D.G. (1992). Helicoidal self-ordering of cellulose microfibrils in aqueous suspension. Int. J. Biol. Macromol. *14*, 170–172.

Roche, E. and Chanzy, H. (1981). Electron microscopy study of the transformation of cellulose I into cellulose III_I in *Valonia*. Int. J. Biol. Macromolecules *3*, 201–206.

Rong, M.Z., Zhang, M.Q., Liu, Y., Yang, G.C. and Zeng, H.M. (2001). The effect of fiber treatment on the mechanical properties of unidirectional sisal-reinforced epoxy composites. Compos. Sci. Technol. *61*, 1437–1447.

Rouilly, A. and Rigal, L. (2002). Agro-materials: a bibliographic review. J. Macromol. Sci. Polym. Rev. *C42*, 441–479.

Rusli, R. and Eichhorn, S.J. (2008). Determination of the stiffness of cellulose nanowhiskers and the fiber-matrix interface in a nanocomposite using Raman spectroscopy. Appl. Phys. Lett. *93*, 033111.

Sakurada, I., Nukushina, Y. and Ito, T. (1962). Experimental determination of the elastic modulus of crystalline regions oriented polymers. J. Polym. Sci. *57*, 651–660.

Sakurada, I., Ito, T. and Nakamae, K. (1964). Elastic moduli of polymer crystals for the chain axial direction. Makromol. Chem. *75*, 1–10.

Sampathrajan, A., Vijayaraghavan, N.C. and Swaminathan, K.R. (1992). Mechanical and thermal properties of particle boards made from farm residues. Bioresour. Technol. *40*, 249–251.

Sandermann, W. (1973). Die "wahren" Dimensionen im Makromolekularen Bereich. Holz Roh-Werkst. *31*, 11.

Sarko, A. and Muggli, R. (1974). Packing analysis of carbohydrates and polysaccharides. III. *Valonia* cellulose and cellulose II. Macromolecules *7*, 486–494.

Sarko, A., Southwick, J. and Hayashi, J. (1976). Packing analysis of carbohydrates and polysaccharides 7. Crystal structure of cellulose III_I and its relationship to other cellulose polymorphs. Macromolecules *9*, 857–863.

Sarko, A. (1987). Cellulose – how much do we know about its structure? In: Wood and Cellulosics: Industrial utilization, biotechnology, structure and properties, J.F. Kennedy, ed. (Ellis Horwood, Chichester, UK), pp. 55–70.

Sasaki, K. and Taylor, I.E.P. (1984). Specific labelling of cell-wall polysaccharides with myo-[2-H-3]-inositol during germination and growth of *Phaseolus-vulgaris L.* Cell Physiology *25*, 989–997.

Sassi, J.F. and Chanzy, H. (1995). Ultrastructural aspects of the acetylation of cellulose. Cellulose *2*, 111–127.

Saxena, I.M. and Brown, R.M.Jr. (2000a). Cellulose biosynthesis: a model for understanding the assembly of biopolymers. Plant Physiol. Biochem. *38*, 57–67.

Saxena, I.M. and Brown, R.M.Jr. (2000b). Cellulose synthases and related enzymes. Current Opinion in Plant Biology *3*, 523–531.

Saxena, I.M. and Brown, R.M.Jr. (2005). Cellulose biosynthesis: Current views and evolving concepts. Ann. Botany *96*, 9–21.

Sisson, W. (1938). The existence of mercerized cellulose and its orientation in *Halicystis* as indicated by x-ray diffraction analysis. Science *87*, 350–351

Sjoström, E. (1981). Wood chemistry fundamentals and applications (Academic Press Inc., New York).

St John Manley, R. (1964). Fine structure of native cellulose microfibrils. Nature *204*, 1155–1157.

Stipanovic, A.J. and Sarko, A. (1976). Packing analysis of carbohydrates and polysaccharides. 6. Molecular and crystal structure of regenerated cellulose II. Macromolecules *9*, 851–857.

Šturcova, A., Davies, G.R. and Eichhorn, S.J. (2005). Elastic modulus and stress-transfer properties of tunicate cellulose whiskers. Biomacromolecules *6*, 1055–1061.

Sugiyama, J., Chanzy, H. and Maret, G. (1992). Orientation of cellulose microcrystals by strong magnetic fields. Macromolecules *25*, 4232–4234.

Tanaka, F. and Iwata, T. (2006). Estimation of the elastic modulus of cellulose crystal by molecular mechanics simulation. Cellulose *13*, 509–517.

Tarchevsky, I.A. and Marchenko, G.N. (1991). Cellulose: Biosynthesis and structure (Springer, Berlin).

Tashiro, K., Kobayashi, M. and Tadokoro, H. (1977). Elastic moduli and molecular structures of several crystalline polymers, including aromatic polyamides. Macromolecules *10*, 413–420.

Tashiro, K. and Kobayashi, M. (1985). Calculation of crystallite modulus of native cellulose. Polym. Bull. *14*, 213–218.

Tashiro, K. and Kobayashi, M. (1991). Theoretical evaluation of three-dimensional elastic constants of native and regenerated celluloses: role of hydrogen bonds. Polymer *32*, 1516–1526.

Taylor, I.E.P. and Wallace, J.C. (1989). The structural association between cellulose and xyloglucan in the primary cell wall of beans. In: Cellulose and wood: chemistry and technology. Proceedings of the Tenth Cellulose Conference, C. Schuerch, ed. (John Wiley and Sons, New York), pp. 273–282.

Tolstoguzov, V. (2004). Why are polysaccharides necessary? Food Hydrocolloids *18*, 873–877.

Treloar, L.R.G. (1960). Calculations of elastic moduli of polymer crystals: III. Cellulose, Polymer *1*, 290–303.

Tze, W.T.Y., Gardner, D.J., Tripp, C.P. and O'Neill, S.C. (2006). Cellulose fiber/polymer adhesion: effects of fiber/matrix interfacial chemistry on the micromechanisms of the interphase. J. Adhes. Sci. Technol. *20*, 1649–1668.

Tze, W.T.Y., O'Neill, S.C., Tripp, C.P., Gardner, D.J., and Shaler, S.M. (2007). Evaluation of load transfer in the cellulosic-fiber/polymer interphase using a micro-Raman tensile test. Wood Fiber Sci. *39*, 184–195.

U.S. Department of Energy, Office of Science. *Genomics: GTL Roadmap*; Available online: http://genomicsgtl.energy.gov/roadmap/ (accessed on August 12, 2012)

Valadez-Gonzalez, A., Cervantes-Uc, J.M., Olayo, R. and Herrera-Franco, P.J. (1999). Effect of fibre surface treatment on the fibre-matrix bond strength of natural fibre reinforced composites. Compos. B *30*, 309–320.

Van Daele, Y., Revol, J.F., Gaill, F. and Goffinet, G. (1992). Characterization and supramolecular architecture of the cellulose-protein fibrils in the tunic of the sea peach *(Halocynthia papillosa,* Ascidiacea, Urochordata). Biol. Cell *76*, 87–96.

Vietor, R.J., Mazeau, K., Lakin, M. and Perez, S. (2000). A priori crystal structure prediction of native cellulose. Biopolymers *54*, 342–354.

Wada, M., Kondo, T. and Okano, T. (2003). Thermally induced crystal transformation from cellulose Iα to Iβ. Polym. J. *35*, 155–159.

Wada, M., Heux, L. and Sugiyama, J. (2004). Polymorphism of cellulose I family: reinvestigation of cellulose IV_I. Biomacromolecules *5*, 1385–1391.

Watanabe, A., Morita, S. and Ozaki, Y. (2007). Temperature-dependent changes in hydrogen bonds in cellulose Iα studied by infrared spectroscopy in combination with perturbation-correlation moving-window two-dimensional correlation spectroscopy: Comparison with cellulose Iβ. Biomacromolecules *8*, 2969–2975.

Wellard, H.J. (1954). Variation in the lattice spacing of cellulose. J. Polym. Sci. *13*, 471–476.

Willstätter, R. and Zechmeister, L. (1913). Zur Kenntnis der Hydrolyse von Cellulose I. Ber. Deutsch. Chem.. Ges. *46*, 2401–2412.

Witter, R.; Sternberg, U.; Hesse, S., Kondo, T., Koch, F-Th. and Ulrich, A.S. (2006). 13C chemical shift constrained crystal structure refinement of cellulose Iα and its verification by NMR anisotropy experiments. Macromolecules *39*, 6125–6132.

Woodcock, C. and Sarko, A. (1980). Packing analysis of carbohydrates and polysaccharides. 11. Molecular and crystal structure of native ramie cellulose. Macromolecules *13*, 1183–1187.

Wool, R.P. and Sun, X.S. (2005). Bio-based polymers and composites (Elsevier, Amsterdam).

Xu, W., Mulhern, P.J., Blackford, B.L., Jericho, M.H. and Templeton, I. (1994). A new atomic force microscopy technique for the measurement of the elastic properties of biological materials. Scanning Microsc. *8*, 499–506.

Yeh, W.-Y. and Young, R.J. (1999). Molecular deformation processes in aromatic high modulus polymer fibres. Polymer *40*, 857–870.

Yu, L.; Dean, K. and Li, L. (2006). Blends and composite from renewable resources. Prog. Polym. Sci. *31*, 576–602.

Zugenmaier, P. (2001). Conformation and packing of various crystalline cellulose fibers. Prog. Polym. Sci. *26*, 1341–1417.

2 Preparation of microfibrillated cellulose

The hierarchical structure of natural fibers can be destructured using a top-down deconstruction strategy consisting in extracting the structural cellulose microfibrils' sub-elements. Destruction of the multi-level organization of natural fibers can be done mechanically by submitting slurries of cellulose fibers to high shearing forces. The ensuing material, called microfibrillated cellulose, is composed of nanosized cellulose fibrils with a high aspect ratio. This chapter describes the different techniques reported in the literature to prepare this new material. Morphological features and induced changes in properties are also reported.

2.1 Fiber fibrillation process

If plant cell walls are subjected to a strong enough mechanical disintegration action, the original fiber structure is ruined and microfibrils or microfibril bundles with diameters in the order of 10–100 nm can be extracted. The length can reach the μm scale. Different top-down strategies have been used to reverse the natural assembly process found in hierarchically structured biomaterials, i.e. to extract nanosized fibers from microsized natural fibers.

Acronym	Terminology	Reference
MFC	Microfibrillated Cellulose	(Herrick et al., 1983; Turbak et al., 1983)
–	Cellulose Microfibrils	(Dufresne et al., 1997; Dinand et al., 1999)
–	Fibrillated Cellulose	(Azizi Samir et al., 2004)
–	Nanofibrillar Cellulose	(Jin et al., 2004)
–	Fibril Aggregates	(Cheng et al., 2007)
–	Nanoscale Cellulose Fibrils	(Pääkkö et al., 2007)
–	Microfibrillated Cellulose Nanofibers	(Henriksson et al., 2007)
–	Cellulose Fibril Aggregates	(Cheng et al., 2007)
CNF	Cellulose Nanofibers	(Abe et al., 2007; Alemdar and Sain, 2008)
–	Cellulose Nanofibrils	(Henriksson et al., 2008; Ahola et al., 2008a; 2008b)
–	Cellulose Microfibers	(Bhattacharya et al., 2008)
–	Microfibril Aggregates	(Abe et al., 2009)
–	Cellulose Microfibril Aggregates	(Abe and Yano, 2009)
–	Cellulose Fibrils	(Cheng et al., 2009a; 2009b))
NFC	Nanofibrillated Cellulose	(Mörseburg and Chinga-Carrasco, 2009; Chinga-Carrasco and Syverud, 2010)
–	Microfibrillar Cellulose	(Spence et al., 2010)

Table 2.1: Different terminologies used in the literature to describe the material resulting from the cellulose fiber fibrillation process.

The terminology microfibrillated cellulose (MFC) was first used by Herrick et al. (1983) and Turbak et al. (1983) from ITT Rayonier Research Center in Shelton, Washington, USA in two companion papers published in the same journal. The research reported in these papers was carried out over a five-year period. The first one was oriented toward the study of the chemical properties (Herrick et al., 1983) while the other one concentrated on physical properties and end-use applications (Turbak et al., 1983). However, different terminologies are used to describe the material resulting from the cellulose fiber fibrillation process sometime leading to misunderstanding and ambiguities. The various terms used in the literature to describe MFC are reported in Table 2.1.

2.1.1 Purification of cellulose

Bleached pulps are often used in order to skip the matrix removal process. If other material is used, it is generally necessary to submit it to a purification step using chemical treatments to remove as much of the non-cellulosic components. This purification step needs to be adapted depending on the source. The biomass is generally first ground to increase the accessibility of the material to further treatments. Dewaxing in a Soxhlet apparatus with a toluene/ethanol or benzene/ethanol mixture is sometime performed.

Then the residue is dispersed in a 2% sodium hydroxide (NaOH) solution. This alkali extraction treatment is performed at 80°C and followed by filtration and washing with water to remove the soluble polysaccharides. The washed product is then bleached with a sodium chlorite ($NaClO_2$) solution in a buffer medium under mechanical stirring following a well-established method (Wise et al., 1946). This treatment removes most of the residual phenolic molecules like lignin or polyphenols and proteins, and the resulting bleached product consists essentially of individualized cells. This treatment has been applied to sugar beet pulp (Dinand et al., 1996; Dufresne et al., 1997; Dinand et al., 1999; Leitner, 2007; Agoda-Tandjawa et al., 2010), potato pulp (Dufresne et al., 2000), cladodes (Malainine et al., 2003; Malainine et al., 2005) and peel of *Opuntia ficus indica* (Habibi et al., 2009), swede root tissue (Bruce et al., 2005), hemp (*Cannabis sativa L.*) (Wang et al., 2007), soybean pods (Wang and Sain, 2007a; Wang and Sain, 2007b), sugarcane bagasse (Bhattacharya et al., 2008), wheat straw and soy hulls (Alemdar and Sain, 2008), sisal (Siqueira et al., 2009), banana rachis (Zuluaga et al., 2009), and rachis of the date palm tree (Bendahou et al., 2010). An alternative method consists in using potassium hydroxide (KOH) instead of sodium hydroxide (NaOH) (Abe et al., 2007; Abe and Yano, 2009; Chen et al., 2011).

Specific alkaline purification treatment of fiber pulp leads to the solubilization of lignin and partial disencrustation of the cellulose microfibrils from the other components, leaving a small amount of hemicellulose and eventually pectin at the microfibril surface. These components are critical as they are responsible not only for the ease in cell wall disruption during mechanical treatment, but also for the specific properties of MFC when homogenized and suspended in water (Dinand et al., 1996). Indeed, residual non-cellulosic components on the surface of MFC are for instance responsible for the stability of aqueous suspensions. The alkaline extraction needs to be carefully controlled to avoid undesirable cellulose degradation and to ensure that the reaction only occurs at the fiber surface, leaving intact nanofibrils for extraction (Bhatnagar and Sain, 2005; Wang and Sain, 2007a).

At this stage, the different cell walls are generally well individualized (Figure 2.1), but the microfibrils are still associated within the cell wall. The resulting bleached material consists of individual cell-ghosts corresponding essentially to the flattened microfibrillar envelopes of the cellulose microfibrils within the cell wall. In order to extract and individualize the microfibrils from the cell walls, a mechanical treatment is required.

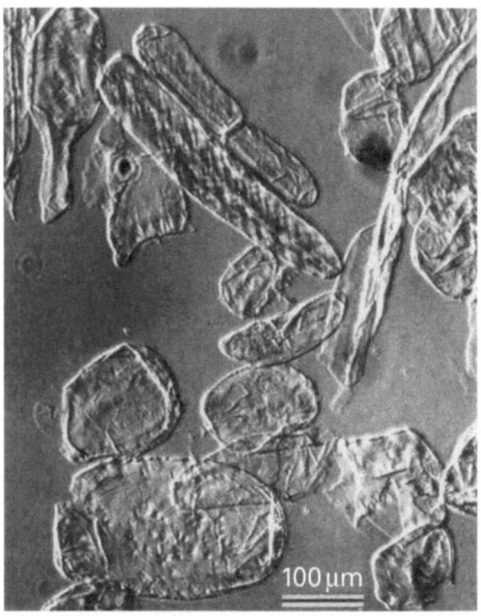

Fig. 2.1: Optical micrograph in Nomarski contrast showing individualized sugar beet cell wall (Dufresne et al., 1997).

Attempts to use unpurified sugar beet (Dufresne et al., 1997) or kraft wood (Spence et al., 2010a; Spence et al., 2010b) pulps were also reported. The motivation of these studies was to investigate the use of MFC containing pectins or aromatic lignin that occurs in nature intimately linked to native cellulose fibers.

2.1.2 High-pressure homogenization

In the pioneering work of Herrick et al. (1983) and Turbak et al. (1983), commercial softwood pulps were used. It was shown that pulps produced by the sulfite pulping process gave the best results. Both never-dried and uncut, and wood pulp fibers cut in the dry state to a fiber length of 0.7 mm and suspended in water were employed. Diluted slurries of cellulose fibers (1–2 wt%) were subjected to repeated high-pressure (55 MPa) homogenizing action. A laboratory Gaulin type milk homogenizer was used. Homogenizing temperature was controlled in the range 70–90°C. In this process, the suspension is passed through a thin slit where it is subjected to large shear forces. During homogenization dilute slurries of fibers are pumped at high pressure and fed through a spring-loaded valve assembly. As this valve opens and closes in rapid succession, the fibers are subjected to a large pressure drop with shearing and impact forces. This combination of forces promotes a high degree of fibrillation of the cellulosic fibers, resulting in the production of MFC. Figure 2.2 illustrates the functional part of a slit homogenizer. The function of the homogenizer is explained in detail by Rees (1974).

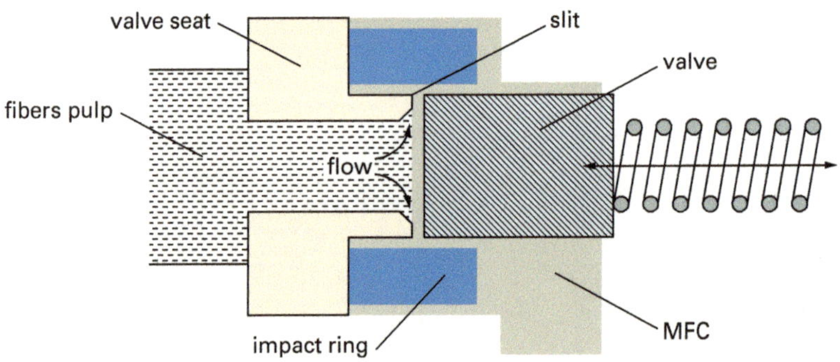

Fig. 2.2: Scheme of the slit homogenizer.

The homogenizer is fitted with a product recycle line so that a given volume of slurry can be passed through the homogenizer valve in a plug-flow manner and the number of passes through the machine can be estimated by measuring the flow rate. After 5 to 10 passes through the homogenizer, the aqueous pulp dispersion becomes creamy, more viscous and translucent. Well-homogenized suspension is a stable hydrogel that does not settle out or separate appreciably from the aqueous phase during 18 months or more when stored at room temperature in a closed container. Such gels have pseudoplastic viscosity properties and are very fluid when stirred at high shear rate (see Chapter 6). During homogenization, pulp fibers are rapidly fibrillated into a high-volume spongy structure, expanded in surface area and opened into their substructural microfibrils. While water is the most convenient and cheapest liquid medium for preparing MFC dispersions with exceptionally smooth characteristics, any polar fluid may be used, and the degree of microfibrillation that is achieved depends on the polarity and swelling properties of the chosen liquid. Satisfactory MFC dispersions have been prepared in glycerin, propylene glycol, dimethyl sulfoxide, dimethylformamide, and mixtures of these with water (Turbak et al., 1983).

An alternative to the homogenizer is the microfluidizer (Microfluidics Inc., USA). The fluid slurry is pumped through z-shaped interaction chambers where it is submitted to high shear forces (Figure 2.3). Pressure can reach levels as high as 40,000 psi, i.e. around 2,760 bars. Within the chamber, there are specially designed fixed geometry micro-channels through which the product accelerates to high velocities, creating desired shear and impact forces as the slurry stream impinges upon itself and on wear-resistant surfaces. As the intensifier pump continues its travel in one direction, a series of check valves allow the product to be drawn into the opposite end of the pump. As the intensifier pump completes its stroke, it reverses direction and the new volume of product is pressurized, repeating the process. This creates a constant flow of product at near constant pressure through the interaction chamber. Upon exiting the interaction chamber, the product may be directed through a heat exchanger, recirculated through the system for further processing or directed externally to the next step in the process. It is necessary to repeat the process several times, in order to improve the degree of fibrillation. Typical applications for the microfluidizer are cell disruption, emulsion generation, liposome generation and particle size reduction into the nanometer range.

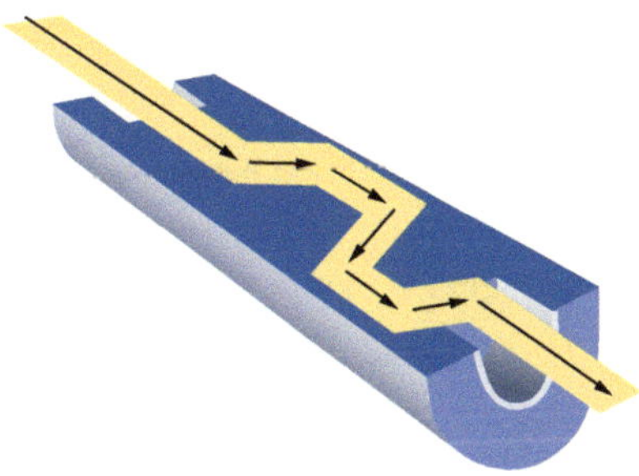

Fig. 2.3: Details of the z-shaped interaction chamber of the microfluidizer (Microfluidics Inc., USA).

2.1.3 Grinding

Mechanical treatment of cellulosic fibers can also be performed through a friction grinding process by grinding discs. The grinder features two nonporous ceramic grinding discs with an adjustable clearance between the upper and lower discs. While the upper grinding disc is fixed, the lower one is rotated at a high speed. These discs have surfaces fitted with bars and grooves against which the fibers are subjected to repeated cyclic stresses. The raw material is fed into a hopper and dispersed by centrifugal force into the clearance between the grinding stones where it is ground into ultra-fine particles, after being subjected to massive compression, shearing and rolling friction forces. During grinding, cellulose fiber fibrillation is obtained by passing the cellulose slurry between the static grinding stone and the rotating grinding stone (Figure 2.4) revolving at ~ 1,500 rpm and designed to give shearing stress to the longitudinal fiber axis of the fibrous material.

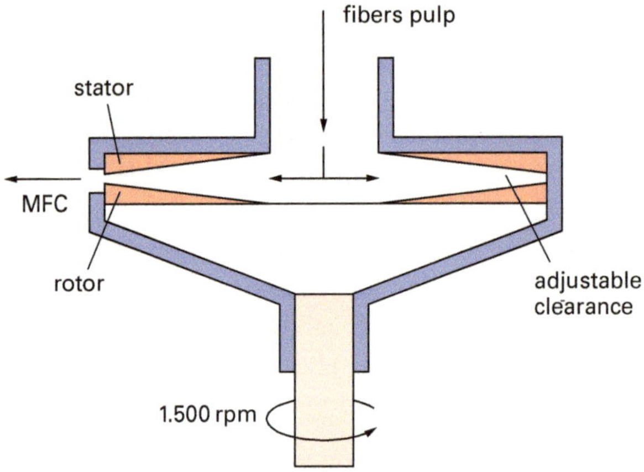

Fig. 2.4: Scheme of the friction grinding process by grinding discs.

Taniguchi and Okamura (1998) obtained MFC with diameters in the range 20–90 nm by a super-grinding method. A variety of natural fibers were used, such as wood pulp fibers, cotton fibers, tunicin, chitosan, silk fibers and collagen. The pulp was passed 10 times through the super-grinding machine and homogeneous, strong and translucent films between 3 and 100 μm thick were obtained by casting the ensuing suspension on plastic plates and drying. Hybrid MFC films were also prepared by mixing slurries of two or more types of natural fibers.

By repeating the grinding treatment ten times, 50–100 nm wide cellulose nanofibers were obtained (Iwamoto et al., 2007). However, degradation of the pulp fibers resulting from the high shearing forces generated by the grinding stones was reported. Because of the complicated multilayered structure of plant fibers and interfibrillar hydrogen bonding, the material obtained by this method consists of aggregated nanofibers with a wide distribution in width (Abe et al., 2007). An efficient extraction process of wood cellulose nanofibers as they exist in the cell wall, with a uniform width of 15 nm, was proposed using a very simple mechanical treatment (Abe et al., 2007). After removal of the matrix substance (lignin and hemicellulose), microscopic observation showed the presence of individualized microfibril bundles approximately 15 nm wide. However, as the sample still maintained the initial cell shape, the slurry was submitted to a mechanical treatment. The grinding treatment was performed in undried state, keeping the material in the water-swollen state, thus avoiding irreversible hydrogen bonding between cellulose bundles. Therefore, it was shown that the never-dried process was effective and enabled the extraction of the natural nanofibers as they exist in wood cell walls, i.e. with a uniform diameter around 15 nm.

Grinding is sometimes used as a pretreatment step of the fiber suspension before submitting it to a subsequent high-pressure homogenization process (Stelte and Sanadi, 2009).

In a recent study (Uetani and Yano, 2011), it was reported that nanofibrillation of never-dried pulp with a high-speed blender yields MFC showing the same degree of fibrillation with less damage to the nanofibrils compared with the pulp treated in a grinder. In addition, the high-speed blender is an open system that enabled observation of the fibrillation mechanism. It was shown that during nanofibrillation many balloons were formed along the pulp during agitation (Figure 2.5). As the balloons extended to the edges, the fibrils individualized rapidly. Moreover, the investigation of the degree of fibrillation in various NaCl solutions suggested that the repulsion due to the electric double layer of the MFC surface plays a critical role in the occurrence of fibrillation via balloons (Uetani and Yano, 2011).

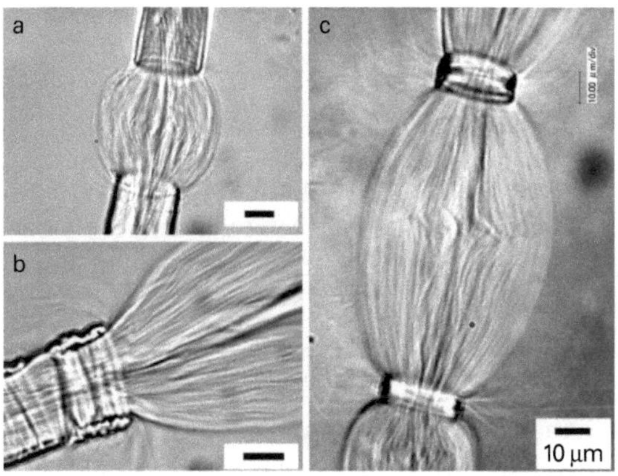

Fig. 2.5: Optical microscope observation of balloons on straw pulp. The interstices of longitudinal S2 secondary wall fibers swell because of some kind of repulsion force (Uetani and Yano, 2011).

2.1.4 Cryocrushing

Cryocrushing is a marginal and little used method because in this procedure the target diameter of microfibrils ranges between 0.1 and 1 µm as reported in a study considering bleached northern black spruce pulp (Chakraborty et al., 2005). Rather large bundles of microfibrils are therefore considered. It consists in immersing refined fibers in liquid nitrogen to freeze the water in the fibers. The high-shear refining step was used to form fibrillation at the surface of the fiber bundles. The frozen pulp is subsequently crushed with a cast iron mortar and pestle. During this step, the ice crystals exert pressure on the cell walls and sufficient energy was expected to be imparted to provoke the liberation of nanofibrils. This material was then soaked in water and homogeneously dispersed using a disintegrator or freeze-dried. It was reported that 75% of the fibrils have diameters up to 1 μm for the freeze-dried material whereas it was 89% for the water-dispersed material (Chakraborty et al., 2005). This difference was ascribed to the process of hornification for the dry sample. Some fibers with a diameter as large as 5 μm were also observed. Reduced diameters were obtained when pretreating the fibers with a fungus (Janardhanan and Sain, 2006). The cryocrushing method was also applied to chemically treated flax, hemp and rutabaga fibers (Wang et al., 2007), soybean stock (Wang and Sain, 2007a), soybean pods (Wang and Sain, 2007b), and wheat straw and soy hulls (Alemdar and Sain, 2008). Nanofibers in the range 10–100 nm are generally obtained. However, it is worth noting that in these four last studies, cryocrushing was rather used as a pretreatment since the material was later submitted to a high-pressure defibrillation process, as already reported (Dufresne et al., 1997). It explains why the lateral dimensions of the ensuing MFC were reduced compared to the earlier works.

2.1.5 High-intensity ultrasonication

Another common laboratory-scale method for cell disruption applies ultrasound (typically 20–50 kHz) to the sample (sonication). This technique has also been marginally reported in the literature for the preparation of MFC. In principle, the high-frequency is generated electronically and the mechanical energy is transmitted to the sample via a metal probe that oscillates with high frequency. The probe is placed into the cell-containing sample and the high-frequency oscillation causes a localized high pressure region resulting in cavitation and impaction, ultimately breaking open the cells. Some disadvantages beset this technique, such as heat generation by the ultrasound process that must be dissipated, high noise levels (most systems require hearing protection and sonic enclosures), yield variability, and free radicals generation that can react with other molecules.

A common sonifier apparatus equipped with a cylindrical titanium alloy probe tip 2.5 cm in diameter was used at a frequency of 20 kHz to prepare nanofibers from various materials such as spider and silkworm silks, and chitin, collagen, cotton, bamboo, ramie and hemp fibers (Zhao et al., 2007). After ultrasonication, the material was cooled to room temperature and nanofibers with uniform diameters in the range 25–120 nm were collected at the bottom of the vessel. The treatment was conducted at different powers and durations and it was reported that the disassembly process became faster when increasing the intensity of the ultrasonication. It was also observed that the defibrillation was faster for fibroin fibers than for collagen, chitin and cellulose fibers.

High-intensity ultrasonication was also reported as a method to prepare MFC from regenerated and native cellulose fibers, as well as microcrystalline cellulose (MCC) (Cheng et al., 2007; Wang and Cheng, 2009; Cheng et al., 2009a; Cheng et al., 2009b). It was argued that high-intensity ultrasonication treatment can produce efficient mechanical oscillating power because of cavitation that includes the formation, expansion, and implosion of microscopic gas bubbles when the liquid molecules absorb ultrasonic energy. The action of hydrodynamic forces of the ultrasound is expected to lead to the defibrillation of lignocellulosic fibers. Six factors were considered, viz. power, temperature, time, concentration of the suspension, fiber size and distance from the ultrasonic probe (Wang and Cheng, 2009). However, only a mixture of microscale and nanoscale fibrils was obtained (Cheng et al., 2009a). The diameter of the ensuing particles was widely distributed between 20–30 nm to several microns, showing that some fibrils were peeled from the fibers whereas some remain on the fiber surface.

MFC has been also prepared by high-intensity ultrasonication from chemically purified cellulose fibers extracted from wood, bamboo, wheat straw and flax (Chen et al., 2011). Except for flax, nanofibers with diameters ranging between 10 and 40 nm were isolated. For flax, it was observed that the nanofibrillation was not uniform and that large aggregates remained after the ultrasonic treatment. It was supposed to be due to the high cellulose content of flax and ensuing low hemicellulose, lignin and other matrix material amount removal during the chemical treatment. This could lead to strong hydrogen bonding forces between the nanofiber bundles that still persist in the purified material. MFC was also prepared from MCC by ultrasonication using different experimental conditions (power and time) (Frone et al., 2011).

2.1.6 Electrospinning

Electrostatic fiber spinning or "electrospinning" is a versatile method to prepare fibers with diameters ranging from several microns down to 100 nm through the action of electrostatic forces (Figure 2.6). In this process a polymer solution is positively charged to high voltage to produce the submicron scale fibers from an orifice to a collector (Huang et al., 2003). At a voltage sufficient to overcome surface tension forces, fine jets of polymer solution or polymer melt shoot out toward a grounded collector. The jet is stretched and elongated before it reaches the target, dries and is collected as an interconnected web of fibers with typical diameter of several hundred nanometers. Electrospinning shares characteristics of both electrospraying and conventional solution dry spinning of fibers. The process is non-invasive and does not require the use of coagulation chemistry or high temperatures to produce solid threads from solution. This makes the process particularly suited to the production of fibers using large and complex molecules. Electrospun fibers have a much thinner diameter (from nanometer to micrometer) than those obtained from conventional spinning processes (melt spinning, solution spinning and so on). The principle of this technique therefore differs notably from others since it does not consist in a top-down deconstruction of natural fibers.

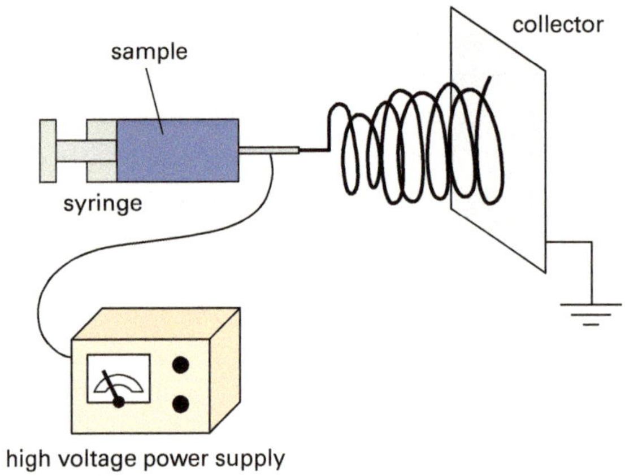

Fig. 2.6: Electrospinning setup.

Difficulties to find suitable solvents to dissolve cellulose and to process it in the melt state make cellulose nanofibers difficult to be directly prepared by electrospinning. It therefore requires either chemical derivatization or physical dissolution in a suitable solvent.

One approach is to first prepare nanofibers of cellulose derivatives. For instance, electrospun fibers of cellulose acetate (Liu and Hsieh, 2002; Liu and Hsieh, 2003; Son et al., 2004; Ma et al., 2005), carboxymethyl cellulose sodium salt, hydroxypropyl methylcellulose, methylcellulose, and enzymatically treated cellulose (Frenot et al., 2007) have been reported. Indeed, cellulose acetate is for instance soluble in many common solvents (especially acetone and other organic solvents). Electrospun cellulose acetate membrane was found to be a nonwoven mesh of fibers with poor mechanical strength because the fibers do not adhere with each other (Ma et al., 2005). Such kind of material cannot be handled or its surface modified. The usual procedure consists in heat treating the nanofiber at a temperature close to but below its melting point (224–230°C) to fuse the nanofibers with each other but maintaining the nanofibrous morphology. Then the electrospun nanofibers are treated in alkaline solution to completely remove the acetyl groups via hydrolysis reaction and to obtain regenerated cellulose nanofibers.

Various systems for dissolving cellulose without any chemical derivatization have been studied and reported, such as N,N-dimethylacetamide (DMAc)/LiCl

(McCormick et al., 1985), dimethyl sulfoxide (DMSO)/triethylamine/SO_2 (Isogai et al., 1987), N-methylmorpholine-N-oxide (NMMO) (Rosenau et al., 2002) and NaOH/urea aqueous solution (Cai and Zhang, 2005). However, the traditional viscose method for obtaining regenerated cellulose is costly and environmentally unfriendly.

Electrospinning of cellulose has also been attempted by using ionic liquids as spinning solvents (Kim et al., 2006; Viswanathan et al., 2006; Xu et al., 2008; Quan et al., 2010). For instance, nonwoven cellulose fibers have been obtained by electrospinning of cellulose in 1-butyl-3-methylinmidazolium chloride (BMIMCl) (Quan et al., 2010). The syringe was enclosed in a constant-temperature chamber (100°C) because of the high melting point of BMIMCl. The electrospun cellulose fibers were collected in a water bath to remove and dissolve quickly BMIMCl. Addition of DMSO during the dissolution of cellulose allowed smooth cellulose fibers of finer diameters to be obtained (500–800 nm). It was ascribed to the DMSO-induced enhanced swelling of cellulose in BMIMCl, so that the viscosity was reduced and the electrospinning facilitated.

2.2 Pretreatments

The flocculating nature of the fibers can cause problems during running through the narrow slit when applying high-pressure homogenization treatment to the slurry of cellulose fibers (Herrick et al., 1983). In addition, this production route is normally connected to high energy consumptions associated with the fiber delamination. Indeed, when a cellulosic pulp fiber suspension is homogenized the procedure is often repeated several times (10–15) in order to increase the degree of fibrillation. With increasing homogenization cycles, energy demand increases and values over 30,000 kWh/ton are not uncommon (Nakagaito and Yano, 2004). Even higher values reaching 70,000 kWh/ton have also been reported (Eriksen et al., 2008). The energy consumption of the mechanical treatment with a microfluidizer has been estimated assuming a processing pressure of 1,500 bar (Zimmermann et al., 2010). Considering that 10 L of cellulose pulp at 1–2 wt% need about 15 min to pass once in the microfluidizer, the crucial parameter having a strong influence on the energy consumption was identified as the number of passes. With four passes, the energy consumption was estimated to 8.5 kW. Its value increased to 14,875 kW for only three passes more with a different source of cellulose. A very precise comparative study of energy consumption and physical properties of MFC produced by different processing methods, viz. homogenizer, microfluidizer and grinder was reported (Spence et al., 2011). For bleached and unbleached kraft hardwood pulps, a comparison of the consumption as a function of the mechanical treatment, the number of passes, the pressure and the speed was reported. It was concluded that homogenizer resulted in MFC with the highest specific surface area and films with the lowest water vapor transmission rate, in spite of high energy consumption. Films produced by microfluidizer and grinder presented superior physical, optical and water interaction properties, suggesting that these materials could be produced in a more economical way for packaging applications. This high energy demand limits the application of MFC to date and the defibrillation of cellulosic fibers is a challenging process to perfect. In addition, the MFC provided solely by mechanical disintegration primarily consists of thick bundles despite high energy input.

Therefore, different pretreatments have been used to limit these problems, e.g. mechanical cutting (Herrick et al., 1983), acid hydrolysis (Boldizar et al., 1987), enzymatic treatment (Henriksson et al., 2007, Pääkkö et al., 2007), and introduction of charged groups e.g. through carboxymethylation (Wågberg et al., 2008) or 2,2,6,6-tetramethylpiperidine-1-oxyl (TEMPO)-mediated oxidation. These pretreatments of cellulose fibers have become popular with the aim of reducing the amount of mechanical energy required to liberate the microfibrils. It has been shown that energy consumption can be heavily decreased using these pretreatments to values around 1,000 kWh/ton (Siró and Plackett, 2010). This energy range is comparable with the one required to produce mechanical pulp and is therefore industrially viable.

2.2.1 Enzymatic pretreatment

The enzymatic hydrolysis of cellulose, particularly hydrogen-bonded and ordered crystalline cellulose, is a complex process (Hayashi et al., 1998). The widely accepted mechanism for cellulose hydrolysis suggests that three different types of enzyme activities work synergistically in a complete cellulase system during this process (Henrissat, 1994; Liu et al., 2009). Based on their structural properties, cellulases can be divided into three groups:

(i) endoglucanases or β-1,4-endoglucanases, which randomly hydrolyze accessible intramolecular β-1,4-glucosidic bonds in cellulose chains, generating oligosaccharides of various lengths and consequently new chain ends (Liu et al., 2009; Zhang et al., 2006);

(ii) exoglucanases (cellobiohydrolases) acting on the chain termini to release soluble cellobiose (cellobiohydrolase) or glucose (glucanohydrolase) as major products (Zhang et al., 2006);

(iii) β-glucosidases which hydrolyze cellobiose to glucose, in order to eliminate cellobiose inhibition (Liu et al., 2009; Zhang et al., 2006).

The use of cellulases for applications in the pulp and paper industry has been extensively reported, mainly for paper recycling or fiber refining (Bajpai, 1999; Spiridon and Popa, 2000; Viikari et al., 2007; Kenealy and Jeffries, 2003). Enzymes are used in fiber processing to degrade or modify hemicelluloses and lignin, while retaining the cellulosic portion. However, only limited works have been published on the use of enzymatic hydrolysis for the preparation of MFC (Pääkkö et al., 2007; Henriksson et al., 2007; Siqueira et al., 2010a; Siqueira et al., 2011).

MFC has been prepared by treating bleached kraft pulp with OS1, a fungus isolated from Dutch Elm trees infected with Dutch elm disease (Janardhnan and Sain, 2006). The fungal culture was added to the fiber suspension in a sterile flask with appropriate amounts of sucrose and yeast extract to support the fungal growth. The enzymatically treated fibers were refined and this pretreatment was shown to have a significant impact on the morphology of the fibers. A shift towards lower fiber diameters was observed for a four-day treatment. The maximum yield of fiber was below 100 nm, whereas it was between 100 and 250 nm for unpretreated fibers. It was shown that fungus OS1 had only a mild activity against cellulose, which is of interest because it minimizes the loss of cellulose during the preparation of MFC.

A combination of mild enzymatic hydrolysis using endoglucanase and high-pressure shear forces to prepare MFC from bleached wood pulp was reported (Henriksson et al., 2007, Pääkkö et al., 2007). The authors praised the milder hydrolysis as provided by enzymes in comparison to the more "aggressive" acid hydrolysis, even if the two concepts and target nanoparticles are obviously totally different. Whereas it was not possible to obtain homogeneous material when using solely the mechanical shearing without enzymatic hydrolysis, because of severe blocking problems during the homogenization step, they found that the enzymatic hydrolysis step successfully facilitates the preparation of MFC (Pääkkö et al., 2007). On the basis of transmission electron (TEM) and atomic force microscopies (AFM), and cross-polarization/magic-angle spinning (CP/MAS) ^{13}C-NMR, it was shown that the resulting MFC mainly consisted of fibrils with a diameter of 5–6 nm and fibril aggregates around 10-20 nm. One effect of high concentration enzymatic pretreatment and mechanical defibrillation is to reduce fiber length and increase the extent of fine material (Henriksson et al., 2007). However, pretreated fibers subjected to the lowest enzyme concentration (0.02%) were also successfully disintegrated while molecular weight and fiber length were well preserved. This preparation technique was also used by other authors (Svagan et al., 2007; López-Rubio et al., 2007; Henriksson et al., 2008; Pääkkö et al., 2008; Sehaqui et al., 2010).

Bleached sisal fibers have been treated with two types of commercial cellulases, viz. an endoglucanase and an exoglucanase (Siqueira et al., 2010a). The treatment was applied following two different procedures, either before (pretreatment) or after (post-treatment) mechanical shearing with a microfluidizer. It was shown that the use of endoglunases allows obtaining a mixture of MFC and stiff rod-like nanoparticles whereas the exoglucanases preserve the web-like structure of MFC regardless of the sequence of the treatments (pretreatment or post-treatment). A concept to integrate the production of MFC and sugar/cellulosic biofuel (ethanol) was reported (Zhu et al., 2011). It was found that cellulase fractionation produced a high-quality glucose stream for biofuel production through yeast fermentation, with efficiency of over

90%. It also resulted in a recalcitrant cellulosic solid fraction that was used to prepare MFC by mechanical homogenization.

2.2.2 Carboxymethylation

Carboxymethylation consists in substituting hydroxyl groups of the glucopyranose monomers that make up the cellulose backbone by carboxymethyl groups (CH_2COOH). It results from a chemical reaction between cellulose and monochloroacetic acid in the presence of sodium hydroxide (Walecka, 1956). Indeed, under alkaline conditions the accessibility of fibers to chemicals is increased by swelling and the hydroxyl group of cellulose shows high activity. Isopropanol is generally used as a suitable solvent. Two consecutive steps of reactions are required consisting in basification and etherification as follow:

$$[C_6H_7O_2(OH)_3]_n + n\,NaOH \rightarrow [C_6H_7O_2(OH)_2ONa]_n + n\,H_2O$$
$$[C_6H_7O_2(OH)_2ONa]_n + n\,ClCH_2COONa \rightarrow [C_6H_7O_2(OH)_2OCH_2COONa]_n + n\,NaCl \quad (2.1)$$

Depending on the degree of substitution, the polar (organic acid) carboxyl groups can render the cellulose soluble and chemically reactive.

MFC has been prepared from cellulose fibers pretreated by carboxymethylation prior to the homogenization step (Wågberg et al., 1987). Carboxymethylated fibers have been mechanically homogenized, dispersed by ultrasonication and centrifuged to prepare MFC. Centrifugation was performed to eliminate contamination from the titanium microtip probe resulting from the slow disintegration/destruction/erosion that occurred during the ultrasonic treatment. The carboxymethylation pretreatment makes the fibrils highly charged and, hence, easier to liberate. Clear dispersions were prepared with a concentration of 1–2 $g \cdot L^{-1}$. Ensuing fibrils have a cross-section between 5 and 15 nm and a length of more than 1 μm with a charge density of about 0.5 $meq \cdot g^{-1}$ (Wågberg et al., 2008). From these data the surface potential, at full dissociation of the carboxyl groups, was calculated with the Poisson-Boltzmann equation assuming that the charges were located on the surface of the fibrils. Values ranging between 200 and 250 mV, depending on the pH, salt concentration and diameter of the fibrils were reported. It was also shown that a pH higher than 10 was required to dissociate all the charges of the MFC. These values were used to calculate the interaction energy between the fibrils in water using the Derjaguin-Landau-Verwey-Overbeek (DLVO) theory. The stability resulting from electrostatic repulsion fibrils was found to be high, providing the carboxyl groups are dissociated.

Enzymatically pretreated MFC is less homogeneous and presents a higher fibril width of 10–30 nm (Aulin et al., 2009). The number of charged groups on the fibril surfaces is also very different. Carboxylated MFC was used to form multilayers with cationic polyelectrolytes (Aulin et al., 2008).

2.2.3 TEMPO-mediated oxidation pretreatment

The most commonly used pretreatment method to prepare MFC is 2,2,6,6-tetramethylpiperidine-1-oxyl (TEMPO)-mediated (or TEMPO-mediated) oxidation of cellulosic fibers. A more complete description of this reaction is given in Chapter 5. This type of nanocellulose is sometimes referred to as TEMPO-oxidized cellulose nanofibers (TOCN).

With this oxidative reaction, it was reported that regenerated cellulose can be completely converted into water-soluble polyglucuronic acid (Isogai and Kato, 1998; Tahiri and Vignon, 2000). For native cellulose fibers, the initial fibrous morphology is maintained and the oxidation reaction proceeds throughout the fibers but occurs only at the surface of the microfibrils, which therefore become negatively charged (Saito et al., 2005). The electrostatic repulsion caused by anionic carboxylate groups between the TEMPO-oxidized cellulose microfibrils should overcome the numerous interfibrillar hydrogen bonds originally present in the wood cell walls. Despite this beneficial surface derivatization, it was not possible to disintegrate the TEMPO-oxidized cellulose fibers into individual microfibrils, probably because the oxidation was achieved on dried fibers that present reduced accessibility (Isogai and Kato, 1998; Tahiri and Vignon, 2000; Saito and Isogai, 2004; Saito et al., 2005; Saito and Isogai, 2005; Montanari et al., 2005).

The individualization of cellulose microfibrils through the mechanical treatment of TEMPO-oxidized never-dried fibers from different sources was first reported in 2006 (Saito et al., 2006). The samples were easily disintegrated into individual microfibrils by a simple moderate mechanical treatment in water. TEMPO-mediated oxidation was performed at room temperature, using sodium bromide, NaBr, and sodium hypochlorite, NaClO, as an additional catalyst and primary oxidant, respectively. *A bulk degree of oxidation of about 0.2 for each anhydroglucose unit of cellulose was necessary for a smooth disintegration of sulfite wood pulp, whereas only small amounts of individual microfibrils were obtained at lower oxidation levels.* This protocol was successfully applied to different sources of cellulose. It was found that at pH 10 optimal conditions were reached, giving cellulose nanofibers with 3–4 nm in width and a few microns in length.

It was shown that when applying a TEMPO/NaClO/$NaClO_2$ system under weakly acidic conditions (pH 6.8, 60°C), no aldehyde groups remain in the oxidized product and no depolymerization of cellulosic chains through β-elimination occurs (Saito et al., 2009). However, the thermal degradation temperature of cellulose was found to strongly decrease from 300 to 200°C upon the introduction of carboxylate groups through the TEMPO-mediated oxidation (Fukuzumi et al., 2009).

Another oxidation route, namely periodate oxidation, to produce dialdehyde cellulose, followed by chlorite oxidation to convert aldehydes into carboxylic groups has been investigated (Tejado et al., 2012). The two successive chemical treatments were carried out to various extents in order to achieve various degrees of oxidation and the relation between the carboxylic content of cellulose fibers and the disintegration energy required to convert them into MFC was analyzed. Compared to TEMPO-mediated oxidation, this oxidation route allows the introduction of a larger amount of carboxylic groups and consequently the study of the effect of the charge content on the disintegration energy over a wider range. The carboxylate content was varied between 1.0 and 3.5 $mmol \cdot g^{-1}$. It was shown that the isolation of MFC can be achieved without the necessity of applying any mechanical energy other than that required to stir fiber suspensions during the chemical treatments if the oxidation is sufficiently high. The mechanism responsible for this "spontaneous" disintegration was shown to be the dissolution of overcharged amorphous domains, which become solubilized upon surpassing a charge threshold set around 3 $mmol \cdot g^{-1}$. However, the length and the crystallinity of the nanoparticles were severely affected.

2.3 Morphology

MFC can be prepared not only from wood, which is the most important source of cellulosic fibers, but from any cellulose source material and the defibrillation is used to delaminate the cell walls of the fibers and to liberate the nanosized fibrils. Non-wood plants, such as agricultural crops and by-products, generally contain less lignin than wood and the bleaching process if therefore simplified. When by-products, such as pulp after juice extraction, are used as raw materials, fewer processing steps to obtain cellulose are required (Bruce et al., 2005). At present, agricultural by-products are either burned, used for low-value products such as animal feed or used in biofuel production. In addition, crop residues can be valuable sources of cellulosic nanoparticles because of their renewability. Moreover, in agricultural fibers, the cellulose microfibrils are less tightly wound in the primary than in the secondary cell wall, resulting in a lower energy consumption to prepare MFC (Dinand et al., 1996).

Production of MFC from various non-wood sources has been reported in the literature, including sugar beet pulp (Dinand et al., 1996; Dufresne et al., 1997; Dinand et al., 1999; Azizi Samir et al., 2004; Leitner et al., 2007; Agoda-Tandjawa et al., 2010), potato pulp (Dufresne et al., 2000), algae (Imai et al., 2003), cladodes, i.e. stems (Malainine et al., 2003; Malainine et al., 2005), and peel of prickly pear fruit (Habibi et al., 2009) of *Opuntia ficus-indica*, a cactus, swede root (Bruce et al., 2005), hemp (*Cannabis sativa L.*) (Wang et al., 2007), soybean pods (Wang and Sain, 2007a; 2007b), sugarcane bagasse (Bhattacharya et al., 2008), wheat straw and soy hulls (Alemdar and Sain, 2008), sisal (Siqueira et al., 2009; 2010b), banana rachis (Zuluaga et al., 2009), and rachis of date palm tree (Bendahou et al., 2010).

Figure 2.7 shows the morphology of MFC obtained from different cellulosic sources. TEM is generally used to investigate the morphology of MFC. For observation, a drop of dilute MFC suspension is deposited onto glow-discharged carbon-coated electron microscopy grids. The excess liquid is absorbed with filter paper, and a drop of 2% uranyl acetate negative stain is added before drying. The excess liquid is blotted, and the remaining film of stain is allowed to dry. Regardless of the source, this material consists of long entangled filaments. Both individual nanofibrils and microfibril bundles can be observed. Larger fragments and unfibrillated fibers are sometimes observed (Andresen et al., 2006; Andresen and Stenius, 2007; Wang and Cheng, 2009). The manufacturing process and source of cellulose influence the particle diameter distribution of the MFC. The high density of hydroxyl groups on

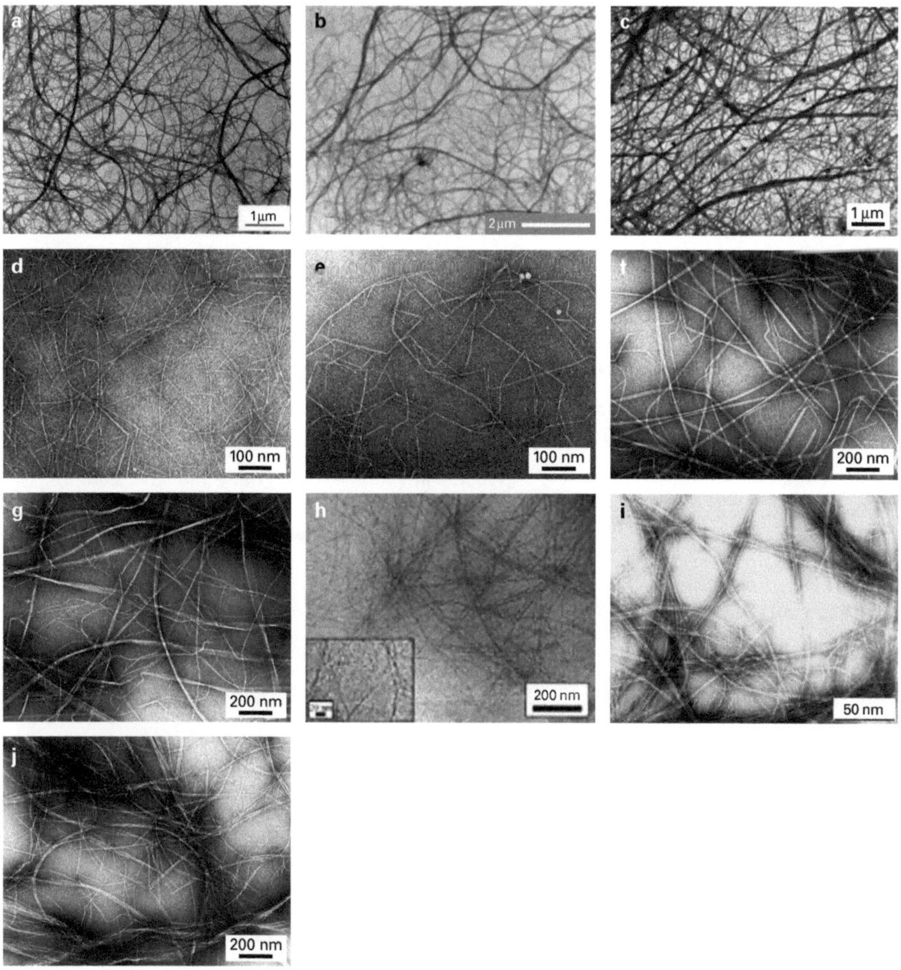

Fig. 2.7: Transmission electron micrographs showing microfibrillated cellulose obtained after high-pressure mechanical treatment: (a) sugar beet pulp (Dufresne et al., 1997), (b) potato pulp (Dufresne et al., 2000), (c) *Opuntia ficus-indica* (Malainine et al., 2003), (d) bleached sulfite wood pulp (Saito et al., 2006), (e) cotton (Saito et al., 2006), (f) tunicin (Saito et al., 2006), (g) bacterial cellulose (Saito et al., 2006), (h) bleached sulfite softwood cellulose pulp (Pääkkö et al., 2007), (i) prickly pear skin (Habibi et al., 2009), and (j) banana rachis (Zuluaga et al., 2009).

Source	Mechanical Treatment	Pretreatment	Width (nm)	Reference
Banana Rachis	High-Pressure Homogenization	None	5–60	(Zuluaga et al., 2009)
Corn Cobs			5–60	(Shogren et al., 2011)
Date Palm Tree Rachis			5–10	(Bendahou et al., 2010)
Empty Fruit Bunches of Oil Palm			5–40	(Jonoobi et al., 2011)
Kenaf			10–90	(Jonoobi et al., 2009)
Lemon Peel			3–10	(Rondeau-Mouro et al., 2003)
Maize Bran			5–20	(Rondeau-Mouro et al., 2003)
Opuntia ficus indica Cladodes			5	(Malainine et al., 2003; 2005)
Peel of Prickly Pear Fruits			2–5	(Habibi et al., 2009)
Potato Pulp			5	(Dufresne et al., 2000)
Radiata Pine (*Pinus radiate D. Don*) Wood			15	(Abe et al., 2007)
Rubberwood			10–90	(Jonoobi et al., 2011)
Sisal			52 ± 15	(Siqueira et al., 2009)
Sugar Beet Pulp			5	(Azizi Samir et al., 2004)
Sugar Beet Pulp			30–100	(Leitner et al., 2009)
Sugar Beet Pulp			2–15	(Agoda-Tandjawa et al., 2010)
Sugarcane Bagasse			30	(Bhattacharya et al., 2008)
Sulfite Softwood Pulp			10–100	(Herrick et al., 1983)
Soybean Stock		Cryocrushing	50–100	(Wang and Sain, 2007a)
Soybean Pods			50–100	(Wang and Sain, 2007b)
Wheat Straw			10–80	(Alemdar and Sain, 2008)
Soy Hulls			20–120	(Alemdar and Sain, 2008)
Bleached Kraft Eucalyptus Pulp		Enzymatic	20	(Zhu et al., 2011)
Bleached Sulfite Pulp			5–20	(Pääkkö et al., 2007; 2008)
Bleached Sulfite Softwood Pulp			30 ± 10	(Svagan et al., 2007)
Bleached Wood Sulfite Pulp			5–30	(Henriksson et al., 2007)
Sulfite Softwood Pulp		Carboxymethylation	10–15	(Aulin et al., 2008)
			5–15	(Wågberg et al., 2008)
Bacterial Cellulose		TEMPO-Mediated Oxidation	3–100	(Saito et al., 2006)
Bleached Sulfite Pulp			3–5	(Saito et al., 2006)
Cotton			3–5	(Saito et al., 2006)
Hardwood Bleached Kraft Pulp			5	(Saito et al., 2009)
Softwood/Hardwood Bleached Kraft Pulp			3–4	(Fukuzumi et al., 2009)
Tunicin			3–20	(Saito et al., 2006)

Source	Mechanical Treatment	Pretreatment	Width (nm)	Reference
Cotton	Grinding	None	20–90	(Taniguchi and Okamura, 1998)
Douglas Fir (*Pseudotsuga menziesii*) Softwood			12–20	(Abe and Yano, 2009)
Potato Pulp			12–55	(Abe and Yano, 2009)
Rice Straw			12–35	(Abe and Yano, 2009)
Tunicin			20–90	(Taniguchi and Okamura, 1998)
Wood Pulp			20–90	(Taniguchi and Okamura, 1998)
Bleached Northern Black Spruce Pulp	Cryocrushing	None	100–1000	(Chakraborty et al., 2005)
		Enzymatic	10–100	(Janardhnan and Sain, 2006)
Bamboo	High-Intensity Ultrasonication	None	10–40	(Chen et al., 2011)
Flax			15–100	(Chen et al., 2011)
Needle Fir (*Abies nephrolepis*) Wood			10–20	(Chen et al., 2011)
Regenerated Cellulose Fibers			30–μms	(Cheng et al., 2009a; 2009b)
Wheat Straw			15–35	(Chen et al., 2011)

Table 2.2: Typical width values for MFC obtained by different methods.

the surface of the cellulosic nanoparticles can also lead to the formation of larger agglomerates (Zimmermann et al., 2004). It has been reported that the presence of hemicelluloses or pectin in MFC pulp limits this agglomeration and tends to facilitate the defibrillation process (Dinand et al., 1999; Hult et al., 2001; Iwamoto et al., 2008). However, removal of non-cellulosic components increases the relative crystallinity of the MFC pulp (Alemdar and Sain, 2008).

Although a combination of microscopic techniques with image analysis can provide information on MFC widths, it is more difficult to determine the length of the particles because of entanglements and difficulties in identifying both ends of individual microfibrils. Indeed, the observation scale for length and diameter are quite different. MFC suspensions are generally not homogeneous and they consist of individual cellulose microfibrils and microfibril bundles. Table 2.2 summarizes some typical width values reported in the literature for MFC obtained by different methods. It is worth noting that these values refer most often to the most individualized fibrils and do not take into account larger bundles. In addition, these values are only indicative since they greatly depend on the experimental conditions, like for instance the pretreatment conditions and/or the number of passes during the defibrillation process.

Solid state nuclear magnetic resonance (NMR) was also used to estimate the lateral fibril dimensions. Lateral sizes of 4 nm for sugar beet pulp MFC (Heux et al., 1999) and 4.6 nm for bleached sulfite softwood cellulose pulp MFC (Pääkkö et al., 2007) have been reported. These values are in good agreement with microscopic observations.

2.4 Degree of fibrillation

Besides direct microscopic observations, other measurements can be carried out to access indirectly the extent of fibrillation.

2.4.1 Turbidity of the suspension

The defibrillation process changes the appearance of the cellulose aqueous suspension. It can be accessed by a simple visual inspection (Seydibeyoğlu and Oksman, 2008). The homogeneity of the suspension increases when increasing the fibrillation treatment. It can also be accessed by measuring the UV-visible transmittance through the suspension (Saito et al., 2006; Besbes et al., 2011). For suspended fibrils thinner that the wavelength, the light scattering is proportional to the mass/length or the cross section area (Carr et al., 1977). For this reason, the suspension becomes more and more transparent at the same solid concentration as the disintegration proceeds.

2.4.2 Viscosity of the suspension

After disintegration, the MFC is typically available as a suspension in polar liquid, usually water. During homogenization, the suspension changes from a low viscosity to a high viscosity medium (Bruce et al., 2005; Iwamoto et al., 2005; Henriksson et al., 2007; Zimmermann et al., 2010) suggesting some applications. This is due to the increase in the Einstein coefficient, which increases with increasing the length to diameter ratio of suspended particles. Rheological studies can thus give information about the fibrillation state of the particles. Normally, a 2 wt% fiber suspension is used. At higher concentrations, the increased viscosity during processing becomes too high so that the suspension cannot be moved forward by the pumping system. The change in suspension viscosity upon defibrillation will be presented in Chapter 6.

2.4.3 Porosity and density

The porosity and density of MFC films is of obvious significance to physical properties. These parameters strongly depend on the structure of the nanoparticles, presence of residual components and technique used to prepare the films. In particular, application of pressure during film formation results in lower porosity and higher density values. The density is expected to increase upon the severity of the homogenizing process, since smaller diameter results in better packing. Density of MFC films can be simply determined from their geometric dimensions and their mass. It is also possible to calculate it from the displacement of the sample when immersed in mercury (Svagan et al., 2007; Henriksson et al., 2008). The use of the technique of porosimetry

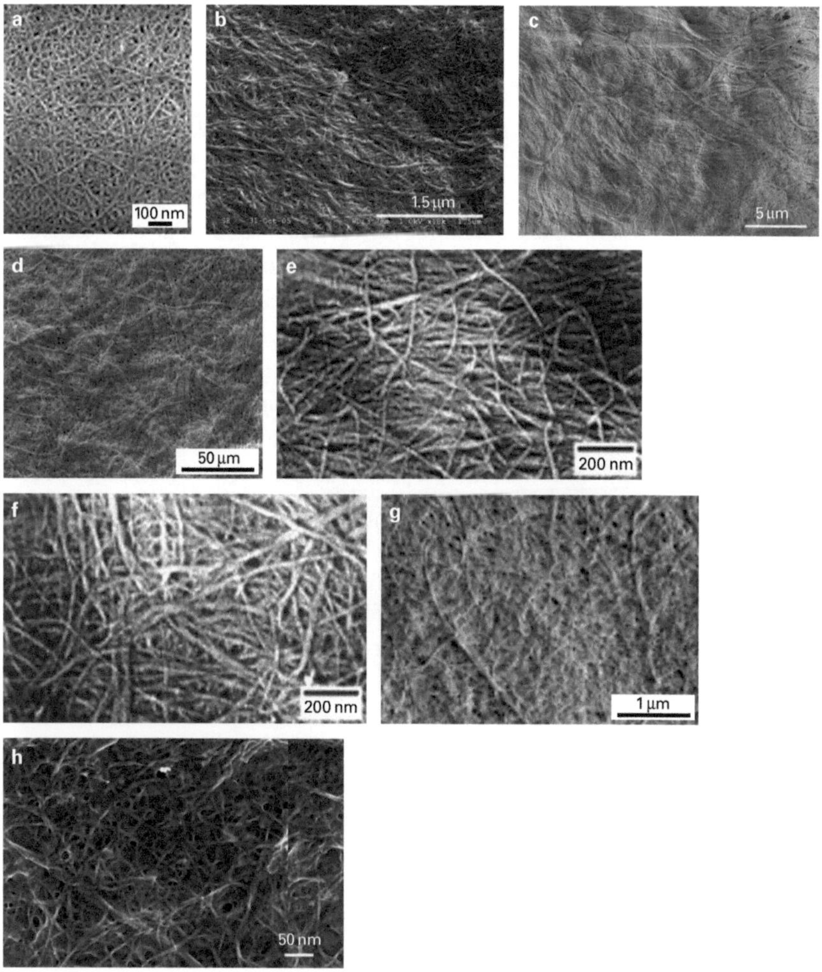

Fig. 2.8: Scanning electron micrographs (SEM-FEG) of microfibrillated cellulose films: (a) wood powder from Radiata pine (*Pinus radiata* D. Don) (Abe et al., 2007), (b) softwood dissolving pulp (Henriksson et al., 2008), (c) kraft pulp (Nakagaito and Yano, 2008a), (d) sisal (Siqueira et al., 2009), (e) rice straw (Abe and Yano, 2009), (f) potato tuber (Abe and Yano, 2009), (g) softwood sulfite pulp (Sehaqui et al., 2010), and (h) sisal (Belbekhouche et al., 2011).

involving the intrusion of a non-wetting liquid (often mercury) at high pressure into the film was also reported (Pääkkö et al., 2008). The pore size can be determined based on the external pressure needed to force the liquid into a pore against the opposing force of the liquid's surface tension. Computerized image analysis of field emission scanning electron microscopy (FEG-SEM) images was performed to estimate the average surface porosity of MFC films (Syverud et al., 2011). A value around 10% was reported. Positron annihilation lifetime spectroscopy (PALS) was also used to determine the pore sizes of wood and tunicin TEMPO-oxidized MFC films at 0% RH (Fukuzumi et al., 2011). This technique is used to study voids and defects in solids. A pore size around 0.47 nm from the film surface to the interior of the film was reported.

Figure 2.8 shows the surface of films obtained by casting-evaporation of MFC extracted from different sources and observed by scanning electron microscopy with field emission gun (SEM-FEG). When the water is removed from the MFC gel, a cellulose microfibril network is formed with interfibrillar hydrogen bonding. The film displays a more or less porous structure similar to a paper sheet and the web-like morphology of MFC is clearly observed. Moreover, microfibril bundles can generally be seen. The film consists of trapped filaments. Non-cellulosic components, such as residual lignin, hemicellulose, pectin, extractive substances and fatty acids, present at the surface of MFC (Dufresne et al., 1997; Bendahou et al., 2010) are supposed to play an important role on the cohesion and therefore properties of MFC films.

The pores are irregular in shape, as expected in a high-density network. The typical pore diameter is in the range 10–50 nm and a porosity value of 28% was reported (Henriksson et al., 2008). Films prepared from a solvent exchange procedure performed on filtered MFC films before drying from less hydrophilic liquids such as methanol, ethanol or acetone show porosities up to 40%. Porosities of MFC aerogels as high as 98% were also reported and it was shown that it can be simply tuned by freeze-drying (Pääkkö et al., 2008). The porosity of MFC films was also estimated from air permeability measurements (Belbekhouche et al., 2011). A value of 35% was reported corresponding to a density around 1.33 $g \cdot cm^{-3}$.

Similar density values (1.31–1.36 $g \cdot cm^{-3}$) have been reported in the literature for MFC isolated from different sources (Abe and Yano, 2009). Nevertheless, lower density values (0.81–1.07 $g \cdot cm^{-3}$ and 1.14 $g \cdot cm^{-3}$) have also been reported (Syverud and Stenius, 2009; Svagan et al., 2007). An increase of density by a factor of about 4 was reported during the fibrillation process for both hard- and softwood fibers (Stelte and Sanadi, 2009). However, filtration and drying under load allow higher densities (1.48–1.53 $g \cdot cm^{-3}$) to be obtained (Yano and Nakahara, 2004; Nogi et al., 2009).

2.4.4 Mechanical properties

Improved mechanical strength and stiffness of solid films obtained from MFC suspensions after water removal are expected compared to untreated fibers. This aspect will be developed in Section 2.5 of the present Chapter.

2.4.5 Water retention

The water retention value (WRV) test provides an indication of fibers' ability to take up water and swell. The WRV is also highly correlated to the bonding ability of the fibers and it is related to the exposed surface area of cellulose (Turbak et al., 1983). The test is carried out by placing a pad of moist fibers or nanofibers in a centrifuge tube that has a fritted glass filter at its base. The centrifuge is accelerated to remove water from the outside surfaces of the fiber. The centrifuged fiber pad is weighed (W_{wet}), dried at 105°C, and then reweighed (W_{dry}). The WRV value corresponds to the ratio of the water content after centrifugation to the dry weight of the sample. It is given by:

$$\mathrm{WRV} = \frac{W_{wet} - W_{dry}}{W_{dry}} \cdot 100 \qquad (2.2)$$

The water retention was found to increase upon defibrillation (Herrick et al., 1983; Turbak et al., 1983; Nakagaito and Yano, 2004; Iwamoto et al., 2005; Cheng et al., 2007; Cheng et al., 2009b; Wang and Cheng, 2009; Spence et al., 2010a). WRV values have been found to decrease when increasing the lignin content (hydrophobic) of the fibers used to prepare MFC (Spence et al., 2010a). The increase in WRV upon fibrillation was shown to be higher for hardwood samples than for softwood samples. A linear increase of both the Young's modulus and bending strength of MFC films as a function of WRV was reported (Yano and Nakahara, 2004). It was suggested that high interactive forces developed between fibrils because of the nanometer web-like network induced by the defibrillation process. After approximately ten high-pressure homogenizing passes, no further increase was observed.

2.4.6 Degree of polymerization

The length of the microfibrils is obviously related to the molecular weight or the degree of polymerization (DP) of constitutive cellulosic chains. It can be estimated from the average intrinsic viscosity value [η], determined from solution viscosity measurements, which is given by:

$$[\eta] = \lim_{c\to 0} \frac{\eta_{sp}}{c} = \lim_{c\to 0} \frac{\eta - \eta_0}{\eta_0 c} \tag{2.3}$$

In this equation, η_{sp} is the specific viscosity, η the solution viscosity, η_o the viscosity of the pure solvent, and c the concentration of the polymer. The specific viscosity expresses the incremental viscosity due to the presence of the polymer in the solution. Normalizing η_{sp} to concentration expresses the capacity of a polymer to increase the solution viscosity, i.e. the incremental viscosity per unit concentration of polymer. Extrapolation at zero concentration gives the intrinsic viscosity which is generally expressed in $mL \cdot g^{-1}$. A practical method for the determination of intrinsic viscosity is with an Ubbelohde viscometer. The intrinsic viscosity is related to the molecular weight M of the polymer through the Mark–Houwink equation:

$$[\eta] = KM^a \tag{2.4}$$

The values of the Mark–Houwink parameters, a and K, depend on the particular polymer-solvent system. Cellulose is generally dissolved in cupriethylenediamine (CED). For cellulose dissolved in CED, $K = 0.42$ and $a = 1$ for $DP < 950$, and $K = 2.28$ and $a = 0.76$ for $DP > 950$ (Marx-Figini, 1978).

Size exclusion chromatography (SEC) can also be used to determine the molecular weight and gives information on its distribution. Dimethylacetamide (DMAc)/LiCl can be used as the solvent system (Henriksson et al., 2007).

It was found that the mechanical isolation process used to prepare MFC induces a decrease of the DP of around 27% (Herrick et al., 1983; Turbak et al., 1983). More recently, similar degradation was reported (Henriksson et al., 2007). A shift in molecular weight distribution towards lower values was observed for enzymatically pretreated MFC, although the higher molecular weight fractions were found to be well preserved. A decrease in viscosities/DPs between 15% and 63% was reported upon fibrillation of cellulose fibers from different sources (Zimmermann et al., 2010). It was found that degradation was higher for starting materials with a high viscosity, whereas those already degraded due to chemical/mechanical pretreatment show lower decreases. A decrease of the DP was also reported for MFC obtained by defibrillation of dissolved pulp fibers using the grinding method (Iwamoto et al., 2007). Different DP of cellulosic chains can also be obtained depending on the enzyme pretreatment process before fibrillation (Henriksson et al., 2008).

2.4.7 Specific surface area

The specific surface area, A_{sp}, of a solid material is a measure of the total surface area per unit of mass. Its value is therefore expected to increase upon defibrillation process of cellulosic fibers. It is usually measured by gas adsorption using the BET

(Brunauer–Emmett–Teller) isotherm. For cellulose fibers, nitrogen is classically used as adsorbate. The specific surface area of the fibers can be calculated from the monolayer adsorbed gas volume, V_m, the molar gas volume, V_M (22.414 $L \cdot mol^{-1}$ for nitrogen) and using the known surface area which one gas molecule covers on the absorbent surface, σ (16 $Å^2$ for nitrogen):

$$A_{sp} = \frac{V_m \cdot N_A \cdot \sigma}{V_M} \tag{2.5}$$

where N_A is Avogadro's constant. However, this technique is not adapted for MFC. Indeed, a low value around 7.5 $m^2 \cdot g^{-1}$ was reported for cellulose fibrils with diameters in the range 10–100 nm (Lu et al., 2008). The actual surface area should be much higher than this value from geometrical considerations, but the sample was freeze-dried for BET analysis causing fibril aggregation. A BET specific surface area of 70 $m^2 \cdot g^{-1}$ was reported for MFC aerogels using an N_2 adsorption-desorption method (Pääkkö et al., 2008). This value and an assumption of cylindrical fibrils suggested a fibril diameter of the order of 30 nm, which agreed quite well with microscopic observations. A specific surface area of 60–120 $m^2 \cdot g^{-1}$ was reported for bacterial cellulose and tunicin from solvent exchange drying of hydrogels and rapid freeze-drying of the suspension by spraying onto a cooled copper plate (Kuga et al., 2002).

Nevertheless, the apparent specific surface area of MFC can be estimated from geometric considerations. A value of 51 $m^2 \cdot g^{-1}$ was reported for MFC around 52 nm wide extracted from sisal fibers (Siqueira et al., 2010b). It was used to estimate the fraction of surface hydroxyl groups to determine grafting efficiency values. The specific surface area of MFC was also determined using the Congo red adsorption method (Spence et al., 2010a), that was earlier applied to cellulose (Ougiya et al., 1998) and chitin nanocrystals (Goodrich and Winter, 2007). Congo red is the sodium salt of benzidinediazo-bis-1-naphthylamine-4-sulfonic acid (formula: $C_{32}H_{22}N_6Na_2O_6S_2$). It has a strong, though apparently non-covalent affinity to cellulose fibers.

The Congo red concentration was determined by measuring the UV-visible absorption at 500 nm. The concentration of Congo red was varied and the maximum adsorbed amount, A_{max}, was derived from Langmuir isotherms. It is related to the specific surface area, A_{sp}, by the following equation:

$$A_{sp} = \frac{A_{max} \cdot N_A \cdot \sigma}{M_W \cdot 10^{21}} \tag{2.6}$$

where N_A is Avogadro's constant, σ the surface area of a single dye molecule (1.73 nm^2), and M_W the molecular weight of Congo red (696 $g \cdot mole^{-1}$). Specific surface area values in the range 30–200 $m^2 \cdot g^{-1}$ were reported for MFC obtained from wood pulp of different compositions (Spence et al., 2010a). No correlation was found between the surface area and lignin content, but for each subclass of pulp fibers (softwood and hardwood) the specific surface area was found to increase with the respective lignin content.

2.4.8 Crystallinity

Cellulose crystallinity in MFC is of interest since it directly affects its physical and mechanical properties. It is generally estimated by X-ray diffraction measurements. Indeed, cellulose is infusible and classical techniques such as differential scanning calorimetry (DSC) and dilatometry cannot be used. The degree of crystallinity, χ_c, is commonly measured as the ratio of the area of diffraction portion from the crystalline part of the sample, A_c, to the whole area of the scattering from the same sample, $A_c + A_a$. The value of A_c can be obtained after an appropriate subtraction of the scattering portion from the amorphous background, A_a:

$$\chi_c = \frac{A_c}{A_c + A_a} \cdot 100 \tag{2.7}$$

However, the deconvolution of crystalline peaks from the amorphous halo can be problematic. The relative degree of cellulose crystallinity or crystallinity index (CrI) can be more easily calculated according to the Segal method (Segal et al., 1959) even if more accurate and sophisticated methods exist. It was shown that the crystallinity index is a time-saving empirical measurement of relative crystallinity. This method is based on the intensity measured for two diffraction peaks on the X-ray diffraction pattern of the cellulosic sample. The CrI value is given by the following empirical equation:

$$\mathrm{CrI} = \frac{I_{002} - I_{Am}}{I_{002}} \cdot 100 \tag{2.8}$$

where I_{002} is the maximum intensity of the (002) lattice diffraction peak and I_{Am} is the intensity scattered by the amorphous part of the sample. The diffraction peak for (002) lattice is located at a diffraction angle around $2\theta = 22.5°$ and the intensity scattered by the amorphous part is measured as the lowest intensity at a diffraction angle around $2\theta = 18.0°$.

Defibrillation was found to induce an increase of crystallinity compared to untreated purified fibers as reported for wood, rice straw and potato tuber (Cheng et al., 2007; Abe and Yano, 2009; Cheng et al., 2009a). It can be explained by the partial degradation and removal of amorphous cellulose during mechanical treatment. However, a decrease of the degree of crystallinity was reported for MFC obtained by defibrillation of dissolved pulp fibers using the grinding method (Iwamoto et al., 2007).

Moreover, the size (width), W, of cellulose crystallites can be estimated from the width of the peak around $2\theta = 22.5°$ at half-height by means of the following semi-empirical equation:

$$W = \frac{0.9 \cdot \lambda}{\Delta\theta \cdot \cos\theta} \tag{2.9}$$

where λ is the wavelength of the X-ray radiation (generally Cu Kα, $\lambda = 1.5418$ Å), $\Delta\theta$ is the half width after curve fitting, and θ is the Bragg angle for the (002) reflection.

Other methods such as Fourier transform infrared (FTIR) (Cheng et al., 2009a) and solid state nuclear magnetic resonance (NMR) (Heux et al., 1999) have been used to estimate the increase of crystallinity of cellulosic materials upon mechanical treatment. From NMR study, it was suggested that some constrained parts of the microfibrils, previously appearing as disordered material, have been cut during the mechanical process. These parts could reorganize into crystals, due to the increased mobility provided by water.

2.5 Mechanical properties of MFC films

Several authors have reported the mechanical properties of MFC thin films (20–200 µm) and demonstrated their high stiffness and their ability for reinforcement. These films can be obtained by casting-evaporation, filtration and hot-pressing or spin-coating of MFC gels. The preparation of films of oriented MFC using a dynamic sheet former was also reported (Syverud and Stenius, 2009).

The simplest and overused technique to access the mechanical properties of MFC films is tensile testing. In this test, the sample is subjected to uniaxial tension upon constant elongation rate until failure. The applied force or tension necessary to induce the elongation rate is recorded as a function of elongation. Stress-strain curves can be plotted from the knowledge of the dimensions of the sample. Properties that are directly determined via a tensile test are ultimate tensile stress or strength, maximum elongation or strain at break, and Young's or tensile modulus. The latter is measured from the initial slope of the stress-strain curve. Mechanical properties obtained for MFC films prepared from different sources are reported in Table 2.3.

The porosity and density of the MFC film are expected to play an important role on the properties, and in particular mechanical properties of the film. The specific Young's modulus and strength can be calculated by dividing the experimental values by the bulk density of the MFC sheet (Dufresne et al., 1997; Abe and Yano, 2009). The experimental testing conditions (temperature, humidity, and cross-head speed) also affect the measured characteristics. The strain at break and Young's modulus were found to increase from 2.1% and 13 GPa, respectively, at 0% RH (relative humidity) (Svagan et al., 2007) to 6.9% and 14.7 GPa, at 50% RH (Henriksson et al., 2008). Sur-

Source	Film Preparation	E (GPa)	ε_R (%)	σ_R (MPa)	Reference
Hardwood	Filtration	6.2	7.0	222	(Fukuzumi et al., 2009)
		6.2–6.5	7.0–11.5	222–312	(Saito et al., 2009)
	Casting	1–3	2	10–60	(Stelte and Sanadi, 2009)
Potato Tuber	Filtration	11.4	–	230	(Abe and Yano, 2009)
Rice Straw	Filtration	11	–	230	(Abe and Yano, 2009)
Softwood	Filtration	16	1.7	250	(Yano and Nakahara, 2004)
	Casting			80–100	(Zimmermann et al., 2004; 2005)
	Filtration	14.0	2.6	104	(Henriksson and Berglund, 2007)
	Casting	13.0	2.1	180	(Svagan et al., 2007)
	Filtration	10.4–13.7	3.3–10.1	129–214	(Henriksson et al., 2008)
		11	–	210	(Abe and Yano, 2009)
		6.9	7.6	233	(Fukuzumi et al., 2009)
		15.7–17.5	5.3–8.6	104–154	(Syverud and Stenius, 2009)
	Casting	2.5	6–11	80	(Stelte and Sanadi, 2009)
	Filtration	13	–	223	(Nogi et al., 2009)
		7.4–10.3	2.8–6.6	122–232	(Sehaqui et al., 2010)
	Casting	4.4–5.4	–	7–90	(Rodionova et al., 2012)
Sugar Beet Pulp	Casting	2.5–3.2	–	–	(Dufresne et al., 1997)
		9.3	3.2	104	(Leitner et al., 2007)
Swede Root Pulp	Filtering	7	–	100	(Bruce et al., 2005)

Table 2.3: Mechanical properties of MFC films obtained from tensile tests: Young's modulus (E), strain at break (ε_R) and strength (σ_R).

prisingly, the moisture did not have a negative effect on the Young's modulus based on these studies. However, Figure 2.9 shows the evolution of the storage modulus of a MFC film as a function of the RH measured during a humidity scan in dynamic mechanical analysis (DMA). When increasing RH from 5 to 90% RH, the storage modulus decreases from about 30 to 18.9 GPa (Aulin et al., 2010). Water present in the amorphous segments acts as a plasticizer, reducing the intermolecular interactions between the MFC fibrils, thus reducing the stiffness of the material.

The tensile modulus of MFC films was found to increase with the duration of the mechanical treatment of the fiber pulp (Dufresne et al., 1997). It was also shown

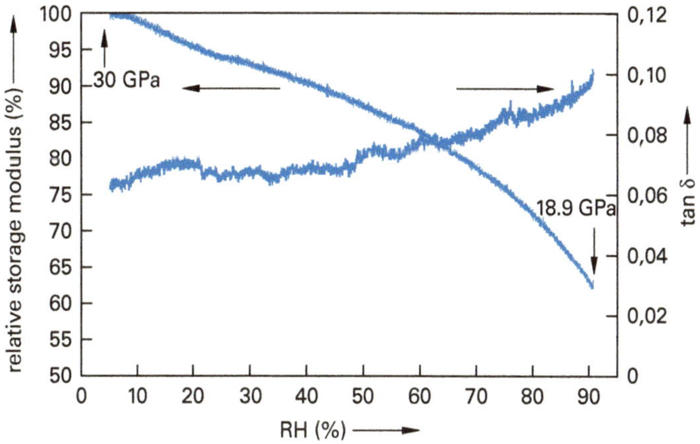

Fig. 2.9: Evolution of the relative storage tensile modulus and tangent of the loss angle of an MFC film obtained from sulfite softwood-dissolving pulp during a relative humidity scan (Aulin et al., 2010).

that MFC films with a thickness of 70 µm presented high strength properties with an average Young's modulus of 10 GPa in spite of high porosity (20–28%) (Henriksson et al., 2008). For MFC extracted from sugar beet pulp, a tensile modulus and strength of 9.3 GPa and 104 MPa, respectively, have been reported (Leitner et al., 2007). The authors reported a clear increase of mechanical properties after high-pressure homogenization. Indeed, the unhomogenized pulp displayed values of 4.6 GPa for the Young's modulus and 73 MPa for the tensile strength. The same tensile strength (104 MPa) was reported for MFC films with a thickness of 21 µm but the value of the tensile modulus was significantly higher (15.7 GPa) (Syverud and Stenius, 2009). Similar results were reported with MFC films having a Young's modulus and a tensile strength of 13 GPa and 180 MPa, respectively (Svagan et al., 2007). The differences observed could be explained by the various sources, or by the process used to produce MFC films. For instance, MFC films were obtained from spruce by vacuum filtering (Syverud and Stenius, 2009), whereas they were obtained from a blend of spruce and pine by casting (Svagan et al., 2007). Films with good mechanical properties displaying a Young's modulus and a tensile strength of 13 GPa and 223 MPa, respectively, were obtained (Nogi et al., 2009).

For MFC extracted from sugar beet pulp, it was found that the elimination of pectins naturally present in the pulp induced a decrease of the tensile modulus in dry atmosphere (Dufresne et al., 1997). Indeed, pectins act as a binder between cellulose microfibrils and improve the mechanism of load transfer when the film is submitted to a mechanical stress. This binding mechanism increases the cohesion of the material because of hydrogen bonding and/or covalent interactions between pectins, remaining hemicelluloses and cellulose microfibrils. However, a decrease of the tensile modulus was reported under a humid atmosphere because of the hydrophilic character of pectins that soften in the presence of moisture. Addition of 2 wt% oxidized starch to the MFC film was reported to increase the bending strength from 250 to 310 MPa and decrease the Young's modulus from 16 to 12.5 GPa (Yano and Nakahara, 2004). The tensile strength of films prepared from wood pulp and tunicin MFC was reported to be about 2.5 and 2.7 times higher than that of print-grade paper and polyethylene, respectively (Taniguchi and Okamura, 1998).

For TEMPO-pretreated MFC, because sodium carboxylate groups abundantly present on the surface of nanofibers cannot form hydrogen bonds, it may be possible to further improve the mechanical properties of MFC films by converting the sodium carboxylate structures to free carboxylate groups by, for example, gaseous acid treatment (Isogai et al., 2011). The tensile strength of self-standing films prepared from TEMPO-oxidized wood cellulose by TEMPO/NaCl/$NaClO_2$ oxidation in water at pH 6.8 was found to be higher than for TEMPO/NaBr/NaClO oxidation in water at pH 10 (Saito et al., 2009). It was ascribed to the higher DP values obtained for cellulose with the former system. However, the elastic modulus of the films was similar, despite the different TEMPO-mediated oxidation methods used. Even after the pulping process lignin remains in the pulp and it can be removed by TEMPO-mediated oxidation. It was found that the decrease in lignin content resulted in improved tensile strength of the MFC film (Rodionova et al., 2012). It was ascribed to enhanced bonding ability of cellulose after removal of lignin.

The mechanical performance of MFC films was compared to that obtained for bacterial cellulose sheets (Nakagaito et al., 2005). Significantly higher modulus and strength were reported for bacterial cellulose sheets. In addition, in contrast to the deforming behavior of MFC films, bacterial cellulose films deformed almost linearly until failure. It was attributed to a high planar orientation of the ribbon-like bacterial cellulose elements when compressed into sheets and to their ultra-fine structure, which allows more extensive hydrogen bonding. Successive loading-unloading experiments were performed on MFC films prepared from softwood dissolving pulp to examine the nature of the plastic region (Henriksson et al., 2008). Both the Young's modulus and yield stress were reported to increase by increasing the number of loading cycles in the plastic range. It was ascribed to a reorganization of the nanofibril network structure when deformed beyond yielding.

2.6 Optical properties of MFC films

The optical properties of MFC films have been mainly investigated by Yano's group in Kyoto. The regular light transmittance of MFC films impregnated with a transparent resin can be used to evaluate indirectly the degree of fibrillation. The procedure consists in first preparing a thin mat of MFC by water removal from the suspension. After drying, the mats are immersed in a transparent thermosetting resin such as acrylic, epoxy or phenol-formaldehyde to fill the cavities of the MFC sheet. The impregnation is performed under vacuum. Impregnated mats are taken out and the resin is finally cured by ultraviolet light. The increase in weight during resin impregnation is used to estimate the fiber content in the final nanocomposite sheet, generally in the range 60–70 wt%.

The regular light transmittance of nanocomposite sheets can be determined using a UV-visible spectrometer. Measurements are performed in the wavelength range 200–1000 nm by placing the specimen 22–25 cm from the entrance port of the integrating sphere.

It is well known that composite materials suffer from increased light scattering, resulting in a loss of transparency, caused by differences in the refractive indices of the materials. However, it is possible to obtain optically transparent composites by combining materials with significantly different refractive indices and reducing the size of the reinforcing phase below the wavelength of the visible light. The regular light transmittance at a 600 nm wavelength, which is in the middle of the visible wavelength range, was measured for MFC sheets prepared by grinding from dissolved pulp fibers and impregnated with acrylic resin as a function of the number of passes through the grinder (Iwamoto et al., 2007). It was found to increase abruptly from 53% for one pass to 80% for five passes. Above five passes through the grinder, the increase rate was linear and more gradual, and the degradation of light transmission compared to the neat acrylic resin (91%) was limited. Similar results were reported for bacterial cellulose films impregnated with acrylic (Nogi and Yano, 2008) or epoxy resin (Yano et al., 2005), and plant fiber-based nanofibers (Iwamoto et al., 2005). However, the regular transmittance of nanocomposite sheets prepared from wood-based MFC impregnated with acrylic resin was higher than that of bacterial cellulose nanocomposites, confirming the more homogeneous nature and lower thickness of MFC (Abe et al., 2007). The type and duration of the fibrillation process was also found to affect the regular light transmittance of MFC sheets impregnated with acrylic resin (Uetani and Yano, 2011).

Films made only from MFC can also be optically transparent if the cellulose nanofibers are densely packed, and the interstices between the fibers are small enough to avoid light scattering (Nogi et al., 2009). However, it was shown that mechanical compression performed on freeze-dried MFC did not result in transparency (Figure 2.10). It was suggested that the nanofibers were deformed under load but recovered after unloading, and the spaces created resulted in light scattering. Films prepared by slow filtration, drying and compression were much more densely packed, and were not optically transparent but translucent (Figure 2.10), probably because of surface light scattering. The films formed by filtration presented a high transparency thanks to a polishing step with emery paper. The transparency of the MFC sheet (thickness 55 µm) reached 71.6% at a wavelength of 600 nm (Figure 2.10). The transmittance at 600 nm of softwood and hardwood TEMPO-oxidized MFC films were found to be around 90% and 78%, respectively (Fukuzumi et al., 2009). The lower light transmittance of hardwood cellulose was ascribed to the presence of xylan that was supposed to interfere in part with complete dispersion of the nanofibrils in water.

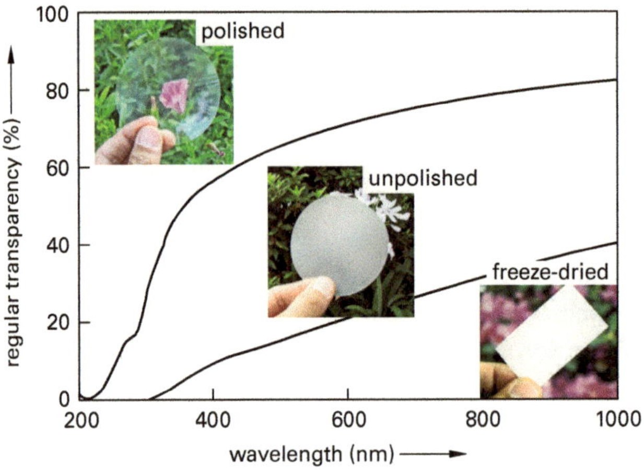

Fig. 2.10: Light transmittance of MFC sheets (Nogi et al., 2009).

2.7 Functionalization of MFC films

MFC films have been used as a support for fluorescent silver nanoclusters (smaller than 2 nm) by dipping the film into a silver nanocluster/poly(methacrylic acid) solution (Díez et al., 2011). Fluorescence and antibacterial activity of the films were reported.

2.8 Conclusions

Several methods of mechanical fibrillation of the biomass can be used to prepare microfibrillated cellulose (MFC). Devices such as the high-pressure homogenizer, microfluidizer, and grinder have been attempted. Moreover, a combination of processes or pretreatments such as refining, cryocrushing, enzymatic treatment, ultrasonication, and chemical treatment (carboxymethylation, TEMPO) have been studied to improve the fibrillation process. MFC is a new material composed of high aspect ratio nanosized cellulose fibrils. It consists of long entanglement filaments made of microfibril bundles. This top-down mechanically-induced destructuration strategy provides new functionalities and potential applications to cellulose either in suspension or in the solid state.

2.9 References

Abe, K., Iwamoto, S. and Yano, H. (2007). Obtaining cellulose nanofibers with a uniform width of 15 nm from wood. Biomacromolecules *8*, 3276–3278.

Abe, K. and Yano, H. (2009). Comparison of the characteristics of cellulose microfibril aggregates of wood, rice straw and potato tuber. Cellulose *16*, 1017–1023.

Agoda-Tandjawa, G., Durand, S., Berot, S., Blassel, C., Gaillard, C., Garnier, C. and Doublier, J.L. (2010). Rheological characterization of microfibrillated cellulose suspensions after freezing. Carbohydr. Polym. *80*, 677–686.

Ahola, S., Salmi, J., Johansson, Laine, J. and Österberg, M. (2008a). Model films from native cellulose nanofibrils. Preparation, swelling and surface interactions. Biomacromolecules *9*, 1273–1282.

Ahola, S., Österberg, M. and Laine, J. (2008b). Cellulose nanofibrils – Adsorption with poly(amideamine) epichlorohydrin studied by QCM-D and application as a paper strength additive. Cellulose *15*, 303–314.

Ahola, S., Turon, X., Österberg, M., Laine, J. and Rojas, O.J. (2008c). Enzymatic hydrolysis of native cellulose nanofibril and other cellulose model films – Effect of surface structure. Langmuir *24*, 11592–11599.

Alemdar, A. and Sain, M. (2008). Isolation and characterization of nanofibers from agricultural residues – Wheat straw and soy hulls. Biores. Technol. *99*, 1664–1671.

Andresen, M., Johansson, L.S., Tanen, B.S. and Stenius, P. (2006). Properties and characterization of hydrophobized microfibrillated cellulose. Cellulose *13*, 665–677.

Andresen, M. and Stenius, P. (2007). Water-in-oil emulsions stabilized by hydrophobized microfibrillated cellulose. J. Disper. Sci. Technol. *28*, 837–844.

Aulin, C., Varga, I., Claesson, P.M., Wågberg, L. and Lindström, T. (2008). Buildup of polyelectrolyte multilayers of polyethyleneimine and microfibrillated cellulose studied by in situ dual-polarization interferometry and quartz crystal microbalance with dissipation. Langmuir *24*, 2509–2518.

Aulin, C., Ahola, S., Josefsson, P., Nishino, T., Hirose, Y., Österberg, M. and Wågberg, L. (2009). Nanoscale cellulose films with different crystallinities and mesostructures – Their surface properties and interaction with water. Langmuir *25*, 7675–7685.

Aulin, C., Gällstedt, M. and Lindström, T. (2010). Oxygen and oil barrier properties of microfibrillated cellulose films and coatings. Cellulose *17*, 559–574.

Azizi Samir, M.A.S., Alloin, F., Paillet, M. and Dufresne, A. (2004). Tangling effect in fibrillated cellulose reinforced nanocomposites. Macromolecules *37*, 4313–4316.

Bajpai, P. (1999). Application of enzymes in the pulp and paper industry. Biotech. Prog. *15*, 147–157.

Belbekhouche, S., Bras, J., Siqueira, G., Chappey, C., Lebrun, L., Khelifi, B., Marais, S. and Dufresne, A. (2011). Water sorption behavior and gas barrier properties of cellulose whiskers and microfibrils films. Carbohydr. Polym. *83*, 1740–1748.

Bendahou, A., Kaddami, H. and Dufresne, A. (2010). Investigation on the effect of cellulosic nanoparticles' morphology on the properties of natural rubber based nanocomposites. Eur. Polym. J. *46*, 609–620.

Besbes, I., Alila, S. and Boufi, S. (2011). Nanofibrillated cellulose from TEMPO-oxidized eucalyptus fibres: Effect of the carboxyl content. Carbohydr. Polym. *84*, 975–983.

Bhatnagar, A. and Sain, M. (2005). Processing of cellulose nanofiber-reinforced composites. J. Reinf. Plast. Compos. *24*, 1259–1268.

Bhattacharya, D., Germinario, L.T. and Winter, W.T. (2008). Isolation, preparation and characterization of cellulose microfibers obtained from bagasse. Carbohydr. Polym. *73*, 371–377.

Boldizar, A., Klason, C., Kubát, J., Näslund, P. and Sáha, P. (1987). Prehydrolyzed cellulose as reinforcing filler for thermoplastics. Intern. J. Polymeric Mater. *11*, 229–262.

Bruce, D.M., Hobson, R.N., Farrent, J.W. and Hepworth, D.G. (2005). High-performance composites from low-cost plant primary cell walls. Compos. Part A: Appl. Sci. Manufact. *36*, 1486–1493.

Cai, J. and Zhang, L.N. (2005). Rapid dissolution of cellulose in LiOH/urea and NaOH/urea aqueous solutions. Macromol. Biosci. *5*, 539–548.

Carr, M.E.Jr., Shen, L.L. and Hermans, J. (1977). Mass-length ratio of fibrin fibers from gel permeation and light scattering. Biopolymers *16*, 1–15.

Chakraborty, A., Sain, M. and Kortschot, M. (2005). Cellulose microfibrils: A novel method of preparation using high shear refining and cryocrushing. Holzforschung *59*, 102–107.

Chen, W.S., Yu, H.P., Liu, Y.X., Hai, Y.F., Zhang, M.X. and Chen, P. (2011). Isolation and characterization of cellulose nanofibers from four plant cellulose fibers using a chemical-ultrasonic process. Cellulose *18*, 433–442.

Cheng, Q., Wang, S., Rials, T.G. and Lee, S.H. (2007). Physical and mechanical properties of polyvinyl alcohol and polypropylene composite materials reinforced with fibril aggregates isolated from regenerated cellulose fibers. Cellulose *14*, 593–602.

Cheng, Q., Wang, S. and Han, Q. (2009a). Novel process for isolating fibrils from cellulose fibers by high-intensity ultrasonication. II. Fibril characterization. J. Appl. Polym. Sci. *115*, 2756–2762.

Cheng, Q., Wang, S. and Rials T.G. (2009b). Poly(vinyl alcohol) nanocomposites reinforced with cellulose fibrils isolated by high intensity ultrasonication. Compos. Part A: Appl. Sci. Manufact. *40*, 218–224.

Chinga-Carrasco, G. and Syverud, K. (2010). Computer-assisted quantification of the multiscale structure of films made of nanofibrillated cellulose. J. Nanopart. Res. *12*, 841–851.

Díez, I., Eronen, P., Österberg, M., Linder, M.B., Ikkala, O. and Ras, R.H.A. (2011). Functionalization of nanofibrillated cellulose with silver nanoclusters: Fluorescence and antibacterial activity. Macromol. Biosci. *11*, 1185–1191.

Dinand, E., Chanzy, H., and Vignon, M.R. (1996). Parenchyma cell cellulose from sugar beet pulp: preparation and properties. Cellulose *3*, 183–188.

Dinand, E., Chanzy, H., and Vignon, M.R. (1999). Suspensions of cellulose microfibrils from sugar beet pulp. Food Hydrocol. *13*, 275–283.

Dufresne, A., Cavaillé, J.Y. and Vignon, M.R. (1997). Mechanical behavior of sheets prepared from sugar beet cellulose microfibrils. J. Appl. Polym. Sci. *64*, 1185–1194.

Dufresne, A., Dupeyre, D. and Vignon, M.R. (2000). Cellulose microfibrils from potato tuber cells: Processing and characterization of starch-cellulose microfibril composites. J. Appl. Polym. Sci. *76*, 2080–2092.

Eriksen, Ø, Syverud, K. and Gregersen, Ø. (2008). The use of microfibrillated cellulose produced from kraft pulp as a strength enhancer in TMP paper. Nord. Pulp Paper Res. J. *23*, 299–304.

Frenot, A., Henriksson, M.W. and Walkenström, P. (2007). Electrospinning of cellulose-based nanofibers. J. Appl. Polym. Sci. *103*, 1473–1482.

Frone, A.N., Panaitescu, D.M., Donescu, D., Spataru, C.I., Radovici, C., Trusca, R. and Somoghi, R. (2011). Preparation and characterization of PVA composites with cellulose nanofibers obtained by ultrasonication. BioResources *6*, 487–512.

Fukuzumi, H., Saito, T., Iwata, T., Kumamoto, Y. and Isogai, A. (2009). Transparent and high gas barrier films of cellulose nanofibers prepared by TEMPO-mediated oxidation. Biomacromolecules *10*, 162–165.

Fukuzumi, H., Saito, T., Iwamoto, S., Kumamoto, Y., Ohdaira, T., Suzuki, R. and Isogai, A. (2011). Pore size determination of TEMPO-oxidized cellulose nanofibril films by positron annihilation lifetime spectroscopy. Biomacromolecules *12*, 4057–4062.

Goodrich, J. and Winter, W. (2007). α-chitin nanocrystals prepared from shrimp shells and their specific surface area measurement. Biomacromolecules *8*, 252–257.

Habibi, Y. Mahrouz, M. and Vignon, M.R. (2009). Microfibrillated cellulose from the peel of prickly pear fruits. Food Chem. *115*, 423–429.
Hayashi, N., Sugiyama, J., Okano, T. and Ishihara, M. (1998). The enzymatic susceptibility of cellulose microfibrils of the algal-bacterial type and the cotton-ramie type. Carbohydr. Res. *305*, 261–269.
Henriksson, M. and Berglund, L.A. (2007). Structure and properties of cellulose nanocomposite films containing melamine formaldehyde. J. Appl. Polym. Sci. *106*, 2817–2824.
Henriksson, M., Henriksson, G., Berglund, L.A. and Lindström, T. (2007). An environmentally friendly method for enzyme-assisted preparation of microfibrillated cellulose (MFC) nanofibers. Eur. Polym. J. *43*, 3434–3441.
Henriksson, M., Berglund, L.A., Isaksson, P., Lindström, T. and Nishino, T. (2008). Cellulose nanopaper structure of high toughness. Biomacromolecules *9*, 1579–1585.
Henrissat, B. (1994). Cellulases and their interaction with cellulose. Cellulose *1*, 169–196.
Herrick, F.W., Casebier, R.L., Hamilton, J.K. and Sandberg, K.R. (1983). Microfibrillated cellulose: morphology and accessibility. J. Appl. Polym. Sci. Polym. Symp. *37*, 797–813.
Heux, L., Dinand, E. and Vignon, M.R. (1999). Structural aspects in ultrathin cellulose microfibrils followed by ^{13}C CP-MAS NMR. Carbohydr. Polym. *40*, 115–124.
Huang, Z.M., Zhang, Y.Z., Kotaki, M. and Ramakrishna, S. (2003). A review on polymer nanofibers by electrospinning and their applications in nanocomposites. Compos. Sci. Technol. *63*, 2223–2253.
Hult, E.L., Larsson, P.T. and Iversen, T. (2001). Cellulose fibril aggregation – An inherent property of kraft pulps. Polymer *42*, 3309–3314.
Imai, T., Putaux, J.L. and Sugiyama, J. (2003). Geometric phase analysis of lattice images from algal cellulose microfibrils. Polymer 44, 1871–1879.
Isogai, A., Ishizu, A. and Nakano, J. (1987). Dissolution mechanism of cellulose in SO_2-amine-dimethylsulfoxide. J. Appl. Polym. Sci. *33*, 1283–1290.
Isogai, A. and Kato,Y. (1998). Preparation of polyglucuronic acid from cellulose by TEMPO-mediated oxidation. Cellulose *5*, 153–164.
Isogai, A., Saito, T. and Fukuzumi, H. (2011). TEMPO-oxidized cellulose nanofibers. Nanoscale *3*, 71–85.
Iwamoto, S., Nakagaito, A.N., Yano, H. and Nogi, M. (2005). Optically transparent composites reinforced with plant fiber-based nanofibers. Appl. Phys. A *81*, 1109–1112
Iwamoto, S., Nakagaito, A.N. and Yano, H. (2007). Nano-fibrillation of pulp fibers for the processing of transparent nanocomposites. Appl. Phys. A-Mater. Sci. Process *89*, 461–466.
Iwamoto, S., Abe, K. and Yano, H. (2008). The effect of hemicelluloses on wood pulp nano-fibrillation and nanofiber network characteristics. Biomacromolecules *9*, 1022–1026.
Iwatake, A., Nogi, M. and Yano, H. (2008). Cellulose nanofiber-reinforced polylactic acid. Compos. Sci. Technol. *68*, 2103–2106.
Janardhanan, S. and Sain, M. (2006). Isolation of cellulose microfibrils – An enzymatic approach. BioResources *1*, 176–188.
Jin, H., Nishiyama, Y., Wada, M. and Kuga, S. (2004). Nanofibrillar cellulose aerogels. Colloids Surf. A: Physicochem. Eng. Aspects *240*, 63–67
Jonoobi, M., Harun, J., Shakeri, A., Misra, M. and Oksman, K. (2009). Chemical composition, crystallinity and thermal degradation of bleached and unbleached kenaf bast (*Hibiscus cannabinus*) pulp and nanofibers. BioResources *4*, 626–639.
Jonoobi, M., Khazaeian, A., Tahir, P.D., Azry, S.S. and Oksman, K. (2011). Characteristics of cellulose nanofibers isolated from rubberwood and empty fruit bunches of oil palm using chemo-mechanical process. Cellulose *18*, 1085–1095.
Kenealy, W.R. and Jeffries, T.W. (2003). Enzyme processes for pulp and paper: A review of recent developments. In: Wood deterioration and preservation: Advances in our changing world.

G. Barry, D.D. Nicholas and T.P. Schultz, eds. (Oxford University Press, Washington, DC), pp. 210–239.

Kim, C.W., Kim, D.S., Kang, S.Y., Marquez, M. and Joo, Y.L. (2006). Structural studies of electrospun cellulose nanofibers. Polymer *47*, 5097–5107.

Kuga, S., Kim, D.Y. and Nishiyama, Y. (2002). Nanofibrillar carbon from native cellulose. Mol. Cryst. Liq. Cryst. *387*, [237]/13–[243]19.

Leitner, J., Hinterstoisser, B., Wastyn, M., Keckes, J. and Gindl, W. (2007). Sugar beet cellulose nanofibril-reinforced composites. Cellulose *14*, 419–425.

Liu, H. and Hsieh, Y. (2002). Ultrafine fibrous cellulose membranes from electrospinning of cellulose acetate. J. Polym. Sci. Part B: Polym. Phys. *40*, 2119–2129.

Liu, H. and Hsieh, Y. (2003). Surface methacrylation and graft copolymerization of ultrafine cellulose fibers. J. Polym. Sci. Part B: Polym. Phys. *41*, 953–964.

Liu, H., Fu, S.Y., Zhu, J.Y., Li, H. and Zhan, H.Y. (2009). Vizualization of enzymatic hydrolysis of cellulose using AFM phase imaging. Enzyme Microb Technol *45*, 274–281.

López-Rubio, A., Lagaron, J.M., Ankerfors, M., Lindström, T., Nordqvist, D., Mattozzi A. and Hedenqvist, M.S. (2007). Enhanced film forming and film properties of amylopectin using micro-fibrillated cellulose. Carbohydr. Polym. *68*, 718–727.

Lu, J., Askeland, P. and Drzal, L.T. (2008). Surface modification of microfibrillated cellulose for epoxy composite applications. Polymer *49*, 1285–1296.

Ma, Z., Kotaki, M. and Ramakrishna, S. (2005). Electrospun cellulose nanofiber as affinity membrane. J. Membr. Sci. *265*, 115–123.

Malainine, M.E., Dufresne, A., Dupeyre, D., Mahrouz, M. and Vignon, M.R. (2003). Structure and morphology of cladodes and spines of *Opuntia ficus-indica*. Cellulose extraction and characterisation. Carbohydr. Polym. *51*, 77–83.

Malainine, M.E., Mahrouz, M. and Dufresne, A. (2005). Thermoplastic nanocomposites based on cellulose microfibrils from *Opuntia ficus indica* parenchyma cell. Compos. Sci. Technol. *65*, 1520–1526.

Marx-Figini, M. (1978). Significance of the intrinsic viscosity ratio of unsubstituted and nitrated cellulose in different solvents. Angew. Makromol. Chem. *72*, 161–171.

McCormick, C.L., Calais, P.A. and Hutchinson, B.H. (1985). Solution studies of cellulose in lithium chloride and N,N-dimethylacetamide. Macromolecules *18*, 2394–2401.

Montanari, S., Roumani, M., Heux, L. and Vignon, M.R. (2005). Topochemistry of carboxylated cellulose nanocrystals resulting from TEMP-mediated oxidation. Macromolecules *38*, 1665–1671.

Mörseburg, K., and Chinga-Carrasco, G. (2009). Assessing the combined benefits of clay and nanofibrillated cellulose in layered TMP-based sheets. Cellulose *16*, 795–806.

Nakagaito, A.N. and Yano, H. (2004). The effect of morphological changes from pulp fiber towards nano-scale fibrillated cellulose on the mechanical properties of high-strength plant fiber based composites. Appl. Phys. A-Mater. Sci. Process *78*, 547–552.

Nakagaito, A.N. and Yano, H. (2005). Novel high-strength biocomposites based on microfibrillated cellulose having nano- orderunit web-like network structure. Appl Phys A-Mater. Sci. Process *80*, 155–159.

Nakagaito, A.N., Iwamoto, S. and Yano H. (2005). Bacterial cellulose: the ultimate nano-scalar cellulose morphology for the production of high-strength composites. Appl. Phys. A-Mater. Sci. Process *80*, 93–97.

Nakagaito, A.N. and Yano, H. (2008a). Toughness enhancement of cellulose nanocomposites by alkali treatment of the reinforcing cellulose nanofibres. Cellulose *15*, 323–331.

Nakagaito, A.N. and Yano, H. (2008b). The effect of fiber content on the mechanical and thermal expansion properties of biocomposites based on microfibrillated cellulose. Cellulose *15*, 555–559.

Nogi, M. and Yano, H. (2008). Transparent nanocomposites based on cellulose produced by bacteria offer potential innovation in the electronics device industry. Adv. Mater. *20*, 1849–1852.
Nogi, M., Iwamoto, S., Nakagaito, A.N. and Yano, H. (2009). Optically transparent nanofiber paper. Adv. Mater. *21*, 1595–1598.
Ougiya, H., Hioki, N., Watanabe, K., Morinaga, Y., Yoshinaga, F. and Samejima, M. (1998). Relationship between the physical properties and surface area of cellulose derived from adsorbates of various molecular sizes. Biosci. Biotechnol. Biochem. *62*, 1880–1884.
Pääkkö, M., Ankerfors, M., Kosonen, H., Nykänen, A., Ahola, S., Österberg, M., Ruokolainen, J., Laine, J., Larsson, P.T., Ikkala, O. and Lindström, T. (2007). Enzymatic hydrolysis combined with mechanical shearing and high-pressure homogenization for nanoscale cellulose fibrils and strong gels. Biomacromolecules *8*, 1934–1941.
Pääkkö, M., Vapaavuori, J., Silvennoinen, R., Kosonen, H., Ankerfors, M., Lindström, T., Berglund, L.A. and Ikkala, O. (2008). Long and entangled native cellulose I nanofibers allow flexible aerogels and hierarchically porous templates for functionalities. Soft Matter *4*, 2492–2499.
Quan, S.L., Kang, S.G. and Chin, I.J. (2010). Characterization of cellulose fibers electrospun using ionic liquid. Cellulose *17*, 223–230.
Rees, L.H. (1974). Evaluating homogenizers for chemical industries. Chem. Eng. *13*, 86–92.
Rodionova, G., Saito, T., Lenes, M., Eriksen, Ø, Gregersen, Ø, Fukuzumi, H. and Isogai, A. (2012). Mechanical and oxygen barrier properties of films prepared from fibrillated dispersions of TEMPO-oxidized Norway spruce and Eucalyptus pulp. Cellulose *19*, 705–711.
Rondeau-Mouro, C., Bouchet, B., Pontoire, B., Robert, P., Mazoyer, J and Buléon, A. (2003). Structural features and potential texturising properties of lemon and maize cellulose microfibrils. Carbohydr. Polym. *53*, 241–252.
Rosenau, T., Potthast, A., Adorjan, I., Hofinger, A., Sixta, H. Firgo, H. and Kosma, P. (2002). Cellulose solution in N-methylmorpholine-N-oxide (NMMO) – degradation processes and stabilizers. Cellulose *9*, 283–291.
Saito, T. and Isogai, A. (2004). TEMPO-mediated oxidation of native cellulose. The effect of oxidation conditions on chemical and crystal structures of the water-insoluble fractions. Biomacromolecules *5*, 1983–1989.
Saito, T. and Isogai, A. (2005). Novel method to improve wet strength of paper. Tappi J. *4*, 3–8.
Saito, T., Shibata, I., Isogai, A., Suguri, N. and Sumikwa, N. (2005). Distribution of carboxylate groups introduced into cotton linters by the TEMPO-mediated oxidation. Carbohydr. Polym. *61*, 414–419.
Saito, T. and Isogai, A. (2006). Introduction of aldehyde groups on surfaces of native cellulose fibers by TEMPO-mediated oxidation. Colloids Surface A *289*, 219–225.
Saito, T., Nishiyama, Y., Putaux, J.L., Vignon, M. and Isogai, A. (2006). Homogeneous suspensions of individualized microfibrils from TEMPO-catalyzed oxidation of native cellulose. Biomacromolecules *7*, 1687–1691.
Saito, T., Hirota, M., Tamura, N., Kimura, S., Fukuzumi, H., Heux, L. and Isogai, A. (2009). Individualization of nano-sized plant cellulose fibrils by direct surface carboxylation using TEMPO catalyst under neutral conditions. Biomacromolecules *10*, 1992–1996.
Segal, L., Creely, J.J., Martin, A.E. and Conrad, C.M. (1959). An empirical method for estimating the degree of crystallinity of native cellulose using the X-ray diffractometer. Text. Res. J. *29*, 786–794.
Sehaqui, H., Liu, A., Zhou, Q. and Berglund, L.A. (2010). Fast preparation procedure for large, flat cellulose and cellulose/inorganic nanopaper structures. Biomacromolecules *11*, 2195–2198.
Seydibeyoğlu, M.Ö. and Oksman, K. (2008). Novel nanocomposites based on polyurethane and micro fibrillated cellulose. Compos. Sci. Technol. *68*, 908–914.
Shogren, R.L., Peterson, S.C., Evans, K.O. and Kenar, J.A. (2011). Preparation and characterization of cellulose gels from corn cobs. Carbohydr. Polym. *86*, 1351–1357.

Siqueira, G., Bras, J. and Dufresne, A. (2009). Cellulose whiskers vs. microfibrils: influence of the nature of the nanoparticle and its surface functionalization on the thermal and mechanical properties of nanocomposites. Biomacromolecules *10*, 425–432.

Siqueira, G., Tapin-Lingua, S., Bras, J., da Silva Perez, D. and Dufresne, A. (2010a). Morphological investigation of nanoparticles obtained from enzymatic and acid hydrolysis of sisal fibers. Cellulose *17*, 1147–1158.

Siqueira, G., Bras, J. and Dufresne, A. (2010b). New process of chemical grafting of cellulose nanoparticles with a long chain isocyanate. Langmuir *26*, 402–411

Siqueira, G., Tapin-Lingua, S., Bras, J., Da Silva Perez, D. and Dufresne, A. (2011). Mechanical properties of natural rubber nanocomposites reinforced with cellulosic nanoparticles obtained from enzymatic and acid hydrolysis of sisal fibers. Cellulose *18*, 57–65.

Siró, I. and Plackett, D. (2010). Microfibrillated cellulose and new nanocomposite materials: A review. Cellulose *17*, 459–494.

Son, W.K., Youk, J.H., Lee, T.S. and Park, W.H. (2004). Electrospinning of ultrafine cellulose acetate fibers: Studies of a new solvent system and deacetylation of ultrafine cellulose acetate fibers. J. Polym. Sci. Part B: Polym. Phys. *42*, 5–11.

Spence, K., Venditti, R., Habibi, Y., Rojas, O. and Pawlak, J. (2010a). The effect of chemical composition on microfibrillar cellulose films from wood pulps: Mechanical processing and physical properties. Biores. Technol. *101*, 5961–5968.

Spence, K., Venditti, R., Rojas, O., Habibi, Y. and Pawlak, J. (2010b). The effect of chemical composition on microfibrillar cellulose films from wood pulps: Water interactions and physical properties for packaging applications. Cellulose *17*, 835–848.

Spence, K., Venditti, R., Rojas, O., Habibi, Y. and Pawlak, J. (2011). A comparative study of energy consumption and physical properties of microfibrillated cellulose produced by different processing methods. Cellulose *18*, 1097–1111.

Spiridon, I. and Popa, V.I. (2000). Application of microorganisms and enzymes in the pulp and paper industry. Cell. Chem. Technol. *34*, 275–285.

Stelte, W. and Sanadi, A.R. (2009). Preparation and characterization of cellulose nanofibers from two commercial hardwood and softwood pulps. Ind. Eng. Chem. Res. *48*, 11211–11219.

Svagan, A.J., Azizi Samir, M.A.S. and Berglund, L.A. (2007). Biomimetic polysaccharide nanocomposites of high cellulose content and high toughness. Biomacromolecules *8*, 2556–2563.

Syverud, K. and Stenius, P. (2009). Strength and barrier properties of MFC films. Cellulose *16*, 75–85.

Syverud, K., Xhanari, K., Chinga-Carrasco, G., Yu, Y. and Stenius, P. (2011). Films made of cellulose nanofibrils: Surface modification by adsorption of a cationic surfactant and characterization by computer-assisted electron microscopy. J. Nanopart. Res. *13*, 773–782.

Tahiri, C. and Vignon, M.R. (2000). TEMPO-oxidation of cellulose: synthesis and characterization of polyglucuronans. Cellulose *7*, 177–188.

Taniguchi, T. and Okamura, K. (1998). New films produced from microfibrillated natural fibres. Polym. Inter. *47*, 291–294.

Tejado, A., Alam, Md.N., Antal, M., Yang, H. and van de Ven, T.G.H. (2012). Energy requirements for the disintegration of cellulose fibers into cellulose nanofibers. Cellulose *19*, 831–842.

Turbak, A.F., Snyder, F.W. and Sandberg, K.R. (1983). Microfibrillated cellulose: A new cellulose product. Properties, uses, and commercial potential. J. Appl. Polym. Sci. Polym. Symp. *37*, 815–827.

Uetani, K. and Yano, H. (2011). Nanofibrillation of wood pulp using a high-speed blender. Biomacromolecules *12*, 348–353.

Viikari, L., Alapuranen, M., Puranen, T., Vehmaanpera, J. and Siika-Aho, M. (2007). Thermostable enzymes in lignocellulose hydrolysis. Adv. Biochem. Engin./Biotechnol. *108*, 121–145.

Viswanathan, G., Murugesan, S., Pushparaj, V., Nalamasu, O., Ajayan, P.M. and Linhard, R.J. (2006). Preparation of biopolymer fibers by electrospinning from room temperature ionic liquids. Biomacromolecules *7*, 415–418.

Wågberg, L., Winter, L., Ödberg, L. and Lindström, T. (1987). On the charge stoichiometry upon adsorption of a cationic polyelectrolyte on cellulosic materials. Colloids and Surf. *27*, 163–173.

Wågberg, L., Decher, G., Norgren, M., Lindström, T., Ankerfors, M. and Axnäs, K. (2008). The build-up of polyelectrolyte multilayers of microfibrillated cellulose and cationic polyelectrolytes. Langmuir *24*, 784–795.

Walecka, J.A. (1956). An investigation of low degree of substitution carboxymethylcelluloses. Tappi *39*, 458–463.

Wang, B. and Sain, M. (2007a). Dispersion of soybean stock-based nanofiber in a plastic matrix. Polym. Inter. *56*, 538–546.

Wang, B. and Sain, M. (2007b). Isolation of nanofibers from soybean source and their reinforcing capability on synthetic polymers. Compos. Sci. Technol. *67*, 2521–2527.

Wang, B., Sain, M. and Oksman, K. (2007). Study of structural morphology of hemp fiber from the micro to the nanoscale. Appl. Compos. Mater. *14*, 89–103.

Wang, S. and Cheng, Q. (2009). A novel process to isolate microfibrils from cellulose fibers by high-intensity ultrasonication, Part 1: Process optimization. J. Appl. Polym. Sci. *113*, 1270–1275.

Wise, L.E., Murphy, M. and D'Addieco, A.A. (1946). Chlorite holocellulose, its fractionation and bearing on summative wood analysis and on studies on hemicelluloses. Paper Trade J. *122*, 35–43.

Xu, S., Zhang, J., He, A., Li, J., Zhang, H. and Han, C.C. (2008). Electrospinning of native cellulose from nonvolatile solvent system. Polymer *49*, 2911–2917.

Yano, H. and Nakahara, S. (2004). Bio-composites produced from plant microfiber bundles with a nanometer unit web-like network. J. Mat. Sci. *39*, 1635–1638.

Yano, H., Sugiyama, J., Nakagaito, A.N., Nogi, M., Matsuura, T., Hikita, M. and Handa, K. (2005). Optically transparent composites reinforced with networks of bacterial nanofibers. Adv. Mat. *17*, 153–155.

Zhang, Y.H.P., Himmel, M.E. and Mielenz, J.R. (2006). Outlook for cellulase improvement: Screening and selection strategies. Biotechnol. Adv. *24*, 452–481.

Zhao, H.P., Feng, X.Q. and Gao, H. (2007). Ultrasonic treatment for extracting nanofibers from nature materials. Appl. Phys. Lett. *90*, 073112.

Zhu, J.Y., Sabo, R. and Luo, X. (2011). Integrated production of nano-fibrillated cellulose and cellulosic biofuel (ethanol) by enzymatic fractionation of wood fibers. Green Chem. *13*, 1339–1344.

Zimmermann, T., Pohler, E. and Geiger, T. (2004). Cellulose fibrils for polymer reinforcement. Adv. Eng. Mat. *6*, 754–761.

Zimmermann, T., Pohler, E. and Geiger, T. (2005). Mechanical and morphological properties of cellulose fibril reinforced nanocomposites. Adv. Eng. Mat. *7*, 1156–1161.

Zimmermann, T., Bordeanu, N. and Strub E. (2010). Properties of nanofibrillated cellulose from different raw materials and its reinforcement potential. Carbohydr. Polym. *79*, 1086–1093.

Zuluaga, R., Putaux, J.L., Cruz, J., Vélez, J., Mondragón, I. and Gañán, P. (2009). Cellulose microfibrils from banana rachis: Effect of alkaline treatments on structural and morphological features. Carbohydr. Polym. *76*, 51–59.

3 Preparation of cellulose nanocrystals

Cellulose nanocrystals are the other important family of cellulosic nanoparticles. Their preparation involves a chemical acid hydrolysis process intended to dissolve amorphous chains from the cellulose fibers and to release crystal domains. Historically, the preparation of such nanoparticles has already been reported. They therefore consist of a sub-entity of cellulose microfibrils having more defined geometrical features of rod-like nanoparticles than microfibrillated cellulose.

3.1 Pioneering works on the acid hydrolysis of cellulose

The extraction or isolation of crystalline cellulosic regions, in the form of monocrystals, is a simple process based on acid hydrolysis. This process is not a synthesis as sometimes found in literature since Mother Nature has already done the job. In the 1940s, Nickerson and Habrle (1947) observed that the degradation induced by boiling cellulose fibers in acidic solution (hydrochloric and sulfuric acids) reached a limit after a certain time of treatment. Boiling 2.45 N hydrochloric acid (HCl) – 0.6 M ferric chloride ($FeCl_3$) and 2.5 N sulfuric acid (H_2SO_4) were used. In the presence of $FeCl_3$, the glucose formed was catalytically oxidized to carbon dioxide and water. The use of $FeCl_3$ was dropped and it was proposed to use 6N or 4N HCl at the boil for hydrolyzing native cellulose or regenerated structures, respectively (Philipp et al., 1947). It was shown that hydrolysis proceeds at a rapid initial rate which coincides with an equally rapid fall in moisture regain and cuprammonium viscosity of the recoverable non-hydrolyzed material. It was suggested that cellulose molecule sections which interlink the crystallites in the chain direction were attacked first. It corresponds to highly disordered or readily accessible cellulose, presumably the amorphous component. This part of the structure, highly hygroscopic, represented relatively small percentages of the intact material. With further hydrolysis, cellulose breakdown continued and regain increased, but viscosity remained practically constant, corresponding to the beginning of the breakdown of the so-called mesomorphous material. At this stage, the acidic attack appeared to be limited to the lateral surfaces of crystallites (Nickerson and Habrle, 1947). However, alternative mechanisms, such as rapid disintegration of the particle once attack is commenced (Millett et al., 1954; Immergut and Rånby, 1956), or end-attack of the crystallites (Sharples, 1958) have been suggested.

Both the amorphous and mesomorphous components were supposed to constitute the intercrystalline cellulose chain network. Thereafter, the rate of hydrolysis remained fairly constant for several hours and was assumed to represent hydrolysis of the crystalline component. Under the conditions employed, sulfuric acid appeared to be only about one sixth as active in hydrolyzing cellulose as hydrochloric acid of equivalent concentration. Crystallite length estimates derived from viscosity measurements ranged from 280 glucose units for cotton to 110 for high tenacity viscose rayon. Values of the same order of magnitude have been obtained by the indirect viscosity measurement and direct electron microscope observation (Battista et al., 1956).

This early work was the basis of the pioneering investigation on the preparation of stable colloidal suspensions of cellulose by controlled sulfuric acid-catalyzed degradation (boiling 2.5N H_2SO_4) of native and mercerized wood cellulose, viscose rayon and cotton in the Institute of Physical Chemistry at the University of Uppsala (Rånby, 1949; Rånby and Ribi, 1950; Rånby, 1951; Rånby, 1952). Rånby's work suggested that cotton and wood cellulose were built up from micellar strings (or crystalline areas), free or in aggregates, of uniform width (~ 7 nm), and could be set free by mechanical treatment such as ultrasonic waves. Hydrolysis was viewed as cutting the micellar strings into short fragments, or micelles, while retaining their width. It consisted of bundles of 100–150 cellulose molecules. The dimensions of the isolated micelles agreed fairly well with the dimensions calculated from the widening of the X-ray diffraction measurements for micelles in the cell wall (Hengstenberg and Mark, 1928). A width around 60 Å and a minimum length of 600 Å were reported. It was indicated that the micelle string of native cellulose was held together only by secondary forces, such as hydrogen bonds and van der Waals forces. Rånby proposed that, in cellulose synthesis, micelles rather than molecules may represent the primary structural elements. This theory was subsequently refined by Frey-Wyssling's experimental observations (Frey-Wyssling, 1954), which provided evidence that elementary microfibrils, separated (and joined) by paracrystalline cellulose, represent the basic building unit. The rod-like morphology of hydrolyzed cellulose was confirmed from electron microscopy observations for jute, cotton, hemp, Fortisan and ramie and it was shown that they had the same crystalline structure as the original fibers (Mukherjee et al., 1952; Mukherjee and Woods, 1953).

Simultaneously, the development by Battista (Battista, 1950; Battista et al., 1956) of the hydrochloric acid-assisted degradation of cellulose fibers derived from high-quality wood pulps, followed by ultrasonic treatment, led to the commercialization of microcrystalline cellulose (MCC). MCC is cellulose which has been degraded to a degree of polymerization where there is little further decrease (so called level-off DP). The term was proposed in lieu of the term "limiting degree of polymerization" (Battista, 1950). This degradation can be achieved by either mechanical disintegration or by hydrolysis of purified cellulose after 15 min in 2.5 N HCl at 105°C (Battista et al., 1956). It is a stable, chemically inactive, highly hygroscopic, and physiologically inert material, and presents reversible absorbency and attractive binding properties (Ardizzone et al., 1999; Battista et al., 1961). MCC therefore offered a significant opportunity for multiple uses in the pharmaceutical industry as a tablet binder, in food applications as a texturizing agent and fat replacer, and also, as an additive in paper and composite applications.

The acid-hydrolysis treatment was also used to investigate crystalline models for cellulose (Chang, 1971; Chang, 1974; Schurz and John, 1975; Yachi et al., 1983).

3.2 Pretreatment of natural fibers

To prepare such cellulose monocrystals, cellulose must be directly hydrolyzed. Thus, apart from pure cellulosic sources such as cotton, bleached wood pulp, bacterial cellulose, and MCC, the biomass is generally first submitted to different pretreatments. Removal of extractives (dewaxing) can be performed using a Soxhlet apparatus as described in TAPPI (Tappi test method T204). The extraction can be carried out using a mixture of toluene and ethanol, or benzene and ethanol in the ratio 2:1.

A purification treatment consisting in an alkali treatment with sodium hydroxide (NaOH) or potassium hydroxide (KOH) followed by a bleaching step using acetate buffer (solution of NaOH and glacial acetate acid) and sodium chlorite ($NaClO_2$) as reported in Section 2.1.1 of Chapter 2 to purify cellulose is performed (Figure 3.1). This preliminary step to obtain pure cellulose fibers is crucial and must be done carefully. The alkali extraction is performed to solubilize most of the pectins and hemicelluloses. Although several bleaching methods exist in the wood industry, bleaching of natural fibers is typically performed with the $NaClO_2$ process over a more or less long period depending on the source of cellulose. The bleaching treatment is performed to break down phenolic compounds or molecules having chromophoric groups present in lignin and to remove the by-products of such breakdown, to whiten the material.

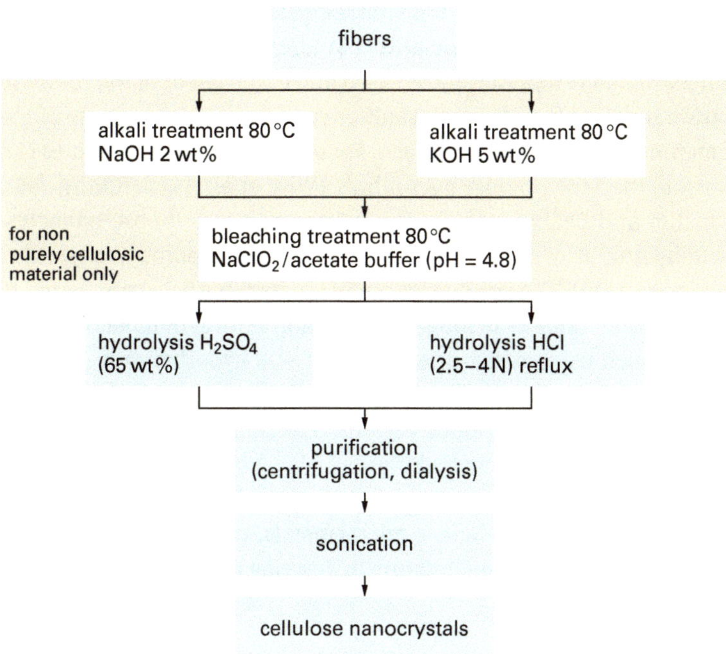

Fig. 3.1: General procedure to obtain cellulose crystallites.

The washing and bleaching steps can be repeated several times for a more effective discoloration of the fibers. Filtration and washing with distilled water is performed between each step.

After removal of non-cellulosic constituents, such as lignin, pectins and hemicelluloses, the bleached material is disintegrated in water, and the resulting suspension is submitted to the hydrolysis treatment with acid.

3.3 Acid hydrolysis treatment

As previously discussed in Chapter 1, cellulose fibers and microfibrils are semicrystalline. This means that apart from crystalline domains, cellulose also occurs in a non-crystalline (amorphous) state. The cellulose amorphous regions are randomly oriented in a spaghetti-like arrangement leading to a lower density compared to nanocrystalline regions (Saxena and Brown, 2005; de Souza Lima and Borsali, 2004). The amorphous regions act as structural defects which are susceptible to acid attack and, under controlled conditions, they may be removed leaving crystalline regions intact (de Souza Lima and Borsali, 2004; Thielemans et al., 2009). This transformation consists of the disruption of amorphous regions surrounding and embedded within cellulose microfibrils. During the acid hydrolysis process, the hydronium ions can penetrate the cellulose chains in the amorphous domains promoting the hydrolytic cleavage of the glycosidic bonds and releasing individual crystallites. It is ascribed to the faster hydrolysis kinetics of amorphous domains compared to crystalline ones. From this process, cellulose monocrystals may be released and extracted from the cellulose substrate. It has been reported that these crystallites can grow in size because of the large freedom of motion after hydrolytic cleavage. For this reason, the crystallites can be larger in dimension than the original microfibrils (Wise et al., 1946; Ranby, 1951; Battista, 1975; Whistler and BeMiller, 1997). The preparation of cellulose nanocrystals by sulfuric acid hydrolysis of dried and never-dried chemical pulps was reported (Kontturi and Vuorinen, 2009). The average length of the nanoparticles was found to be fairly similar, but a higher number of longer crystals and a lower number of shorter crystals were present when using never-dried pulps. It was hypothetically ascribed to tensions building in individual microfibrils upon drying, resulting in irreversible supramolecular changes in the amorphous regions. The tense amorphous regions were deemed as more susceptible to acid hydrolysis.

Different descriptors have been used in the literature to designate these crystalline rod-like nanoparticles, even cellulose microcrystals, despite their nanoscale dimensions. The main terms used in the literature to describe these nanoparticles are reported in Table 3.1.

Acronym	Terminology	Reference
–	Cellulose Micelles	(Rånby, 1949)
–	Level-off D.P. Cellulose Product	(Battista and Smith, 1961)
Wh	Whiskers	(Helbert et al., 1996; Dufresne, 2008)
–	Cellulose Crystallites	(Dong et al., 1996)
–	Cellulose Microcrystals	(Araki et al., 2001)
CNC	Cellulose Nanocrystals	(Grunert and Winter, 2002; Paralikar et al., 2008; Mangalam et al., 2009)
CNW	Cellulose Nanowhiskers	(Oksman et al., 2006; Petersson et al., 2007; Habibi et al., 2008; Braun et al., 2008; Rojas et al., 2009)
–	Nanocellulose	(Morán et al., 2008)
NCC	Nanocrystalline Cellulose	(Bai et al., 2009)

Table 3.1: Different terminologies used in the literature to describe the material resulting from the cellulose fiber fibrillation process.

The nomenclature that will be used further (cellulose nanocrystals) is in agreement with the TAPPI standard recommendation (TAPPI, 2011).

3.3.1 Sources of cellulose

The final properties of cellulose nanocrystals depend on the origin of the cellulosic fibers. Nanocrystals extracted from tunicate and algae sources are several micrometers in length since the cellulose microfibrils in tunicate and algae are highly crystalline. Nanocrystals from bacterial cellulose also have dimensions similar to those obtained from tunicate and algae, while nanocrystals of smaller dimensions are obtained from cotton and wood cellulose for instance. Therefore, specific hydrolysis and extraction procedures have been developed depending on the source of cellulose. These conditions are reported in Table 3.2 for the most common sources that have been investigated in the literature.

Source	Acid	Time (min)	Temperature (°C)	Acid-to-Pulp Ratio ($mL \cdot g^{-1}$)	Reference
Acacia Pulp	H_2SO_4 64%	45	45	17.5	(Pu et al., 2007)
Alfa	H_2SO_4 65%	15	60	25	(Ben Elmabrouk et al., 2009)
Algal (*Valonia*)	H_2SO_4 67%	30 + 30	RT + 70	–	(Revol, 1982; Hanley et al., 1992)

Source	Acid	Time (min)	Temperature (°C)	Acid-to-Pulp Ratio (mL · g^{-1})	Reference
Bacterial Cellulose	H_2SO_4 12–65%	60–180	40–104	–	(Roman and Winter, 2004)
	H_2SO_4 60%	60	51	70	(Hirai et al., 2009)
Bamboo	HNO_3 30%	1440	50	10	(Liu et al., 2010)
Banana	H_2SO_4 64–76%	240–480	45–65	–	(Elanthikkal et al., 2010)
Bio-Residue from Wood Bioethanol Production	H_2SO_4 63.5%	130	44	9.8	(Oksman et al., 2011)
Bleached Soft-wood Kraft Pulp	H_2SO_4 65%	10	70	10	(Araki et al., 1998)
	H_2SO_4 60%	30	60	12.5	(Orts et al., 1998)
	HCl 4N	225	80	30	(Araki et al., 1999)
	H_2SO_4 64%	25–45	45	8.75–17.5	(Beck-Candanedo et al., 2005)
		45	45	8.75	(Roman and Gray, 2005)
Capim Dourado	H_2SO_4 65%	60	50	17–25	(Siqueira et al., 2010a)
Cassava Bagasse	H_2SO_4 6.5M	40	60	20	(Teixeira et al., 2009)
Cladophora	HCl 4N	240	80	–	(Kim et al., 2000)
Coconut Husk Fibers	H_2SO_4 64%	120–180	45	0.1–1	(Rosa et al., 2010)
		30	45		(Fahma et al., 2011)
Cotton	H_2SO_4 64%	60	45	8.75	(Dong et al., 1998)
	H_2SO_4 60%	30	60	12.5	(Orts et al., 1998)
Cottonseed Linter	H_2SO_4 64%	240	45	8.75	(Lu et al., 2005)
	H_2SO_4 30%	360	60	8.75	(Wang et al., 2006)
	HCl 2.5M	20	105	–	(Braun et al., 2008)
Curaúa	H_2SO_4 60%	75	45	20	(Corrêa et al., 2010)
Date Palm Tree (Rachis/ Leaflets)	H_2SO_4 65%	45	45	–	(Bendahou et al., 2009)

Source	Acid	Time (min)	Temperature (°C)	Acid-to-Pulp Ratio (mL · g^{-1})	Reference
Eucalyptus Wood Pulp	H_2SO_4 64%	25	45	8.75	(Beck-Candanedo et al., 2005)
				9	(de Mesquita et al., 2010)
Flax	H_2SO_4 64%	240	45	8.3	(Cao et al., 2007)
Grass of Korea (*genus Zoysia*)	H_2SO_4 65%	15	55	–	(Pandey et al., 2008)
Hemp	H_2SO_4 64%	240	45	8.3	(Cao et al., 2008)
Kenaf	H_2SO_4 65%	20–120	45	–	(Kargarzadeh et al., 2012)
Luffa cylindrica	H_2SO_4 65%	40	50	17–25	(Siqueira et al., 2010b)
MCC	H_2SO_4 44–65%	10–120	40–80	6.7–20	(Bondeson et al., 2006)
	HBr 1.5–2.5M	240	100	40	(Lee et al., 2009)
Mengkuang Leaves	H_2SO_4 60%	45	45	–	(Sheltami et al., 2012)
Mulberry	H_2SO_4 64%	30	60	20	(Li et al., 2009)
Pea Hull	H_2SO_4 64%	240–1440	45	8.3	(Chen et al., 2009)
Ramie	H_2SO_4 64%	240	45	8.75	(Lu et al., 2006)
	H_2SO_4 65%	30	55	–	(Habibi and Dufresne, 2008)
Rice Husk	H_2SO_4 10 mol · L^{-1}	40	50	–	(Johar et al., 2012)
Rice Straw	H_2SO_4 64%	30–45	45	8.75	(Lu and Hsieh, 2012)
Sisal	H_2SO_4 65%	15	60	25	(Garcia et al., 2006)
	H_2SO_4 60%	30	45	–	(Morán et al., 2008)
	H_2SO_4 65%	40	50	17–25	(Siqueira et al., 2009)
Sugar Beet Pulp	H_2SO_4 60%	35	40	–	(Azizi Samir et al., 2004)
Sugarcane Bagasse	H_2SO_4 6M	30–75	45	20	(Teixeira et al., 2011)
Tunicin	H_2SO_4 55%	20	60	–	(Anglès and Dufresne, 2000)
Wheat Straw	H_2SO_4 65%	60	25	–	(Helbert et al., 1996)

Table 3.2: Hydrolysis conditions for the preparation of nanocrystals from different cellulosic fibers.

It was shown that the use of MFC instead of native cellulose microfibers to be submitted to the acid hydrolysis treatment allows obtaining cellulose nanocrystals with a similar morphology but reducing the acid concentration by at least 10 wt% (Siqueira et al., 2010c). In this study, the nanoparticles were obtained from sisal using sulfuric acid.

3.3.2 Nature of the acid

Different strong acids have been shown to successfully degrade cellulose fibers. Sulfuric and hydrochloric acids have been extensively used. However, phosphoric (Koshizawa, 1960; Okano et al., 1999), hydrobromic (Lee et al., 2009; Filipponen, 2010) and nitric acids (Liu et al., 2010) have also been reported for the preparation of crystalline cellulosic nanoparticles. The use of an acid mixture composed of hydrochloric and an organic acid (acetic or butyric) was shown to give narrower diameter polydispersity indices (Braun and Dorgan, 2009). Nevertheless, sulfuric acid has been extensively investigated and appears to be the most effective. The general procedure which can be adopted for any cellulosic source to extract cellulose crystallite is reported in Figure 3.1. However, it needs to be adapted depending on the source of cellulose.

Cellulose nanocrystals were prepared from MCC by using an organic acid (maleic acid) instead of a mineral acid and a sono-chemical-assisted hydrolysis process (Filson et al., 2009). The procedure followed consisted in introducing 25% (wt/vol) of MCC in a solution of maleic acid (50 mM to 100 mM) and heated gently between 15 and 35°C. Finally, a centrifugation step was added to remove the excess acid and the resulting mixture was sonicated. By this method the yield was improved by 10% compared to the conventional H_2SO_4 method.

One of the main reasons for using sulfuric acid as a hydrolyzing agent is that if the cellulose nanocrystals are prepared using hydrochloric acid, their ability to disperse is limited and the suspension is unstable tending to flocculate. Indeed, sulfuric acid reacts with the surface hydroxyl groups via an esterification process allowing the grafting of anionic sulfate ester groups ($-OSO_3^-$) (Rånby, 1949). They are randomly distributed on the cellulosic nanoparticle surface. The presence of these negatively charged sulfate ester groups induces the formation of a negative electrostatic layer covering the nanocrystals and promotes their dispersion in water. The high stability of sulfuric acid-hydrolyzed cellulose nanocrystals results therefore from an electrostatic repulsion between individual nanoparticles. Acid hydrolysis with hydrochloric acid does not produce as many negative surface charges on cellulose nanocrystals, resulting in less stable NCC suspensions. For instance, conductimetric titration has been conducted on cellulose nanocrystals prepared by hydrolyzing bleached softwood kraft pulp with 65% sulfuric acid or 4N hydrochloric acid (Araki et al., 1998). It was shown that the H_2SO_4-treated sample had a surface charge of 84 $mmol \cdot kg^{-1}$

of strong acid and 26 $mmol \cdot kg^{-1}$ of weak acid, ascribed to sulfate ester and carboxyl groups, respectively. No strong acid groups and a small amount of weak acid groups (< 18 $mmol \cdot kg^{-1}$) were present in the HCl-treated sample. The presence of these strongly ionizable sulfur-containing groups on the nanoparticle surfaces was responsible of a monotonically repulsive interaction between an amorphous cellulose sphere and cellulose nanocrystal thin film (Notley et al., 2006).

However, the introduction of charged sulfate groups also has its drawbacks since it compromises the thermostability of the nanocrystals. With increasing applications of cellulose nanocrystals as fillers in polymer nanocomposites, this issue is important since typical processing temperatures for thermoplastics often exceed 200°C. Indeed, inorganic salts and acids are known to act as flame retardants in the pyrolysis and combustion of cellulose, i.e. they increase the char yields at the expense of flammable tars by promoting the dehydration reactions (Tang and Neill, 1964). Consequently, impregnation of cellulose with sulfuric acid has been shown to lower the onset of thermal degradation (Parks, 1971; Julien et al., 1993; Kim et al., 2001), decrease the yield of levoglucosan (Parks, 1971), and increase the yield of carbon (Julien et al., 1993).

The thermal degradation and thermal stability of cellulose nanocrystals obtained from bacterial cellulose have been investigated using thermogravimetric analysis (TGA) (Roman and Winter, 2004). The authors observed that during hydrolysis, the sulfate group content, determined by potentiometric titration, could be increased by increasing the reaction time, reaction temperature, acid concentration and acid-to-pulp ratio. Increasing sulfate groups led to degradation at lower temperatures and a broader temperature range was observed as compared to unhydrolyzed samples. Also increasing the number of sulfate groups on nanocrystals increased the amount of charred residue indicating that sulfate groups are flame-retardant in nature. The degradation was described as a two-step process, viz. low temperature process and high temperature process. The low temperature process involves the degradation of most accessible amorphous regions which are also highly sulfated. The high temperature process involves the degradation of less accessible interior crystalline regions that are comparatively less sulfated. For highly sulfated samples, an extra step involves the degradation of sulfated ends of crystalline regions. Using the Broido method, the authors calculated the activation energies for both degradation processes and concluded that more sulfate groups led to lower activation energies for degradation. It was also suggested that sulfuric acid may be instrumental in catalyzing the degradation process. Direct catalysis by the acidic molecules could be a possible mechanism. Therefore, it was concluded that using sulfuric acid produces sulfated nanocrystals which are susceptible to thermal degradation. At the detriment of nanoparticle aggregation, sulfation of nanocrystals can be reduced using shorter reaction times, lower acid-to-pulp ratio and lower acid concentrations. It was also suggested that hydrochloric acid should be used for hydrolysis since there is minimum sulfation during hydrolysis. Similar conclusions were drawn for cellulose nanocrystals extracted from curaua fibers (Corrêa et al., 2010). However, such nanocrystals have higher aggregation tendencies without sufficient negative surface charges.

3.3.3 Effect and optimization of extraction conditions

Typical procedures currently used for the preparation of cellulose nanocrystals from biomass consist in subjecting pure cellulosic material to strong acid hydrolysis under strictly controlled conditions of temperature, agitation, and time. The shape and size of nanocrystals are more or less fixed by the source of cellulose. This means that different dimensions of nanocrystals can be obtained from different sources of cellulose even under similar experimental conditions. However, the properties of cellulose nanocrystals for a given raw material are affected by the degree of acid hydrolysis, and the exact hydrolysis conditions. The current accepted explanation depicts the process of acid hydrolysis of cellulose as a heterogeneous process. It involves the diffusion of acid into the fiber, and the subsequent cleavage of glycosidic bonds transforming the microfibrils into short nanocrystals. Both steps depend on the nature of the acid and its concentration, hydrolysis temperature and time, and possibly other factors.

The hydrolysis conditions must be mild enough to avoid complete hydrolysis to glucose. For sulfuric acid, the best concentration was found to be in the range 60–70% (Revol et al., 1994). Typically, the concentration of sulfuric acid in hydrolysis reactions to obtain cellulose nanocrystals is around 65 wt% (Table 3.2). It was also shown that in these conditions, the pre-acid-hydrolysis treatment did not affect the dimensions of cellulose nanocrystals (Fahma et al., 2011). However, the temperature can range from room temperature up to 70°C and the corresponding time can vary from 15 min to 24 h depending on the reaction temperature. Obviously, if increasing the temperature of the acid hydrolysis reaction, shorter times should be used. For hydrochloric acid-catalyzed hydrolysis, the acid concentration ranges between 2.5 and 4 N and the reaction is usually performed at reflux temperature.

It was reported in the early works on the degradation of cellulose by sulfuric acid that more drastic acid treatments, involving longer times or higher temperatures, were effective in breaking down cellulose fragments fairly successfully (Mukherjee and Woods, 1953). Attempts were made to fractionate the colloidal suspensions, and trial runs on an ultracentrifuge showed that this should be possible. Electrodialysis was found to be more effective as a cleaning than as a fractionating technique, and simple electrophoresis was even less efficient. It was proposed that relatively mild hydrolysis conditions might cause the crystallization of cellulose chains simultaneously with chain splitting (Ingersoll, 1946; Howsmon, 1949; Hermans and Weidinger, 1949). The mechanism of crystallization which accompanies hydrolysis was shown to be controlled by two interdependent processes, viz. hydrolysis kinetic and crystal growth (Battista, 1950). This effect was found to be much more pronounced for regenerated cellulose than for native celluloses.

Dong et al. (1998) have investigated the properties of cotton nanocrystals prepared from Whatman No. 1 filter paper powder and were among the first researchers studying the effect of hydrolysis conditions on the properties of ensuing nanoparticles. They optimized the acid hydrolysis conditions by varying the hydrolysis temperature and hydrolysis time. The concentration of sulfuric acid was fixed at 64% and the ratio of the filter paper to acid was 1:8.75 $g \cdot mL^{-1}$. It was observed that at a low hydrolysis temperature, 26°C, an 18 h reaction time was required to produce nanocrystals. At 65°C, the hydrolysis could not easily be controlled, side reactions such as dehydration occurred since a yellow color appeared at the early stage of the reaction (10 min), and the sample became black after 1 h. At an intermediate temperature (45°C), the hydrolysis proceeded smoothly and yielded an ivory-white suspension after 1 h reaction. The effect of hydrolysis time at 45°C on the length of cellulose nanocrystals is shown in Figure 3.2. When acid treatment is continued for a longer time period, shorter nanocrystals are produced. The initial rapid decrease in the early stages of the hydrolysis slowed down in the later stages.

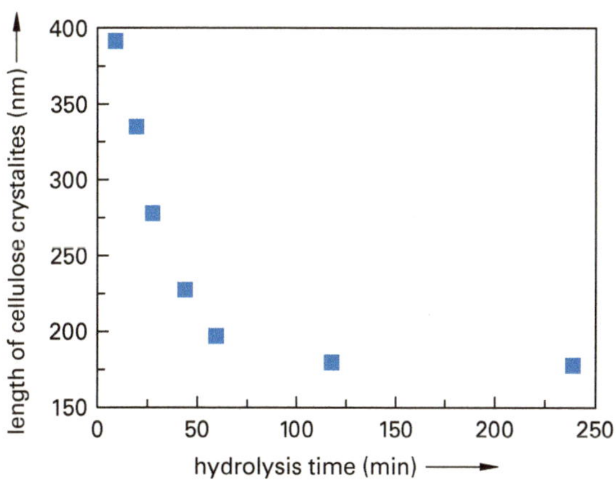

Fig. 3.2: Evolution of the length of cellulose nanocrystals extracted from cotton and prepared by H_2SO_4 hydrolysis as a function of hydrolysis time at 45°C (Dong et al., 1998).

The acid concentration was also found to affect the morphology of nanocrystals as reported by Azizi Samir et al. (2004). In this study, cellulosic nanoparticles were prepared from homogenized sugar beet pulp. Two sulfuric acid concentrations in water were used, namely 20 and 60 wt%, and hydrolysis was carried out at 40°C for 35 min under strong stirring. It was observed that both the length and the diameter of the microfibrils decrease as the concentration of acid increases (Figure 3.3). For the strongly hydrolyzed material (Figure 3.3(c)), most of the amorphous domains were hydrolyzed and straight nanocrystals around 210 nm long were obtained.

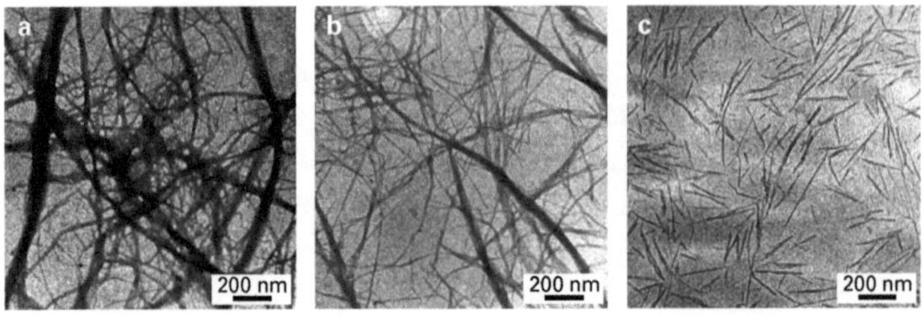

Fig. 3.3: Transmission electron micrographs of a dilute suspension of sugar beet cellulose microfibrils: (a) unhydrolyzed, and H_2SO_4-hydrolyzed at 40°C for 35 min using an acid concentration of (b) 20 wt%, and (c) 60 wt% (Azizi Samir et al., 2004).

The properties of wood cellulose nanocrystals produced by sulfuric acid hydrolysis from two wood sources, viz. softwood (black spruce) and hardwood (eucalyptus), have been studied (Beck-Candanedo et al., 2005). Keeping the reaction time, reaction temperature and acid-to-pulp ratio constant, the authors observed that the two wood cellulose suspensions possessed similar dimensions and surface charge. The critical concentration required to form anisotropic liquid phases was similar for both kinds of suspensions. The authors considered the reaction time, whose range is fairly narrow, as one of the most important parameters to consider in the acid hydrolysis of wood pulp. Too short a reaction time will yield only large undispersible fibers and aggregates, whereas too long a reaction time will digest the cellulose completely to yield its component sugar molecules. It was observed that in the production of softwood suspensions, longer reaction times resulted in nanocrystals with shorter lengths, smaller diameters and lower polydispersity in the length of the nanocrystals. Higher acid-to-pulp ratios also produced shorter crystal length in the softwood suspensions with narrower length distribution. Combining both effects, i.e. a higher acid-to-pulp ratio and longer reaction time, produced shorter nanocrystals with smaller polydispersity and increased critical concentration to form an anisotropic phase (confirmed by Onsager's phase separation theory). The authors also observed that surface charge and sulfur content of the nanocrystals can be decreased by the application of heat, which de-esterifies the sulfate groups (Beck-Candanedo et al., 2005).

A series of cellulose nanocrystals was extracted from pea hull fibers by sulfuric acid hydrolysis (64%) with different hydrolysis times up to 24 h at 45°C (Chen et al., 2009). For increasing reaction times, it was found also that both the length and the diameter of the nanoparticles decreased, while the aspect ratio (ratio of the length to the diameter) first increased, reaching the highest value for a reaction time of 8 h, and then decreased. For cellulose nanocrystals extracted from kenaf bast fibers, it was found that the diameter tended to decrease more or less continuously from about 13 to 11 nm, whereas the length did not display any logical evolution when increasing the duration of the hydrolysis process from 20 to 120 min (Kargarzadeh et al., 2012).

Interestingly, the polydispersity in diameter significantly decreased with the hydrolysis reaction time.

In a slightly different approach, an attempt to find optimized sulfuric acid hydrolysis conditions to prepare cellulose nanocrystals from microcrystalline cellulose (MCC) derived from Norway spruce (*Picea abies*) was also reported (Bondeson et al., 2006). The objective of this study was to find a rapid and high-yield process to obtain aqueous suspension of cellulose nanocrystals. Five important parameters were considered for optimizing the reaction conditions, namely starting concentration of MCC, concentration of sulfuric acid, reaction time, temperature, and sonication time. Response surface methodology was used to determine the effect of the five parameters in order to maximize yield and minimize reaction time. The time and temperature of hydrolysis, as well as acid concentration, were found to be very important single factors in the process. The model predicted that an increase in sulfuric acid concentration, reaction time and reaction temperature would decrease the size of MCC but not the nanocrystal yield significantly. Neither the starting concentrations nor the sonication times had any significant effect on the MCC size or the nanocrystal yield. However, the model predicted that sulfuric acid concentration and reaction time work synergistically in increasing the nanocrystal yields. The optimum reaction conditions suggested by the model were as follows: sulfuric acid concentration = 63.5 wt%, reaction time = 130.3 min, reaction temperature = 44°C, MCC starting concentration = 10.2 $g \cdot 100mL^{-1}$, and duration of sonication = 29.6 min. These optimal reaction conditions predicted by the model produced a low experimental yield of 30%, which was attributed to the degradation of amorphous and crystalline regions during hydrolysis.

The effect of acid strength (1.5M and 2.5M) on the physico-chemical, thermal and morphological properties of cellulose nanocrystals prepared from MCC by acid hydrolysis with hydrobromic acid (HBr) has been reported (Lee et al., 2009). The conditions of the preparation are collected in Table 3.2. The increase in acid concentration induced an increase of crystallinity and thermal stability, and a decrease of the molecular weight of the nanoparticles. The properties of coconut husk fibers cellulose nanocrystals produced by sulfuric acid hydrolysis have been studied (Rosa et al., 2010). Keeping the reaction temperature (45°C) and acid concentration (64 wt%) constant, the authors varied the pulp-to-acid ratio (1–10 $g \cdot mL^{-1}$) and reaction time (120–180 min) to access the effect of preparation conditions on the thermal and morphological properties of the nanoparticles. No correlation was observed.

The effect of hydrolysis conditions of kraft pulp fibers on the degree of polymerization, extent of sulfation, morphology, solid-state characteristics and yield of ensuing cellulose nanoparticles was reported (Hamad and Hu, 2010).

It was shown that the hydrolysis of amorphous cellulosic chains can be performed simultaneously with the esterification of accessible hydroxyl groups to produce surface functionalized nanocrystals in a single step (Braun and Dorgan, 2009). The reaction was carried out in an acid mixture composed of hydrochloric and an organic acid (acetic and butyric). Resulting nanocrystals were of similar dimensions compared to those obtained by hydrochloric acid hydrolysis alone. Narrower diameter polydispersity indices indicate that surface groups aid the individualization of the nanocrystals. The ensuing surface-modified cellulose nanocrystals were dispersible in ethyl acetate and toluene indicating increased hydrophobicity and presumably higher compatibility with hydrophobic polymers.

Aspect ratio-controlled cellulose nanocrystals have been prepared by sulfuric acid hydrolysis of bacterial cellulose produced by *Gluconacetobacter xylinum* by varying the hydrolysis times (Yun et al., 2010). The authors conducted the acid hydrolysis with 65 wt% sulfuric acid at 70°C and various hydrolysis times, i.e. 1, 2 and 3 h. Increasing the hydrolysis time decreased the average length of the resulting nanocrystals to 1380, 840 and 540 nm for hydrolyzing times of 1, 2 and 3 h, respectively. However, their widths remained unaffected around 30 nm and therefore their aspect ratio was reduced to values of 46, 28 and 18, respectively. Also, the sulfur content of the nanoparticles increased with increasing hydrolysis time. The authors suggested that since bacterial cellulose is highly crystalline, acid cannot penetrate the crystalline bundles and hydrolysis occurs due to a surface reaction mechanism. For this reason, the hydrolysis rate was relatively slow and the aspect ratio could be controlled by the hydrolysis time.

3.4 Other processes

Acid hydrolysis is the classical way to prepare cellulose nanocrystals. However, other processes allowing the release of crystalline domains from cellulosic fibers have been recently reported in the literature.

3.4.1 Enzymatic hydrolysis treatment

As reported in Chapter 2, cellulases are composites of endoglucanases, exoglucanases, and cellobiohydrolases. These enzymes act synergistically in the hydrolysis of cellulose. Endoglucanase randomly attacks and hydrolyzes the amorphous regions while exoglucanase attacks the cellulosic chain from either the reducing or non-reducing ends. Cellobiohydrolases hydrolyze cellulose from either the C1 or the C4 ends using a protein in each case, into cellobiose sub-units. Enzymatic hydrolysis is therefore more specific than acid hydrolysis and from the environmental point of view it also displays the advantage to be less corrosive compared to chemical treatments (Henriksson et al., 2007). The enzymatic treatment was also reported to have the advantage of causing less material loss in comparison with other nanocellulose preparation processes such as acid hydrolysis or mechanical fibrillation (Woehl et al., 2010).

The selectivity of endoglucanase towards amorphous regions leaving the crystalline domains intact has been explored to prepare cellulose nanocrystals from recycled pulp using microwave assisted enzymatic hydrolysis (Filson et al., 2009). Ultrasonication was further used to separate the individual crystallites. Rod-like nanoparticles with lengths and widths ranging between 100 nm and 1.8 μm, and 30 and 80 nm, respectively, were obtained from microscopic observations. Dynamic light scattering studies suggested lengths ranging from 100 nm to 3.5 μm. A trimodal frequency length distribution with an average value around 155, 820, and 3,540 nm was observed. The possibility of controlled hydrolysis of MCC using the fungus *Trichoderma reesei* was also explored (Satyamurthy et al., 2011). The penetration of fungus into the ordered regions of MCC during incubation resulted in reduced crystallinity (around 10%) of ensuing nanocrystals compared to acid hydrolyzed-nanoparticles

The preparation of cellulosic nanoparticles using different processing routes, viz. a combination of mechanical shearing, acid and enzymatic hydrolysis, has been reported (Siqueira et al., 2010c). Bleached sisal fibers and two different enzymes, mostly with either an endoglucanase activity or an exoglucanase activity, have been used. The enzymatic treatment was performed on fibers (before mechanical treatment) or on MFC (after mechanical treatment) and the effect of the enzyme concentration was also investigated. Depending on the treatments applied to the pulp and their sequence, a broad range of cellulosic nanoparticles including MFC, nanocrystals and a mixture of both has been obtained. These different nanoparticles induced totally different reinforcing effects when dispersed in a polymeric matrix (Siqueira et al., 2011). A latex of natural rubber was chosen for the study because it allows a more distinct segregation of the results owing to its rubbery amorphous state and hence low modulus value. The two enzymes display different hydrolysis efficiency. The enzyme with the endoglucanase activity induced a much higher DP reduction than the one with an exoglucanase activity. The former allowed cutting the cellulosic chains in specific sites to release smaller nanoparticles. This effect was accentuated when the enzyme concentration was increased. Individual cellulose nanocrystals, 277 nm long and 8.7 nm wide, similar to those obtained by H_2SO_4 hydrolysis, were obtained.

3.4.2 TEMPO oxidation

TEMPO (2,2,6,6-tetramethylpiperidin-1-oxyl)-mediated oxidation has been applied to cellulose fibers under various conditions. Among the three hydroxyl groups borne on C2, C3 and C6 by the glucose unit, the hydroxyl group of C6 is much more reactive than that of C2 and C3. Therefore, the TEMPO system selectively oxidizes C6 primary hydroxyl groups to carboxyl groups. This reaction occurs on the surface of cellulose fibers and in amorphous regions. As the carboxyl content is increased to a certain amount, cellulose begins to disperse in aqueous solution but the crystalline domains remain intact and can therefore be released.

When applying 4-acetamide-TEMPO/NaClO/ $NaClO_2$ oxidation system at 60°C and pH 4.8 for 1–5 days to never-dried mercerized wood cellulose, nanocrystals with the cellulose II structure and carboxylate contents of 2.0–2.4 $mmol \cdot g^{-1}$ have been obtained in the yield 67–77% after disintegration of the oxidized product in water by sonication (Hirota et al., 2010). The dispersion consisted of individual (but slightly aggregated) nanocrystals 4–7 nm in width and 100–200 nm in length. In this study, two types of cellulose nanocrystal samples were prepared. The first one was TEMPO-mediated oxidation of mercerized cellulose and the second one was TEMPO-mediated oxidation of acid hydrolyzed cellulose and mercerized cellulose. By comparing the light transmittance and rheological properties of both kinds of nanocrystals, the authors concluded that cellulose nanocrystals obtained from mercerized cellulose and TEMPO-mediated oxidation revealed better water-dispersibility due to the incorporation of a higher number of carboxylate groups (during oxidation of C6 hydroxyl groups on the crystallite structures of mercerized cellulose). This sample also exhibited highest transmittance, shear stress and viscosity values as compared to the acid hydrolyzed sample due to higher carboxylate content on the nanocrystals thereby producing higher dispersibility and lower number of bundles in water.

High carboxylic acid content cellulose nanocrystals were also prepared from cotton linter pulp by direct ultrasonic-assisted TEMPO oxidation using the TEMPO/NaBr/NaClO system (Qin et al., 2011). Microscopic observations revealed the presence of cellulose nanocrystals 5–10 nm in width and 200–400 nm in length. Stable aqueous suspensions were obtained and a high carboxyl acid content around 1.66 $mmol \cdot g^{-1}$ was reported.

3.4.3 Hydrolysis with gaseous acid

A recent patent reports the hydrolysis of cellulosic materials in the presence of an acid in the gas phase (Kontturi, 2011). In this study, it was shown that gaseous HCl induced the decrease of the degree of polymerization of cellulose to low values at room temperature within 30 min. Nanocrystals were then obtained by mechanical grinding and/or ultrasound treatment of the treated substrate. The principle is based on the fact that there is a thin aqueous layer on the surface of cellulosic fibers. When the gaseous acid adsorbs at the surface of the fibers, a high local acid concentration is reached, leading to a high breakdown rate of amorphous domains. The moisture content of the cellulosic substrate can be between 1% and 80%, but the process was faster with dried fibers because of more tension dissipated in dry amorphous regions. This process can avoid several environmentally harmful and time-consuming steps of the classical acid hydrolysis of cellulose. Indeed, large amounts of water are not necessary for rinsing the sample, the recycling of the acid is easier and the dialysis step can be omitted. The yield was also found to be higher than with liquid acid because of a possible crystallization of cellulose during the gaseous process.

3.4.4 Ionic liquid

Cellulose nanocrystals have been prepared by treating MCC with 1-butyl-3-methylimidazolium hydrogen sulfate ($bmimHSO_4$) ionic liquid (Man et al., 2011). MCC was mixed with $bmimHSO_4$ and treated for 1 h at different temperatures in the range 70–90°C. The ensuing suspension was sonicated and centrifugated for purification. The ionic liquid was found to react with MCC in a similar manner to acids in acid hydrolysis causing hydrolytic cleavage of glycosidic bonds between anhydroglucose units. The ionic liquid was not consumed during treatment and can be thus regenerated. Needle-like nanocrystals with length and diameter of 50–300 nm and 14–22 nm, respectively, were obtained.

3.5 Post-treatment of hydrolyzed cellulose

3.5.1 Purification of the suspension

The suspension resulting from the acid hydrolysis treatment of cellulose is subsequently cooled at room temperature (eventually diluted with ice cubes) and is sometimes filtered over a small-pore fritted glass filter. It is diluted with water, and washed with successive centrifugations. Part of the excess acid and water-soluble fragments can be removed during these centrifugations steps. Dialysis against distilled water is then performed to remove any free acid molecules from the dispersion. This procedure ensures that all ionic materials are removed except the H^+ counterions associated with the sulfate groups at the surface of the nanocrystals. It can be verified by determining the neutrality of the dialysis effluent. Disintegration of aggregates and complete dispersion of the nanocrystals is obtained by a sonication step. This treatment significantly heats up the suspension. These suspensions are generally much diluted because of the formation of a gel for low nanoparticle contents, as for MFC. The aqueous dispersion is stored in the refrigerator after possible filtration to remove residual aggregates and adding a few drops of chloroform to prevent bacterial growth. The solid content of the suspension can be determined gravimetrically by weighing alicots of the suspension before and after drying.

3.5.2 Fractionation

Additional steps of post-treatment of hydrolyzed cellulose nanocrystals have been reported in the literature. Complications in size heterogeneity are inevitable owing to the diffusion-controlled nature of the acid hydrolysis process (Fleming et al., 2001). The aqueous suspensions of cellulose nanocrystals can be separated into isotropic and anisotropic phases by increasing the concentration (by water evaporation).

Evidence of this fractionation was observed by transmission electron microscopy observations (Dong et al., 1998). It was found that the smaller nanoparticles lie in the isotropic phase (top) whereas the longer ones are in the anisotropic phase (bottom) (Hirai et al., 2009).

Tunicin nanocrystal suspensions have been fractionated by ultracentrifugation using a saccharose gradient between 40 and 67 wt% at 26,000 g for 40 min (de Souza Lima and Borsali, 2002). Visual inspection of the suspension evidenced the fractionation process. Figure 3.4(a) shows the cellulose nanocrystal samples before and after ultracentrifugation. Before centrifugation, clearly separate black and clear phases were observed. The latter corresponded to cellulose nanocrystals that were deposited on the surface of the saccharose gradient. After centrifugation, the presence of different bands corresponding to the different distribution in length of the nanoparticles was reported, suggesting the success of the fractionation method. The different fractions were collected from different positions in the centrifugation tube to give narrower samples for further uses. Transmission electron microscopy (TEM) analysis (Figure 3.4(b)) on each fraction confirmed that the fractionated samples had narrow polydispersity and reduced nanocrystal aggregation.

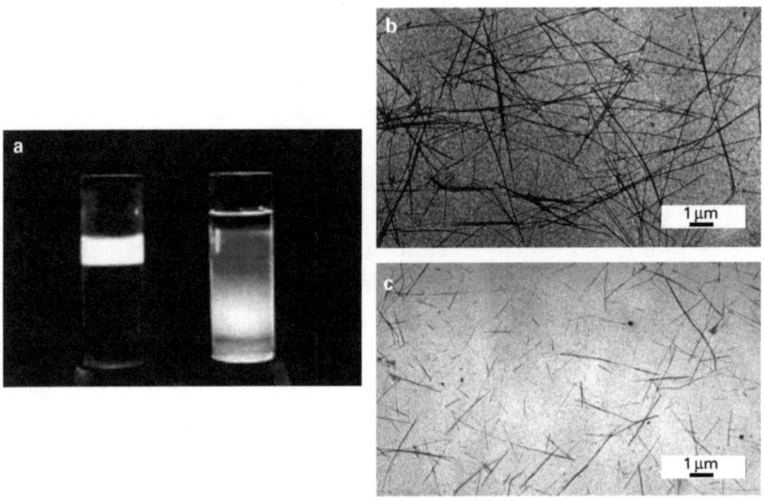

Fig. 3.4: H_2SO_4-hydrolyzed tunicin nanocrystals (a) photographs of the suspension before (left) and after (right) the ultracentrifugation process using a saccharose gradient, and (b) transmission electron micrographs of the nanocrystals before (top) and after (bottom) the ultracentrifugation process (de Souza Lima and Borsali, 2002).

Cellulose nanocrystals with a narrow size distribution were also successfully obtained from MCC using differential centrifugation technique (Bai et al., 2009). In this investigation, MCC was hydrolyzed with sulfuric acid to produce nanocrystal suspensions which were further subjected to centrifugal treatment. Six different nanoparticle fractions were produced by using differential angular velocities, i.e. by creating relative centrifugal force (RCF) for each fraction at a constant centrifugation time. Initially, the suspension was subjected to 500 rpm for 10 min followed by discarding of the precipitate that contained long nanocrystals. Next, the suspension left from the first centrifugation was subjected to 1,000 rpm for 10 min. The precipitate obtained was identified as the 1st fraction. Next, the suspension from 1,000 rpm centrifugation was subjected to 1,500 rpm for 10 min and the precipitate obtained was identified as the 2nd fraction. Again the resulting suspension was further subjected to centrifugation and the whole process was continued until six different fractions were obtained (up to an angular velocity of 3,500 rpm). The authors also derived an inverse relation between the volume of nanocrystals and the RCF applied. It was shown that increasing the RCF in multiple steps would separate the nanocrystals of different volumes if the centrifugation time is kept constant. On studying the volume and length distributions for all nanoparticle fractions, the authors concluded that as RCF was increased, the volume of the individual nanocrystal fractions became smaller, the volume range was narrower, the lengths of particles in a certain fraction were smaller and the length distributions became narrower as well.

3.5.3 Yield

Despite its importance for any potential application, the yield of cellulose nanocrystals prepared from cellulosic materials has been reported in quite a few studies. The importance of this parameter seems to be limited for most investigations performed at the lab scale. The time consuming preparation of these nanoparticles and associated low yield has been stressed in some publications (Pandey et al., 2008; Petersson et al., 2007) even if no quantitative values have been reported.

If the cellulose crystallites extracted by acid hydrolysis were the same in size and structure as the original fibers, then the yield should correspond to the proportion of crystalline domains of the cellulosic material. However, rather lower yields, around 30%, are generally obtained (Mukherjee and Woods, 1953). It was supposed that continued action of the acid does ultimately cause complete dissolution of cellulose in the crystallites, or in an important fraction of them. If this does not take place by lateral erosion, it was hypothesized that there was a progressive shortening of the particles up to a critical length.

However, some yield data have been occasionally reported in the literature. Some are collected in Table 3.3. It is worth noting that these values are only indicative since they strongly depend on the preparation conditions and post-treatment filtration. For instance, it was found that the yield of cellulose nanocrystals prepared from cotton decreased when increasing the sulfuric acid hydrolysis temperature or time, as expected (Dong et al., 1998). A suspension with a yield of 89.8 wt% was obtained when hydrolysis was performed at 26°C for 1 h, whereas it was only 34.4 wt% when the reaction was conducted at the same temperature for 18 h. When using suitable conditions (45°C, 1h) for hydrolysis with H_2SO_4 64% and pulp-to-acid ratio 1:8.75, the yield of nanocrystals was 43.5 wt%.

Source	Yield (wt%)	Reference
Bio-Residue from Wood Bioethanol Production	48*	(Oksman et al., 2011)
Capim Dourado	14**	(Siqueira et al., 2010a)
Cotton	43.5*	(Dong et al., 1998)
Date Palm Tree (Rachis)	15**/34*	(Bendahou et al., 2009)
Kenaf	23–59*	(Kargarzadeh et al., 2012)
MCC	30*	(Bondeson et al., 2006)
MCC	21*	(Bai et al., 2009)
Mengkuang Leaves	28*	(Sheltami et al., 2012)
Recycled Pulp	3.4–38.2*	(Filson et al., 2009)
Rice Straw	4.83–6.43*	(Lu and Hsieh, 2012)
Sisal	30*	(Garcia de Rodriguez et al., 2006)
Softwood Kraft Pulp	31–39*	(Hamad and Hu, 2010)
Tunicin	75*	(Dagnon et al., 2012)

* with respect to the cellulose content.
** with respect to the initial amount of dried fibers.

Table 3.3: Yield of cellulose nanocrystals.

For cellulose nanocrystals obtained from recycled pulp by an enzymatic-mediated process, it was observed that microwave heating produced nanocrystals with higher yield compared to conventional heating since microwave heating is more selective and also reduces reaction time (Filson et al., 2009). The yield of cellulose nanocrystals in the various treatments using conventional and microwave heating ranged between 3.4% and 29.0%, and 4.9% and 38.2%, respectively. After varying the reaction times, temperature, enzyme-to-pulp ratio and mode of heating, microwave-assisted reaction at 50°C was suggested as an optimum condition for enzymatic hydrolysis of recycled pulp.

3.6 Morphology

Pioneering microscopic observations of the single particles obtained from acid-catalyzed degradation of cellulose suggested that they were tabular, and of remarkably uniform width, although they show a very wide range of lengths (Mukherjee and Woods, 1953). The thickness of the particles was found to be about a quarter of its width. For a particular type of cellulose fiber it was observed that the particles have specific lateral dimensions, which were not affected very much by the severity of the acid treatment, which, as far as the individual particles were concerned, seemed only to cause a shortening in length. For nanoparticles obtained from different types of native cellulose, widths and thicknesses ranging from 85 to 155 Å, and from 20 to 35 Å, respectively, were reported. The (101) crystallographic planes were found to be parallel to their largest faces.

The precise shape and dimensions of cellulose nanocrystal can be directly accessed by microscopic observations. Scanning electron microscopy (SEM) and atomic force microscopy (AFM) can be used. Nevertheless AFM is well-known to overestimate the width of the nanocrystals due to the tip-broadening effect. SEM is useful to perform a rapid examination in order to obtain an overview of the sample. The resolution is simply insufficient for more detailed information. Field emission SEM (SEM-FEG) can be used to improve the spatial resolution and minimize sample charging and damage.

In contrast, transmission electron microscopy (TEM) is the best-suited method to identify individual nanocrystals and determine their morphology. For these observations, a drop of a dilute cellulose nanocrystal suspension is deposited and allowed to dry on a carbon-coated grid, previously irradiated with a UV lamp. The grid is negatively stained with 2% uranyl acetate in water to enhance the contrast and dried. Uranyl acetate ($UO_2(CH_3COO)_2 \cdot 2H_2O$) is the acetate salt of uranium. This compound is a nuclear fuel derivative, and its use and possession are sanctioned by international law since it is both radioactive and toxic. It is extensively used as a negative stain in electron microscopy. As seen previously, cellulose nanocrystals can be prepared from any botanical source containing cellulose. Figure 3.5 shows some TEM observations of cellulose fragments obtained by acid hydrolysis from different sources.

The constitutive cellulosic fragments occur as elongated rod-like or needle-shaped particles designated by different appellations as reported in Table 3.1. The term whiskers, which is sometimes used in the literature, refers to specialized long stiff hairs that grow near the mouth of animals such as cats or mice and are usually employed for tactile sensation. Cellulose nanocrystals resemble whiskers because of the tapering at the ends of the nanoparticles, most likely a result of the acid hydrolysis process. When the electron probe is scanned along these rods, the persistence of the spot diffractogram displays the monocrystalline nature of the cellulosic fragment (Favier et al., 1995a). Each rod can therefore be considered as a cellulosic crystal with no apparent defect. The precise physical dimensions of the nanocrystals depend on several factors, including the source of the cellulose, the exact hydrolysis conditions and ionic strength. They generally present a relatively high distribution in length because of the diffusion-controlled nature of the acid hydrolysis. The typical geometrical characteristics for nanocrystals derived from different species and reported in the literature are collected in Table 3.4.

The length is generally of the order of a few hundreds nanometers and the width is of the order of a few nanometers. However, nanoparticles resulting from hydrolysis/sonication treatment are constituted of a few laterally bound elementary crystallites

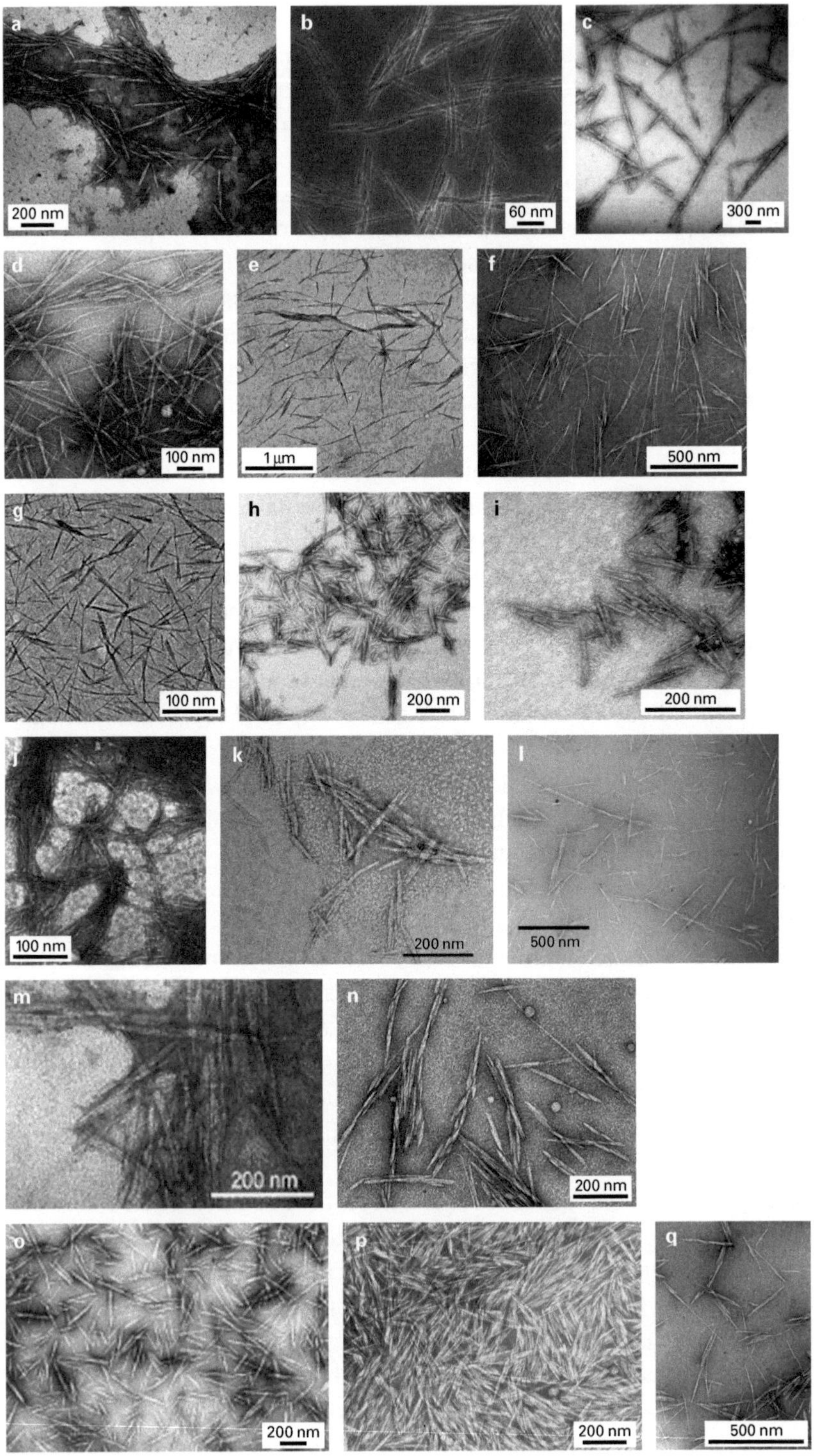
a
200 nm
b
60 nm
c
300 nm
d
100 nm
e
1 μm
f
500 nm
g
100 nm
h
200 nm
i
200 nm
j
100 nm
k
200 nm
l
500 nm
m
200 nm
n
200 nm
o
200 nm
p
200 nm
q
500 nm

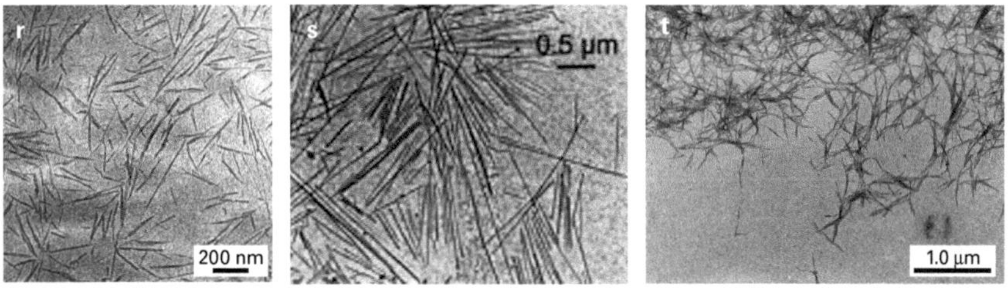

Fig. 3.5: Transmission electron micrographs from a dilute suspension of (a) acacia pulp (Pu et al., 2007), (b) alfa (Ben Elmabrouk et al., 2009), (c) bacterial cellulose (Grunert and Winter, 2002), (d) banana rachis (Zuluaga et al., 2007), (e) bleached softwood kraft pulp (Araki et al., 1998) (the scale bar correspond to 500 nm) (Foto: Jun Araki from Shishun University, Japan), (f) Capim dourado (Siqueira et al., 2010a), (g) cotton (Fleming et al., 2000) (the scale bar corresponds to 400 nm), (h) Curaúa (Corrêa et al., 2010), (i) eucalyptus wood pulp (de Mesquita et al., 2010), (j) grass of Korea (Pandey et al., 2008), (k) kenaf (Kargarzadeh et al., 2012), (l) *Luffa cylindrica* (Siqueira et al., 2010b), (m) MCC (Bondeson et al., 2006), (n) mengkuang leaves (Sheltami et al., 2012), (o) ramie (Habibi et al., 2008), (p) rice straw (Lu and Hsieh, 2012), (q) sisal (Siqueira et al., 2009), (r) sugar beet pulp (Azizi Samir et al., 2004), (s) tunicin (Anglès and Dufresne, 2000), and (t) wheat straw (Helbert et al., 1996) nanocrystals.

that are not separated by conventional acid hydrolysis and sonication process (Elazzouzi-Hafraoui et al., 2008). The length polydispersity was found to have a constant value, whereas the polydispersity for the nanocrystal diameter appears to depend on the acid used for isolation (Braun et al., 2008). When sulfuric acid is used, a smaller polydispersity in diameter is obtained when compared to hydrochloric acid. This was attributed to the increased separation of individual nanocrystals due to electrostatic charges resulting from the introduction of sulfate ester groups when sulfuric acid is used (Braun et al., 2008). For bacterial cellulose, the reason for the non-uniform fragmentation was possibly ascribed to the predominance of the cellulose Iα allomorph or to morphological differences in the microfibrils (Yamamoto and Horii, 1994; Roman and Winter, 2004).

The cross-sectional shape of the Iβ cellulose microfibrils of *Halocynthia papillosa*, a species of tunicate, was reported to be in the form of a parallelogram (Daele et al., 1992; Helbert et al., 1998), whereas Iα-rich algal cellulose microfibrils are square (Revol, 1982; Sugiyama et al., 1985). However, it has been shown that acid hydrolysis used to extract the nanocrystals from tunicate partially erodes the parallelogram-shaped cross-section and the resulting cross-section is a distorted hexagon (Helbert et al., 1998). Ribbon-like morphology was reported for nanocrystals from tunicate (Elazzouzi-Hafraoui et al., 2008). For Iα-rich algal cellulose, it has been shown that *Valonia* microfibrils have a square cross-section (~ 20 nm by ~ 20 nm), whereas *Micrasterias* microfibrils have a rectangular cross-section (~ 5 nm by ~ 20–30 nm) (Revol, 1982; Sugiyama et al., 1985). AFM measurements performed on cotton cellulose nanocrystals revealed a rectangular cross-section with an average width of 20 nm

Source	L (nm)	D (nm)	L/D	Reference
Acacia Pulp	100–250	5–15	20–50	(Pu et al., 2007)
Alfa	200±20	10±2	20	(Ben Elmabrouk et al., 2009)
Algal (*Valonia*)	> 1,000	10–20	∞	(Revol, 1982; Hanley et al., 1992)
Bacterial Cellulose	100-Several 1,000	5–10 · 30–50	–	(Tokoh et al., 1998; Grunert and Winter, 2002)
Banana Rachis	500–1,000	5	–	(Zuluaga et al., 2007)
Bio-Residue from Wood Bioethanol Production	Several 100	10–20	–	(Oksman et al., 2011)
Bleached Softwood Kraft Pulp	180–280	5.0	33–47	(Orts et al., 1998)
	180±80	3.5±0.5	50	(Araki et al., 1999)
	105–141	4.5–5	23.3–28.2	(Beck-Candanedo et al., 2005)
Capim Dourado	300±93	4.5±0.86	67	(Siqueira et al., 2010a)
Cassava Bagasse	360–1,700	2–11	–	(Teixeira et al., 2009)
Cladophora	–	20 · 20	–	(Kim et al., 2000)
Coconut Husk Fibers	80–500	6	39±16	(Rosa et al., 2010)
	–	2.24–2.38	–	(Fahma et al., 2011)
Cotton	177–390	7	–	(Dong et al., 1998)
	170	–	10	(Ebeling et al., 1999)
	100–300	5–15	10	(Teixeira et al., 2010)
Cottonseed Linter	350±70 170–490	40±8 40–60	9 –	(Lu et al., 2005)
Curaúa	80–170	6–10	13–17	(Corrêa et al., 2010)
Date Palm Tree (rachis/leaflets)	260/180	6.1	43/30	(Bendahou et al., 2009)
Eucalyptus Wood Pulp	147±7 145	4.8±0.4 6	30.6 24	(Beck-Candanedo et al., 2005) (de Mesquita et al., 2010)
Flax	100–500	10–30	15	(Cao et al., 2007)
Grass of Korea (*genus Zoysia*)	200–700	10–60	–	(Pandey et al., 2008; Pandey et al., 2009)
Hemp	Several 1,000	30–100	–	(Wang et al., 2007)
Kenaf	158	12	13	(Kargarzadeh et al., 2012)
Luffa cylindrica	242	5.2	47	(Siqueira et al., 2010c)

Source	L (nm)	D (nm)	L/D	Reference
MCC	150–300	3–7	–	(Kvien et al., 2005)
Mengkuang Leaves	50–400	5–25	10–20	(Sheltami et al., 2012)
Mulberry	400–500	20–40	–	(Li et al., 2009)
Pea Hull	240–400	7–12	34	(Chen et al., 2009)
Ramie	350–700 150–250	70–120 6–8		(Lu et al., 2006; Habibi and Dufresne, 2008; Habibi et al., 2008; de Menezes et al., 2009)
Recycled Pulp	100–1,800	30–80	–	(Filson et al., 2009)
Rice Husk	–	15–20	10–15	(Johar et al., 2012)
Rice Straw	117–270	5–30	–	(Lu and Hsieh, 2012)
Sisal	250 215	4 5	60 43	(Garcia et al., 2006) (Siqueira et al., 2009; Siqueira et al., 2010a)
Sugar Beet Pulp	210	5	42	(Azizi Samir et al., 2004)
Sugarcane Bagasse	200–310	2–6	64	(Teixeira et al., 2010b)
Tunicin	100–Several 1,000	10–20	67	(Favier et al., 1995a; Favier et al., 1995b)
Wheat straw	150–300	5	45	(Helbert et al., 1996)

Table 3.4: Geometrical characteristics of cellulose nanocrystals from various sources: length (L), cross section (D) and aspect ratio (L/d).

and average height of 8.9 nm (Ureña-Benavides et al., 2011). Moreover, a 180° twist with a pitch of 700 nm was observed along the crystal length (Hanley et al., 1997).

However, several factors may cause deviations from the idealized cellulose nanocrystal structure. Three categories of deviations have been identified, viz. percent crystallinity, $I\alpha/I\beta$ ratio and lattice defects (Moon et al., 2011). Numerous evidence from microscopy and light scattering techniques reveal that most cellulose crystals have more or less rod-like and cylindrical shape (de Souza Lima and Borsali, 2004; Bai et al., 2009). The cross-sections of microfibrils observed by TEM were found to be square, whereas their AFM topography showed a rounded profile due to convolution with the shape of the AFM tip (Hanley et al., 1992). AFM images of the surface of highly crystalline cellulose microfibrils showed periodicities along the microfibril axis of 1.07 and 0.53 nm that were supposed to correspond to the fiber and glucose unit repeat distances, respectively. Small angle neutron scattering (SANS) technique revealed that cellulose nanocrystals are twisted rods (Orts et al., 1998).

An important parameter for cellulosic nanocrystals is the aspect ratio, which is defined as the ratio of the length to the width. The value of this parameter for nanocrystals derived from different species is reported in Table 3.4. It determines the anisotropic phase formation and reinforcing properties. It can be affected of course by preparation conditions and possibility of aggregation of individual nanoparticles. The aspect ratio varies between 10 for cotton and 67 for tunicin or capim dourado (golden grass). An attempt to correlate the intrinsic viscosity of cellulose nanocrystal suspensions to the aspect ratio of constituting nanoparticles was reported (Boluk et al., 2011). The intrinsic viscosity was determined for suspensions containing different concentrations of NaCl and extrapolated to 1 nm Debye length to calculate the intrinsic viscosity of electroviscous effect-free hard rods. The corresponding aspect ratio was determined from Simha's equation. Even if of the same order of magnitude, the obtained value (41) was higher than the one obtained from microscopic observations (30).

3.7 Degree of hydrolysis

As for MFC (Chapter 2), in addition to direct microscopic observations, other measurements can be used to access indirectly the extent of hydrolysis.

3.7.1 Birefringence of the suspension

Polarized light is generally used for the observation and the identification of oriented structures (in particular crystalline structures). This technique gives a quick indication of the successful preparation of well-dispersed cellulose nanocrystals obtained by sulfuric acid hydrolysis. Theoretically, a beam of unpolarized light entering a crystal will be divided into two beams, vibrating in different planes perpendicular to each other (polarized rays). Cellulose nanocrystals display this property with a strong absorption of one of the polarized rays, i.e. they possess two refractive indices. The light emerging from such a material will be plane-polarized light. This is the property of birefringence. Birefringence of cellulose nanocrystal suspensions can be observed with two linear polarizers crossed at 90° (cross-nicols or crossed-polars). The presence of negative charges on the cellulose nanocrystal surface induces an electrostatic repulsion which can be directly observed.

In 1959, the birefringent character of acid-treated cellulose and chitin crystallites was reported (Marchessault et al., 1959). Figure 3.6 shows a photograph of an aqueous suspension of cellulose nanocrystals observed between cross-nicols and showing the formation of birefringent domains. The birefringence of cellulose nanocrystal aqueous suspensions can be observed through a pair of cross-nicols. This birefringence results from two origins: (i) a structural form anisotropy of cellulose ($\Delta n \sim 0.05$) and (ii) a flow anisotropy resulting from alignment of the nano-rods under flow generally operated before observation. The birefringence magnitude corresponds to

Δn (= $n_e - n_o$), with n_e and n_o being the extraordinary and ordinary index of refraction, respectively. The index n_e is the indice of refraction of light with linear polarizations parallel to the optical axis (axis of symmetry). The index n_o is the indice of refraction of light with linear polarizations perpendicular to the optical axis.

Fig. 3.6: Photograph of an aqueous suspension of capim dourado cellulose nanocrystals (0.50 wt%) observed between cross-nicols showing the formation of birefringent domains (Siqueira et al., 2010a).

This shear-induced alignment was studied for cotton nanocrystals in colloidal aqueous suspension using small-angle X-ray scattering (SAXS) experiments (Ebeling et al., 1999). It was shown that this alignment is shear rate dependent, and it is completely reversible. The presence of planar domains of randomly oriented nanocrystals which align at low shear rates and are broken up at higher shear rates enabling alignment of the individual nanocrystals was suggested. The nanocrystals appeared to be randomly aligned prior to shearing and horizontally aligned along the shear direction when the shear rate exceeded 5 s^{-1}. Ordering of diluted suspensions of cellulose nanocrystals was also evidenced using static and dynamic light scattering (de Souza Lima and Borsali, 2002). Numerous peaks in the scattered intensity were observed, which were interpreted in terms of electrostatic repulsive interactions between nanoparticles. These interactions are of long-range order, and the arrangement of nanocrystals was supposed to be in cylindrical/hexagonal packing. Addition of salt to the suspension induced destruction of this order and leads to nanoparticle flocculation, demonstrating the role of the electrostatic interactions between nanocrystals to control both stability and order.

3.7.2 Viscosity of the suspension

As for MFC, the viscosity of the suspension increases when the cellulosic substrate is defragmented into individual nanocrystals. This aspect will be detailed in Chapter 6.

3.7.3 Porosity and density

The observation of the porosity of films obtained from cellulose nanocrystals is much less referenced than for MFC. However, the density of these films is expected to increase upon the severity of the hydrolysis treatment, since smaller diameters result in better packing. Figure 3.7 shows the surface of a film obtained by casting-evaporation of cellulose nanocrystals extracted from sisal and observed by scanning electron microscopy with field emission gun (SEM-FEG). Needle-like associated nanoparticles are clearly observed and the film consists of a mat of H-bonded cellulosic nanocrystals. The porosity and density of this film were found to be around 62% and 0.9 g cm^{-3}, respectively (Belbekhouche et al., 2011). These values are significantly different from those reported for an MFC film obtained from the same cellulosic source (sisal), viz. 35% and 1.33 g cm^{-3}, respectively (Belbekhouche et al., 2011). The porosity of these films was estimated from air permeability measurements.

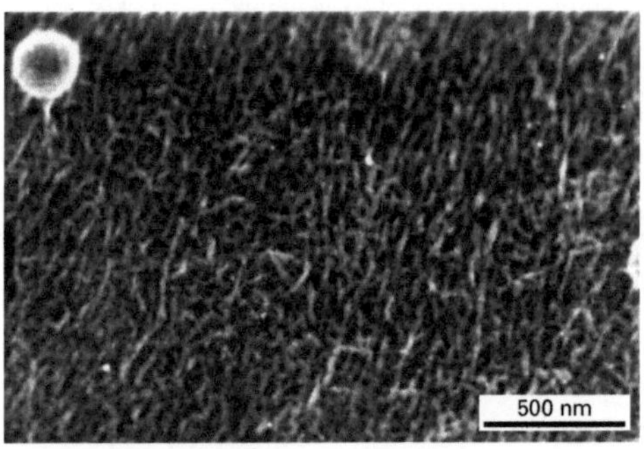

Fig. 3.7: Scanning electron micrograph (SEM-FEG) of a sisal cellulose nanocrystal film (Belbekhouche et al., 2011).

3.7.4 Mechanical properties

Improved mechanical strength and stiffness of solid films obtained from cellulose nanocrystal suspensions after water removal are expected compared to unhydrolyzed fibers. This aspect will be developed in Section 3.8 of the present chapter.

3.7.5 Degree of polymerization

As the hydrolysis proceeds, the degree of polymerization (DP) of the cellulose molecules is expected to decrease. It was found that regardless of the hydrolysis conditions, the same leveling-off values of the DP were reached (Battista, 1950; Battista et al., 1956). Similar conclusions were reported for hydrolysis of different pulps at 60°C and 80°C using HCl and H_2SO_4 with concentrations ranging from 0.5 to 4M (Håkansson and Ahlgren, 2005). Obviously a longer time was needed under milder conditions to reach the leveling-off DP. A good correlation between DP and length of the nanocrystals measured by microscopic observations was reported (Battista et al., 1956). This leveling-off DP was also found to match the periodicity observed for ramie cellulose microfibrils in small angle neutron scattering studies (Nishiyama et al., 2003). It was considered that the microfibrils have 4–5 disordered residues every 300 residues.

An interesting study reported the effect of extraction conditions on the DP of ensuing nanoparticles using kraft pulp fibers as the starting raw material (Hamad and Hu, 2010). The DP of the cellulose nanocrystals was determined from intrinsic viscosity measurements as reported in Chapter 2 (Section 2.4.6) and correlated with the yield of the hydrolysis process (Figure 3.8). At low acid concentration (16 wt%), a significant decrease of the DP was reported from 680 to 180 (initial DP of the raw material was 1178) as the hydrolysis temperature was raised from 45°C to 85°C, but the yield decreased only slightly from 93.7% to 91.5%. The low material loss in these hydrolysis conditions was attributed to the preferential hydrolysis and dissolution of the hemicelluloses from the pulp.

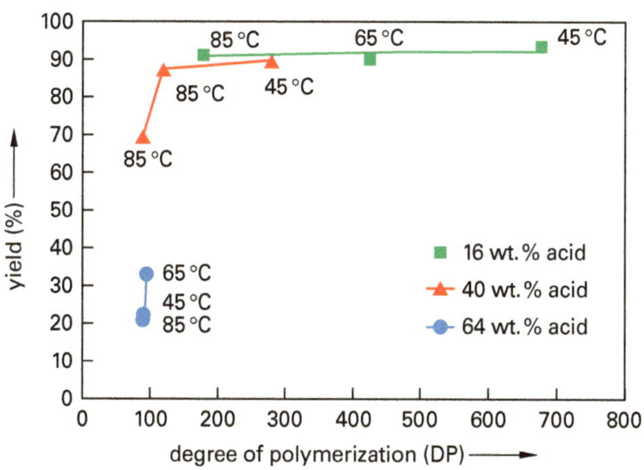

Fig. 3.8: Evolution of the yield as a function of the degree of polymerization of cellulose nanocrystals obtained by acid hydrolysis of softwood kraft pulp (initial DP = 1178) at various sulfuric acid concentrations and temperature for 25 min (Hamad and Hu, 2010).

When increasing the acid concentration to 40 wt%, a similar behavior was reported as when increasing the temperature from 45°C to 65°C, whereas a further increase of temperature to 85°C induced a small decrease in DP and a significant drop in the yield (Hamad and Hu, 2010). The authors suggested that when the crystalline cellulose material reached a DP around 90, the "easier-to-hydrolyze" amorphous material had a sufficiently low DP and/or high sulfation to dissolve. Because no sulfation was detected from elemental analysis for insoluble cellulose, the yield drop was ascribed to the lowered DP of amorphous cellulose. At higher acid concentrations (64 wt%) no more evolution of the DP was observed and the yield was found to be much lower. A higher level of sulfation was observed for the material hydrolyzed in these conditions and it was supposed to increase its water solubility and then decrease the yield of insoluble material.

3.7.6 Specific surface area

Aggregation of cellulose nanocrystals upon drying makes the experimental determination of their specific surface area difficult. However, it should obviously strongly increase after hydrolysis of the amorphous domains of cellulose and release of crystalline nano-domains. Determination of the specific surface area from gas adsorption isotherm is often incorrect because of the irreversible aggregation of the nanoparticles upon drying. Therefore, the drying method of water-suspended cellulose nanocrystals significantly affects preservation of their large surface area.

The specific surface area of cellulose nanocrystals can be estimated from the average geometrical dimensions of the nanoparticles, assuming a rod-like geometry and a density ρ of 1.5 or 1.6 $g \cdot cm^{-3}$ for crystalline cellulose. The contribution of the ends of the cylinder is assumed to be negligible. The specific area, A_{sp}, of cellulose nanocrystals can be determined using the following equation:

$$A_{sp} = \frac{4}{\rho \cdot d} \tag{3.1}$$

where d corresponds to the diameter of the cellulosic rods. Values of 170 (Dufresne, 2000) and 533 $m^2 \cdot g^{-1}$ (Siqueira et al., 2010d) have been reported for nanocrystals extracted from tunicin and sisal, respectively. The corresponding diameters were 15 and 5 nm, respectively, and the density of cellulose crystalline was assumed to be 1.5 $g \cdot cm^{-3}$. Figure 3.9 shows the evolution of the specific surface area as a function of the diameter of rod-like nanoparticles. A sharp increase is observed when the diameter falls below a value of 20 nm, which corresponds to the diameter range of cellulose nanocrystals.

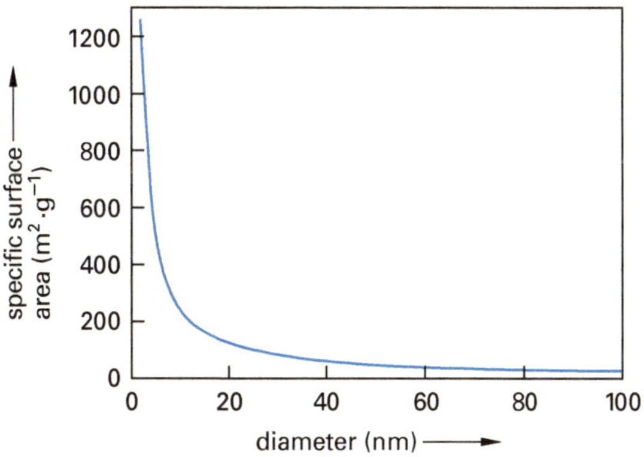

Fig. 3.9: Evolution of the specific surface area of rod-like nanoparticles as a function of their diameter, assuming a density of 1.5 g · cm^{-3} for crystalline cellulose.

3.7.7 Level of sulfation

The hydrolysis treatment of cellulose with sulfuric acid can result in the introduction of sulfate esters at the surface of cellulose nanocrystals, leading to improved electrostatic stabilization of the suspensions. The level of sulfation can be determined by conductimetric titration with base. Sodium hydroxide (NaOH) can be added continuously to the suspension of cellulose nanocrystals and the change in conductivity is monitored by a conductimeter. Standardization of NaOH against volumetric standard sulfuric acid should be performed. It was found, as shown in Figure 3.10, that the conductivity of H_2SO_4-hydrolyzed suspensions first decreased because of the neutralization of strong acid surface groups (Araki et al., 1998). The horizontal portion after the first inflection point corresponds to the neutralization of weak acid groups (probably carboxylic acid) (Figure 3.10). After complete neutralization, the conductivity increased due to excess NaOH. A surface charge of 84 mmol · kg^{-1} of strong acid and 26 mmol · kg^{-1} of weak acid, ascribed to sulfate ester and carboxyl groups, respectively, were reported. These values agree well with the surface charge density of 0.155 e.nm^{-2} reported elsewhere (Dong et al., 1996). A surface charge of $1.71 \cdot 10^{-4}$ mol · g^{-1} was also reported, which, according to the size of cotton nanocrystals used in the study, was equivalent to sulfonating 30% of the available primary hydroxyl groups (Ureña-Benavides et al., 2011). Sulfur content and surface charge can be determined from conductimetric titrations assuming a cylindrical shape of nanocrystals and density of 1.6 g · cm^{-3} for crystalline cellulose. These parameters were found to increase when increasing the strength of the acid hydrolysis treatment (Roman and Winter, 2004; Beck-Candanedo et al., 2005). For HCl-treated cellulose, no strong acid groups and a small amount of weak acid were observed as shown in Figure 3.10 (Araki et al., 1998).

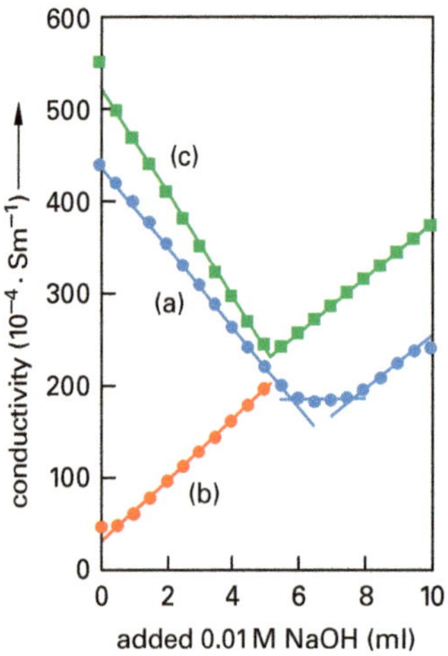

Fig. 3.10: Conductometric titration curves of (a) 698 mg of H_2SO_4-treated bleached softwood kraft pulp nanocrystal suspension, (b) 298 mg of HCl treated nanocrystal and (c) 265 mg of HCl treated nanocrystal with initial addition of 5 ml of 0.01 M HCl (Araki et al., 1998).

The sulfur content can be also determined by elemental analysis (Dong et al., 1998; Hamad and Hu, 2010) or fluorescent X-ray analysis (Araki et al., 1999). The surface charge can also be evaluated by determining the zeta-potential of the suspension (Corrêa et al., 2010; Elanthikkal et al., 2010).

3.7.8 Crystallinity

During hydrolysis, the crystallinity of the extracted insoluble cellulose material is expected to increase because of the accessibility and selective hydrolysis of amorphous regions of cellulose. Sharper diffraction peaks for hydrolyzed cellulose are a good indication of a higher degree of crystallinity. It was shown that the Segal approach (Segal et al., 1959), commonly adopted in the lignocellulosic field, was incapable of differentiating materials of patently dissimilar solid-state characteristics (Hamad and Hu, 2010). The deconvolution approach (see Chapter 2, Section 2.4.8) was suggested as a more robust method to analyze the solid-state characteristics of lignocellulosics. Figure 3.11(a) shows the evolution of the DP of acid-hydrolyzed softwood kraft pulp as a function of the crystallinity of the insoluble residue. At high sulfuric acid concentration (64 wt%), crystallinity reached a maximum at around 90%. At the same time, the crystallite size was found to decrease (Figure 3.11(b)). Table 3.5 shows the values of crystallinity reported in the literature for hydrolyzed cellulose from different sources. However, these values should only be considered an indication because of difficulties in determining them and differences in experimental conditions.

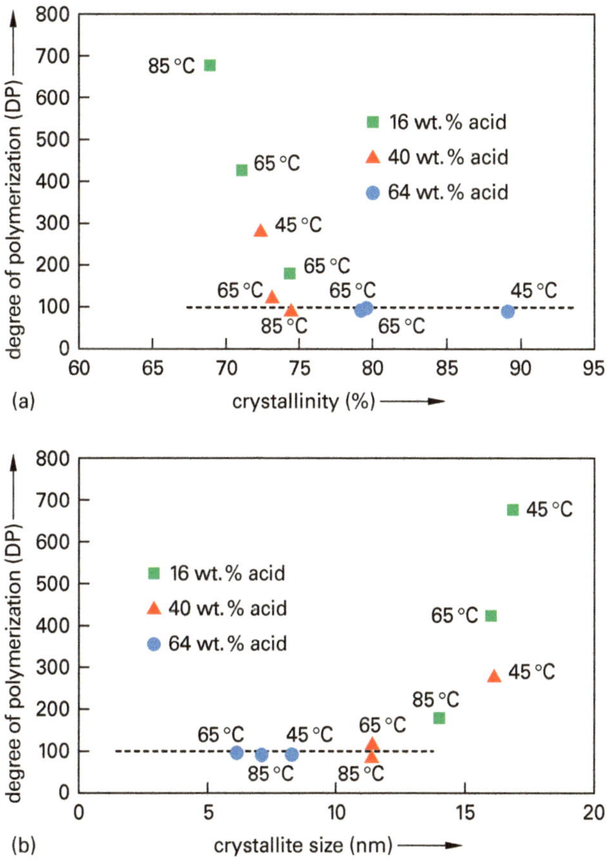

Fig. 3.11: Evolution of the degree of polymerization as a function of (a) crystallinity, and (b) crystallite size of cellulose nanocrystals obtained by acid hydrolysis of softwood kraft pulp (initial DP = 1178) at various sulfuric acid concentrations and temperatures for 25 min (Hamad and Hu, 2010).

Source	Method	Crystallinity (%)	Reference
Bacterial Cellulose	Deconvolution	85	(Roman and Winter, 2004)
Bamboo	Deconvolution	46	(Liu et al., 2010)
Bio-Residue from Wood Bio-ethanol Production	Segal	75	(Oksman et al., 2011)
Capim Dourado	Segal	91	(Siqueira et al., 2010a)
Cassava Bagasse	Segal	54	(Teixeira et al., 2009)
Coconut Husk	ns*	50–57	(Fahma et al., 1011)
	Deconvolution	62–66	(Rosa et al., 2010)
Cotton Linter	Segal	84	(Qin et al., 2011)
Curaúa	Deconvolution	72–87	(Corrêa et al., 2010)
Kenaf	Segal	75–82	(Kargarzadeh et al., 2012)
Luffa cylindrica	Segal	96	(Siqueira et al., 2010b)
MCC	Segal	87–92	(Lee et al., 2009)
Mulberry	ns*	73	(Li et al., 2009
Rice Husk	Segal	59	(Johar et al., 2012)
Rice Straw	Segal	86–91	(Lu and Hsieh, 2012)
Sisal	Segal	75	(Morán et al., 2008)
Sofwood Kraft Pulp	Segal	80–86	(Hamad and Hu, 2010)
	Deconvolution	69–89	(Hamad and Hu, 2010)
Sugarcane Bagasse	Deconvolution	70–87	(Teixeira et al., 2011)

* non-specified

Table 3.5: Crystallinity of cellulose nanocrystals from various sources.

3.8 Mechanical properties of nanocrystal films

Intrinsic mechanical properties of cellulose nanocrystals are obviously important parameters for their reinforcement capability. However, it will be seen in Chapter 9 that the outstanding mechanical properties of cellulose nanocrystal reinforced polymer nanocomposites can be well understood from the formation of a continuous nanoparticle network. The formation of this rigid network, resulting from strong interactions between nanocrystals, is assumed to be governed by a percolation mechanism. In this approach, an important parameter is the modulus of the percolating cellulosic nanoparticles, which is obviously different from the one of individual nanocrystals. This modulus can be assumed to be similar, in principle, to the one of a paper sheet for which the hydrogen bonding forces provide the basis of its stiffness. However, because of the nanoscale effect stronger interactions are expected. This modulus can be experimentally determined from tensile tests performed on films prepared by water evaporation of a suspension of nanocrystals.

Very few data are reported in the literature. For tunicin (Favier et al., 1995a; Favier et al., 1995b), and wheat straw cellulose nanocrystals (Helbert et al., 1996), the tensile modulus was around 15 GPa and 6 GPa, respectively. For tunicin nanocrystals lower values (5 GPa; 2 GPa) were also reported (Ljungberg et al., 2005; Capadona et al., 2007). These values are much higher than those obtained for films made of chitin nanocrystals. Tensile modulus values around 0.5 and 2 GPa were reported for chitin nanocrystals extracted from squid pen (Paillet and Dufresne, 2001) and *Riftia* tubes (Morin and Dufresne, 2002), respectively. As suggested from these values and from the literature (Capadona et al., 2007), the reinforcing capacity of cellulose nanocrystals is related to the aspect ratio of the nanoparticles. A benchmarking of cellulose nanocrystals from different sources and impact of the geometrical characteristics on the specific mechanical properties of the percolating network was reported (Bras et al., 2011). Even if other factors are controlling the mechanical properties of this network, such as porosity, density, orientation and organization of the nanoparticles, it seems that a weak correlation exists between the aspect ratio of the nanocrystals and mechanical stiffness of the mat as shown in Figure 3.12. The stiffness of the film was found to increase with the aspect ratio of the nanocrystals.

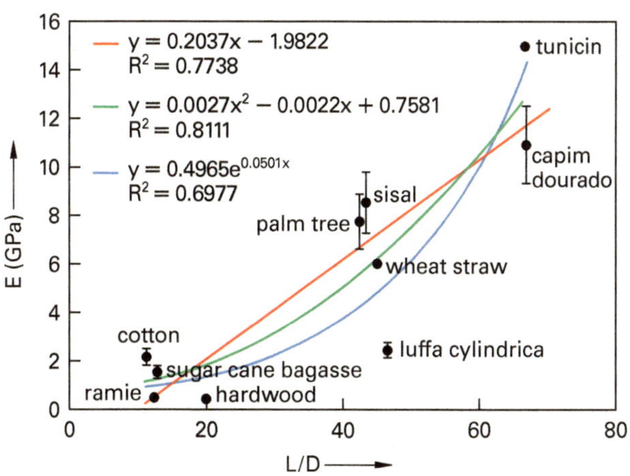

Fig. 3.12: Evolution of the Young's modulus of the cellulosic nanocrystals film determined from tensile tests as a function of the aspect ratio of the constituting nanocrystals (Bras et al., 2011).

3.9 Conclusions

Individual cellulose nanocrystals are produced by breaking down cellulose fibers and isolating the crystalline domains. Strong acid hydrolysis, a process described nearly 60 years ago, is generally used to isolate cellulose nanocrystals. The current accepted explanation depicts this process of acid hydrolysis as a heterogeneous process that involves the diffusion of acid into the cellulose fibers, followed by cleavage of glycosidic bonds. Sulfuric acid has been extensively investigated and appears to be the most effective. High surface area and aspect ratio needle-like nanoparticles are obtained from this procedure. The physical dimensions that are obtained for cellulose nanocrystals are determined by both the source of cellulose and the hydrolysis conditions that are used during extraction.

3.10 References

Anglès, M.N. and Dufresne, A. (2000). Plasticized starch/tunicin whiskers nanocomposites. 1. Structural analysis. Macromolecules *33*, 8344–8353.

Araki, J., Wada, M., Kuga, S. and Okano, T. (1998). Flow properties of microcrystalline cellulose prepared by acid treatment of native cellulose. Colloid Surface A *142*, 75–82.

Araki, J., Wada, M., Kuga, S. and Okano, T. (1999). Influence of surface charge on viscosity behaviour of cellulose microcrystal suspension. J. Wood Sci. *45*, 258–261.

Araki, J., Wada, M. and Kuga, S. (2001). Steric stabilization of a cellulose microcrystal suspension by poly(ethylene glycol) grafting. Langmuir *17*, 21–27.

Ardizzone, S., Dioguardi, F.S., Mussini, T., Mussini, P.R., Rondinini, S., Verceli, B. and Vertova, A. (1999). Microcrystalline cellulose powders: structure, surface features and water sorption capability. Cellulose, *6*, 57–69.

Azizi Samir, M.A.S., Alloin, F., Paillet, M. and Dufresne, A. (2004). Tangling effect in fibrillated cellulose reinforced nanocomposites. Macromolecules *37*, 4313–4316.

Bai, W., Holbery, J. and Li, K. (2009). A technique for production of nanocrystalline cellulose with a narrow size distribution. Cellulose *16*, 455–465.

Battista, O.A. (1950). Hydrolysis and crystallization of cellulose. Ind. Eng. Chem. *42*, 502–507.

Battista, O.A., Coppick, S., Howsmon, J.A., Morehead, F.F. and Sisson, W.A. (1956). Level-off degree of polymerization – relation to polyphase structure of cellulose fibres. Ind. Eng. Chem. *48*, 333–335.

Battista, O.A., Hill, D. and Smith, P.A. (1961). Level-off D.P. cellulose products. US Patent 2,978,446.

Battista, O.A. (1975). Microcrystalline polymer science (MacGraw-Hill, New York).

Beck-Candanedo, S., Roman, M. and Gray, D.G. (2005). Effect of reaction conditions on the properties and behaviour of wood cellulose nanocrystals suspensions. Biomacromolecules *6*, 1048–1054.

Belbekhouche, S., Bras, J., Siqueira, G., Chappey, C., Lebrun, L., Khelifi, B., Marais, S. and Dufresne, A. (2011). Water sorption behavior and gas barrier properties of cellulose whiskers and microfibrils films. Carbohydr. Polym. *83*, 1740–1748.

Ben Elmabrouk, A., Thielemans, W., Dufresne, A. and Boufi, S. (2009). Preparation of poly(styrene-co-hexylacrylate)/cellulose whiskers nanocomposites via miniemulsion polymerization. J. Appl. Polym. Sci. *114*, 2946–2955.

Bendahou, A., Habibi, Y., Kaddami, H. and Dufresne, A. (2009). Physico-chemical characterization of palm from *Phoenix dactylifera* – L, preparation of cellulose whiskers and natural rubber-based nanocomposites. J. Biobased Mat. Bioenergy *3*, 81–90.

Boluk, Y., Lahiji, R., Zhao, L. and McDermott, M.T. (2011). Suspension viscosities and shape parameter of cellulose nanocrystals (CNC). Colloid Surface A *377*, 297–303.

Bondeson, D. Mathew, A. and Oksman, K. (2006). Optimization of the isolation of nanocrystals from microcrystalline cellulose by acid hydrolysis. Cellulose *13*, 171–180.

Bras, J., Viet, D., Bruzzese, C. and Dufresne, A. (2011). Correlation between stiffness of sheets prepared from cellulose whiskers and nanoparticles dimensions. Carbohydr. Polym. *84*, 211–215.

Braun, B., Dorgan, J.R. and Chandler, J.P. (2008). Cellulosic nanowhiskers. Theory and application of light scattering from polydisperse spheroids in the Rayleigh-Gans-Debye regime. Biomacromolecules *9*, 1255–1263.

Braun, B. and Dorgan, J.R. (2009). Single-step method for the isolation and surface functionalization of cellulosic nanowhiskers. Biomacromolecules *10*, 334–341.

Cao, X., Dong, H and Li, C.M. (2007). New nanocomposite materials reinforced with flax cellulose nanocrystals in waterborne polyurethane. Biomacromolecules *8*, 899–904.

Cao, X., Chen, Y., Chang, P.R., Stumborg, M. and Huneault, M.A. (2008). Green composites reinforced with hemp nanocrystals in plasticized starch. J. Appl. Polym. Sci. *109*, 3804–3810.

Capadona, J.R., Van den Berg, O., Capadona, L., Schroeter, M., Tyler, D., Rowan, S.J. and Weder, C. (2007). A versatile approach for the processing of polymer nanocomposites with self assembled nanofibre templates. Nature Nanotechnology *2*, 765–769.

Chang, M. (1971). Folding chain model and annealing of cellulose. J. Polym. Sci.: Part C *36*, 343–362.

Chang, M.M.Y. (1974). Crystallite structure of cellulose. J. Polym. Sci. Polym. Chem *12*, 1349–1374.

Chen, Y., Liu, C., Chang, P.R., Cao, X. and Anderson, D.P. (2009). Bionanocomposites based on pea starch and cellulose nanowhiskers hydrolyzed from pea hull fibre: Effect of hydrolysis time. Carbohydr. Polym. *76*, 607–615.

Corrêa, A.C., Teixeira, E.M., Pessan, L.A. and Mattoso, L.H.C. (2010). Cellulose nanofibers from curaua fibers. Cellulose *17*, 1183–1192.

Daele, Y.V.; Revol, J.F.; Gaill, F. and Goffinet, G. (1992). Characterization and supramolecular architecture of the cellulose-protein fibrils in the tunic of the sea peach (*Halocynthia papillosa*, Ascidiacea, Urochordata). Biol. Cell *76*, 87–96.

Dagnon, K.F., Shanmuganathan, K., Weder, C. and Rowan, S.J. (2012). Water-triggered modulus changes of cellulose nanofiber nanocomposites with hydrophobic polymer matrices. Macromolecules 45, 4707–4715.

de Mesquita, J.P., Donnici, C.L. and Pereira, F.V. (2010). Biobased nanocomposites from layer-by-layer assembly of cellulose nanowhiskers with chitosan. Biomacromolecules *11*, 473–480.

de Souza Lima, M.M. and Borsali, R. (2002). Static and dynamic light scattering from polyelectrolyte microcrystal cellulose. Langmuir *18*, 992–996.

de Souza Lima, M.M. and Borsali, R. (2004). Rodlike cellulose microcrystals: Structure, properties and applications. Macromol. Rapid Commun. *25*, 771–787.

Dong, X.M., Kimura, T., Revol, J.F. and Gray, D.G. (1996). Effects of ionic strength on the isotropic-chiral nematic phase transition of suspensions of cellulose crystallites. Langmuir *12*, 2076–2082.

Dong, X.M., Revol, J.F. and Gray, D.G. (1998). Effect of microcrystallite preparation conditions on the formation of colloid crystals of cellulose. Cellulose *5*, 19–32.

Dufresne, A. (2000). Dynamic mechanical analysis of the interphase in bacterial polyester/ cellulose whiskers natural composites. Compos. Interfaces *7*, 53–67.

Dufresne, A. (2008). Cellulose-based composites and nanocomposites. In: Monomers, polymers and composites from renewable resources, A. Gandini and M.N. Belgacem, eds. (Elsevier, Great Britain), pp. 401–418.

Ebeling, T., Paillet, M., Borsali, R., Diat, O., Dufresne, A., Cavaillé, J.Y. and Chanzy, H. (1999). Shear-induced orientation phenomena in suspensions of cellulose microcrystals, revealed by small-angle X-ray scattering. Langmuir *15*, 6123–6126.

Elanthikkal, S., Gopalakrishnapanicker, U., Varghese, S. and Guthrie, J.T. (2010). Cellulose microfibres produced from banana plant wastes: Isolation and characterization. Carbohydr. Polym. *80*, 852–859.

Elazzouzi-Hafraoui, Nishiyama, Y., Putaux, J.L., Heux, L., Dubreuil, F. and Rochas, C. (2008). The shape and size distribution of crystalline nanoparticles prepared by acid hydrolysis of native cellulose. Biomacromolecules *9*, 57–65.

Fahma, F., Iwamoto, S., Hori, N., Iwata, T. and Takemura, A. (2011). Effect of pre-acid-hydrolysis treatment on morphology and properties of cellulose nanowhiskers from coconut husk. Cellulose *18*, 443–450.

Favier, V., Canova, G.R., Cavaillé, J.Y., Chanzy, H., Dufresne, A. and Gauthier, C. (1995a). Nanocomposites materials from latex and cellulose whiskers. Polym. Adv. Technol. *6*, 351–355.

Favier, V., Chanzy, H. and Cavaillé, J.Y. (1995b). Polymer nanocomposites reinforced by cellulose whiskers. Macromolecules *28*, 6365–6367.

Filson, P.B., Dawson-Andoh, B.E. and Schwegler-Berry, D. (2009). Enzymatic-mediated production of cellulose nanocrystals from recycled pulp. Green Chem. *11*, 1808–1814.

Filson, P.B. and Dawson-Andoh, B.E. (2009). Sono-chemical preparation of cellulose nanocrystals from lignocellulose derived materials. Biores. Technol. *100*, 2259–2264.

Fleming, K., Gray, D.G., Prasannan, S. and Matthews, S. (2000). Cellulose crystallites: A new and robust liquid crystalline medium for the measurement of residual dipolar couplings. J. Am. Chem. Soc. *122*, 5224–5225.

Fleming, K., Gray, D.G. and Matthews, S. (2001). Cellulose crystallites. Chem. Eur. J. *7*, 1831–1835.

Frey-Wyssling, A. (1954). The fine structure of cellulose microfibrils. Science *119*, 80–82.

Garcia de Rodriguez, N.L., Thielemans, W. and Dufresne, A. (2006). Sisal cellulose whiskers reinforced polyvinyl acetate nanocomposites. Cellulose *13*, 261–270.

Grunert, M. and Winter, W. T. (2002). Nanocomposites of cellulose acetate butyrate reinforced with cellulose nanocrystals. J. Polym. Envir. *10*, 27–30.

Habibi, Y. and Dufresne, A. (2008). Highly filled bionanocomposites from functionalized polysaccharide nanocrystals. Biomacromolecules *9*, 1974–1980.

Habibi, Y., Goffin, A.L., Schiltz, N., Duquesne, E., Dubois, P. and Dufresne, A. (2008). Bionanocomposites based on poly(ε-caprolactone)-grafted cellulose nanocrystals by ring opening polymerization. J. Mat. Chem. *18*, 5002–5010.

Håkansson, H. and Ahlgren, P. (2005). Acid hydrolysis of some industrial pulps: Effect of hydrolysis conditions and raw material. Cellulose, *12*, 177–183.

Hamad, W.Y. and Hu, T.Q. (2010). Structure-process-yield interrelations in nanocrystalline cellulose extraction. Can. J. Chem. Eng. *88*, 392–402.

Hanley, S.J., Giasson, J., Revol, J.F. and Gray, D.G. (1992). Atomic force microscopy of cellulose microfibrils: Comparison with transmission electron microscopy. Polymer *33*, 4639–4642.

Hanley, S.J., Revol, J.F., Godbout, L. and Gray, D.G. (1997). Atomic force microscopy and transmission electron microscopy of cellulose microfibrils from *Micrasterias denticulata*; Evidence for a chiral helical microfibril twist. Cellulose *4*, 209–220.

Helbert, W., Cavaillé, J.Y. and Dufresne, A. (1996). Thermoplastic nanocomposites filled with wheat straw cellulose whiskers. Part I: Processing and mechanical behavior. Polym. Comp. *17*, 604–611.

Helbert, W., Nishiyama, Y., Okano, T. and Sugiyama, J. (1998). Molecular imaging of *Halocynthia papillosa* cellulose. J. Struct. Biol. *124*, 42–50.

Hengstenberg, J. and Mark, H. (1928). Über die Form und die Größe der Mizelle von Zellulose und Kautschuk. Z. Krist., *69*, 271–284.

Henriksson, M., Henriksson, G., Berglund, L.A. and Lindström, T. (2007). An environmentally friendly method for enzyme-assisted preparation of microfibrillated cellulose (MFC) nanofibers. Eur. Polym. J. *43*, 3434–3441.
Hermans, P.H. and Weidinger, A. (1949). Change in crystallinity upon heterogeneous acid hydrolysis of cellulose fibers. J. Polym. Sci. *4*, 317–322.
Hirai, A., Inui, O., Horii, F. and Tsuji, M. (2009). Phase separation behaviour in aqueous suspensions of bacterial cellulose nanocrystals prepared by sulphuric acid treatment. Langmuir *25*, 497–502.
Hirota, M., Tamura, N., Saito, T. and Isogai, A. (2010). Water dispersion of cellulose II nanocrystals prepared by TEMPO-mediated oxidation of mercerized cellulose at pH 4.8. Cellulose *17*, 279–288.
Howsmon, J.A. (1949). Water sorption and the poly-phase structure of cellulose fibers. Textile Research J. *19*, 152–162.
Immergut, E.A. and Rånby, B.G. (1956). Heterogeneous acid hydrolysis of native cellulose fibers. Ind. Eng. Chem. *48*, 1183–1189.
Ingersoll, H.G. (1946). Fine structure of viscose rayon. J. Appl. Phys. *17*, 924–939.
Johar, N. Ahmad, I. and Dufresne, A. (2012). Extraction, preparation and characterization of cellulose fibres and nanocrystals from rice husk. Ind. Crops *37*, 93–99.
Julien, S., Chornet, E. and Overend, R.P. (1993). Influence of acid pretreatment (H_2SO_4, HCl, HNO_3) on reaction selectivity in the vacuum pyrolysis of cellulose. J. Anal. Appl. Pyrolysis. *27*, 25–43.
Kargarzadeh, H., Ahmad, I., Abdullah, I., Dufresne, A., Zanariah Zainuddin, S.Y., and Sheltami, R.M. (2012). Effect of hydrolysis conditions on the morphology, crystallinity and thermal stability of cellulose nanocrystals extracted from kenaf bast fibres. Cellulose *19*, 855–866.
Kim, U.J., Kuga, S., Wada, M., Okano, T. and Kondo T. (2000). Periodate oxidation of crystalline cellulose. Biomacromolecules *1*, 488–492.
Kim, D.Y., Nishiyama, Y., Wada, M. and Kuga, S. (2001). High-yield carbonization of cellulose by sulfuric acid impregnation. Cellulose *8*, 29–33.
Kontturi, E. and Vuorinen, T. (2009). Indirect evidence of supramolecular changes within cellulose microfibrils of chemical pulp fibers upon drying. Cellulose *16*, 65–74.
Kontturi, E. (2011). Process for preparing micro- and nanocrystalline cellulose. Patent WO2011/114005.
Koshizawa, T. (1960). Degradation of wood cellulose and cotton linters in phosphoric acid. Kami Pa Gikyoshi *14*, 455.
Kvien, I., Tanem, B.S. and Oksman, K. (2005). Characterization of cellulose whiskers and their nanocomposites by atomic force and electron microscopy. Biomacromolecules *6*, 3160–3165.
Lee, S.Y., Mohan, D.J., Kang, I.A., Doh, G.H., Lee, S. and Han, N.O. (2009). Nanocellulose reinforced PVA composite films: Effects of acid treatment and filler loading. Fibers and Polymers *10*, 77–82.
Li, R., Fei, J., Cai, Y., Li, Y., Feng, J. and Yao, J. (2009). Cellulose whiskers extracted from mulberry: A novel biomass production. Carbohydr. Polym. *76*, 94–99.
Ljungberg, N., Bonini, C., Bortolussi, F., Boisson, C., Heux, L. and Cavaillé, J.Y. (2005). New nanocomposite materials reinforced with cellulose whiskers in atactic polypropylene: Effect of surface and dispersion characteristics. Biomacromolecules *6*, 2732–2739.
Liu, D., Zhong, T., Chang, P.R., Li, K. and Wu, Q. (2010). Starch composites reinforced by bamboo cellulosic crystals. Bioresource Technol. *101*, 2529–2536.
Lu, P. and Hsieh, Y.-L. (2012). Preparation and characterization of cellulose nanocrystals from rice straw. Carbohydr. Polym. *87*, 564–573.
Lu, Y, Weng, L. and Cao X. (2005). Biocomposites of plasticized starch reinforced with cellulose crystallites from cottonseed linter. Macromol. Biosci. *5*, 1101–1107.

Lu, Y., Weng, L. and Cao, X. (2006). Morphological, thermal and mechanical properties of ramie crystallites – Reinforced plasticized starch biocomposites. Carbohydr. Polym. *63*, 198–204.

Man, Z., Muhammad, N., Sarwono, A., Bustam, M.A., Kumar, M.V. and Rafiq, S. (2011). Preparation of cellulose nanocrystals using an ionic liquid. J. Polym. Environ. *19*, 726–731.

Mangalam, A. P., Simonsen, J. and Benight, A. (2009). Cellulose/DNA hybrid nanomaterials. Biomacromolecules *10*, 497–504.

Marchessault, R.H., Morehead, F.F. and Walter, N.M. (1959). Liquid crystal systems from fibrillar polysaccharides. Nature *184*, 632–633.

Millett, M.A., Moore, W.E. and Saeman, J.F. (1954). Preparation and properties of hydrocelluloses. Ind. Eng. Chem. *46*, 1493–1497.

Moon, R.J., Martini, A., Nairn, J., Simonsen, J. and Youngblood, J. (2011). Cellulose nanomaterials review: Structure, properties and nanocomposites. Chem. Soc. Rev. *40*, 3941–3994.

Morán, J.I., Alvarez, V.A., Cyras, V.P. and Vásquez, A. (2008). Extraction of cellulose and preparation of nanocellulose from sisal fibers. Cellulose *15*, 149–159.

Morin, A. and Dufresne, A. (2002). Nanocomposites of chitin whiskers from *Riftia* tubes and poly(caprolactone). Macromolecules *35*, 2190–2199.

Mukherjee, S.M., Sikorski, J. and Woods, H.J. (1952). Electron-microscopy of degraded cellulose fibres. J. Text. Inst. *43*, T196–T201.

Mukherjee, S.M. and Woods, H.J. (1953). X-ray and electron microscope studies of the degradation of cellulose by sulphuric acid. Biochim. Biophys. Acta *10*, 499–511.

Nickerson, R.F. and Habrle, J.A. (1947). Cellulose intercrystalline structure: Study by hydrolytic methods. Ind. Eng. Chem. *39*, 1507–1512.

Nishiyama, Y., Kim, U.J., Kim, D.Y., Katsumata, K.S., May, R.P. and Langan, P. (2003). Periodic disorder along ramie cellulose microfibrils. Biomacromolecules *4*, 1013–1017.

Notley, S.M., Eriksson, M., Wågberg, L., Beck, S. and Gray, D.G. (2006). Surface forces measurements of spin-coated cellulose thin films with different crystallinity. Langmuir *22*, 3154–3160.

Okano, T., Kuga, S., Wada, M., Araki, J. and Ikuina J. *Fine cellulose particle and its production.* (1999). 1999-343301.

Oksman, K., Mathew, A. P., Bondeson, D. and Kvien, I. (2006). Manufacturing process of cellulose whiskers/polylactic acid nanocomposites. Compos. Sci. Technol. *66*, 2776–2784

Oksman, K., Etang, J.A., Mathew, A.P. and Jonoobi, M. (2011). Cellulose nanowhiskers separated from a bio-residue from wood bioethanol production. Biomass & Bioenergy *35*, 146–152.

Orts, W.J., Godbout, L., Marchessault, R.H. and Revol, J.F. (1998). Enhanced ordering of liquid crystalline suspensions of cellulose microfibrils: A small angle neutron scattering study. Macromolecules *31*, 5717–5725.

Paillet, M. and Dufresne, A. (2001). Chitin whisker reinforced thermoplastic nanocomposites. Macromolecules *34*, 6527–6530.

Pandey, J.K., Lee, J.W., Chu, W.S., Kim, C.S., Lee, C.S. and Ahn, S.H. (2008). Cellulose nanowhiskers from grass of Korea. Macromol. Res. *16*, 396–398.

Pandey, J.K., Chu, W.S., Kim, C.S., Lee, C.S. and Ahn, S.H. (2009). Bio-nano reinforcement of environementally degradable polymer matrix by cellulose whiskers from grass. Composites: Part B *40*, 676–680.

Paralikar, S. A., Simonsen, J. and Lombardi, J. (2008). Poly (vinyl alcohol)/cellulose nanocrystals barrier membranes. J. Membr. Sci. *320*, 248–258.

Parks, E.J. (1971). Thermal analysis of modified cellulose. Tappi J. *54*, 537–544.

Petersson, L., Kvien, I. and Oksman, K. (2007). Structure and thermal properties of poly(lactic acid)/ cellulose whiskers nanocomposites materials. Compos. Sci. Technol. *67*, 2535–2544

Philipp, H.J., Nelson, M.L. and Ziifle, H.M. (1947). Crystallinity of cellulose fibers as determined by acid hydrolysis. Textile Research J. *17*, 585–596.

Pu, Y., Zhang, J., Elder, T., Deng, Y., Gatenholm, P. and Ragauskas, A.J. (2007). Investigation into nanocellulosics versus acacia reinforced acrylic films. Composites: Part B *38*, 360–366.

Qin, Z.Y., Tong, G.L., Chin, Y.C.F. and Zhou, J.C. (2011). Preparation of ultrasonic-assisted high carboxylate content cellulose nanocrystals by TEMPO oxidation. BioResources *6*, 1136–1146.

Rånby, B.G. (1949). Aqueous colloidal solutions of cellulose micelles. Acta Chem. Scand. *3*, 649–650.

Rånby, B.G. and Ribi, E. (1950). Über den Feinbau der Zellulose. Experimentia *6*, 12–14.

Rånby, B.G. (1951). III. Fibrous macromolecular systems. Cellulose and muscle. The colloidal properties of cellulose micelles. Faraday Soc. *11*, 158–164.

Rånby, B.G. (1952). The cellulose micelles. Tappi *35*, 53–58.

Revol, J.F. (1982). On the cross-sectional shape of cellulose crystallites in *Valonia ventricosa*. Carbohydr. Polym. *2*, 123–134.

Revol, J.F., Godbout, L., Dong, X.M., Gray, D.G., Chanzy, H. and Maret, G. (1994). Chiral nematic suspensions of cellulose crystallites; Phase separation and magnetic field orientation. Liq. Cryst. *16*, 127–134.

Rojas, O.J., Montero, G.A. and Habibi, Y. (2009). Electronspun nancomposites from polystyrene loaded with cellulose nanowhiskers. J. Appl. Polym. Sci. *113*, 927–935.

Roman, M. and Winter, W.T. (2004). Effect of sulphate groups from sulphuric acid hydrolysis on the thermal degradation behaviour of bacterial cellulose. Biomacromolecules *5*, 1671–1677.

Roman, M. and Gray D.G. (2005). Parabolic focal conics in self-assembled solid films of cellulose nanocrystals. Langmuir *21*, 5555–5561.

Rosa, M.F., Medeiros, E.S., Malmonge, J.A., Gregorski, K.S., Wood, D.F., Mattoso, L.H.C., Glenn, G., Orts, W.J. and Imam, S.H. (2010). Cellulose nanowhiskers from coconut husk fibers: Effect of preparation conditions on their thermal and morphological behaviour. Carbohydr. Polym. *81*, 83–92.

Satyamurthy, P., Jain, P., Balasubramanya, R.H. and Vigneshwaran, N. (2011). Prepapration and characterization of cellulose nanowhiskers from cotton fibres by controlled microbial hydrolysis. Carbohydr. Polym. *83*, 122–129.

Saxena, I.M. and Brown, R.M.J. (2005). Cellulose biosynthesis: Current views and envolving concepts. Annals of Botany *96*, 9–21.

Schurz, J. and John, K. (1975). Long periods in native and regenerated celluloses. Cell. Chem. Technol. *9*, 493–501.

Segal, L., Creely, J.J., Martin, A.E. and Conrad, C.M. (1959). An empirical method for estimating the degree of crystallinity of native cellulose using the X-ray diffractometer. Text. Res. J. *29*, 786–794.

Sharples, A. (1958). The hydrolysis of cellulose and its relation to structure. Part 2. Trans. Faraday Soc. *54*, 913–917.

Sheltami, R.M., Ahmad, I., Abdullah, I., Dufresne, A. and Kargarzadeh, H. (2012). Extraction of cellulose nanocrystals from mengkuang leaves (Pandanus tectorius). Carbohydr. Polym. *88*, 772–779.

Siqueira, G., Bras, J. and Dufresne, A. (2009). Cellulose whiskers versus microfibrils: Influence of the nature of the nanoparticle and its surface functionalization on the thermal and mechanical properties of nanocomposites. Biomacromolecules *10*, 425–432.

Siqueira, G., Abdillahi, H., Bras, J. and Dufresne, A. (2010a). High reinforcing capability cellulose nanocrystals extracted from *Syngonanthus nitens* (Capim Dourado). Cellulose *17*, 289–298.

Siqueira, G., H., Bras, J. and Dufresne, A. (2010b). *Luffa cylindrica* as a lignocellulosic source of fiber, microfibrillated cellulose, and cellulose nanocrystals. BioResources *5*, 727–740.

Siqueira, G., Tapin-Lingua, S., Bras, J., da Silva Perez, D. and Dufresne, A. (2010c). Morphological investigation of nanoparticles obtained from combined mechanical shearing, and enzymatic and acid hydrolysis of sisal fibers. Cellulose *17*, 1147–1158.

Siqueira, G., Bras, J. and Dufresne, A. (2010d). New process of chemical grafting of cellulose nanoparticles with a long chain isocyanate. Langmuir *26*, 402–411.

Siqueira, G., Tapin-Lingua, S., Bras, J., da Silva Perez, D. and Dufresne, A. (2011). Mechanical properties of natural rubber nanocomposites reinforced with cellulosic nanoparticles obtained from combined mechanical shearing, and enzymatic and acid hydrolysis of sisal fibers. Cellulose *18*, 57–65.

Sugiyama, J., Harada, H., Fujiyoshi, Y. and Uyeda, N. (1985). Lattice images from ultrathin sections of cellulose microfibrils in the cell wall of *Valonia macrophysa* Kütz. Planta *166*, 161–168.

Tang, W.K. and Neill W.K. (1964). Effect of flame retardants on pyrolysis and combustion of a-cellulose. J. Polym. Sci.: Part C *6*, 65–81.

TAPPI Test Method T204. Solvent Extractives of Wood and Pulp. In: Tappi Test Methods, (Technical Association of the Pulp and Paper Industry, Atlanta, GA).

TAPPI (2011). Workshop on International Standards for Nanocellulose, Arlington, USA, June 9, 2011.

Teixeira, E.M., Pasquini, D., Curvelo, A.A.S., Corradini, E., Belgacem, M.N. and Dufresne, A. (2009). Cassava bagasse cellulose nanofibrils reinforced thermoplastic cassava starch. Carbohydr. Polym. *78*, 422–431.

Teixeira, E.M., Bondancia, T.J., Teodoro, K.B.R., Corrêa, A.C., Marconcini, J.M. and Mattoso, L.H.C. (2011). Sugar cane bagasse whiskers: Extraction and characterizations. Ind. Crops Prod. *33*, 63–66.

Thielemans, W., Warbey, C.R. and Walsh, D.A. 2009. Permselective nanostructured membranes based on cellulose nanowhiskers. Green Chem. *11*, 531–537.

Ureña-benavides, E.E., Ao, G., Davis, V.A. and Kitchens, C.L. (2011). Rheology and phase behavior of lyotropic cellulose nanocrystal suspensions. Macromolecules *44*, 8990–8998.

Wang, Y., Cao, X. and Zhang, L. (2006). Effects of cellulose whiskers on properties of soy protein thermoplastics. Macromol. Biosci. *6*, 524–531.

Wang, B., Sain, M. and Oksman, K. (2007). Study of structural morphology of hemp fiber from the micro to the nanoscale. Appl. Compos. Mater. *14*, 89–103.

Wise, L.E., Murphy, M. and D'Addieco, A.A. (1946). Chlorite holocellulose, its fractionation and bearing on summative wood analysis and on studies on hemicelluloses. Paper Trade Journal *122*, 35–43.

Whistler, R.L. and BeMiller, J.M. (1997). Cellulosics. In: Carbohydrate chemistry for food scientists, (American Association of Cereal Chemists, Inc., Saint Paul, MN), Vol. 7 p. 155.

Woehl, M.A., Canestraro, C.D., Mikowski, A., Sierakowski, M.R., Ramos, L.P. and Wypych, F. (2010). Bionanocomposites of thermoplastic starch reinforced with bacterial cellulose nanofibres: Effect of enzymatic treatment on mechanical properties. Carbohydr. Polym. *80*, 866–873.

Yachi, T., Hayashi, J., Takai, M. and Shimizu, Y. (1983). Supermolecular structure of cellulose: stepwise decrease in LODP and particle size of cellulose hydrolyzed after chemical treatment. J. Appl. Polym. Sci.: Appl. Polym. Symp. *37*, 325–343.

Yamamoto, H. and Horii, F. (1994). *In situ* crystallization of bacterial cellulose. 1. Influences of polymeric additives, stirring and temperature on the formation of celluloses Ia and Ib as revealed by cross polarization magic angle spinning (CP/MAS) ^{13}C NMR spectroscopy. Cellulose *1*, 57–66.

Yun, Y.S., Cho, S.Y. and Jin, H.J. (2010). Flow-induced liquid crystalline solutions prepared from aspect ratio-controlled bacterial cellulose nanowhiskers. Mol. Cryst. Liq. Cryst. *519*, 141–148.

Zuluaga, R., Putaux, J.L., Restrepo, A., Mondragón, I. and Ganan, P. (2007). Cellulose microfibrils from banana farming residues: Isolation and characterization. Cellulose *14*, 585–592

4 Bacterial cellulose

Cellulose is the major component of plant biomass in which it plays a structural role. However, it is also found as microbial extracellular polymer. Bacterial or microbial cellulose belongs to specific products of primary metabolism and constitutes mainly a protective coating. It is an extracellular product of vinegar bacteria which was described by Pasteur as "a sort of moist skin, swollen, gelatinous and slippery ..." (Ring, 1982). One of the most important features of bacterial cellulose is its chemical purity, which distinguishes this cellulose from that of plants, usually associated with hemicelluloses and lignin. The removal of the latter is inherently difficult. However, cellulose synthesized by bacteria showed identical molecular structure to that made by plants.

4.1 Production of cellulose by bacteria

Microbial cellulose is a form of cellulose that is produced by bacteria. The cellulose mat associated with the production of vinegar, Kombucha tea and nata de coco has been observed and used for centuries, even though not isolated and formerly recognized as secondary metabolite of the bacterium. It was first confirmed as cellulose in 1886 (Brown, 1886). While working with *Bacterium aceti* (acetic acid bacteria), the author identified a gelatinous solid mass in the course of vinegar fermentation, which was referred to as "vinegar plant" or "mother". This material was reported to be tough to tear, with the touch and feel of animal tissue. It was found to be chemically equivalent to cell-wall cellulose.

Bacterial cellulose (BC) is a product of microbes' primary metabolic processes. It is synthesized by bacteria from the genera *Zoogloea*, *Sarcina* (Canale-Parola and Wolfe, 1960), *Salmonella*, *Rhizobium* (Napoli et al., 1975), *Pseudomonas* (Spiers et al., 2003), *Escherichia*, *Agrobacterium* (Matthysse et al., 1995), *Aerobacter*, *Achromobacter*, *Azotobacter*, *Alcaligenes*, and *Acetobacter*. An overview of BC producers is reported in Table 4.1. From a commercial point of view, only the *Acetobacter* species produce enough cellulose. The most studied and most used BC-producing bacterium specie is *Acetobacter xylinum*, formerly known as *Acetobacterium xylinum* and *Bacterium xylinodes*, reclassified as the genus *Gluconacetobacter*. It is found wherever the fermentation of sugars and plant carbohydrates takes place, in unpasteurized or unsterilized fruit juices, flowers, rotting fruits, and alcoholic beverages. It converts ethanol to acetic acid. This non-photosynthetic organism can procure glucose, sugar, glycerol, or other organic substrates and convert them into pure cellulose (Brown et al., 1976).

Genus	Cellulose structure
Acetobacter	Extracellular pellicle composed of ribbons
Achromobacter	Fibrils
Aerobacter	Fibrils
Agrobacterium	Short fibrils
Alcaligenes	Fibrils
Pseudomonas	No distinct fibrils
Rhibozium	Short fibrils
Sarcina	Amorphous cellulose
Zoogloea	Not well defined

Table 4.1: Bacterial cellulose producers (adapted from Jonas and Farah, 1998).

There have been conflicting reports on the exact biological function of microbial cellulose. The most familiar form of BC is that of a pellicle on the top of a static cultured growth media. Therefore, it has been hypothesized that cellulose-forming bacteria use the produced cellulose fleece as a designed architecture for living in different environments and protection against drying, enemies, and lack of oxygen and food (Cannon and Anderson, 1991; Klemm et al., 2011). It could act as a floatation device bringing the bacteria to the oxygen-rich air-media interface, even if experiments conducted on submerged oxygen-permeable silicone tubes show that cellulose grows well submerged if enough oxygen is present (Yoshino et al., 1996). Its function seems to be not only to hold the bacteria in the aerobic environment but also to serve a diversity of functions, such as protection from ultraviolet light, retention of moisture, and colonization of substrates (Williams and Cannon, 1989). Pure *Gluconacetobacter* strains can be bought from international collections of micro-organisms. It includes the strains ATCC 23769, 10145, 53582, AX5 and many others (Klemm et al., 2001). It was shown that oxygen is vital for cellulose production.

Unlike the complex formation in plants, *A. xylinum* produces a well-defined extracellular mesh of cellulose (Aschner and Hestrin, 1946; Ross et al., 1991). It is generally accepted that the synthesis of fibrillar cellulose by *A. xylinum* is an extracellular process. It occurs between the outer membrane and cytoplasma membrane by cellulose-synthesizing complexes (or terminal complexes) linearly arranged in association with pores at the surface of the bacterium (Cannon and Anderson, 1991; Jonas and Farah, 1998) as schematized in Figure 4.1. First, glucan chain aggregates consisting of approximately 10–100 glucan chains are elongated or extruded from the complex. These sub-elementary fibrils leave the pores which are located about 10 nm apart in a distinct array on the surface (Zaar, 1979), and are then assembled to form microfibrils. A single cell of *Acetobacter* has a linear row of pores from which glucan chain aggregates are spun. These microfibrils have a rectangular cross-section of $10–20 \cdot 30–40$ Å^2. The final step consists in the formation of a ribbon of crystalline cellulose from the tight assembly of many secreted microfibrils in the culture medium (Forge and Preston, 1977). Production occurs mostly at the interface of liquid and air. These ribbons are 3–4 nm thick and about 70–130 nm wide (Brown et al., 1976; Zaar, 1977; Yamanaka et al., 2000). The length ranges from 1 to 20 μm and they form a dense 3D nanofiber network or reticulated structure, stabilized by extensive hydrogen bonding. The degree of polymerization (DP) of BC is usually between 2,000 and 6,000, whereas it is about 13,000 to 14,000 for plant cellulose (Jonas and Farah, 1998). In some cases it reaches even 11,000 or 14,000 (Watanabe et al., 1998). This entangled mesh of nanofibrils produces a gelatinous membrane known as pellicle consisting of pure cellulose and cells entrapped within it (Figure 4.2).

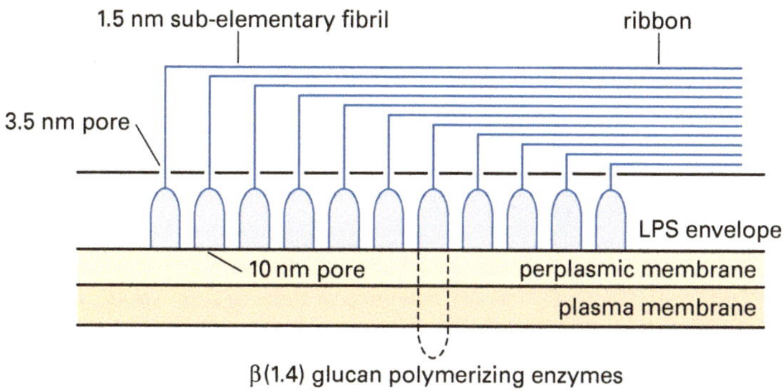

Fig. 4.1: Schematic model for the formation process of the normal ribbon assembly in the *Acetobacter xylinum* system (adapted from Vandamme et al., 1997).

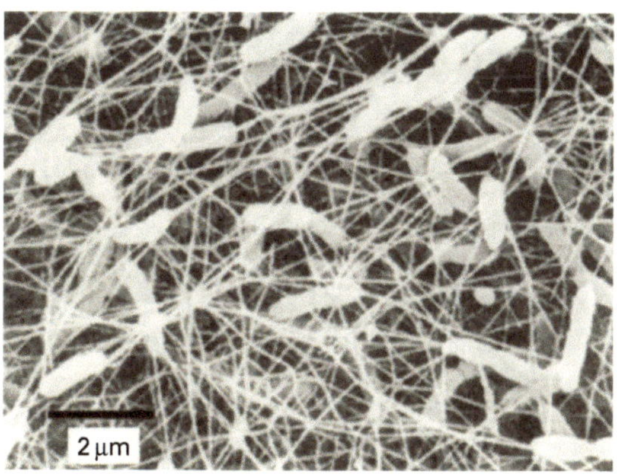

Fig. 4.2: Scanning electron micrograph of the surface of a freeze-dried bacterial cellulose pellicle (Iguchi et al., 2000).

Figure 4.3 shows the formation of nanofibers and ribbons from rod-like *Gluconacetobacter* bacteria. Each cell acts as a nano-spinneret, producing a bundle of nanometric fibrils. The lateral dimension of BC increases as bacteria grow and as the population increases. During bacterial growth, new production sites for BC become available and the BC ribbon widens. At cell division, BC production sites are distributed between two daughter cells, which also increase BC ribbon width (Zaar, 1977).

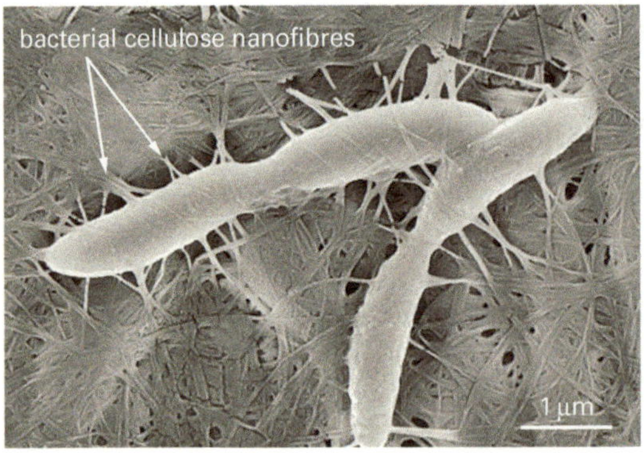

Fig. 4.3: Gluconacetobacter bacteria forming cellulose nanofibers and ribbons (http://www.vtnews.vt.edu/articles/2008/11/2008-693.html; access date: May 20, 2012).

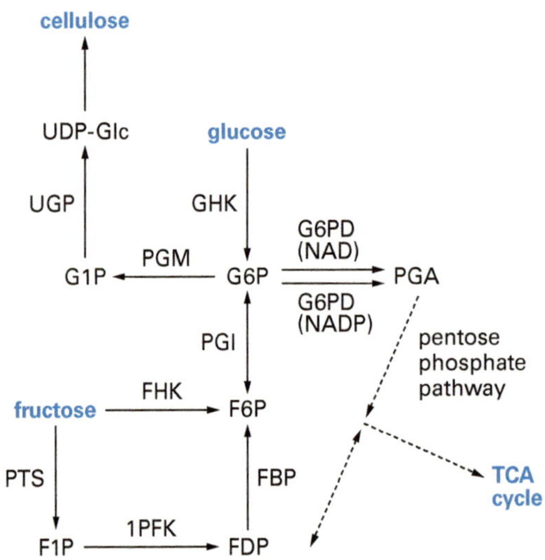

Fig. 4.4: Pathways of carbon metabolism in *Acetobacter xylinum* (adapted from Tonouchi et al., 1996). Abbreviations: 1PFK, 1-phosphofructokinase; EMP, Embden-Myerhoff pathway; F1P, fructose-1-phosphate; F6P, fructose-6-phosphate; FBP, fructose bis-phosphatase; FDP, fructose diphosphate; FHK, fructose hexokinase; G1P, glucose-1-phosphate; G6P, glucose-6-phosphate; G6PD, glucose-6-phosphate dehydrogenase; GHK, glucose hexokinase; PGI, phosphoglucose isomerase; PGM, phosphoglucomutase; PTS, phosphotransferase system; TCA cycle, tricarboxylic acid cycle or Krebs cycle; UDP, uridine-diphosphate; UGP, UDP-glucose pyrophosphorylase;

Determination of the biosynthetic pathway of carbon metabolism in *A. xylinum* has been intensively investigated. It is schematically represented in Figure 4.4. The conversion of the glucose substrate into cellulose involves at least four steps of biochemical reaction (Klemm et al., 2001). First, glucose is converted to glucose-6-phosphate (G6P) by the enzyme gluconase. It is followed by the isomerization of glucose-6-phosphate into glucose-1-phosphate (G1P) by phosphoglucomutase (PGM). Then, glucose-1-phosphate is converted into uridine-diphosphate (UDP)-glucose-1-phosphate, before being converted into pyrophosporylase. In *A. xylinum* this enzyme is activated by cyclic nucleotide (c-di-GMP). The cellulose synthase activator c-di-GMP is synthesized in *A. xylinum* by the enzyme diguanylate cyclise, and its concentration is regulated by the action of phosphodiesterases. Finally, UDP-Glc is polymerized into cellulose by cellulose synthase (Tal et al., 1998).

4.2 Influence of carbon source

Several animals, fungi, and bacteria can assemble cellulose. However, these organisms are devoid of photosynthetic capacity and usually require glucose or some organic substrate synthesized by a photosynthetic organism to assemble their cellulose. Some bacteria can utilize methane or sulfur substrates to produce glucose and other organic substrates for cellulose.

A carbon source consisting of low-molecular weight sugar such as D-glucose is required for cellulose production by bacteria. D-glucose acts not only as an energy source but also as a cellulose precursor. Various modifications have been made to improve the mechanical, chemical, structural and biological properties and production yield of BC. *A. xylinum* is the most prolific cellulose-producing bacterium found in nature. A typical single cell can convert up to 108 glucose molecules per hour into cellulose. As many as a million cells can be packed into a large liquid droplet, each converting up to 108 glucose molecules per hour into cellulose, and it is considered that the product should virtually be made before one's eyes. However, bacteria can grow under controllable conditions and produce cellulose from various carbon sources, such as monosaccharides, disaccharides, starch, alcohols, and organic acids (Jonas and Farah, 1998).

Incorporation of mannitol (Oikawa et al., 1995a) and arabitol (Oikawa et al., 1995b) in the culture medium was found to be able to produce 3.8 and 6.2 times, respectively, more cellulose than glucose did. The use of ethanol as the sole carbon source in the culture medium has been shown to be ineffective. However, when ethanol is included with a suitable carbon source such as glucose, an increase of cellulose production has been reported in some cases (Tarr and Hibbert, 1931; Son et al., 2001; Naritomi et al., 1998a; Krystynowicz et al., 2002; Park et al., 2003). The inclusion of lactate (Naritomi et al., 1998b), pyruvate, aldehyde, acetate, citrate and succinate in the culture medium has also been shown to increase the cellulose yield (Dudman, 1959; Matsuoka et al., 1996). Similar results were obtained with the plant inhibitor, 1-methylcyclopropene (Hu and Catchmark, 2010). An increase of crystallinity and Iα content has been observed when using lignosulfonate in static culture at a concentration 1 wt% to stimulate cellulose synthesis (Keshk and Sameshima, 2006). An increase of BC productivity of almost 57% was also observed. The incorporation of carboxymethylcellulose (CMC) in the culture medium led to contradictory reports. Some studies show an increase in cellulose yield with a saturation around 1% CMC (Cheng et al., 2009), possibly attributed to an increase in the solubility of BC in the presence of the highly soluble CMC, which may bind to the surface of BC fibrils. Others report a decrease or no significant impact on the yield (Chao et al., 2001; Tantratian et al., 2005).

Conflicting results have also been reported for additives such as agar, xanthan, sodium alginate and acetic acid (Chao et al., 2001; Bae et al., 2004; Bae and Shoda, 2005; Zhou et al., 2007). Increased productivity of BC was reported by adding water soluble agar in concentration 0.1–1 wt% in the culture medium (Bae et al., 2004). Optimized culture conditions were determined using a response surface analysis (Bae and Shoda, 2005). Differences in pore size, degree of polymerization, crystallinity, fiber widths and mechanical strength have been observed by inclusion in the culture medium of cellulose-producing bacteria of different additives such as acetyl glucomannan (Tokoh et al., 1998), various antibiotics (Yamanaka et al., 2000), glucose-phosphate (Basta and El-Saied, 2009), polyethylene glycol 400, polyethylene glycol 4000, β-cyclodextrin (Hessler and Klemm, 2009), Tween 80, urea, fluorescent brightener 28 (Huang et al., 2010) and hydroxypropylmethyl cellulose (HPMC) (Huang et al., 2011). Phosphate containing BC was found to be an environmentally friendly paper additive for the production of functionalized paper with improved strength, fire retardation and filler loading (Basta and El-Saied, 2009). In situ modification of BC with fluorescent brightener 28, HPMC and CMC during fermentation improved the rehydration ability of BC (Huang et al., 2010; Huang et al., 2011).

4.3 Culture conditions

Macroscopic morphology of BC strictly depends on culture conditions. Most works are based on the so-called Hestrin–Schramm medium (Hestrin and Schramm, 1954). It contains 2 wt% glucose, 0.5% yeast extract, 0.5 wt% peptone, 0.27 wt% anhydrous disodium phosphate and 0.15 wt% citric acid monohydrate. The pH is adjusted to around 5.0 using 1 N acetic acid

Acetobacter are Gram-negative, aerobic, rod-like micro-organisms of unusual acid tolerance. It is accepted that the optimal pH range for cellulose production by *A. xylinum* is below 7 (Hestrin and Schramm, 1954). The pH generally ranges between 4 and 7 (Masaoka et al., 1993; Ishikawa et al., 1995; Oikawa et al., 1995a; Oikawa et al., 1995b; Watanabe and Yamanaka, 1995; Matsuoka et al., 1996; Hwang et al., 1999). Optimal growth temperature conditions range between 25 and 30°C (Cannon and Anderson, 1991), even if most studies used 28–30°C (Geyer et al., 1994; Ishikawa et al., 1995; Oikawa et al., 1995a; Oikawa et al., 1995b; Borzani and de Souza, 1995; Watanabe and Yamanaka, 1995; Iguchi et al., 2000). Oxygen is an important factor for cellulose production. It was shown that O_2 tensions affect both the cellulose production and performance of the membrane (Watanabe and Yamanaka, 1995). In the initial stage, the bacteria increase their population by consumption of oxygen initially dissolved in the medium. At the same time, cellulose is synthesized in the liquid phase. Only bacteria present in the vicinity of the upper side of the pellicle surface at the air/pellicle interface, and therefore associated with oxygen, can maintain their cellulose-producing activity whereas those below the surface of the pellicle are dormant. This means that the producing cells as well as nutritive sources must be transported through the pellicle to its surface (Borzani and de Souza, 1995). Progressively, the system becomes turbid and, after a while, a white pellicle appears on the surface whose thickness increases steadily with time, reaching over 25 mm in four weeks as shown in Figure 4.5 (Iguchi et al., 2000).

Fig. 4.5: Bacterial cellulose layers grown with different culture time (maximum 4 weeks) (Iguchi et al., 2000).

A. xylinum is generally grown statically because otherwise no uniform pellicle can be produced. In static culture conditions, bacteria produce cellulose in the form of pellicles or mats on the surface of nutrient broth at the oxygen-rich air-liquid interface (Hestrin and Schramm, 1954; Hestrin, 1963). In these conditions, bacteria are cultivated in Erlenmeyer flasks containing Hestrin–Schramm medium at 28°C for 10–14 days. The advantage of the traditional static culture is its simplicity. However, the number of pellicles formed on the surface of the media is smaller than in other methods because the growth of cellulose-producing bacteria is relatively slow (Son et al., 2001). This is ascribed to the formation of an effective barrier between atmospheric oxygen on one side and the nutrient on the other. It reduces the rate of oxygen penetration through the pellicles to the cell (Dudman, 1960).

Agitated culture conditions include shake culture and rotating disk system. In these conditions, BC can be produced in flasks and in a bioreactor. These methods are more efficient for BC production and are preferred for large scale production. In agitated culture conditions, BC is produced in a form of irregular granules, stellate and fibrous strands, well dispersed in culture broth (Vandamme et al., 1998; Czaja et al., 2004). Moreover, differences in the 3D nanofiber network, crystallinity index and cellulose Iα content have been reported. The production of cellulose in the form of characteristic spheres obtained in agitated culture is shown in Figure 4.6. A hypothetical mechanism of sphere formation and cell arrangement in the agitated culture has been proposed (Czaja et al., 2004). During agitation, it was observed that cells were stacked together in organized groups around the outer surface of the cellulose spheres.

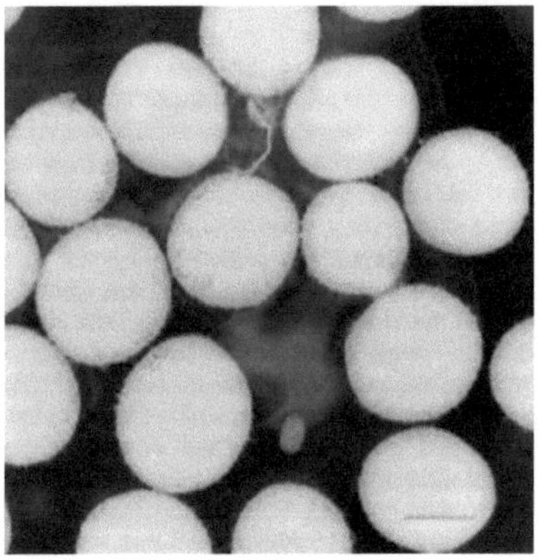

Fig. 4.6: Cellulose spheres formed in the agitated culture conditions (Czaja et al., 2004). Scale bar = 5 mm.

Microscopic observations revealed morphological differences between agitated-BC and static-BC (Johnson and Neogi, 1989). Nanofibrils produced under static conditions are more extended and piled above one another in a crisscrossing manner, and have a larger cross-sectional width. The cellulose pellicle obtained in agitated culture consists of entangled and curved nanofibrils. These morphological differences contribute varying crystallinity. The crystallinity index and size of crystallites were found to be lower for BC synthesized in agitated culture (Watanabe et al., 1998; Czaja et al., 2004). In addition, a significant portion of cellulose II was reported. Cellulose synthesized from agitated culture conditions was also characterized by a lower Iα mass fraction than cellulose that was produced statically. A broad range of new concepts based on a combination of static and agitated cultivation has been investigated (Klemm et al., 2011).

Bacteria and residues from the culture medium entrapped within the cellulose network can be removed using a suitable purification method of the isolated never-dried BC pellicle. It consists in treating it with a diluted sodium hydroxide solution under reflux for 10–120 min depending on the thickness of the pellicle (Embuscado et al., 1996; Klemm et al., 2005). Optimal alkali isolation conditions of BC were determined using a response surface methodology (Embuscado et al., 1996). No detectable damage of the cellulose occurred under these conditions.

4.4 In situ modification of bacterial cellulose

Because BC is produced as very fine fibrils from in situ control of cellulose formation, its morphology can therefore be tuned or tailor-designed during biogenesis. The constituents of the culture medium and the inclusion of additives in the growth medium affect cellulose production, as the assembly of cellulose is susceptible to chemical and physical influences by the compounds present during synthesis and aggregation (Uhlin et al., 1995). The structure, morphology and physical properties can be affected by the presence of an additive in the culture medium, effectively creating in situ modifications. The additive can bind directly to the cellulose during production and interfere with the crystallization, or co-crystallize with cellulose. It may also interfere with the bacterial cells themselves, thereby altering the cellulose production indirectly.

Some studies have considered addition of cellulose derivatives, hemicelluloses, and pectins in the culture medium to highlight the biogenesis of cellulose and plant cell walls (Haigler et al., 1982; Uhlin et al., 1995; Yamamoto et al., 1996; Hirai et al., 1998). The compatibility between the polymer and the cellulose obviously governs the alterations that take place in the cellulose in terms of microfibril and ribbon dimensions, crystalline allomorphs, and crystallinity index. Properties of BC can also be altered by manipulating its biosynthesis so as to synthesize copolymers or miscible blends. For instance, BC/chitosan copolymers have been obtained by biosynthesizing BC in a chitosan-enriched medium (Ciechanska, 2004). Improved water absorbance and mechanical properties were reported.

Preparation of nanocomposites by in situ methods has been reported by exploiting the competitive adsorption of a host water-soluble polymer present in the culture medium. Finely dispersed BC/poly(ethylene oxide) (PEO) nanocomposite materials with different morphologies and BC contents ranging from 15 to 59 wt% have been obtained by adding PEO to the growth culture medium of *A. xylinum* (Brown and Laborie, 2007). When increasing the BC content, a decrease of the nanofiber size and aggregation in larger bundles was observed indicating that PEO mixed with the cellulose on the nanometer scale. Bundles of nanofiber reached sizes in the 121–770 nm range when adding 5% PEO in the culture medium, indicating that nanofibers were bonded together by PEO (Figure 4.7). The fine dispersion of BC nanofibers hindered the crystallization of PEO, lowering its melting point and crystallinity although remaining bacterial cell debris also contributed to the melting point depression. Moreover, improved thermal stability and mechanical properties of PEO were reported.

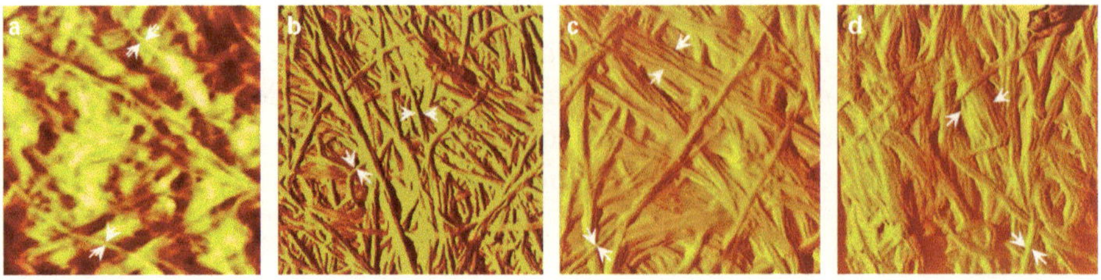

Fig. 4.7: AFM topographical images (3 · 3 μm^2) of compression-molded BC/PEO nanocomposites obtained in (a) Hestrin–Schramm (HS) medium, (b) HS medium with 1% PEO, (c) HS medium with 3% PEO, and (d) HS medium with 5% PEO (Brown and Laborie, 2007). Arrows indicate measurements of nanofiber bundles.

Cellulose derivatives and poly(vinyl alcohol) (PVA) have been added in the culture medium with the aim to prepare nanocomposites with improved water retention ability and ion absorption (Seifert et al., 2003). Increased water absorption capacities were observed by the addition of methylcellulose and CMC, whereas the opposite trend was reported by the addition of PVA. BC/PVA nanocomposites were obtained using a 5% PVA solution in the culture medium (Gea et al., 2010). It was shown that the preparation of BC/PVA by this in situ process did not affect significantly the morphology of the cellulosic fibrils. Gelatinized starch was also included in the culture medium at a concentration of 2 wt% to obtain BC/starch nanocomposites (Grande et al., 2009). It was observed that the crystallinity of BC was preserved and that starch formed a layer covering the cellulosic fibrils. In all these cases, the BC was mostly dispersed and the nanocomposites displayed improved mechanical properties.

4.5 Bacterial cellulose hydrogels

BC fibrils aggregate due to hydrogen bonding and van der Waals forces and are immobilized in a stable well-defined cellulose network. These forces cause the fibrils to interact, and they are held apart by adsorbed water layers. One of the unique features of this pure cellulose membrane is that it is very strong in the never-dried state, and it can hold hundreds of times its weight in water. Indeed, large amounts of water, up to 99%, can be incorporated leading to form-stable hydrogels. The hydrophilicity of BC pellicle is usually explained by the presence of pore structures and tunnels (Klemm et al., 2001). Water plays a key role as a spacer element and as a stabilizing agent of the network and pore structure because of H-bond interactions with cellulose.

The shape of the hydrogel can be designed by choosing the appropriate reactor form and function (static or agitated cultivation) (Klemm et al., 2011). Under classical static conditions in glass or polypropylene containers with circular or square profiles, the BC products are formed in the shape of the reactor (Gatenholm and Klemm,

2010). The BC hydrogel bodies adopt the dimensions of the cultivation vessel. Figure 4.8(a) shows a typical BC fleece consisting of the BC nanofiber network and soaked with water in its native hydrated state formed in a circular reactor. Thinner films can be obtained after shorter cultivation times. Tubular hollow bodies can be prepared by matrix-type reactors (Gatenholm and Klemm, 2010). In this case, the BC tubes are not formed directly in the reactor vessel but on a glass or silicon cylinder as a matrix/template placed in the culture medium inside of the reactor. The tubes can be simply removed and purified. Depending on the dimension of the matrix, BC tubes with diameters ranging from 0.5 mm up to centimeter-scale and with lengths of 1 to more than 20 cm can be prepared. BC tubes of different dimensions are shown in Figure 4.8(b).

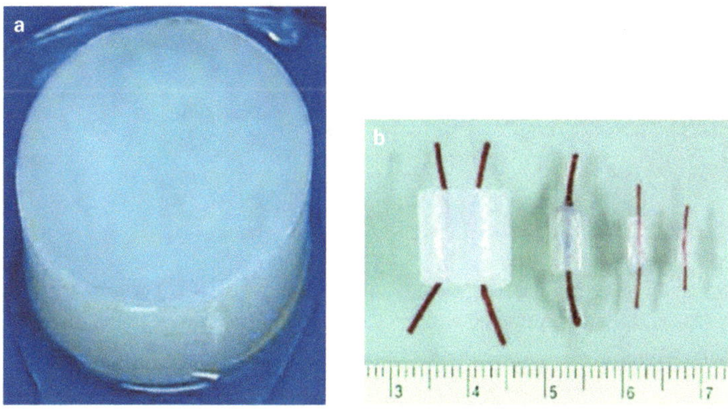

Fig. 4.8: (a) Typical bacterial cellulose fleece formed in a reactor of circular profile (diameter 3 cm and thickness 2 cm), and (b) bacterial cellulose tubes of different dimensions (Gatenholm and Klemm, 2010). In panel (b) ruler units are in centimeters.

For biomedical applications it is highly desirable to fine-tune the properties of the scaffold or implant to match the properties of the material that it intends to regenerate or replace. For instance, BC has been soaked into hydroxyapatite to develop a composite scaffold for bone regeneration (Hong et al., 2006; Wan et al., 2006). Phosphorylated BC pellicles have been pre-incubated for 3 days in a $CaCl_2$ solution to obtain a homogeneous calcium phosphate layer. The ensuing $CaCl_2$-treated phosphorylated BC was then immersed in a simulated body fluid at 37°C for 7 or 14 days to initiate biomimetic calcium phosphate formation on the surface of BC network. Formation of homogeneous layers on the surface of BC fibrils was observed, whose thickness increased with increased soaking time (Figure 4.9). BC has also been augmented by immersion in solutions of polyacrylamide and gelatin, yielding hydrogels with improved toughness (Yasuda et al., 2005). Similarly, immersion of bacterial cellulose into PVA has yielded hydrogels with a wide range of mechanical properties of interest for cardiovascular implants (Millon et al., 2006). The objective of this study was to develop a PVA-based hydrogel that not only mimics the non-linear mechanical properties displayed by cardiovascular tissues, but also their anisotropic behavior.

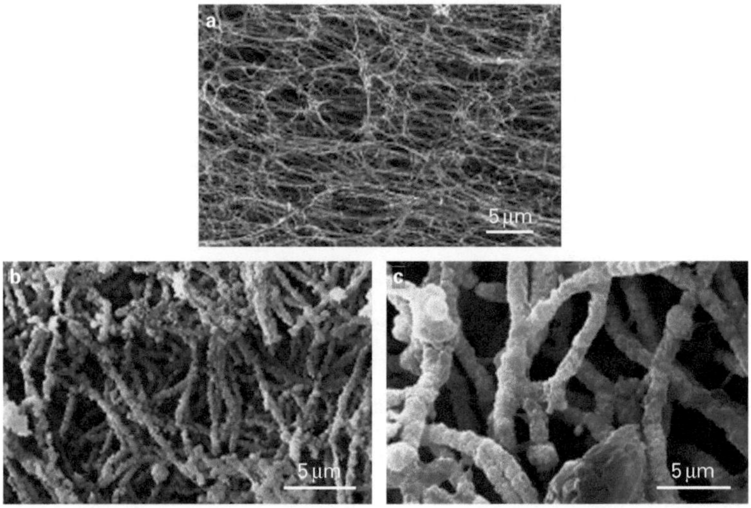

Fig. 4.9: Scanning electron micrographs of (a) freeze-dried phosphorylated bacterial cellulose and hydroxyapatite coated bacterial cellulose obtained after (b) 7 days and (c) 14 days immersion in a simulated body fluid (Wan et al., 2006).

4.6 Bacterial cellulose films

BC hydrogels can be dried to obtain solid films. Various methods, such as hot-pressing, critical point drying, freeze-drying, solvent exchange procedure from water to ethanol or acetone, and air-drying under normal or high pressure, have been reported in the literature. The drying process greatly influences the morphology and properties of the films. Mild drying conditions preserve the stability of the fibril network leading to the formation of highly porous aerogels (Figure 4.10(b)). These materials present low density and high re-swelling capacity. Alternatively, during pressure-induced drying the nanofibril network collapses resulting in high density films (Figure 4.10(a)). The mechanical properties of the latter are obviously improved. BC sheets can be obtained by sandwiching the native pellicle between stainless-steel meshes to squeeze out water followed by hot-pressing. The mechanical properties of BC films were found to correlate with the compression pressure applied for film formation (Retegi et al., 2010).

Fig. 4.10: Scanning electron micrographs of partial acetylated bacterial cellulose films with different degrees of substitution (DS): (a) original BC obtained by direct drying; (b) original BC obtained by freeze-drying; (c) DS 0.04; (d) DS 0.09; (e) DS 0.19; (f) DS 0.37; (g) DS 1.44; (h) DS 2.77 (Kim et al., 2002). Acetylated samples have been obtained by direct drying and the scale bar for all SEMs corresponds to 3 μm.

It was reported that extensive purification of BC via alkaline and oxidative agents improved the modulus of BC films (Yamanaka et al., 1989; Nishi et al., 1990). This was ascribed to the removal of impurities, such as protein, nucleic acid from the bacteria and media remaining after culture, promoting contact between individual cellulose nanofibrils and increasing hydrogen bonding in the network (Yamanaka et al., 1989). It was also observed that adding a cell division inhibitor or an organic reducing agent improved the modulus of BC sheet by artificially changing the form of cellulose-producing bacteria to various ribbon-shaped nanofibrils (Ishihara and Yamanaka, 2003). Orientation of cellulose fibrils was induced by deforming uniaxially never-dried BC fleeces and keeping under strain during the drying process (Bohn et al., 2000). Improved mechanical strength and modulus of ensuing films were reported. It was also observed that the stretchability and thus the orientability of the fibrils were strongly improved by loosening the inter-fibrillar H-bonds. This was achieved by drawing the samples in sodium hydroxide solutions with concentrations in the range 8–10 wt%.

BC was also partially acetylated by the fibrous acetylation method to decrease its hydrophilicity (Kim et al., 2002). Wet BC pellicles were squeezed in a polyethylene cloth and soaked in anhydrous acetic acid. The microfibrillar morphology and cellulose I structure were preserved up to moderate degrees of acetylation. It was observed that acetylation proceeded from the surface of the nanofibrils, leaving the core portion unreacted. Figure 4.10 shows the SEM observation of original BC films (obtained by direct drying, shown in panel (a), and freeze-drying, shown in panel (b)) and acetylated samples (direct drying, panels (c)–(h)). Because of the mode of drying (direct drying), densely coagulated films are observed up to DS 0.19 (Figures 4.10(c)–(e)). The nanofibrillar units became gradually visible from DS 0.37 (Figure 4.10(f)) and up, and are fully developed for the DS 2.77 sample (Figure 4.10(h)). This change is considered to result from increased hydrophobicity of the acetylated surfaces.

BC-based composites can be prepared by either in situ or post-modification. The in situ modification method consists in adding the polymer directly into culture medium, whereas in the post-modification method the BC gel is impregnated with the polymer solution. Hot-pressing is the following step to obtain solid composite films. Both techniques have been used to prepared PVA-based composites (Gea et al., 2010). Although the initial PVA concentration was similar, a maximum content of PVA of 1.3% was observed for the in situ formed composite compared to impregnated composites that achieved a content of 3.7% PVA. Compared to pure BC films, a decrease of the Young's modulus and increase in toughness was observed, especially for in situ grown samples. It has been ascribed to the partial disruption of hydrogen bonds naturally occurring in BC films. Enhanced transparency was also reported. No significant decrease of the mechanical properties was reported upon adding gelatinized starch into the culture medium with regard to pure BC films (Grande et al., 2009).

The potential of transparent resin reinforced with BC nanofibers for optically functional materials exhibiting low thermal expansion was reported (Yano et al., 2005; Nogi and Yano, 2008). Dried BC sheets were impregnated with different neat resins and either cured by ultraviolet light or hot-pressed. Due to the size effect, the nanofiber network led to a very low loss of transparency of the original resin, even at high fiber content, coupled with high mechanical strength and high flexibility (Figure 4.11(a)). The ensuing composites were found to be able to be folded or bent without damage (Nogi and Yano, 2008) as shown in Figures 4.11(b)–(c) and to have the potential to shift the electronics display industry from the traditional batch process to the much-anticipated more cost-effective continuous roll-to-roll manufacturing process. It was also shown that reinforcement with acetylated BC nanofibrils displays reduced moisture absorption and improved thermal degradation resistance regarding transparency (Nogi et al., 2006).

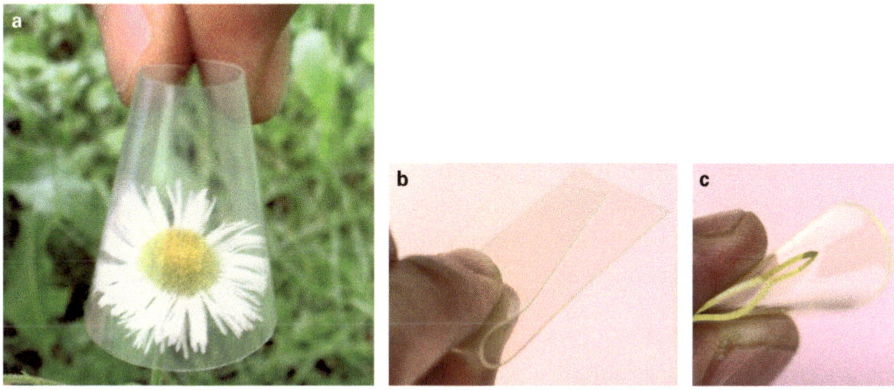

Fig. 4.11: (a) Flexibility and transparency of a 65 μm thick acrylic resin nanocomposite film reinforced with 60 wt% BC nanofibers (Yano et al., 2005); (b) foldable transparent 0.7 mm thick acrylic resin nanocomposite sheet reinforced with 5 wt% BC nanofibers; and (c) the more fragile neat acrylic resin sheet of the same thickness (Nogi and Yano, 2008).

Fibrils of BC can also act as a template for the in situ synthesis of ferrites for magnetization of the film (Sourty et al., 1998a; Sourty et al., 1998b) or for the insertion of nanoparticles. Ferrites were synthesized in situ in two different neutral cellulose gels: a never-dried bacterial cellulose membrane and a never-dried cast film using N-methylmorpholine-N-oxide as solvent (Sourty et al., 1998a). TEM micrographs showed the presence of ferrites in two different shapes, acicular and equiaxial, respectively hydrated ferric oxides (FeOOH) and the spinel oxides: maghemite (γ-Fe_2O_3) or magnetite (Fe_3O_4). Thin sections of BC showed these particles to be located along the cellulose microfibrils, which were assumed to provide a site for their nucleation. Room temperature magnetization curves showed all samples to be superparamagnetic. Impregnation of BC with silver nitrate and the subsequent reduction with water-soluble sodium borohydride, which forms silver metal as nanoparticles on the supporting BC fibrils, was reported (Maneerung et al., 2008). The unique structure and the high oxygen (ether and hydroxyl) density of BC fibrils were found to constitute an effective nanoreactor for in situ synthesis of silver nanoparticles. The freeze-dried silver nanoparticle-impregnated BC exhibited a strong antimicrobial activity against both *Staphylococcus aureus* (Gram-positive bacteria) and *Escherichia coli* (Gram-negative bacteria), which are general bacteria found on a contaminated wound. The addition of silica particles to the culture medium led to the formation of BC-nanosilica composites (Yano et al.,

2008). This method allowed incorporation of 50% silica in BC, whereas it was limited to 10% when immersing the BC hydrogel directly in different concentrations of silica sols to induce diffusion of the particles into the fibrillar network. It was reported that the treatment of BC fleeces with titanium tetraisopropoxide followed by its hydrolysis resulted in titania-coated cellulose nanofrils (Yamanaka et al., 2000). Upon removal of BC by heating up to 500°C, TiO_2 nanotubes were obtained.

4.7 Applications of bacterial cellulose

The main distinguishing features of BC over its plant counterpart are:
- Its finer and more intricate structure.
- Absence of foreign compound (hemicellulose or lignin) to be removed.
- Longer and higher tensile strength fibers (higher crystallinity).
- It can be grown on a variety of substrates and to virtually any shape.
- The quality of the pellicle can be tailored by changing the nature of the culture medium used and the strain of *A. xylinum*. Cellulose can be directly modified during assembly (delayed crystallization by introduction of dyes into culture medium, control of molecular weight and crystallinity of the cellulose during assembly).
- Possibility to orientate cellulose during synthesis to obtain uniaxially strengthened membranes.
- Possibility to genetically modify the cellulose product (direct synthesis of cellulose derivatives, such as cellulose acetate, carboxymethylcellulose, methyl cellulose, etc. ...., control of cellulose crystalline allomorph (cellulose I or cellulose II), control of the molecular weight).
- Higher absorbency in the hydrated state leading to higher capacity to hold water (selective porosity, higher wet strength and surface-to-volume carrier capacity).

The principal disadvantages for a more intensive large scale commercial use are:
- Its high price, resulting from the high price of the substrate and low volumetric yields, which is roughly 100 times higher than plant cellulose.
- Lack of large scale production capacity.
- The timely expansion and maintenance of the cell culture for production.

Few industries currently use BC, but a deeper knowledge of this material and exploitation of its properties could change this trend. Its high purity and unusual physicochemical properties offer a wide range of special applications. A few representative industry sectors and potential products are described below.

The Philippine dessert, nata de coco, is the most well-known application in the food industry. The unique gel-like properties of BC can be exploited for thickeners in ice cream and salad dressing. Combined with its complete indigestibility in the human intestinal tract, it also makes it an attractive food base for diet foods, such as low calorie ice creams, snack and candies. Kombucha elixir or Manchurian tea is a fermented product which is consumed by a growing number of individuals for improved health needs. *Acetobacter* along with yeasts are cultured in a medium containing tea extract and sugar. The cellulose is not consumed in this product and the fermentation extract is utilized.

BC can be used for audio speaker diaphragm products. Acoustic membrane made from BC in high-end earphones was exploited by Sony (Nishi et al., 1990). The requirements for this application are high values of the Young's modulus, sound propagation velocity and relatively high internal loss. Traditionally, diaphragms of loudspeakers are made from paper (so-called cone paper) because it can be processed into lightweight diaphragm. The unique dimensional stability of microbial cellulose gives rise to a sound transducing membrane which maintains higher sonic velocity over wide frequency ranges, thus being the best material to meet the rigid requirements for optimal sound transduction. Moreover, higher modulus values and comparable high values of the internal loss were reported for BC sheets compared to cone paper.

The pulp and paper industry is another area of potential applications for BC. BC has been investigated as a binder in papers and because it consists of extremely small clusters of cellulose microfibrils this property greatly adds to strength and durability of pulp when integrated into paper. Archival document repair and high strength specialty paper, such as paper base for long-lived currency, have been suggested.

Other non-food applications include cosmetic products (skin cream, astringents, base for artificial nails, thickener and strengthener for fingernail polish), oil spill cleanup sponge, absorptive base for toxic material removal, mineral and oil recovery, and water purification via ultrafilters and reverse osmosis membranes.

A promising area of applications is healthcare. Indeed, BC is biocompatible and non-toxic making it a good candidate material for medical applications (Klemm et al., 2001; Gatenholm and Klemm, 2010; Klemm et al., 2011). Potential applications include wound dressings, and scaffolds for tissue engineering, soft tissue replacement and artificial blood vessels, but also drug delivery agents and artificial skin substrate. BC was fabricated as an in vitro scaffold and implanted as a substitute for a small diameter artery (Schumann et al., 2009).

4.8 Conclusions

Can bacterial or microbial cellulose compete with traditional cellulose sources? This question remains unanswered until commercial scale up and fermentation development become mature. It is a unique family of nanocellulose and it offers some specific features compared to cellulose nanoparticles extracted from the biomass. Some specific or niche applications that cannot be reached with more conventional nanocellulosic materials extracted plant-based materials can be considered. However, until now the most restrictive drawback is related to the current production costs.

4.9 References

Aschner, M. and Hestrin, S. (1946). Fibrillar structure of cellulose of bacterial and animal origin. Nature *157*, 659.

Bae, S., Sugano, Y. and Shoda, M. (2004). Improvement of bacterial cellulose production by addition of agar in a jar fermentor. J. Biosci. Bioeng. *97*, 33–38.

Bae, S. and Shoda, M. (2005). Statistical optimization of culture conditions for bacterial cellulose production using Box-Behnken design. Biotechnol. Bioeng. *90*, 20–28.

Basta, A.H. and El-Saied, H. (2009). Performance of improved bacterial cellulose application in the production of functional paper. J. Appl. Microbiol. *107*, 2098–2107.

Bohn, A., Fink, H.P., Ganster, J. and Pinnow, M. (2000). X-ray texture investigations of bacterial cellulose. Macromol. Chem. Phys. *201*, 1913–1921.

Borzani, W. and de Souza, S. (1995). Mechanism of the film thickness increasing during the bacterial production of cellulose on non-agitated liquid media. Biotechnol. Lett. *17*, 1271–1272.

Brown, A.J. (1886). On acetic ferment which forms cellulose. J. Chem. Soc. *49*, 432–439.

Brown, E.E. and Laborie, M.P.G. (2007). Bioengineering bacterial cellulose/poly(ethylene oxide) nanocomposites. Biomacromolecules *8*, 3074–3081.

Brown, R.M.Jr., Willison, J.H.M. and Richardson, C.L. (1976). Cellulose biosynthesis in *Acetobacter xylinum*: 1. Visualization of the site of synthesis and direct measurement of the *in vivo* process. Proc. Nat. Acad. Sci. USA *73*, 4565–4569.

Canale-Parol, E. and Wolfe, R.S. (1960). Studies on *Sarcina Ventricula* I. Stock culture, J. Bacteriol. *79*, 857–862.

Cannon, R.E. and Anderson, S.M. (1991). Biogenesis of bacterial cellulose. Crit. Rev. Microbiol. *17*, 435–447.

Cheng, K.C., Catchmark, J.M. and Demirci, A. (2009). Effect of different additives on bacterial cellulose production by *Acetobacter xylinum* and analysis of material property. Cellulose *16*, 1033–1045.

Ciechanska, D. (2004). Multifunctional bacterial cellulose/chitosan composite materials for medical applications. Fibres Text. East. Eur. *12*, 69–72.

Czaja, W., Romanovicz, D. and Brown, R.M. Jr. (2004). Structural investigations of microbial cellulose produced in stationary and agitated culture. Cellulose *11*, 403–411.

Dudman, W.F. (1959). Cellulose production by *Acetobacter-acetigenum* in defined medium. J. Gen. Microbiol. *21*, 327–337.

Dudman, W. F. (1960). Cellulose production by *Acetobacter* strain in submerged culture. J. Gen. Microbiol. *22*, 25–39.

Embuscado, M.E., De Miller, J.N. and Marks, J.S. (1996). Isolation and partial characterization of cellulose produced by *Acetobacter xylinum*. Food Hydrocolloids *10*, 75–82.

Forge, A. and Preston, R.H. (1977). An Electron Microscope Examination of *Acetobacter xylinum* Showing the Ultrastructure of Cells and the Association Cellulose Microfibrils. Ann. Bot. *41*, 437–446.

Gatenholm, P. and Klemm, D. (2010). Bacterial nanocellulose as a renewable material for biomedical applications. MRS Bull. *35*, 208–213.

Gea, S., Bilotti, E., Reynolds, C.T., Soykeabkeaw, N. and Peijs, T. (2010). Bacterial cellulose-poly(vinyl alcohol) nanocomposites prepared by an *in-situ* process. Mater. Lett. *64*, 901–904.

Geyer, U., Klemm, D. and Schmauder, H.P. (1994). Kinetics of the utilization of different C sources and the cellulose formation by *Acetobacter xylinum*. Acta Biotechnol. *14*, 261–266.

Grande, C.J., Torres, F.G., Gomez, C.M., Troncoso, O.P., Canet-Ferrer, J. and Martinez-Pastor, J. (2009). Development of self-assembled bacterial cellulose-starch nanocomposites. Mat. Sci. Eng. C *29*, 1098–1104.

Haigler, C.H., White, A.R., Brown, R.M.J. and Cooper, K.M. (1982). Alteration of *in vivo* cellulose ribbon assembly by carboxymethylcellulose and other cellulose derivatives. J. Cell Biol. *94*, 64–69.

Hestrin, S. and Schramm, M. (1954). Synthesis of cellulose by Acetobacter xylinum. 2. Preparation of freeze-dried cells capable of polymerizing glucose to cellulose. Biochem. J. *58*, 345–352.

Hestrin, S. (1963). Bacterial cellulose. Meth. Carbohydr. Chem. *3*, 4–9.

Hessler, N. and Klemm, D. (2009). Alteration of bacterial nanocellulose structure by *in situ* modification using polyethylene glycol and carbohydrate additives. Cellulose *16*, 899–910.

Hirai, A., Tsuji, M., Yamamoto, H. and Horii, F. (1998). *In situ* crystallization of bacterial cellulose III. Influences of different polymeric additives on the formation of microfibrils as revealed by transmission electron microscopy. Cellulose *5*, 201–213.

Hong, L., Wang, Y.L., Jia, S.R., Huang, Y., Gao, C. and Wan, Y.Z. (2006). Hydroxyapatite/bacterial cellulose composites synthesized via a biomimetic route. Mater. Lett. *60*, 1710–1713.

Hu, Y. and Catchmark, J.M. (2010). Influence of 1-methylcyclopropene (1-MCP) on the production of bacterial cellulose biosynthesized by *Acetobacter xylinum* under the agitated culture. Lett. Appl. Microbiol. *51*, 109–113.

Huang, H.C., Chen, L.C., Lin, S.B., Hsu, C.P. and Chen, H.H. (2010). In situ modification of bacterial cellulose network structure by adding interfering substances during fermentation. Bioresour. Technol. *101*, 6084–6091.

Huang, H.C., Chen, L.C., Lin, S.B. and Chen, H.H. (2011). Nano-biomaterials application: *In situ* modification of bacterial cellulose structure by adding HPMC during fermentation. Carbohydr. Polym. *83*, 979–987.

Hwang, J.W., Yang, Y.K., Hwang, J.K., Pyun, Y.R. and Kim, Y.S. (1999). Effects of pH and dissolved oxygen on cellulose production by *Acetobacter xylinum* BRC5 in agitated culture. J. Biosci. Bioeng. *88*, 183–188.

Iguchi, M., Yamanaka, S. and Budhiono, A. (2000). Review. Bacterial cellulose – A masterpiece of nature's arts. J. Mater. Sci. *35*, 261–270.

Ishihara, M. and Yamanaka, S. (2003). Modified bacterial cellulose. US Patent 662419 B2.

Ishikawa, A., Masuoka, M., Tsuchida, T. and Yoshinaga, F. (1995). Increase in cellulose production by sulfa guanidine-resistant mutants derived from *Acetobacter xylinum* subsp. *sucrofermentans*. Biosci. Biotech. Biochem. *59*, 2259–2262.

Johnson, D.C. and Neogi, A.N. (1989). Sheeted products formed from reticulated microbial cellulose. US Patent 4,863,565.

Jonas, R. and Farah, L.F. (1998). Production and application of microbial cellulose. Polym. Degrad. Stab. *59*, 101–106.

Keshk, S. and Sameshima, K. (2006). Influence of lignosulfonate on crystal structure and productivity of bacterial cellulose in a static culture. Enzyme Microb. Technol. *40*, 4–8.

Kim, D.Y., Nishiyama, Y. and Kuga, S. (2002). Surface acetylation of bacterial cellulose. Cellulose *9*, 361–367.

Klemm, D., Schumann, D., Udhardt, U. and Marsch, S. (2001). Bacterial synthesized cellulose – Artificial blood vessels for microsurgery. Prog. Polym. Sci. *26*, 1561–1603.

Klemm, D., Heublein, B., Fink, H.P. and Bohn, A. (2005). Cellulose: Fascinating biopolymer and sustainable raw material. Angew. Chem. Int. Ed. *44*, 3358–3393.

Klemm, D., Kramer, F., Moritz, S., Lindström, T., Ankerfors, M., Gray, D. and Dorris, A. (2011). Nanocelluloses: a new familly of nature-based materials. Angew. Chem. Int. Ed. *50*, 5438–5466.

Krystynowicz, A., Czaja, W., Wiktorowska-Jezierska, A., Goncalves-Miskiewicz, M., Turkiewicz, M. and Bielecki, S. (2002). Factors affecting the yield and properties of bacterial cellulose. J. Ind. Microbiol. Biot. *29*, 189–195.

Maneerung, T., Tokura, S. and Rujiravanit, R. (2008). Impregnation of silver nanoparticles into bacterial cellulose for antimicrobial wound dressing. Carbohydr. Polym. *72*, 43–51.

Masaoka, S., Ohe, T. and Sakota, N. (1993). Production of cellulose from glucose by *Acetobacter xylinum*. J. Ferment. Bioeng. *75*, 18–22.

Matsuoka, M., Tsuchida, T., Matsushita, K., Adachi, O. and Yoshinaga, F. (1996). A synthetic medium for bacterial cellulose production by *Acetobacter xylinum* subsp sucrofermentans. Biosci. Biotechnol. Biochem. *60*, 575–579.

Matthysse, A.G., Thomas, D. and White, A.R. (1995). Mechanisms of cellulose synthesis in *Agrobacterium tumefaciens*. J. Bacteriol. *177*, 1076–1081.

Millon, L.E., Mohammadi and H., Wan, W.K. (2006). Anisotropic polyvinyl alcohol hydrogel for cardio-vascular applications. J. Biomed. Mater. Res. B *79*, 305–311.

Napoli, C., Dazzo, F. and Hubbell, D. (1975). Production of cellulose microfibrils by *Rhizobium*. Appl. Microbiol. *30*, 123–131.

Naritomi, T., Kouda, T., Yano, H. and Yoshinaga, F. (1998a). Effect of ethanol on bacterial cellulose production from fructose in continuous culture. J. Ferment. Bioeng. *85*, 598–603.

Naritomi, T., Kouda, T., Yano, H. and Yoshinaga, F. (1998b). Effect of lactate on bacterial cellulose production from fructose in continuous culture. J. Ferment. Bioeng. *85*, 89–95.

Nishi, Y., Uryu, M., Yamanaka, S., Watanabe, K., Kitamura, N, Iguchi, M. and Mitsuhashi, S. (1990). The structure and mechanical properties of sheets prepared from bacterial cellulose. J. Mat. Sci. *25*, 2997–3001.

Nogi, M., Abe, K., Handa, K., Nakatsubo, F., Ifuku, S. and Yano, H. (2006). Property enhancement of optically transparent bionanofiber composites by acetylation. Appl. Phys. Lett. *89*, 233123-1–233123-3.

Nogi, M. and Yano, H. (2008). Transparent nanocomposites based on cellulose produced bt bacteria offer potential innovation in the electronics device industry. Adv. Mat. *20*, 1849–1852.

Park, J.K., Jung, J.Y. and Park, Y.H. (2003). Cellulose production by *Gluconacetobacter hansenii* in a medium containing ethanol. Biotechnol. Lett. *25*, 2055–2059.

Oikawa, T., Ohtori, T. and Ameyanea, M. (1995a). Production of cellulose from D-mannitol by *Acetobacter xylinum* KU-1. Biosci. Biotech. Biochem. *59*, 331–332.

Oikawa, T., Morino, T. and Ameyanea, M. (1995b). Production of cellulose from D-arabitol by *Acetobacter xylinum* KU-1. Biosci. Biotech. Biochem. *59*, 1564–1565.

Retegi, A., Gabilondo, N., Peña, C., Zuluaga, R., Castro, C., Gañan, P., de la Caba, K. and Mondragon, I. (2010). Bacterial cellulose films with controlled microstructure-mechanical property relationships. Cellulose *17*, 661–669.

Ring, G.J.F. (1982). Study of polymerization kinetics of bacterial cellulose through gel permeation chromatography. In: Cellulose and other natural polymer systems, R.M. Brown, ed. (Plenum, New York), pp. 299–325.

Ross, P., Mayer, R. and Benziman, M. (1991). Cellulose biosynthesis and function in bacteria. Microbiol. Mol. Biol. Rev. *55*, 35–58.

Schumann, D.A., Wippermann, J., Klemm, D.O., Kramer, F., Koth, D., Kosmehl, H., Wahlers, T. and Salehi-Gelani, S. (2009). Artificial vascular implants from bacterial cellulose: Preliminary results of small arterial substitutes. Cellulose *16*, 877–885.

Seifert, M., Hesse, S., Kabrelian, V. and Klemm, D. (2003). Controlling the water content of never dried and reswollen bacterial cellulose by the addition of water-soluble polymers to the culture medium. J. Polym. Sci. A *42*, 463–470.

Son, H.J., Heo, M.S., Kim, Y.G. and Lee, S.J. (2001). Optimization of fermentation conditions for the production of bacterial cellulose by a newly isolated Acetobacter sp A9 in shaking cultures. Biotechnol. Appl. Biochem. *33*, 1–5.

Sourty, E., Ryan, D.H. and Marchessault, R.H. (1998a). Characterization of magnetic membranes based on bacterial and man-made cellulose. Cellulose *5*, 5–17.

Sourty, E., Ryan, D.H. and Marchessault, R.H. (1998b). Ferrite-loaded membranes of microfibrillar bacterial cellulose prepared by *in situ* precipitation. Chem. Mater. *10*, 1755–1757.

Spiers, A.J., Bohannon, J., Gehrig, S.M. and Rainey, P.B. (2003). Biofilm formation at the air-liquid interface by the *Pseudomonas fluorescens* SBW25 wrinkly spreader requires an acetylated form of cellulose. Mol. Microbiol. *50*, 15–27.

Tal, R., Wong, H.C., Calhoon, R., Gelfand, D., Fear, A., Volman, G., Mayer, R., Ross, P., Amikam, D., Weinhouse, H., Cohen, A., Sapir, S., Ohana, P. and Benziman, M. (1998). Three cdg operons control cellular turnover of cyclic di-GMP in *Acetobacter xylinum*: Genetic organization and occurrence of conserved domains in isoenzymes. J. Bacteriol. *180*, 4416–4425.

Tantratian, S., Tammarate, P., Krusong, W., Bhattarakosol, P. and Phunsri, A. (2005). Effect of dissolved oxygen on cellulose production by *Acetobacter* sp. J. Sci. Res. Chulalongkorn Univ. *30*, 179–186.

Tarr, H.L.A. and Hibbert, H. (1931). Studies on reactions relating to carbohydrates and polysaccharides. XXXV. Polysaccharide synthesis by the action of *Acetobacter xylinus* on carbohydrates and related compounds. Can. J. Res. *4*, 372–388.

Tokoh, C., Takabe, K., Fujita, M. and Saiki, H. (1998). Cellulose synthesized by *Acetobacter xylinum* in the presence of acetyl glucomannan. Cellulose *5*, 249–261.

Tonouchi, N., Tsuchida, T., Yoshinaga, F., Beppu, T. and Horinouchi, S. (1996). Characterization of the biosynthetic pathway of cellulose from glucose and fructose from *Acetobacter xylinium*. Biosci. Biotech. Biochem. *60*, 1377–1379.

Uhlin, K.I., Atalla, R.H. and Thompson, N.S. (1995). Influence of hemicelluloses on the aggregation patterns of bacterial cellulose. Cellulose *2*, 129–144.

Vandamme, E.J., De Baets, S., Vanbaelen, A., Joris, K. and De Wulf, P. (1998). Improved production of bacterial cellulose and its application potential. Polym. Degrad. Stab. *59*, 93–99.

Wan, Y.Z., Hong, L., Jia, S.R., Huang, Y., Zhu, Y., Wang, Y.L. and Jiang, H.J. (2006). Synthesis and characterization of hydroxyapatite/bacterial cellulose nanocomposites. Compos. Sci. Technol. *66*, 1825–1832.

Watanabe, K. and Yamanaka, S. (1995). Effects of oxygen tension in the gaseous phase on production and physical properties of bacterial cellulose formed under static culture conditions. Biosci. Biotech. Biochem. *59*, 65–68.

Watanabe, K., Tabuchi, M., Morinaga, Y. and Yoshinaga, F. (1998). Structural features and properties of bacterial cellulose produced in agitated culture. Cellulose *5*, 187–200.

Williams, S. and Cannon, R. (1989). Alternative environmental roles for cellulose produced by *Acetobacter xylinum*. Appl. Environ. Microbiol. *55*, 2448–2452.

Yamamoto, H., Horii, F. and Hirai, A. (1996). *In situ* crystallization of bacterial cellulose. II. Influences of different polymeric additives on the formation of celluloses I_a and I_b at the early stage of incubation. Cellulose *3*, 229–242.

Yamanaka, S., Watanabe, K. and Kitamura, N. (1989). The structure and mechanical properties of sheets prepared from bacterial cellulose. J. Mat. Sci. *24*, 3141–3145.

Yamanaka, S., Ishihara, M. and Sugiyama, J. (2000). Structural modification of bacterial cellulose. Cellulose *7*, 213–225.

Yano, H., Sugiyama, J., Nakagaito, A.N., Nogi, M., Matsuura, T., Hikita, M. and Handa, K. (2005). Optically transparent composites reinforced with networks of bacterial nanofibers. Adv. Mat. *17*, 153–155.

Yano, S., Maeda, M., Nakajima, M., Hagiwara, T. and Sawaguchi, T. (2008). Preparation and mechanical properties of bacterial cellulose nanocomposites loaded with silica nanoparticles. Cellulose *15*, 111–120.

Yasuda, K., Gong, J.P., Katsuyama, Y., Nakayama, A., Tanabe, Y., Kondo, E., Ueno, M. and Osada, Y. (2005). Biomechanical properties of high-toughness double network hydrogels. Biomaterials *26*, 4468–4475.

Yoshino, T., Asakura, T. and Toda, K. (1996). Cellulose production by *Acetobacter pasteurianus* on silicone membrane. J. Ferment. Bioeng. *81*, 32–36.

Zaar, K. (1977). The biogenesis of cellulose by *Acetobacter Xylinum*. Eur. J. Cell Biol. *16*, 1–15.

Zaar, K. (1979). Visualisation of pores (export sites) correlated with cellulose production in the envelope of the gram-negative bacterium *Acetobacter xylinum*. J. Cell Biol. *80*, 773–777.

Zhou, L.L., Sun, D.P., Hu, L.Y., Li, Y.W. and Yang, J.Z. (2007). Effect of addition of sodium alginate on bacterial cellulose production by *Acetobacter xylinum*. J. Ind. Microbiol. Biot. *34*, 483–489.

5 Chemical modification of nanocellulose

Extensive studies have been conducted on the utilization of cellulose nanocrystals, microfibrillated cellulose and bacterial cellulose as reinforcements in polymer nanocomposites. Efficiency of nanofiller dispersion in the matrix and adequacy of nanofiller-matrix interfacial interactions are widely known to critically affect material physical and mechanical properties. Because of its inherent hydrophilicity, it is difficult to achieve a good dispersion level of nanocellulose reinforcements in water-insoluble or non-water dispersible polymer matrices. This limitation stems primarily from the high affinity of nanocellulose for water and its inability to disperse readily in organic solvents. Cellulosic nanoparticles are characterized by hydrogen bonding-induced aggregation which is magnified because of their nanosized and fibrillar structures. To prevent self-aggregation (sometimes referred to as hornification) and promote efficient dispersion in non-aqueous media, the surface of cellulosic nanoparticles can be modified with hydrophobic compounds using covalent and non-covalent coupling techniques. A wide variety of chemical modification techniques, including coupling hydrophobic small molecules, grafting polymers and oligomers, and adsorbing hydrophobic compounds to surface hydroxyl groups of cellulosic nanoparticles, can be employed. Besides the problem of dispersibility with polymeric matrices, improved nanofiller-matrix interaction is expected to enhance the stress transfer from the matrix to the dispersed phase and then improve the load-bearing capability of the material. Moreover, the surface of the nanoparticles can be tailored to impart specific functionality to nanocellulose.

5.1 Reactivity of cellulose

According to its molecular structure, cellulose is an active chemical because of the presence of three hydroxyl groups in each anhydroglucose unit. These hydroxyl groups are therefore mainly responsible for the reactions of cellulose. As a carbohydrate, the chemistry of cellulose is primarily the chemistry of alcohols and it forms many of the common derivatives of alcohols, such as esters, ethers, etc. However, despite the fact that cellulose has a chemical similarity with sugars bearing three hydroxyl groups its reactivity should not be simply regarded as that of a trihydric alcohol. Indeed, native cellulose is a high molecular weight polymeric substance. Moreover, if the soluble form of cellulose is not considered, i.e. if the solid cellulosic substrate is suspended in a non-swelling liquid reaction medium, the reactions of cellulose occur under heterogeneous conditions.

Assuming the numbering system shown in Figure 5.1 for carbon atoms in anhydroglucose unit of cellulose, the hydroxyl group at the 6 position acts as a primary alcohol whereas the hydroxyl groups in the 2 and 3 positions behave as secondary alcohols. Indeed, it can be seen from Figure 5.1 that the carbon atom which carries the hydroxyl group in the 6 position is only attached to one alkyl group, while the carbons with the hydroxyl groups in the 2 and 3 positions are joined directly to two alkyl groups. Under heterogeneous conditions, the reactivity of these three hydroxyl groups can be affected by their inherent chemical reactivity, by steric effects produced by the reacting agent and by steric effects derived from the supramolecular structure of cellulose (Wakelyn, 1998). The relative reactivity has been established from esterification studies. It has been reported that the hydroxyl group at the 6 position can react ten times faster than the other OH groups (Hebeish and Guthrie, 1981). This high reactivity is generally ascribed to its isomerization. The reactivity of the hydroxyl group in the 2 position has, in the meantime, been found to be twice that of one of the hydroxyl group at the 3 position.

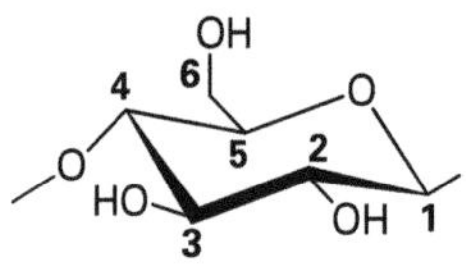

Fig. 5.1: The numbering system for carbon atoms in anhydroglucose unit of cellulose.

The hydroxyl groups of cellulose can be partially or fully reacted with various chemicals to provide derivatives with useful properties. These derivatives form the basis for much of the industrial technology of cellulose in use today. Because of the strong hydrogen bonds that occur between cellulose chains, cellulose does not melt or dissolve in common solvents. Thus, it is difficult to convert the short fibers from wood pulp into the continuous filaments needed for artificial silk, an early goal of cellulose chemistry. Several different cellulose derivatives were examined as early routes to artificial silk, but only two, the acetate and xanthate esters, are of commercial importance for fibers today.

An important parameter to characterize the extent or efficiency of reaction of cellulose is the degree of substitution (DS). It indicates the average number of hydroxyl groups of anhydroglucose moiety that are substituted after reaction. The DS value ranges between 0 and 3 and the maximum value of 3 indicates that all three hydroxyl groups are substituted. For composite applications, a non-swelling reaction medium should be used so that the reaction is restricted to the surface of the fiber/nanoparticle keeping its integrity. Therefore, rather than expressing the grafting efficiency referring to the number of grafted moieties per anhydroglucose unit regardless of their localization in the fiber/nanoparticle, it is preferable to refer to the fraction of substituted hydroxyl groups among those that are able to react (the OH groups at the surface).

The fraction of accessible hydroxyl groups is not easy to determine but can be roughly estimated from the specific surface area of the fiber/nanoparticle. For microscopic cellulosic fibers, it was shown that only 2% of hydroxyl groups are accessible at the surface for chemical modification (Trejo-O'Reilly et al., 1997). This calculation was based on the morphological characteristics of the fibers and taking into account that swelling was negligible in such non-swelling solvents. The determination of the specific surface area and then accessibility and reactivity of MFC and cellulose nanocrystals is more difficult to access by classical techniques because drying induces aggregation phenomenon of the nanoparticles leading to an underestimated value of the specific surface area as reported in Chapters 2 and 3. Therefore, geometrical considerations obtained from microscopic observations are generally used (Siqueira et al., 2009). The number of cellulose chains close packed to form MFC or nanocrystals can then be estimated from crystallographic data for native cellulose (Gardner and Blackwell, 1974). For instance, it was shown that cellulose nanocrystals extracted from wheat straw (Helbert et al., 1996) or sisal (Siqueira et al., 2009) both with a width around 5 nm consist of close packing of about 40 chains. For sisal MFC with a width of 52 nm, the number of close packed cellulose chains was estimated around 4,000 (Siqueira et al., 2009). In the case of tunicin nanocrystals with a square section equal to $15 \cdot 15\ nm^2$, it was estimated that 344 unit cells and 688 chains were packed, since there are two chains per unit cell (Favier et al., 1997). Assuming a square cross section for cellulosic nanoparticles, the number and fraction of cellulosic chains laying on the surface can be estimated. However, even for these surface cellulosic chains, not all hydroxyl groups are accessible, since some are oriented toward the inner of the nanoparticle. It was assumed that only one-third and one-half of hydroxyl groups from cellulosic chains at the surface of nanocrystals and MFC, respectively, can react (Siqueira et al., 2009). Indeed, due to higher crystallinity of nanocrystals, and knowing that primary hydroxyl on C6 is the most reactive, it was assumed that the fraction of hydroxyl groups available on the surface was 1/3 for nanocrystals and 1/2 for MFC. Therefore, the DS experimentally determined should be divided by the fraction of cellulose chain at the surface of the nanoparticles and by a factor 3 for nanocrystals and by a factor 2 for MFC, to refer to the fraction of substituted hydroxyl groups that are able to react (the OH groups at the surface).

A higher reactivity of cellulosic fibers and DS with a more homogeneous substituent distribution can be obtained using different activation treatments (Roy et al., 2009). It includes (i) opening surface cannulae, internal pores and cavities, and interfibrillar interstices, (ii) disrupting fibrillar aggregations, in order to make available additional surface, (iii) disturbing the crystalline order, and (iv) altering the crystal modification, thus changing the hydrogen bonding scheme and the relative availability of reactive hydroxyl groups (Krassig, 1985; Krassig, 1993). The most frequently used activation method consists in swelling cellulose in solutions of acids, bases, salts and some organic solvents (Roy et al., 2009).

5.2 Surface chemistry of cellulose nanoparticles

Because of their nanoscale dimensions, cellulose nanoparticles display a high surface area generally of the order of few 100 $m^2 \cdot g^{-1}$ and their ample surface hydroxyl groups (2–3 $mmol \cdot g^{-1}$) enable targeted surface modification to introduce virtually any desired surface functionality (Eyley and Thielemans, 2011). However, the surface chemistry of cellulose nanoparticles is primary governed by the extraction procedure used to prepare these nanoparticles from the native cellulosic substrate (Moon et al., 2011). Figure 5.2 shows the surface chemistries provided by the most common extraction methods.

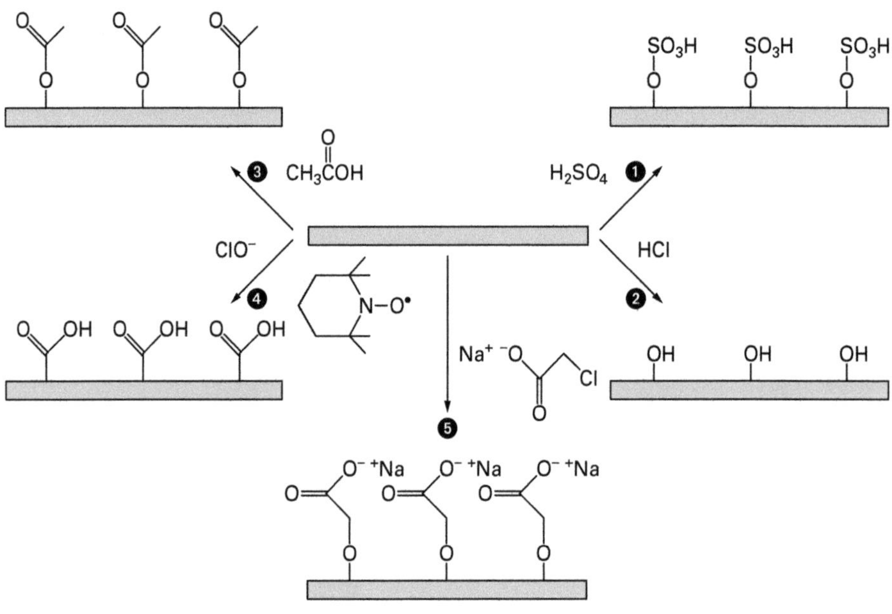

Fig. 5.2: Distinctive surface chemistries provided by the most common extraction methods of cellulose nanoparticles: sulfuric acid hydrolysis provides sulfate ester groups (route ❶), hydrochloric acid hydrolysis provides hydroxyl groups (route ❷), acetic acid hydrolysis provides acetyl groups (route ❸), TEMPO mediated hypochlorite treatment (route ❹) and carboxymethylation (route ❺) provide carboxylic acid groups.

Cellulose nanocrystals are generally prepared by sulfuric acid hydrolysis inducing the formation of sulfate ester groups on the surface of the nanoparticles. These charged moieties, whose density increases with the severity (acid concentration, time, temperature) of the treatment, are responsible for the high stability of the H_2SO_4-prepared nanocrystal aqueous suspensions. Hydrolysis of cellulose with hydrochloric acid preserves the hydroxyl groups of the native fibers but leads to less stable suspensions. More marginally, acetic acid mixed with hydrochloric acid was used to prepare functionalized cellulose nanocrystals in a single step (Braun and Dorgan, 2009). Fischer esterification of hydroxyl groups induced simultaneously the hydrolysis of amorphous cellulose chains and isolation of functionalized nanoparticles with an acetylated surface.

MFC is generally prepared with purely mechanical methods. Hydroxylated surfaces similar to native cellulose are therefore obtained. However, as reported in Chapter 2, several strategies have been proposed to decrease the energy consumption for fibrillation. One of these methods, increasingly used, is TEMPO (2,2,6,6-tetramethylpiperidine-1-oxyl radical)-mediated oxidation of cellulose coupled with mild mechanical disintegration treatment (Isogai et al., 2011). The primary oxidant, such as hypochlorite, selectively oxidizes primary alcohol groups of cellulose. This oxidation pretreatment allows leaving carboxylic acid groups at the surface of fibrillated cellulose. Carboxymethylation was also used as pretreatment prior to mechanical defibrillation to decrease the energy associated with the process (Wågberg et al., 2008). The ensuing carboxymethylated MFC material also bear carboxyl groups. Water-redispersible MFC in powder form was also prepared from refined, bleached beech pulp by carboxymethylation and mechanical disintegration (Eyholzer et al., 2010). The sequence of the treatments was changed and it was observed that samples that were first carboxymethylated formed a more transparent suspension in water and showed a more homogeneous network.

It has been shown that when cotton cellulose nanocrystals were chemically modified directly after acid hydrolysis followed by dialysis, reproducibility between batches was problematic (Labet and Thielemans, 2011). It was suspected to result from a variable surface composition. Given the inherent purity of acid hydrolysis extracted nanoparticles, this issue was believed to be due to the presence of adsorbed species at the surface of the nanocrystals blocking reactive sites. Soxhlet extraction with different solvents of cellulose nanocrystals from several batches was performed and extracted impurities were analyzed by nuclear magnetic resonance (^{1}H and ^{13}C NMR) and mass spectrometries. Dichloromethane, ethanol, ethyl acetate, heptane, methanol and isopropanol were used as solvents and it was shown that a variety of adsorbed species can be removed by Soxhlet extraction. It was concluded from this study that a 24 h ethanol extraction is required and sufficient for the purification of cellulose nanocrystals and this extraction was recommended to be performed before any surface modification attempt, from a simple surface esterification to grafting polymers from the surface.

The hydroxyl groups from the cellulose nanoparticle surface need to be fully accessible for chemical reactions. This means that their surface should remain clean and that aggregation of the nanoparticles should be avoided. A medium-driven surface adaptation phenomenon of cellulose to minimize its surface free energy was evidenced using MFC (Johansson et al., 2011). Changes of surface properties, nanoscale structure and surface accessibility subsequent to solvent exchanges or drying in air were investigated. In this study, aqueous MFC suspensions were solvent exchanged to acetone and then to either toluene or dimethyl acetamide (DMAc). Due to the amphiphilic character of both DMAc and cellulose, MFC remained well dispersed in DMAc while extensive aggregation of the nanoparticles and loss of both the reactivity and the nanostructure occurred when solvent exchanging to non-polar toluene. Passivation of the cellulose surface when exposed to non-hydrophilic media was evidenced. Actually, in water, an excellent hydrogen bonding solvent, the surface hydroxyl groups stay clean and accessible whereas upon drying, MFC aggregates and accumulates air-borne contaminants to adapt to the non-polar medium. These contaminants were evidenced through the C—C carbon peak in X-ray photoelectron spectroscopy (XPS) data and contact angle measurements. As a result, it was shown that silylation conducted in cellulose compatible DMAc gave a significantly higher DS compared to the silylation conducted in toluene.

5.3 Non-covalent surface chemical modification of cellulose nanoparticles

The surface of cellulose nanoparticles can be modified and tuned by physical interactions or adsorption of molecules or macromolecules onto their surface. This method appears much simpler than chemical grafting since no chemical reaction is involved.

5.3.1 Adsorption of surfactant

The surface of cellulose nanoparticles can be tuned for instance by using surfactants as stabilizing agents. Surfactants are usually amphiphilic organic compounds, i.e. compounds containing both hydrophobic groups (so-called tails) and hydrophilic groups (so-called heads). Therefore, a surfactant molecule contains both a water insoluble (or oil soluble) component and a water soluble component. The hydrophilic end of the surfactant molecule may adsorb on the surface of cellulose nanoparticles whereas the hydrophobic end may extend out providing a non-polar surface and lowering the surface tension of the nanoparticle. The ensuing surfactant-coated nanoparticles can be easily dispersed in non-polar organic media.

An anionic surfactant consisting of acid phosphate ester of ethoxylated nonylphenol has been applied to cotton and tunicin nanocrystals (Heux et al., 2000), as well as nanocrystals extracted from MCC (Kvien et al., 2005; Bondeson and Oksman, 2007; Fortunati et al., 2012) and cotton linters (Elazzouzi-Hafraoui et al., 2009). Owing to the high specific surface area of these nanoparticles, the ratio of adsorbed surfactant to cellulose was found to be 0.7 for cotton and 1.5 for tunicin. Small angle neutron scattering (SANS) experiments have shown that a thin layer around 15 Å was formed at the surface of the nanoparticles by the surfactant molecules (Bonini et al., 2002). A similar value was observed from UV-visible spectrophotometry for cotton linter nanocrystals (Elazzouzi-Hafraoui et al., 2009).

Nonionic sorbitan monostearate surfactant was also used to disperse cellulose nanocrystals in polystyrene-based electrospun composite fibers (Kim et al., 2009; Rojas et al., 2009). Model cellulose surfaces based on cellulose nanocrystals extracted from bleached cotton, sisal and ramie fibers were prepared by the Langmuir–Schaeffer technique (Habibi et al., 2010). A cationic surfactant, namely dioctadecyldimethylammonium bromide, was used to prepare cellulose-surfactant complexes that allowed transfer of the nanocrystal from the air/liquid interface in the aqueous suspension to hydrophobic solid substrates. Strong interactions between negatively H_2SO_4-prepared cellulose nanocrystals and the cationic surfactant by way of electrostatic interactions were expected.

Cellulose films prepared from carboxymethylated MFC were modified by coating with various amounts of a fluorosurfactant, namely perfluorooctadecanoic acid ($C_{17}F_{35}COOH$) (Aulin et al., 2008). When MFC was prepared using a TEMPO-mediated oxidation pretreatment, films obtained from these nanofibrils had anionic sites on their surface. Therefore, their surface can be easily modified with a cationic surfactant applied directly on the film surface. N-hexadecyl trimethylammonium bromide (cetyltrimethylammonium bromide CTAB) dissolved in water was deposited on the surface of MFC films (Syverud et al., 2011). The adsorbed layer of CTAB was found to increase the hydrophobicity of the film without affecting its mechanical properties significantly. Because of the antiseptic properties of CTAB, new applications can be considered. CTAB, as well as didodecyl- (DDDAB) and dihexadecyl ammonium bromide (DHDAB) were used to control the water repellency of cellulose nanofibrils (Xhanari et al., 2011). In this study, the surfactant was directly added to MFC in water suspension. Because DDDAB and DHDAB are insoluble in water at room temperature, it was necessary to solubilize them by heating at 75°C. Small aggregates (admicelles) of surfactant were shown to form on the surface of the nanofibrils, well below critical micelle concentrations.

5.3.2 Adsorption of macromolecules

The basic principle is similar to the one provided by a surfactant. This means that to be adsorbed on the surface of cellulose nanoparticles, the macromolecule should bear both hydrophilic and hydrophobic moieties.

Surface modification of cotton cellulose nanocrystals using a xyloglucan oligosaccharide-based triblock copolymer was reported (Zhou et al., 2009). By mimicking lignin-carbohydrate copolymers, a xyloglucan oligosaccharide-poly(ethylene glycol) (PEG)-polystyrene (PS) triblock copolymer was synthesized by coupling prefabricated linear blocks and adsorbed onto the surface of nanoparticles. PEG was used as a compatible block between the carbohydrate and hydrophobic PS blocks. A solution of the copolymer in dimethylformamide (DMF) was mixed with the aqueous suspension of nanocrystals and the mixture was subsequently freeze-dried. Ensuing nanocrystals showed good dispersion abilities in non-polar liquids.

High molecular weight polyoxyethylene (PEO) dissolved in water was adsorbed on the surface of cotton nanocrystals (Ben Azouz et al., 2012). Adsorption was suspected from the rheological behavior of the PEO solution whose viscosity decreased when the nanoparticles were added. When saturation of the nanocrystal surface was reached, the viscosity of the suspension was found to increase. These PEO-coated cellulose nanocrystals were successfully extruded with hydrophobic polymer, enhancing both the dispersibility and thermal stability of the nanoparticles.

5.4 Esterification, acetylation and acylation

The hydrophobization of cellulose surface has often been achieved through the well-known aptitude of cellulose to undergo esterification reactions. The cellulose esterification process basically uses acid anhydrides or acyl chlorides as acetylating agents. Esterification is a reaction that introduces an ester functional group COO on the surface of cellulose nanoparticles by condensation of a carboxylic acid group COOH and alcohol group OH. Acetylation is a reaction that introduces an acetyl functional group $COCH_3$ on the surface of cellulose nanoparticles. It is the basic reaction involved in the preparation of the cellulose derivative, cellulose acetate, which is of commercial importance for the preparation of fibers and film base. Native cellulose can be gradually converted into cellulose acetate and cellulose triacetate upon the addition of acetic anhydride in the presence of a small amount of catalyst, such as sulfuric or perchloric acid. Commercially, the largest amount of cellulose acetate is produced by the so-called "acetic acid process". In this heterogeneous process, the cellulose is first swelled in acetic acid and then acetylated with acetic anhydride in the presence of sulfuric acid or perchloric acid as catalysts. Acetic acid is recovered as a by-product of the reaction as shown in Figure 5.3.

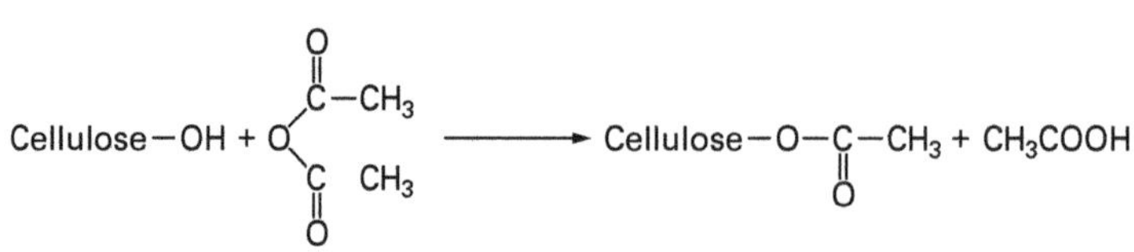

Fig. 5.3: Acetylation of cellulose with acetic anhydride.

This reaction has been conducted on different nanocellulosic substrates. Some of these studies are collected in Table 5.1. Several methods have been used to achieve acetylation and esterification of cellulose nanoparticles. Based on a non-swelling

Source of Cellulose	**Nano-particle**	**Reagent**	**Objective of the Modification**	**Reference**
Bacterial Cellulose	–	Acetic anhydride	Enhancement of properties of acrylic resin	(Nogi et al., 2006; Ifuku et al., 2007)
		Palmitoyl acid (gas phase)	Hydrophobization	(Berlioz et al., 2009)
		Acetic, hexanoic and dodecanoic acids	Hydrophobization	(Lee et al., 2011)
Bleached Sulfite Wood Pulp	MFC	Acetic anhydride	Dispersion in chloroform and compatibilization with PLA	(Tingaut et al., 2010)
Cotton Linters	CNC	Acetic acid and butyric acid	Dispersion in ethyl acetate and toluene	(Braun and Dorgan, 2009; Sobkowicz et al., 2009)
		Vinyl acetate	Dispersion in THF	(Çetin et al., 2009)
		Acetic anhydride	Dispersion in chloroform	(Lin et al., 2011)
Kenaf Bast Fibers	MFC	Acetic anhydride	Dispersion in acetone and ethanol	(Jonoobi et al., 2010)
MCC	CNC	Acetic anhydride	Redispersibility in water	(Wang et al., 2007)
Norway Spruce Kraft Pulp	MFC	Acetic anhydride	Hydrophobization of films	(Rodionova et al., 2011)
Ramie	CNC	Hexanoyl chloride, lauroyl chloride and stearoyl chloride	Extrusion with LDPE	(de Menezes et al., 2009)
Tunicin		Iso-octadecenyl succinic anhydride and n-tetradecenyl succinic anhydride	Dispersion in solvents of low polarity	(Yuan et al., 2006)
		Acetic anhydride and acetic acid	Ultrastructural aspect of acetylation	(Sassi and Chanzy, 1995)
		Palmitoyl acid (gas phase)	Hydrophobization	(Berlioz et al., 2009)
Valonia		Acetic anhydride and acetic acid	Ultrastructural aspect of acetylation	(Sassi and Chanzy, 1995)

Table 5.1: Acetylation of cellulosic nanoparticles.

reaction mechanism, the reaction only occurred on the cellulose chains located at the surface of the nanoparticles. The limitation on the extent of acetylation lies in the susceptibility and accessibility of the surface.

The ultrastructural investigation of well-characterized tunicin and *Valonia* nanocrystals subjected to homogeneous and heterogeneous acetylation was reported (Sassi and Chanzy, 1995). TEM observations and diffraction contrast images of the nanoparticles at various stages of the reaction showed that the acetylation proceeded by a reduction of the diameters of the nanocrystals, while their lengths were reduced to a lower extent as shown by comparing Figures 5.4(a) and 5.4(b). It was supposed that the nanocrystals break down laterally but not longitudinally upon acetylation (peeling effect). A model of acetylation based on a non-swelling reaction mechanism that affects only the cellulosic chains located at the surface of the nanoparticle was proposed (Figure 5.4(c)). For homogeneous acetylation, the partially acetylated molecules were found to be sucked into the acetylating medium when sufficiently soluble whereas in heterogeneous conditions, the cellulose acetate remained insoluble and surrounded the crystalline core of unreacted cellulose.

Based on earlier studies carried out with chitin (Gopalan Nair et al., 2003) and starch (Angellier et al., 2005) nanocrystals, alkenyl succinic anhydride (ASA) was used to confer high hydrophobicity to cellulose nanocrystals (Yuan et al., 2006). In both former works, organic solvents were used to avoid hydrolysis of ASA. Acylation

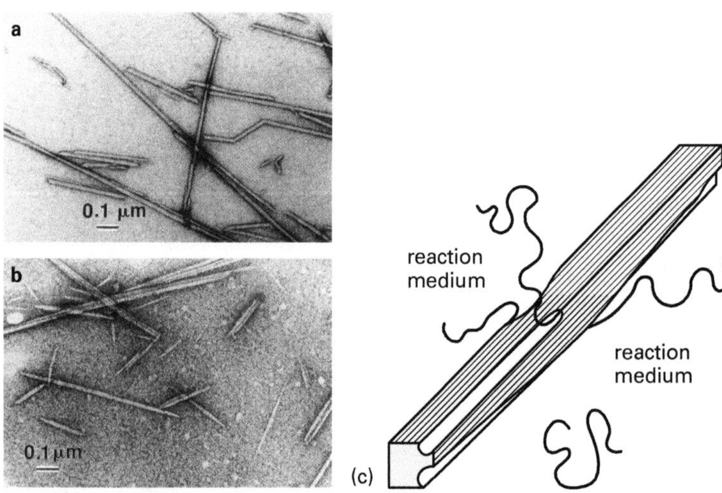

Fig. 5.4: Transmission electron micrographs from a dilute suspension of tunicin nanocrystals (a) before and (b) after partial acetylation (DS = 0.17), and (c) schematic drawing describing the onset of the acetylation of a typical cellulose crystal. A chain that is sufficiently acetylated to become soluble has left the crystal. Three chains in the process of acetylation are partially lifted from the crystal. The crystal is indented by a series of grooves that correspond to the missing cellulose chains (Sassi and Chanzy, 1995).

of chitin nanocrystals was achieved in a dioxane system with 4-(dimethylamino) pyridine as the catalyst and the reaction was carried out for 1 week at 70°C (Gopalan Nair et al., 2003). The toluene/ (dimethylammo)pyridine system was used to modify starch nanocrystals (Angellier et al., 2005). ASA is widely used as a sizing agent in papermaking, where it is applied to pulp fibers in aqueous systems. ASA aqueous emulsions were mixed with tunicin nanocrystal suspensions, freeze-dried and heated at 105°C to induce the esterification of hydroxyl groups of cellulose. From this environmentally-friendly procedure, highly hydrophobic nanocrystals were obtained with low reagent consumption and esterified nanoparticles were dispersed into medium- to low-polarity solvents, i.e. DMSO and 1,4-dioxane. Nanocrystals with different dispersibility could be obtained by controlling the heating time. Surface acetylation of cellulose nanocrystals (obtained from MCC) was undertaken by transesterification of vinyl acetate in the presence of potassium carbonate as catalyst (Çetin et al., 2009). By progressively increasing the reaction times, it was observed that the crystalline structure of the nanoparticles was destroyed. Improved stability of acetylated nanocrystals in tetrahydrofurane (THF) was reported with increased acetylation.

To avoid complex surface functionalization routes, the simultaneous hydrolysis and acetylation of cellulose nanocrystals in a single step has been reported (Braun and Dorgan, 2009). Using a mixture of hydrochloric acid (HCl) and organic acid (acetic and butyric were both demonstrated), amorphous cellulose chains were hydrolyzed and cellulose nanocrystals were functionalized by a Fischer esterification process of hydroxyl groups in a one-pot reaction methodology (Figure 5.5). It was shown that about half of the surface hydroxyl groups were esterified using this procedure. This approach is also quite versatile as the chain length of the covalently grafted surface group is easily controlled through the choice of organic acid employed in the reaction, the only restriction being that it has some miscibility with water. The presence of acetate and butyrate groups affected the hydrophilicity of cellulose nanocrytals making their aqueous suspensions unstable, but they possessed better dispersibility in ethyl acetate and toluene.

Esterification of ramie cellulose nanocrystals by reacting organic fatty acid chlorides with different lengths of the aliphatic chain (C6 to C18) under reflux was reported (de Menezes et al., 2009). The crystalline core of the nanoparticles was found to be unaffected while the grafting density was high enough to allow crystallization of grafted chains when using backbones of 18 carbon atoms. Crystallization of the grafted chains was evidenced from X-ray diffraction and differential scanning calorimetry (DSC) experiments. The DS was found to decrease with increasing carbon chain length of the organic acid used. Similar results were reported for bacterial cellulose (BC) modified using an esterification reaction with acetic acid, hexanoic acid or dodecanoic acid (Lee et al., 2011). Apart from the solution process, a gas-phase process using evaporation of large excess of palmitoyl chloride to achieve a surface to core esterification was investigated (Berlioz et al., 2009). The method was developed for freeze-dried tunicin nanocrystals and bacterial cellulose microfibrils dried

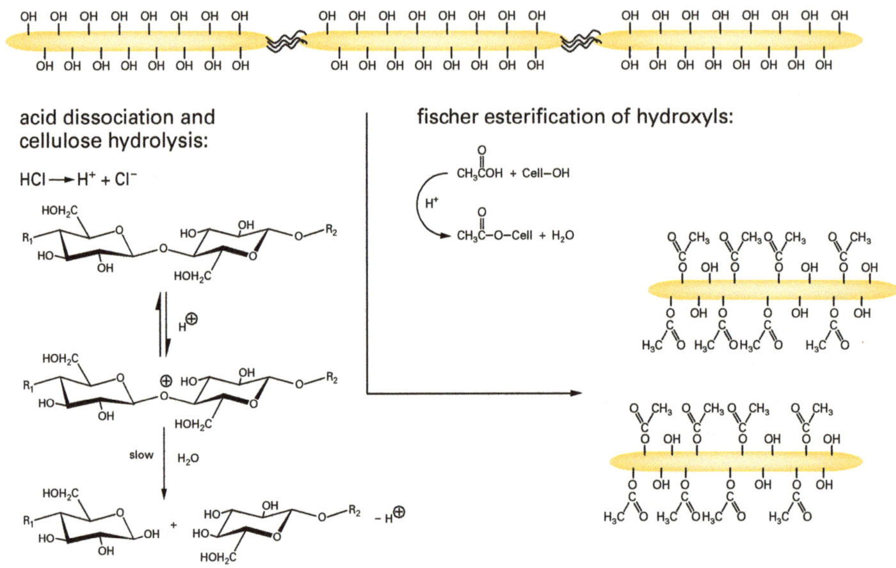

Fig. 5.5: Reaction scheme illustrating the simultaneous occurrence of cellulose hydrolysis and esterification of hydroxyl groups using a mixture of acetic and hydrochloric acid (Braun and Dorgan, 2009).

by the critical point method. This procedure can be extended to esterification of different fatty acid chlorides. The experimental conditions, nature and conditioning of cellulose were found to be important factors controlling the extent of esterification and morphology of the grafted nanoparticles. It was observed that the esterification proceeded from the surface of the cellulosic substrate to the crystalline core. For moderate DS, the surface was fully grafted whereas the cellulose core remained unmodified, and under certain conditions, an almost total esterification could be achieved, leading to highly substituted cellulose esters.

5.5 Cationization

Treatment with cations can be performed to render cellulose nanoparticles cationic by introducing positive charges on their surface. This strategy applied to cellulose fibers is largely used in the pulp and paper industry. A one-step method was reported to introduce positive charges on the surface of cellulose nanocrystals through the grafting of weak or strong ammonium-containing groups, such as epoxypropyltrimethylammonium chloride (EPTMAC) (Hasani et al., 2008). The surface cationization procedure was conducted via a nucleophilic addition of alkali-activated cellulose hydroxyl groups to the epoxy moiety of EPTMAC. Under mild alkaline cationization conditions the original morphology of the nanocrystals was preserved and their integrity was maintained. This functionalization process reversed the surface charge but the charge density was similar. However, functionalized nanocrystals made better monolayers on anionic mica when observed by AFM. Moreover, stable aqueous suspensions of modified nanocrystals showed unexpected thixotropic gelling properties. Shear birefringence was observed, but thixotropic gels inhibit the formation of chiral nematic liquid crystalline phase, because no liquid crystalline chiral nematic phase separation was detected.

5.6 Silylation

Silylation consists in the introduction of substituted silyl groups R_3Si on the surface of cellulose nanoparticles. Many organofunctional silanes have been developed as coupling agents for glass-reinforced polymer composites. The general formula of an organosilane RR'R"SiX shows two classes of functionality. The X functional group is involved in the reaction with the cellulosic substrate (Figure 5.6). The bond between X and the silicon atom in coupling agents is therefore replaced by a bond between the cellulosic substrate and the silicon atom. X is a hydrolyzable group, typically alkoxy, acyloxy, amine, or chlorine. The most common alkoxy groups are methoxy and ethoxy, which give methanol and ethanol as by-products during coupling reactions. Chlorosilanes generate hydrogen chloride as a by-product during the coupling reactions.

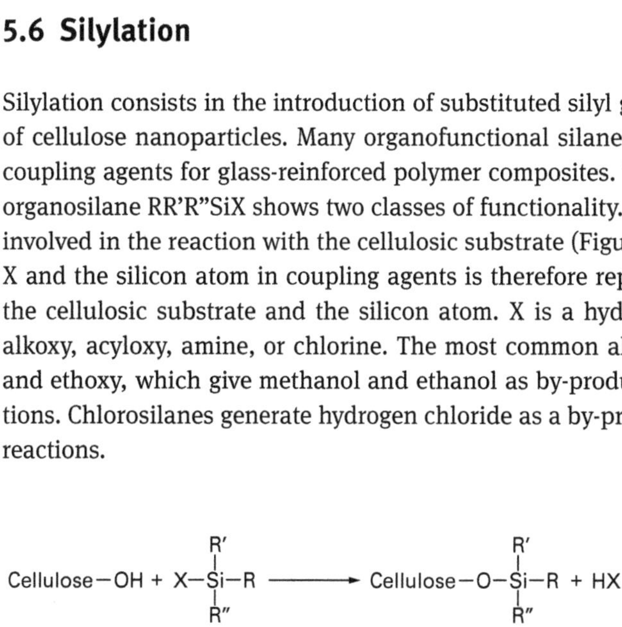

Fig. 5.6: Silylation of cellulose.

This reaction has been conducted on different nanocellulosic substrates. Some of these studies are collected in Table 5.2.

The first partial silylation of cellulose nanoparticles was conducted using tunicin nanocrystals and a series of alkyldimethylchlorosilanes, with alkyl moieties ranging from isopropyl to n-butyl, n-octyl and n-dodecyl (Goussé et al., 2002). It was shown that with a surface DS of the order of 0.6–1, the nanocrystals kept their morphological integrity, but due to their surface silylation, they became readily dispersible in solvents of low polarity such as THF. The resulting suspensions did not flocculate and appeared birefringent when viewed between cross polars. However, for surface DS higher than 1, the core of the nanocrystals also became silylated leading to the loss of their crystal character and it was no longer possible to obtain birefringent suspensions. Cotton nanocrystals functionalized by partial silylation through reaction with

Source of Cellulose	**Nano-particle**	**Silane**	**DS**	**Objective of the Modification**	**Reference**
Bacterial Cellulose	CNC	Hexamethyl disilazane	0.49	Dispersion in acetone and compatibilization with acetate butyrate	(Grunert and Winter, 2002a; 2002b)
Bleached Birch Kraft Pulp	MFC		0.03–0.9	Evidence of medium-driven reactivity of cellulose	(Johansson et al., 2011)
Bleached Spruce Sulfite Cellulose		Chlorodimethyl isopropylsilane	0.6–1	Hydrophobization	(Andresen et al., 2006)
			0–1.1	Stabilization of emulsions	(Andresen and Stenius, 2007)
		Octadecyldimethyl (3-trimethoxysilyl-propyl) ammonium chloride	–	Antimicrobial activity	(Andresen et al., 2007)
Cotton	CNC	N-dodecyldimethyl-chlorosilane	–	Dispersion in THF and chloroform and compatibilization with PLA	(Pei et al., 2010)
Kraft Pulp	MFC	3-aminopropyltri-ethoxysilane and 3-glycidoxypropyl-trimethoxysilane	–	Dispersion in acetone and compatibilization with epoxy	(Lu et al., 2008)
Ramie	CNC	Aminopropyltrietho-xysilane, n-propyl-trimethoxysilane, methacryloxypro-pyltrimethoxysilan, acryloxypropyltrime-thoxysilane	–	Compatibilization with PLA	(Raquez et al., 2012)
Sugar Beet Pulp	MFC	Isopropyl dimethyl-chlorosilane	0–0.36	Dispersion in methyl oleate	(Goussé et al., 2004)
Tunicin	CNC	Isopropyl, n-butyl, n-octyl and n-dode-cyldimethylchloro-silane	0–1	Dispersion in THF	(Goussé et al., 2002)

Table 5.2: Silylation of cellulosic nanoparticles.

n-dodecyldimethylchlorosilane were homogeneously dispersed in PLA and found to increase the crystallization rate of the matrix (Pei et al., 2010).

The preparation of bacterial cellulose nanocrystals topochemically trimethylsilylated was also reported (Grunert and Winter, 2002a). Trimethylsilylation of the crystal surface was performed heterogeneously with hexamethyldisilazane in formamide. As determined by inductively coupled plasma spectrometry the average degree of substitution of these silylated nanocrystals was 0.49, resulting in a molecular weight increase of 22% with respect to unmodified cellulose (Grunert and Winter, 2002b). In other words, 18% of the silylated crystals weight was due to silyl groups. Resulting nanoparticles were dispersed in acetone to process nanocomposites with a cellulose acetate butyrate matrix.

Suspensions of MFC resulting from the homogenization of sugar beet pulp were surface silylated with isopropyl dimethylchlorosilane (Goussé et al., 2004). When mild silylation conditions were applied, the microfibrils retained their morphology, but could be dispersed in a non-flocculating manner into organic solvents. Permanent antimicrobial MFC films were also prepared by grafting quaternary ammonium compound octadecyldimethyl(3-trimethoxysilylpropyl) ammonium chloride (ODDMAC) by a simple adsorption-curing process (Andresen et al., 2007). MFC and ODDMAC were mixed in a mixture of methanol and water (90:10 w/w). The mixture was cast, drained, dried and finally cured for 2 h at 100°C. Hydrophobization of MFC was also obtained by grafting 3-aminopropyltriethoxysilane (APS) and 3-glycidoxypropyltrimethoxysilane (GPS) (Lu et al., 2008). MFC and coupling agents were mixed in acetone, and the mixture was filtered and dried. The reaction was carried out for 2 h at 120°C. Better and stronger adhesion between MFC and the epoxy polymer used as matrix was observed for the treated fibers, which resulted in better mechanical properties of the composite materials.

5.7 Carbamination

Cellulose nanoparticles (both cellulose nanocrystals and MFC) extracted from sisal fibers were chemically modified with n-octadecyl isocyanate ($C_{18}H_{37}NCO$) (Siqueira et al., 2009). The surface chemical modification was carried out in toluene. Never-dried nanoparticles were used and a solvent exchange procedure from water to toluene was performed. The aqueous suspensions of cellulose nanoparticles were solvent exchanged to acetone and then to dry toluene by several successive centrifugation and redispersion operations. Contrarily to nanocrystals, it was found that MFC did not disperse homogeneously in toluene, probably because of the possibility of entanglements between the microfibrils and differences in surface charge. To overcome this problem, a process consisting in an in-situ solvent exchange procedure during the reaction was proposed. Compared to cellulose nanocrystals, a higher grafting density was necessary to achieve dispersion of MFC in an apolar and aprotic liquid medium.

Hexamethylene diisocyanate was also used to modify the surface characteristics of MFC (Stenstad et al., 2008). Water from the MFC suspension was replaced by acetone and then by THF. The reaction was performed for 2 h at 50°C. The results indicated that the treatments can be designed so that the strength properties or solubility of the MFC are not significantly changed.

5.8 TEMPO-mediated oxidation

Transformation of hydroxyl groups from the surface of cellulose nanoparticles into carboxylic groups can be conducted using 2,2,6,6-tetramethylpiperidine-1-oxyl (TEMPO) reagent (Isogai et al., 2011). TEMPO-mediated oxidation of cellulosic fibers is the most commonly used pretreatment to prepare MFC (see Chapter 2, Section 2.2.3). This type of nanocellulose is sometimes referred to as TEMPO-oxidized cellulose nanofibers (TOCN). Catalytic oxidation using TEMPO consists in a selective conversion of hydroxyl groups to aldehyde and carboxylate functional groups under aqueous and mild conditions.

Oxidation is broadly defined as the interaction between oxygen molecules and all the different substances they may contact. Technically, it is more precisely defined as the loss of at least one electron when two or more substances interact. These substances may or may not include oxygen. (2,2,6,6-tetramethylpiperidin-1-oxyl), or (2,2,6,6-tetramethylpiperidin-1-oxidany) or TEMPO is a stable nitroxyl radical (Figure 5.7) that was found to catalyze the oxidation of primary alcohol groups to aldehydes in aqueous media (De Nooy et al., 1995). This heterocycle is a red-orange, sublimable solid with a melting point around 36–38°C, and it is widely used as a radical trap, structural probe for biological systems, reagent in organic synthesis, or mediator in controlled free radical polymerization. The actual oxidant is the N-oxoammonium salt. In a catalytic cycle with sodium hypochlorite and sodium bromide as the stoichiometric oxidant, hypochlorous acid generates the N-oxoammonium salt from TEMPO.

The TEMPO-mediated oxidation reaction of cellulose is schematized in Figure 5.8. The TEMPO/NaBr/NaClO system in water at pH 10–11 is expected to oxidize the C6 primary hydroxyls of cellulose to C6 carboxylate groups. NaClO is added to the aqueous suspension of cellulose in the presence of catalytic amounts of TEMPO and NaBr at room temperature. The C6 primary hydroxyl groups of cellulose are thus entirely and selectively converted to carboxylate groups via C6 aldehyde groups and only NaClO and NaOH are consumed (Saito and Isogai, 2006). TEMPO-mediated oxidation of cellulose nanocrystals involves a topologically confined reaction sequence, and as a consequence of the 2-fold screw axis of the cellulose chain, only half of the accessible hydroxymethyl groups are available to react, whereas the other half are buried within the crystalline particle (Habibi et al., 2006).

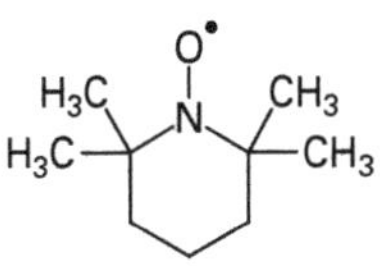

Fig. 5.7: (2,2,6,6-tetramethylpiperidin-1-oxyl), or (2,2,6,6-tetramethylpiperidin-1-oxidany) or TEMPO radical.

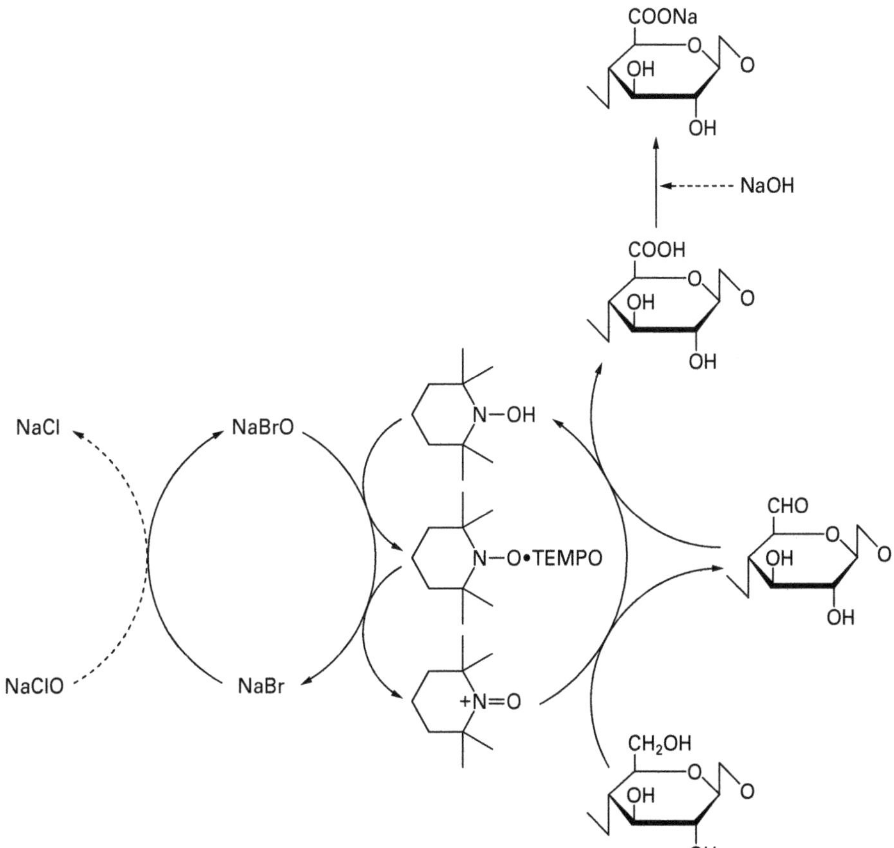

Fig. 5.8: Scheme of TEMPO-mediated oxidation of cellulose (Saito and Isogai, 2006).

However, undesirable side reactions under such alkaline conditions, like significant depolymerization of discoloration or the oxidized cellulose resulting from the presence of residual aldehyde groups, are unavoidable (Shibata and Isogai, 2003; Saito and Isogai, 2004). Indeed, these aldehyde groups are thermally unstable and cause discoloration of the oxidized cellulose when heated or dried over 80°C in the presence of residual acid. Moreover, they disturb the individualization of cellulose microfibrils in water due to the partial formation of hemiacetal linkages between fibrils (Saito and Isogai, 2006). To avoid these side reactions, a TEMPO/NaClO/$NaClO_2$ system under neural or slightly acidic conditions was proposed as an alternative and applied (Saito et al., 2009). Catalytic amounts of TEMPO and NaClO are combined with $NaClO_2$ as the primary oxidant.

Carboxylation of cellulose nanocrystals with the TEMPO/NaBr/NaClO at pH 10–11 and room temperature was used as an intermediate step to promote grafting of poly(ethylene glycol) chains (Araki et al., 2001). The carboxyl content of the carboxylated sample was 915 $mmol \cdot kg^{-1}$. A homogeneous dispersion in water was obtained. TEMPO-mediated oxidation was also conducted on cotton linters cellulose nanocrystals and sugar beet MFC (Montanari et al., 2005). Most oxidized specimens tended to limit their aggregation and promote their individualization. Moreover, TEMPO-mediated oxidation was found to induce cleavage in the amorphous zones of MFC and a decrease of the crystal size was observed. By applying various oxidation conditions to HCl-prepared tunicin nanocrystals, it was found that the degree of oxidation can be tuned (Habibi et al., 2006). With a degree of oxidation up to 0.1, the morphological integrity of the nanocrystals was kept and the surface hydroxymethyl groups were selectively converted to carboxylic groups, thus imparting a negative surface charge to the nanoparticles. Carboxyl content around 300 $\mu mol \cdot g^{-1}$ was found to be necessary to significantly reduce the number of passes to obtain MFC gels (Besbes et al., 2011a; Besbes et al., 2011b).

TEMPO oxidation is often used to selectively activate the primary hydroxyl groups of cellulose nanoparticles by their conversion to carboxylic acids. Further reactions can be carried out using the activated TEMPO-oxidized nanoparticles. For instance, two separate populations of complementary single stranded DNAs were grafted to carboxylated cellulose nanocrystals and the populations were then combined to hybridize the DNA and bond the cellulosic nanoparticles (Mangalam et al., 2009). Grafting was performed using carbodiimide chemistry.

5.9 Polymer grafting

Surface chemical modification of cellulose nanoparticles can be achieved by covalently attaching small molecules as seen previously, but also polymers. Long chain surface chemical modification of cellulose nanoparticles consist in grafting agents bearing a reactive end group and a long "compatibilizing" tail. The general objective of this chemical modification is of course to increase the apolar character of the nanoparticle. In addition, it can yield some extraordinary possibilities. The surface grafted chains can act as binding sites for active agents in drug delivery systems or for toxins in purifying and treatment systems. These grafted chains may also be able to interdiffuse, upon heating, to form the polymer matrix phase. The covalent linkage between reinforcement and matrix will result in near-perfect stress transfer at the interface with exceptional mechanical properties of the composite as a result. Moreover, if the grafted chains and the matrix are the same, better compatibilization can be obtained thanks to the formation of a continuous interphase between the cellulose phase and the polymeric matrix. The possibility to obtain chain entanglements as well as for co-crystallization between covalently linked chains at the surface of

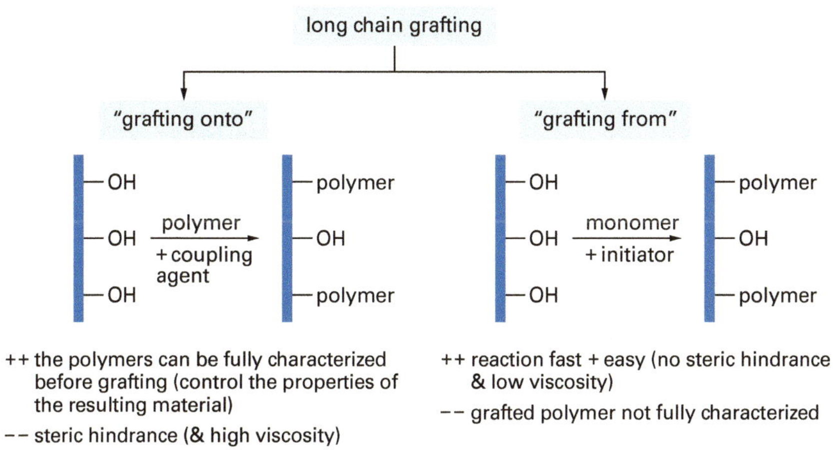

Fig. 5.9: Schematic representation of the "grafting onto" and "grafting from" approaches (Dufresne, 2010).

	Source of Cellulose	Nano-particle	Polymer	Method	Objective of the Modification	Reference
Grafting Onto	Cotton	CNC	PEG	Carboxylation-amidation	Dispersion in water and non-aqueous solvents	(Araki et al., 2001)
			PEO	Alkaline epoxyde ROP	Steric stabilization in water	(Kloser and Gray, 2010)
			PS, P*t*BuA	SI-ATRP	Dispersion in organic solvents	(Harrisson et al., 2011)
	Cotton Linters	CNC	PEO-co-PPO	Peptidic coupling reaction	Providing thermosensitive properties	(Azzam et al., 2010)
	Ramie	CNC	PCL	Isocyanate-mediated reaction	Dispersion in dichloromethane and compatibilization with PCL	(Habibi and Dufresne, 2008)
					Dispersion in DMF and compatibilization with PCL	(Zoppe et al., 2009)

	Source of Cellulose	Nano-particle	Polymer	Method	Objective of the Modification	Reference
Grafting From	Bleached Birch Pulp	MFC	PGMA, PMMA, PHEMA, PBuA, PEA,	Cerium-induced reaction	Functionalization	(Littunen et al., 2011)
	Bleached Spruce Sulfite Cellulose	MFC	PGMA	Cerium-induced reaction	Functionalization	(Stenstad et al., 2008)
	Bleached Wood Sulfite Pulp	MFC	PCL	Sn(Oct)$_2$-catalyzed ROP	Control of the molecular weight of the grafted polymer	(Lönnberg et al., 2008)
	Cotton	CNC	PS	SI-ATRP	Investigation of liquid crystalline phases behavior in DMF	(Yi et al., 2008)
			PMMAZO	SI-ATRP	Investigation of liquid crystalline phases behavior in chlorobenzene	(Xu et al., 2008)
			PDMAEMA	SI-ATRP	Investigation of liquid crystalline phases behavior in water	(Yi et al., 2009)
			PS	SI-ATRP	Pollutant removal from water	(Morandi et al., 2009)
			PAA	SI-LRP	Preparation of well-defined polymer brush architectures	(Majoinen et al., 2011)
	Cotton Linters	CNC	PCL	Sn(Oct)$_2$-catalyzed ROP	Dispersion in dichloromethane and compatibilization with PLA	(Lin et al., 2009)
	Cottonseed Linter Pulp	CNC	WPU	In situ polymerization	Direct processing of WPU-based nanocomposites	(Cao et al., 2009)
	Ramie	CNC	PCL	Sn(Oct)$_2$-catalyzed ROP	Dispersion in dichloromethane and compatibilization with PCL	(Habibi et al., 2008)
			Poly (NiPAAm)	SI-SET-LRP	Functionalization	(Zoppe et al., 2010)
			PCL	Sn(Oct)$_2$-catalyzed ROP	Compatibilization with PCL	(Goffin et al., 2011a)
			PLA	Sn(Oct)$_2$-catalyzed ROP	Compatibilization with PLA	(Goffin et al., 2011b)

Table 5.3: Polymer grafting of cellulosic nanoparticles.

the nanoparticle with those from the matrix should further improve the mechanical properties of the final composite.

Two main different approaches can be used to graft polymers on surfaces, viz. "grafting onto" or "grafting from". These two strategies are schematized in Figure 5.9. The "grafting onto" approach consists in mixing the cellulosic nanoparticles with an existing polymer and a coupling agent to attach the polymer to the nanoparticle surface. In this approach, one cannot expect high grafting densities because of steric hindrance and blocking of reactive sites by the already grafted polymer chains. Moreover, the viscosity of the reaction medium is high because of the presence of macromolecular chains. However, its main advantage is that the properties of the resulting material are perfectly controlled since the molecular weight of the attached polymer can be characterized before grafting. The "grafting from" approach consists in mixing the cellulosic nanoparticles with a monomer and an initiator agent to induce the polymerization of the monomer from the nanoparticle surface. Because of the lower viscosity of the medium and limitation of steric hindrance, this strategy has proven to be a very effective way to create high grafting densities on the surface. However, it is difficult to control and determine precisely the molecular weight of the grafted polymer. Some of these studies are collected in Table 5.3.

5.9.1 Polymer grafting using the "grafting onto" approach

Cellulose nanocrystals obtained by acid hydrolysis of ramie fibers were subjected to isocyanate-mediated reaction to graft polycaprolactone (PCL) chains with various molecular weights (Mn = 10,000 $g \cdot mol^{-1}$ and Mn = 42,500 $g \cdot mol^{-1}$) on their surface (Habibi and Dufresne, 2008). Never-dried nanocrystals were used for the grating and a solvent exchange procedure from water to toluene was performed. For that, aqueous suspension with the desired amount of cellulose nanocrystals (1 wt%) was solvent exchanged to acetone and then to dry toluene by several successive centrifugation and redispersion operations. Sonication was performed after each solvent exchange step to avoid aggregation. However, the suspension in toluene was not stable in time. PCL grafting onto cellulose nanocrystals involved a three-step process. The first step required the reaction of the polymer on isocyanate functionality of phenylisocyanate. The second step involved the reaction of the polymer, now protected, with one isocyanate functionality of toluene 2,4-diisocyanate (2,4-TDI). During the third step, the unreacted second isocyanate functionality of 2,4-TDI was then reacted with the surface hydroxyl groups of the cellulose nanocrystals to graft the polymer chain onto the nanoparticles. Each mixture was washed with toluene, with successive centrifugations and then washed with dichloromethane by successive centrifugations to remove ungrafted polymer chains. The washed modified nanocrystals were subsequently Soxhlet extracted with dichloromethane before drying in a convective oven. The reaction scheme is shown in Figure 5.10.

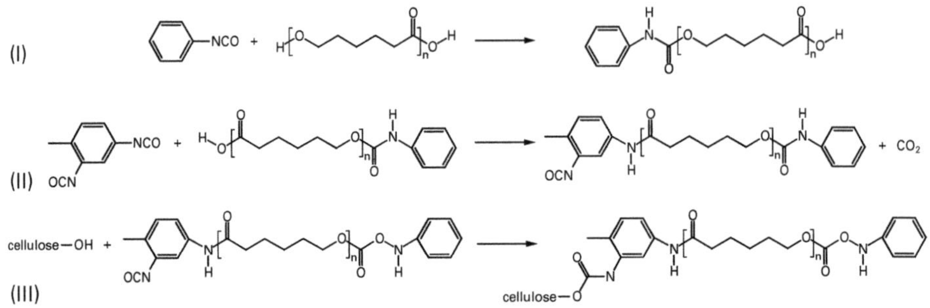

Fig. 5.10: Reaction scheme of the grafting of PCL onto the cellulose nanocrystals surface: (I) capping of the second hydroxyl group of PCL chain end, (II) reaction of the PCL chain with 2,4-TDI to give rise to the grafting agent, and (III) grafting onto the cellulose nanocrystal (adapted from Labet et al., 2007).

Grafted polymeric chains were found to form a crystalline structure at the surface of the nanoparticles as evidenced from X-ray diffraction and differential scanning calorimetry experiments. Nanocomposite films were processed from both unmodified and PCL-grafted nanoparticles, and PCL as matrix using a casting/evaporation technique. It was shown that mechanical properties of resulting films were notably different. Compared to unmodified nanoparticles, the grafting of PCL chains on the surface resulted in lower modulus values but significantly higher strain at break.

Poly(ethylene glycol) chains with a terminal amino group on one end (PEG-NH_2, Mw = 1,000) have been grafted onto the surface of hydrochloric acid-prepared cellulose nanocrystals using a carboxylation-amidation procedure (Araki et al., 2001). Amidation was performed using a water-soluble carbodiimide, 1-ethyl-3-(3-dimethylaminopropyl) carbodiimide (EDC), and N-hydroxysuccinimide (NHS). The freeze-dried PEG-grafted nanocrystals could be redispersed in water or non-aqueous solvents and displayed drastically enhanced dispersion stability evidenced through resistance to addition of 2 M sodium chloride. The amount of bound PEG was 0.2–0.3 $g \cdot g^{-1}$ of cellulose. In order to achieve higher steric instead of electrostatic stabilization, aqueous suspensions of poly(ethylene oxide) (PEO)-grafted cotton nanocrystals were prepared (Kloser and Gray, 2010). Desulfation of sulfuric acid-prepared nanocrystals was carried out with sodium hydroxide and the nanoparticles were functionalized with α-epoxy,ω-methoxy-terminated PEO (Mw = 2,086 $g \cdot mol^{-1}$) under alkaline conditions. Preformed amine-terminated polystyrene and poly(*tert*-butyl acrylate) (P*t*BuA) were grafted onto oxidized cellulose nanocrystals (Harrisson et al., 2011). Thermogravimetric analysis (TGA) was used to estimate the grafting density at around 60–64%.

Grafting of thermosensitive amine-terminated statistical Jeffamine copolymers (PEO-co-PPO) onto the surface of cellulose nanocrystals was achieved by a peptidic coupling reaction (Azzam et al., 2010). TEMPO-oxidized nanocrystals were used and the polymer grafting reaction was performed either in water or in DMF. The grafting density was sufficiently high to induce a steric stabilization of the nanocrystals that prevented flocculation at high ionic strength and made them surface-active. Amidation of TEMPO-oxidized nanocrystals was also performed with 4-amino TEMPO, a nitroxide radical containing a terminal amino group (Follain et al., 2010). The amine coupling brought a low polarity to the nanoparticles enabling the modification of cellulose suspensions into hydrophobic materials.

5.9.2 Polymer grafting using the "grafting from" approach

Several studies reported the preparation of PCL-grafted cellulose nanoparticles using the "grafting from" strategy. PCL is traditionally prepared by the ring-opening polymerization (ROP) of cyclic ε-caprolactone monomer. The catalytic ROP of lactones is the most common synthesis route today and can be carried out in solution by cationic, anionic or coordination-insertion mechanisms depending on the catalyst used. Stannous octoate ($Sn(Oct)_2$) is the most commonly used catalyst in ROP due to its high effectiveness and low toxicity (Storey and Sherman, 2002; Kowalski et al., 2005). In ROP, the hydroxyl groups on the cellulose nanoparticle surface act as initiator and the ratio of monomer to initiating groups determines the DP of the grafted polymer chains (Figure 5.11(a)).

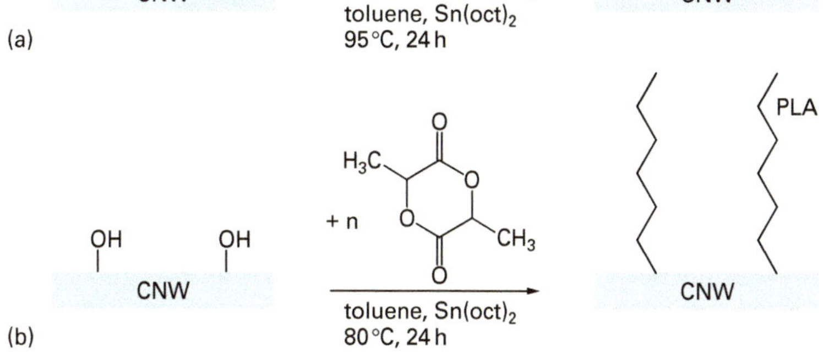

Fig. 5.11: Ring opening polymerization of (a) caprolactone (Goffin, 2010) and (b) L-lactide (Goffin et al., 2011b) as initiated from the surface of cellulose nanocrystals (CNW).

This approach was used to prepare PCL-grafted ramie (Habibi et al., 2008) and native linter (Lin et al., 2009) cellulose nanocrystals. PCL was grafted by $Sn(Oct)_2$-catalyzed ROP.

In the former study (Habibi et al., 2008), the H_2SO_4-hydrolyzed cellulose nanocrystals were neutralized using a 1 wt% NaOH solution and never-dried nanoparticles were used by exchanging water to acetone and then to dry toluene by successive centrifugation and redispersion operations. The grafting reaction was performed at 95°C for 24 h and stopped by adding a few drops of dilute hydrochloric acid solution. The grafting efficiency was evidenced by the long-term stability of suspension of PCL-grafted cellulose nanocrystals in toluene. After PCL grafting, the structural and morphological integrity of the cellulose nanocrystals did not appear to have been affected by $Sn(Oct)_2$-catalyzed polymerization and grafting as shown by comparing the TEM observation of ungrafted and grafted nanoparticles (Figure 5.12). The cellulose nanocrystal content in the recovered nanohybrid sample was determined by gravimetry and estimated to 15 wt%. The nanocrystals were less individualized than native ones and were believed to aggregate as a result of sulfate groups being removed from their surface. Furthermore, the presence of hydrophobic PCL chains on the nanocrystals likely triggers the particle aggregation upon drying. Nanocomposites with high filler content were prepared from neat and PCL-grafted cellulose nanocrystals and high molecular weight PCL as matrix using a casting/evaporation technique from dichloromethane. These PCL-grafted cellulose nanocrystals were also used as "masterbatches" by melt blending with a PCL matrix (Goffin et al., 2011a).

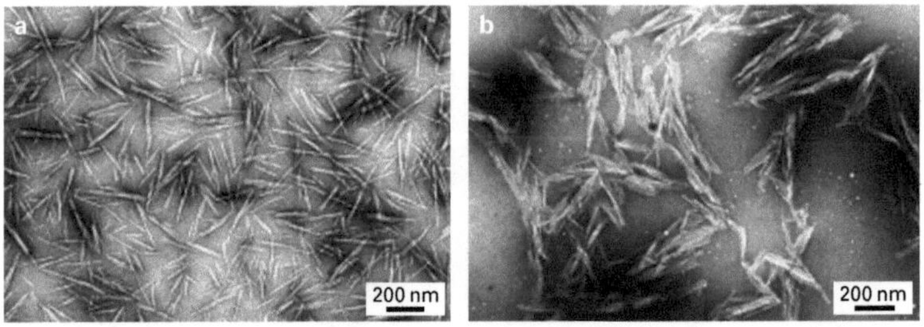

Fig. 5.12: Transmission electron micrograph of ramie cellulose nanocrystals: (a) ungrafted and (b) recovered after ROP/grafting reactions and Soxhlet extraction (Habibi et al., 2008).

In the second investigation (Lin et al., 2009), cellulose nanocrystals were freeze-dried and ROP was carried out using microwave irradiation. Purification and removal of residual monomer, catalyst and homopolymer were obtained by dispersion of the grafted nanocrystals in CH_2Cl_2 and precipitation by methanol. Processing of nanocomposites was performed in CH_2Cl_2 using poly(lactic acid) (PLA) as matrix and casting/evaporation.

A similar approach was used to graft PCL on the surface of MFC. Freeze-dried MFC was mixed with ε-CL monomer and grafting reaction was conducted with a catalytic amount of $Sn(Oct)_2$ at 95°C for 18–20 h (Lönnberg et al., 2008). Because the number of initiating hydroxyl groups on the surface of MFC was unknown, a co-initiator, benzyl alcohol, which is known to be an efficient initiator for ROP of lactones in the presence of $Sn(Oct)_2$, was added to the polymerization system to control the polymerization. Therefore, both free PCL and PCL-grafted MFC were formed simultaneously in the reaction medium. The soluble PCL formed was isolated from the mixture via filtration after dispersion of the reaction product in THF. By changing the amount of added free initiator to monomer, the amount of PCL on the MFC surface was changed to optimize the graft length. Different theoretical lengths of the PCL chain, i.e. DP 300, 600 and 1200, were investigated. The experimental molecular weights of free PCL formed during the grafting reaction were estimated from NMR and size exclusion chromatography (SEC). As expected, the obtained values were significantly lower than the theoretical ones since the theoretical molecular weight was calculated from the ratio of added monomer to free initiator, whereas the experimental value depends on the added monomer to free initiator, as well as the number of initiating groups on the MFC surface. TGA was used to estimate the composition of PCL-grafted MFC and PCL contents of 16%, 19% and 21% were reported, depending on the amount of free initiator to the system. Crystallization of grafted PCL was observed, but because of lower mobility of these chains compared to free PCL, a lower melting point and degree of crystallinity, as well as longer crystallization time were reported.

Molecular dynamics simulation was used to estimate the work of adhesion based on surface energies with different amounts of grafted caprolactone (Bergenstråhle et al., 2008). The types of interactions were principally Coulomb interactions, although the importance of weak Lennard-Jones interactions was larger when the surrounding medium was caprolactone instead of water. Hydrogen bonds were formed extensively between the grafted caprolactone and the surrounding medium.

Another biodegradable polyester, viz. polylactic acid (PLA), was chemically grafted on the surface of cellulose nanocrystals using the "grafting from" approach. $Sn(Oct)_2$–catalyzed ROP of L-lactide was initiated from the hydroxyl groups available at the nanoparticle surface as shown in Figure 5.11(b) to yield PLA-grafted cellulose nanocrystal nanohybrids. The cellulose nanocrystals were neutralized using a 1 wt% NaOH solution and never-dried nanoparticles were used by exchanging water to acetone and then to dry toluene by successive centrifugation and redispersion operations. The grafting reaction was performed at 80°C for 24 h and stopped by adding a few drops of dilute hydrochloric acid solution. The surface-modified cellulose nanocrystals were recovered by precipitation with cold methanol, filtered, and dried at 40°C under vacuum until constant weight. The cellulose nanocrystal content in the recovered nanohybrid sample was determined by gravimetry and estimated to 14 wt%. They were subsequently extruded and injection-molded with PLA.

The cerium (IV) ion is a powerful oxidation agent for alcohol containing 1,2-glycol groups. The mechanism of ceric ion reaction involves the formation of a chelate complex that decomposes to generate free radicals on the cellulose backbone. Epoxy functionality was introduced onto the surface of MFC by oxidation by cerium (IV) followed by graft polymerization of glycidyl methacrylate (GMA) (Stenstad et al., 2008). The cerium-induced grafting reaction was performed using ammonium cerium (IV) nitrate at 35°C in dilute (0.1 wt%) aqueous MFC suspensions. The length of the grafted polymeric chain was varied by regulating the amount of glycidyl methacrylate between 2 and 40 $\mu mol \cdot mg^{-1}$ MFC. Significant degradation of the cellulose chains could occur because of the formation of radicals in the reaction involving ammonium cerium nitrate. However, it was shown that the treatment resulted in only a slight reduction in the molecular weight of cellulose. In the same study (Stenstad et al., 2008), coupling of MFC with maleic anhydride was shown to introduce vinyl groups that could be used as a starting point for grafting reactions for monomers that are insoluble in water, as an alternative to the cerium-induced grafting method. MFC was also grafted in aqueous solution using a redox-initiated free radical polymerization with two acrylates and three methacrylates (Littunen et al., 2011). Cerium ammonium nitrate was used as initiator. The graft copolymerization was dominant over homopolymerization for all monomers. The highest graft yield was obtained with butyl acrylate (BuA) and glycidyl methacrylate (GMA) with 80 wt%. However, it was shown that BuA formed very long chains while for GMA relatively short chains were obtained forming a dense coating.

Cellulose nanocrystal reinforced waterborne polyurethane (WPU) nanocomposites have been prepared via one-pot polymerization, surface grafting, and processing (Cao et al., 2009). Polycaprolactone diol and isophorone diisocyanate (IPDI) were reacted to prepare a WPU prepolymer. During the reaction, cellulose nanocrystals dispersed in DMF were added to promote the reaction between hydroxyl groups from the nanoparticle surface and isocyanate on the end of the WPU prepolymer. Nanocomposites were processed by casting and evaporation. The grafted chains were able to form a crystalline structure on the surface of the nanoparticles and induced the crystallization of the matrix.

Grafting of polymer chains on cellulose nanoparticles can also be performed via living (or controlled) radical polymerization (LRP). Two steps are involved in LRP. The first one is the initial formation of initiating sites for LRP, in which an initiator is immobilized on the nanoparticle. The second step involves the reaction of the initiator-modified nanoparticles with a monomer to induce the polymerization. Copper-mediated LRP is generally chosen for its versatility with respect to monomer choice and ease of synthesis. Depending on the reaction conditions, two mechanisms, atom transfer radical polymerization (ATRP) and single electron transfer-living radical polymerization (SET-LRP), are distinguished.

Cellulose nanocrystals were grafted with styrene by ATRP (Yi et al., 2008). The surface hydroxyl groups were esterified with 2-bromoisobutyrylbromide to yield

2-bromoisobutyryloxy groups, which were used to initiate the polymerization of polystyrene (PS). The immobilization of ATRP initiator was conducted using freeze-dried nanocrystals redispersed in THF at room temperature for 24 h and the polymerization was performed at 110°C for 12 h. The grafted PS chains were cleaved from the nanoparticles under acidic conditions and found to represent 68 wt% of the nanohybrid. They had a number-average molecular weight of 74,700 $g \cdot mol^{-1}$. The chiral-nematic self-ordering of the grafted nanocrystals was investigated. A similar method was used to graft poly[6-(4-(4-methoxyphenylazo)phenoxyl)hexyl methacrylate] (PMMAZO) (Xu et al., 2008) and poly(N,N-dimethylaminoethyl methacrylate) (PDMAEMA) (Yi et al., 2009). For PMMAZO, the grafting percentage was determined from DSC measurements and found to be 74.6% (Xu et al., 2008). The number-average molecular weight of grafted PDMAEMA was estimated at around 10,200 $g \cdot mol^{-1}$ by cleaving from the cellulose backbone under acidic conditions (Yi et al., 2009). Surface-initiated ATRP was also used to prepare a range of PS-grafted nanocrystals with different graft lengths (theoretical DP = 27–171) and tailor grafting density by controlling the final content of initiating sites (Morandi et al., 2009). The grafted nanoparticles exhibited the capacity to absorb the equivalent of 50% of their weight of 1,2,4-trichlorobenzene from water proving their potential for pollutant removal applications.

Cellulose nanocrystals were grafted with thermoresponsive poly(N-isopropylacrylamide) (poly(NiPAAm) brushes via surface-initiated SET-LRP under various conditions at room temperature (Zoppe et al., 2010). Cu(I) was rapidly disproportionated to Cu(0) and Cu(II) yielding simple catalyst removal. It was shown that by increasing the initiator and monomer contents, increased amounts of poly(NiPAAm) were grafted from the surface of cellulose nanocrystals. Saponification of ester linkages was performed to cleave the polymer brushes from the cellulose nanoparticles. Increased molecular weights of polymer brushes were obtained when increasing the initiator and monomer contents. The observed effect of the initiator content was explained by local heterogeneities shifting the SET-LRP equilibrium to the active state. However, a broad molecular weight distribution was observed. Well-defined polymer brush architectures were prepared from cellulose nanocrystal surfaces to yield high grafting density polyelectrolyte brushes of poly(acrylic acid) (PAA) of different lengths by applying the Cu-mediated surface-initiated LRP (Majoinen et al., 2011). Chemical vapor deposition (CVD) was used as a pretreatment method before performing esterification in solution and obtaining full functionalization of the cellulose nanocrystal surface hydroxyl groups keeping the integrity of the cellulose crystal. Well-defined poly(*tert*-butyl acrylate) (P*t*BA) brushes with high grafting density were first synthesized from the nanocrystal surfaces, followed by the acid hydrolysis of their tertiary alkyl functionalities to provide PAA brushes.

5.10 Click chemistry

Click chemistry is tailored to generate substances quickly and reliably by joining small units together. One of the most popular reactions within the click chemistry concept is the azide alkyne Huisgen cycloaddition using a copper catalyst at room temperature.

A method for the grafting of amine-terminated monomers onto surface-modified cellulose nanocrystals followed by click chemistry was reported (Filpponen and Argyropoulos, 2010). Initially, the primary hydroxyl groups on the surface of the nanoparticles were selectively activated to carboxylic acids using TEMPO-mediated hypohalite oxidation. Next, compounds carrying terminal amine functionality were grafted onto the surface of the nanocrystals using these reactive sites for the amidation reaction. The grafted amine compounds contained terminal alkyne or azide functionalities providing the essential precursors to click chemistry. Unique nanoplatelet gels were obtained from this method. Production of a heterogeneous system with facile anion exchange capability was also reported (Eyley and Thielemans, 2011). Imidazoliumbromide salt was grafted onto cellulose nanocrystals using click chemistry. The first step consisted in a chlorination to activate the nanocrystals. The chlorinated nanocrystals were subsequently azidated using sodium azide in DMF. The ionic liquid 1-methyl-3-propargylimidazolium bromide ([MPIM][Br]) was synthesized and grafted onto the azidated nanocrystals in an aqueous solution using Cu(II) sulfate as a pre-catalyst and sodium ascorbate as the reductant. The crystallinity index of the nanoparticle was found to decrease slightly from 0.89 to 0.80 during the chlorination step but remained constant through the other modification reactions.

5.11 Fluorescently labeled nanocellulose

Fluorescent labeling corresponds to the process of covalently attaching a fluorophore to another molecule, such as a protein or nucleic acid. This is generally accomplished using a reactive derivative of the fluorophore that selectively binds to a functional group contained in the target molecule. Fluorescence techniques have been extensively used to study the cellular uptake and biodistribution of nanoparticulate delivery systems, by tracking the localization of the fluorophores. The fluorophore absorbs light energy of a specific wavelength and re-emits energy at a longer wavelength. The most commonly labeled molecules are antibodies, proteins, amino acids and peptides which are then used as specific probes for detection of a particular target.

It is believed that cellulose nanocrystals are promising candidates for applications in nanomedicine (Dong and Roman, 2007). To enable the use of fluorescence techniques in in vitro and in vivo studies, cellulose nanocrystals can be labeled with fluorophores. Fluorescein-5'-isothiocyanate (FTIC) was used to label cellulose nanocrystals prepared from softwood sulfite pulp via a three-step reaction involving epoxy activation on the nanoparticle surface, opening of the epoxy rings with ammonium hydroxide, and coupling of FITC molecules to the primary amino groups (Dong and Roman, 2007). The reaction route is shown in Figure 5.13(a). The FITC content grafted to cellulose nanocrystals was determined by UV/Vis spectroscopy and was estimated to be 0.03 mmol · g^{-1} of cellulose, equivalent to 5 FITC moieties per 1,000 anhydroglucose units. Figures 5.13(b) and 5.13(c) show aqueous suspensions of unlabeled and FITC-labeled cellulose nanocrystals, respectively. The unlabeled suspension was colorless and slightly opaque, whereas the FITC-labeled suspension appeared clear and yellow.

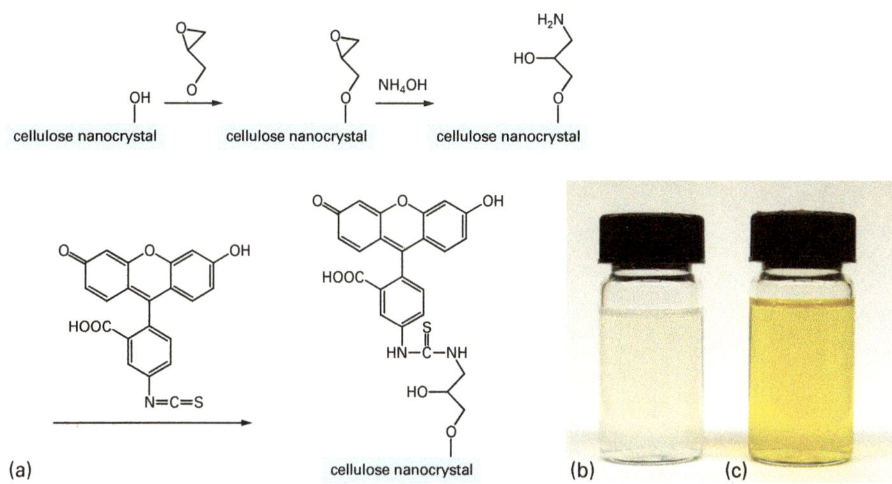

Fig. 5.13: (a) Reaction route for surface fluorescently labeled cellulose nanocrystals with FITC, and aqueous suspensions of (b) unlabeled cellulose nanocrystals (0.8 wt%) and (c) FITC-labeled cellulose nanocrystals (0.5 wt%) (Dong and Roman, 2007).

Cellulose nanocrystals prepared from cotton were converted into ratiometric pH-sensing nanoparticles by dual fluorescent labeling using a one-pot procedure (Nielsen et al., 2010). FTIC and rhodamine B isothiocyanate (RBITC) were reacted with the free hydroxyl groups of cellulose nanocrystals in a basic solution forming a thiocarbamate bond. The average amount of FITC and RBITC attached to the cellulose nanocrystals was estimated by UV/Vis and fluorescence spectroscopy, respectively. Values of 2.8 μmol · g^{-1} and 2.1 μmol · g^{-1} of cellulose were reported for FITC and RBITC, respectively. In the same study, a versatile three-step procedure schematized in Figure 5.14 was reported, extending the number of fluorophores available for grafting. An amine group was introduced via esterification followed by a thiol-ene click reaction. The esterification was carried out by reacting the nanocrystals with methacrylic acid using N,N'-diisopropylcarbodiimide/4-dimethylaminopyridine as catalyst to introduce a double bond. The ensuing nanoparticles were reacted with cysteamine in methanol to introduce the primary amine. Two different pH sensitive fluorophores 5-(and-6)-carboxyfluorescein succinimidyl ester (FAM-SE) and Oregon Green 488 carboxylic acid, succinimidyl ester (OG-SE) were conjugated to two separate batches of cellulose nanocrystals along with the reference fluorophore 5-(and-6)-carboxytetramethylrhodramine succinimidyl ester (TAMRA-SE). The average amount of dye grafted to the nanoparticles was 10.4 μmol · g^{-1} FAM-SE, 4.7 μmol · g^{-1} TAMRA-SE, and 7.3 μmol · g^{-1} OG-SE, 4.2 μmol · g^{-1} TAMRA-SE, respectively.

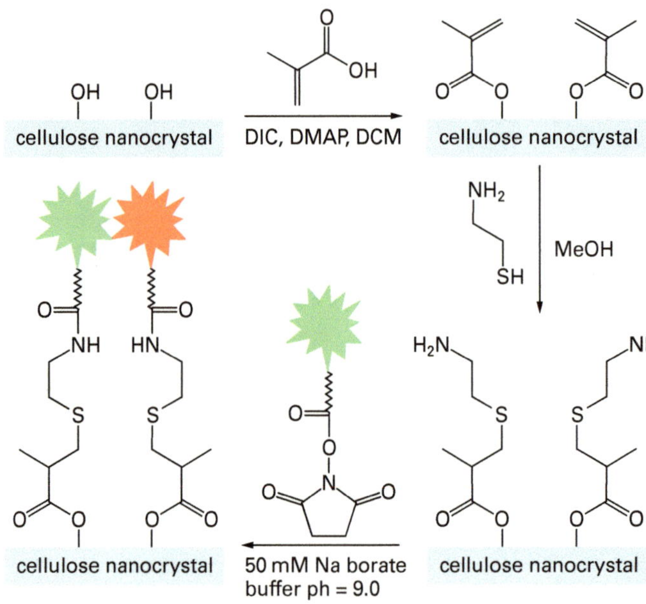

Fig. 5.14: Fluorescent labeling of cellulose nanocrystals with succinimidyl ester dyes (Nielsen et al., 2010).

Probing of cellular uptake and cytotoxicity was conducted for FITC and RBITC labeled cellulose nanocrystals (Mahmoud et al., 2010). No noticeably cytotoxic effect was observed, rendering modified cellulose nanocrystals as promising candidates for bioimaging and drug delivery systems.

Amino acids are the building blocks of proteins and peptides and are involved in the development of a wide range of biocompatible applications and architecture. TEMPO-oxidized MFC was coupled with fluorescent amino acids using a two-step procedure (Barazzouk and Daneault, 2011). First, MFC was activated by N-ethyl-N'-(3-dimethylaminopropyl) carboiimide hydrochloride, forming a stable active ester in the presence of N-hydroxysuccinimide. Then, the active ester was reacted with the amino groups on the amino acids, forming an amide bond between MFC and amino acids. After coupling with amino, MFC clearly showed the characteristic absorption and fluorescence features of the grafted amino acid. Anchoring of L-leucine moieties to cellulose nanocrystals allowing the creation of a binding site to which drugs or targeting molecules can be attached was also reported (Cateto and Ragauskas, 2011). The chemical modification of cellulose nanocrystals with L-leucine amino acid by esterification reaction was conducted through a two-step process involving the reaction between Fmoc-L-leucine and cellulose and removal of Fmoc-protecting group.

5.12 Evidence of surface chemical modification

Regardless of the grafting strategy adopted, extensive washing, Soxhlet extraction or solubilization-centrifugation should be performed before characterization to remove ungrafted species. Several techniques are generally simultaneously used to evidence the surface chemical grafting of the nanoparticles.

5.12.1 X-ray diffraction analysis

It is important to verify that the X-ray diffraction pattern remains unchanged after chemical modification compared to the pristine sample and still displays the characteristics of cellulose I. The X-ray diffraction pattern of cellulose I is characterized by the main diffraction signals at 2θ values of 15°, 16°, 22.5° and 34° attributed to the diffraction planes 101, $10\bar{1}$, 002 and 040, respectively. If only the surface of the nanoparticle is chemically modified, the initial crystallinity should be retained. If the grafted chains are able to crystallize at the surface of the nanoparticle, extra diffraction peaks could be observed, corresponding to the crystalline brush surrounding the nanoparticle (Habibi and Dufresne, 2008; de Menezes et al., 2009).

5.12.2 Dispersion in organic solvent

A simple and valuable experiment to evidence the surface chemical modification efficiency consists in trying to suspend the modified nanoparticles in an organic solvent and comparing the stability of the suspension to neat nanoparticles. The unmodified nanoparticle suspension should settle more rapidly than modified nanoparticles. However, in some cases and particularly for polymer-grafted nanoparticles, one cannot exclude that free (ungrafted) species can interact with cellulose and behave as a surface-compatibilizer. A simple physical mixture of pristine cellulose nanoparticles and the surface chemical modifier can be prepared in the same proportion and observed under the same conditions. Examples are provided in Figure 5.15. After stopping the stirring agitation, all suspensions are turbid. However, differences appear in the long-term dispersion stability. Only the polymer-grafted suspensions remain turbid and homogeneous while the supernatant of other suspensions becomes clear and colorless. Polymer chains grafted on cellulose nanocrystal surface are highly solvated maintaining the nanoparticles in stable suspension in the organic solvent. However, this method also depends on many parameters such as DS.

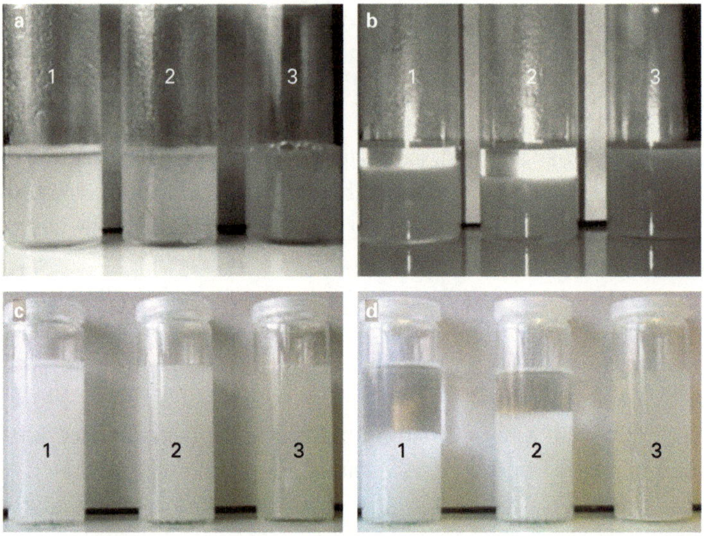

Fig. 5.15: (a), (b) Suspensions in toluene of (1) neat cellulose nanocrystal, (2) PCL and cellulose nanocrystals physical mixture, and (3) PCL-grafted cellulose nanocrystals (Habibi et al., 2008). Pictures recorded (a) immediately after stopping the stirring agitation and (b) 15 min later. (c), (d) Suspensions in chloroform of (1) neat cellulose nanocrystal, (2) PLA and cellulose nanocrystals physical mixture, and (3) PLA-grafted cellulose nanocrystals (Goffin et al., 2011b). Pictures recorded (c) immediately after stopping the stirring agitation and (d) 72 h later.

Another similar method consists in mixing the pristine or modified materials with two immiscible solvents, both with different polarities and densities and to observe with which solvent they are the best wetted. This gives a qualitative indication about the affinity between these two substances (Angellier et al., 2005).

5.12.3 Contact angle measurements

The contact angle is the angle at which a liquid/*vapor* interface meets a *solid* surface. If the liquid molecules are strongly attracted by the molecules of the solid surface (for example, water on hydrophilic cellulose) then a drop of the liquid tends to spread out on the solid surface, leading to a low contact angle value. Typically, the contact angle of water on a cellulose nanocrystal film surface ranges between 13° and 29° (Anglès and Dufresne, 2000; Eriksson et al., 2007; Aulin et al., 2009; Dankovich and Gray, 2011). Roughness of the cellulosic substrate can explain the dispersion in values. Upon hydrophobization of the nanoparticles, weaker attractions between water and solid molecules result in higher contact angles.

The dispersive and polar components of the surface energy of cellulose nanoparticles before and after chemical modification can be evaluated by applying the Owens–Wendt approach (Owens and Wendt, 1969). According to the Owens–Wendt model, the adhesion work is assumed to verify the following equation:

$$W_a = 2\sqrt{\gamma_S^d}\sqrt{\gamma_L^d} + 2\sqrt{\gamma_S^p}\sqrt{\gamma_L^p} \tag{5.1}$$

where $\gamma_L{}^d$, $\gamma_L{}^p$, $\gamma_S{}^d$ and $\gamma_S{}^p$ are the dispersive and polar surface tension of the liquid and the solid, respectively. According to the Young–Dupré equation, the adhesion work is also given by:

$$W_a = \gamma_L(1 + \cos\theta) \tag{5.2}$$

Thereby,

$$\gamma_L(1 + \cos\theta) = 2\sqrt{\gamma_S^d}\sqrt{\gamma_L^d} + 2\sqrt{\gamma_S^p}\sqrt{\gamma_L^p} \tag{5.3}$$

by plotting,

$$\frac{\gamma_L(1 + \cos\theta)}{2\sqrt{\gamma_L^d}} = f\left(\frac{\sqrt{\gamma_L^p}}{\sqrt{\gamma_L^d}}\right) \tag{5.4}$$

$\sqrt{\gamma_S^p}$ and $\sqrt{\gamma_S^d}$ are deduced from the slope and the Y-axis point, respectively, of the straight line that is obtained. The surface energy of cellulose nanoparticles is given by the following equation:

$$\gamma_S = \gamma_S^d + \gamma_S^p \tag{5.5}$$

For the static angle values, at least three standard liquids of different polarity should be used, namely purely dispersive probes such as hexadecane and α-bromonaphtalene and liquids with increasing polar character like diiodomethane, formamide and distilled water. Upon hydrophobization, the dispersive component of the surface energy of cellulose remains roughly constant while the polar component decreases (de Menezes et al., 2009).

5.12.4 Gravimetry

The grafting efficiency can be determined by weighting the grafted cellulose nanoparticles (Habibi et al., 2008; Goffin et al., 2011b, Littunen et al., 2011) even if it is not strictly speaking an experimental technique. The following parameters were derived from gravimetric measurements for polymer-grafted cellulose nanoparticles using the "grafting onto" approach (Littunen et al., 2011):

$$\begin{aligned} \text{Conversion} &= \frac{m_G + m_P}{m_M} \cdot 100\% \\ \text{Graft yield} &= \frac{m_G}{m_C} \cdot 100\% \\ \text{Graft efficiency} &= \frac{m_G}{m_G + m_P} \cdot 100\% \\ \text{Polymer weight fraction} &= \frac{m_G}{m_G + m_C} \cdot 100\% \end{aligned} \tag{5.6}$$

where m_G, m_P, m_M and m_C correspond to the mass of the grafted polymer, homopolymer, fed monomer and cellulose, respectively. However, to apply this method one should ensures that the weight gain does not result from a solvent uptake when using a swelling solvent.

5.12.5 Fourier transform infrared (FTIR) spectroscopy

FTIR is a spectroscopic technique classically used to identify and study chemical compounds. It is based on how infrared radiation is absorbed by the compounds' chemical bonds. The mid-infrared, approximately 4000–400 cm^{-1} (2.5–25 μm), is used to study the fundamental vibrations and associated rotational-vibrational structure. A low amount of dried cellulose nanoparticles (typically 1 wt%) is generally ground into a fine powder with KBr and pressed as pellets.

The FTIR spectrum of unmodified cellulose nanoparticles shows characteristic bands, viz. the hydrogen bonded OH stretching at ca. 4000–2995 cm^{-1}, the CH stretching at 2900 cm^{-1}, the OH bending of adsorbed water at 1635 cm^{-1}, the CH_2 bending at 1430 cm^{-1}, the CH bending at 1380 cm^{-1}, the C—O stretching at 1058 and 1035 cm^{-1}, the CH bending or CH_2 stretching at 900 cm^{-1}, which indicates the amorphous structure, e.g. in MFC, and the OH out-of-plane bending at 687 cm^{-1} (Klemm et al., 1998).

New bands are observed for chemically modified cellulose nanoparticles. The assignment of these new bands linked to the new functional groups introduced at the surface of the nanoparticles can be performed using an IR spectroscopy correlation table.

5.12.6 Elemental analysis

Elemental analysis is used to quantify the mass fraction of carbon, hydrogen, oxygen and other atoms. It is based on atomic absorption of the investigated element. Compared to the pristine cellulose nanoparticle, the chemical modification changes the composition of the sample. The DS value can be obtained from this technique according to:

$$DS = \frac{6 \cdot Mc - C \cdot M_{AGU}}{M_g \cdot C - M_{Cg}} \tag{5.7}$$

where C is the relative carbon content in the sample, and $6 \cdot Mc$, M_{AGU}, M_g and M_{Cg} correspond to the carbon molecular weight of one anhydroglucose unit (72.07), mass of anhydroglucose unit (162.14), mass of the grafted molecule, and carbon mass of the grafted molecule, respectively.

5.12.7 X-ray photoelectron spectroscopy (XPS)

XPS is a powerful tool to investigate chemical changes resulting from surface modification. In XPS experiments, photons of a specific energy are used to excite the electronic states of atoms below the surface of the sample. Electrons ejected from the surface are energy-filtered before the intensity for a defined energy is recorded by a detector. Since core level electrons in solid-state atoms are quantified, the resulting energy spectra exhibit resonance peaks characteristic of the electronic structure for atoms at the sample surface. The escape depth of the ejected electrons is limited and ejected electrons from depths greater than 10 nm have a low probability of leaving the surface without undergoing an energy loss event, and therefore contribute to the background signal rather than well-defined photoelectric peaks. It therefore eliminates from the analysis the bulk of unmodified bulk material. Quantitative XPS analysis can be performed by calculating the atomic concentrations from the photoelectron peak areas using Gaussian –Lorentzian deconvolutions. Low resolution spectrum for cellulose shows that carbon and oxygen atoms are the main components. The corresponding peaks are observed around 287 and 534 eV, respectively. A small amount of sulfur is also detected when using H_2SO_4-hydrolysed nanocrystals.

The carbon 1s spectrum can be resolved into different contributions of bonded carbon which reflect their local environments, namely carbon without oxygen bonds C1 (C—C and C—H_x), carbon with one oxygen bond C2 (C—O), carbon with two oxygen bonds C3 (O—C—O) and carbon with three oxygen bonds C4 (O—C=O). The carbon-without-oxygen-bond contribution in the C1s emission is always set at 285.0 eV (Watts and Wolstenholme, 2003). From the formula of pure cellulose (Figure 5.1), the theoretical O/C ratio is 0.83, the contributions of C1 and C4 should be zero and the ratio of carbon atoms with two bonds to oxygen relative to carbon atoms with one bond to oxygen (C3/C2) should be 0.2. However, small amounts of C—C/ C—H_x (C1) and O—C═O (C4) carbon are always present in XPS spectra of pure cellulose nanoparticles although the cellulose molecule does not contain such carbon bonds. The presence of these non-cellulosic carbons has been explained by accumulation of air-borne contaminants and hemicelluloses (Johansson et al., 2011) as detailed in Section 5.2 of the present chapter.

After chemical modification, the proportions of carbon atoms involved in various chemical bonds are expected to change and new types of carbon bonds can appear depending on the grafted molecule. An example is given in Figure 5.16 that shows the deconvolution of the C1s peak for cellulose nanocrystals extracted from ramie fibers and chemically modified by grafting organic acid chlorides with different lengths of the aliphatic chain by an esterification reaction (de Menezes et al., 2008). The intensity of the C1 peak referring to aliphatic chains (C—H) considerably increased for modified samples compared to the pristine cellulose nanocrystals. The surface coverage by the alkane chains (θ_{SC}) can be estimated from the C1 relative area according to:

$$\theta_{SC} = A_{C1m} - A_{C1um} \tag{5.8}$$

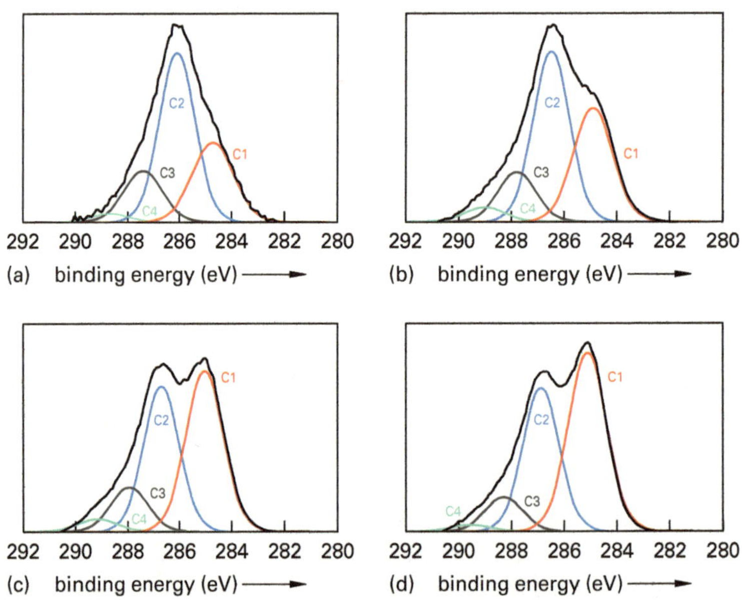

Fig. 5.16: Deconvolution of the C1S XPS signal into its constituent contributions (C1: C—C and C—H_x, C2: C—O, C3: O—C—O, and C4: O—C═O) for (a) unmodified ramie cellulose nanocrystals, and chemically modified with (b) hexanoyl chloride, (c) lauroyl chloride, and (d) stearoyl chloride (de Menezes et al., 2009).

where A_{C1m} and A_{C1um} correspond to the areas for the modified and unmodified samples, respectively.

5.12.8 Time of flight mass spectrometry (TOF-MS)

TOF-MS is a surface-sensitive analytical method that uses a pulsed ion beam to remove molecules from the very outermost surface (atomic monolayers) of a sample. These ions are accelerated by an electric field. Ions with the same electric charge acquire the same kinetic energy but their velocity depends on the mass/charge ratio. This ratio determines the time (time of flight) taken by a charged molecule to reach a detector located at a known distance. Two ionization techniques are considered, viz. matrix-assisted laser desorption/ionization (MALDI-TOF-MS) and secondary ion mass spectrometry (TOF-SIMS).

MALDI-TOF-MS was used to show that PEO chains grafted onto the surface of cellulose nanocrystals had the same atomic mass units fragments than the starting polymer before grafting (Kloser and Gray, 2010). It was concluded that the PEO chains were covalently bonded to the nanocrystal surface. This technique was also used to determine the molecular weights of PAA chains grafted on cellulose nanocrystals (Majoinen et al., 2011).

5.12.9 Solid-state NMR spectroscopy

The solid state NMR spectrum of unmodified cellulose nanoparticles exhibits a pattern typical of native cellulose I with specific signals (Azzam et al., 2010). The region from 60 to 70 ppm is assigned to the hydroxymethyl C6 carbons, split in a 62–63 ppm contribution from the disordered chains mainly at the surface of the nanoparticles and 65 ppm signal arising from the crystalline core. The region between 70 and 80 ppm is attributed to C2, C3, and C5 carbons that are undistinguishable. The region from 80 to 91 ppm is assigned to C4 carbons, where the wide signal at 83–84 ppm corresponds to the C4 carbons of disordered cellulose chains and the sharp signals at 88–89 ppm to crystalline ones. The region around 105 ppm is ascribed to C1 carbons. For oxidized cellulose nanoparticles, an additional signal around 174 ppm characteristic of the carboxylic acid is observed (Montanari et al., 2005). Upon grafting, important changes occur in the NMR spectrum of cellulose nanoparticles and new resonances characteristic of the grafted species appear. According to the attribution and area of each signal, the DS of modified nanoparticles can be estimated by the ratio of corresponding areas.

5.12.10 Thermogravimetric analysis (TGA)

Thermal stability is a crucial factor, especially when the cellulose nanoparticles are used as a reinforcement for polymers, because melt processing requires elevated temperatures. Compared to inorganic nanoparticles, many biomaterials suffer from inferior thermal stability. It is therefore important that alterations of the surface chemistry do not decrease the onset temperature of thermal degradation. TGA is typically performed by heating the sample from room temperature to 600–900°C at a given heating rate in a nitrogen, air or helium atmosphere.

TGA has been used to estimate the amount of PCL that was grafted on the surface of MFC (Lönnberg et al., 2008). Both MFC and free PCL displayed a monomodal shape derivative curve (DTGA) with only one maximum degradation rate temperature. On the contrary, the DTGA curves of PCL-grafted MFC had a bimodal shape with two different maxima and curve fitting was used to determine the sample composition. H_2SO_4-prepared cellulose nanocrystals display a reduced thermal stability because of the sulfate groups formed during the acid hydrolysis treatment. It was shown that their degradation temperature can be increased either by physical adsorption of macromolecules (Ben Azouz et al., 2012) or surface chemical grafting (Yi et al., 2009).

5.12.11 Differential scanning calorimetry (DSC)

Because of strong H-bonding interactions between cellulose chains, no thermal transition is observed before the degradation of the material. In some cases, a melting endotherm is observed for polymer-grafted cellulose nanoparticles (Habibi and Dufresne, 2008; Habibi et al., 2008; Lönnberg et al., 2008; Cao et al., 2009). It is attributed to the grafted polymeric chains that form a crystalline structure at the surface of the nanoparticles. Compared to the typical values of the melting point reported for the free polymer, the temperature position of this melting endotherm is lower because of the restricted size of these crystallites. As the molecular weight of the grafted chains increased, their melting temperature increased as shown for PCL-grafted MFC (Lönnberg et al., 2008). Because of the restricted mobility of grafted chains compared to free polymer chains, a lower degree of crystallinity and longer crystallization time were observed (Lönnberg et al., 2008).

5.13 Conclusions

Cellulose is inherently hydrophilic making it incompatible with organic liquid media and hydrophobic polymer matrices. However, the reactive hydroxyl groups of cellulose can be advantageously exploited to tune the surface properties of cellulose nanoparticles. Several strategies, including non-covalent and covalent bonding, have been proposed in the literature to hydrophobize these nanoparticles. It allows broadening the range of liquid media in which cellulose nanoparticles can be suspended and polymer matrices that can be used in association with them to process nanocomposite materials. Most of the studies reported on the surface chemical modification involve cellulose nanocrystals and covalent bonding. The common surface chemical modifications of cellulose nanocrystals are summarized in Figure 5.17 (Lin et al., 2012). They can be categorized into three distinctive groups, namely (1) substitution of hydroxyl groups with small molecules (as indicated with red arrows in Figure 5.17), (2) polymer grafting based on the "grafting onto" strategy with different coupling agents (as indicated with blue arrows in Figure 5.17), and (3) polymer

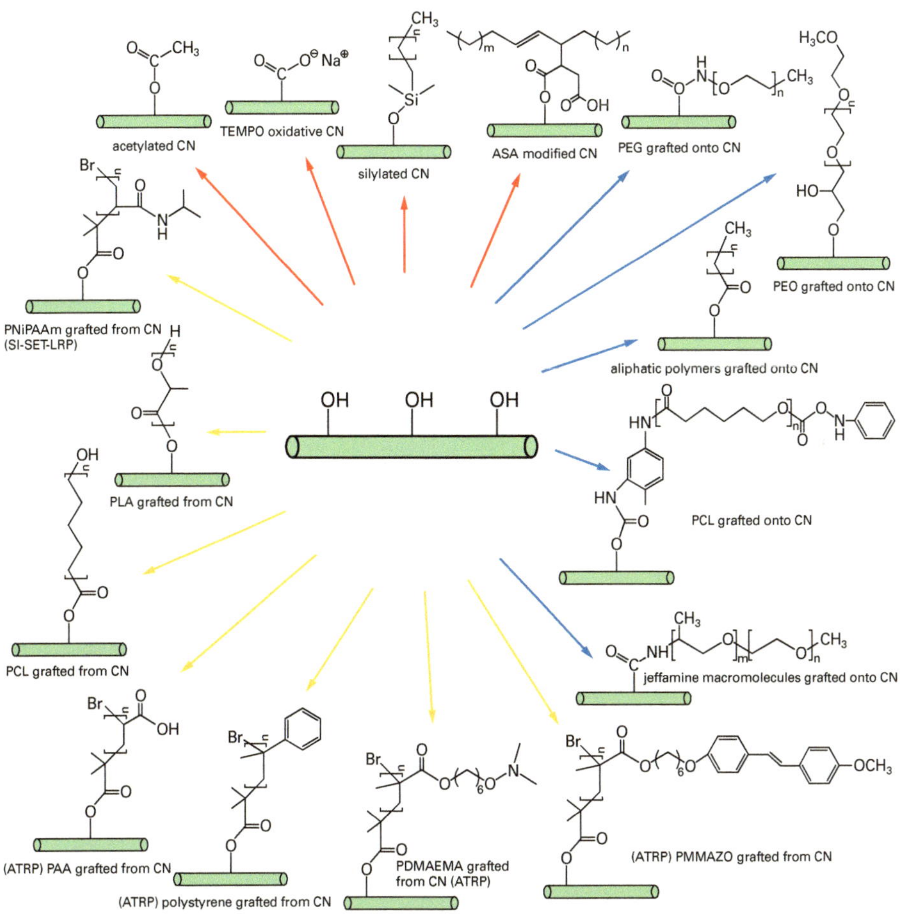

Fig. 5.17: Common surface covalent chemical modifications of cellulose nanocrystals. PEG: poly(ethylene glycol); PEO: poly(ethylene oxide); PLA: poly(lactic acid); PAA: poly(acrylic acid); PNiPAAm: poly(*N*-isopropylacrylamide); PDMAEMA: poly(*N*,*N*-dimethylaminoethyl methacrylate) (Lin et al., 2012).

grafting based on the "grafting from" approach with a radical polymerization involving ring opening polymerization (ROP), atom transfer radical polymerization (ATRP) and single-electron transfer living radical polymerization (SET-LP) (as indicated with yellow arrows in Figure 5.17).

5.14 References

Andresen, M., Johansson, L.S., Tanem, B.S., and Stenius, P. (2006). Properties and characterization of hydrophobized microfibrillated cellulose. Cellulose *13*, 665–677.

Andresen, M. and Stenius, P. (2007). Water-in-oil emulsions stabilized by hydrophobized microfibrillated cellulose. J. Disper. Science Technol. *28*, 837–844.

Andresen, M., Stenstad, P., Møretrø, T., Langsrud, S., Syverud, K., Johansson, L.S. and Stenius, P. (2007). Nonleaching antimicrobial films prepared from surface-modified microfibrillated cellulose. Biomacromolecules *8*, 2149–2155.

Angellier, H., Molina-Boisseau, S., Belgacem, M.N. and Dufresne, A. (2005). Surface chemical modification of waxy maize starch nanocrystals. Langmuir *21*, 2425–2433.

Anglès, M.N. and Dufresne, A. (2000). Plasticized starch/tunicin whiskers nanocomposites. 1. Structural analysis. Macromolecules *33*, 8344–8353.

Araki, J., Wada, M. and Kuga, S. (2001). Steric stabilization of a cellulose microcrystal suspension by poly(ethylene glycol) grafting. Langmuir *17*, 21–27.

Aulin, C., Shchukarev, A., Lindqvist, J., Malmström, E., Wågberg, L. and Lindström, T. (2008). Wetting kinetics of oil mixtures on fluorinated model cellulose surfaces. J. Colloid Interface Sci. *317*, 556–567.

Aulin, C. Ahola, S., Josefsson, P., Nishino, T., Hirose, Y., Österberg, M. and Wågberg, L. (2009). Nanoscale cellulose films with different crystallinities and mesostructures – Their surface properties and interaction with water. Langmuir *25*, 7675–7685.

Azzam, F., Heux, L., Putaux, J.L. and Jean, B. (2010). Preparation by grafting onto, characterization, and properties of thermally responsive polymer-decorated cellulose nanocrystals. Biomacromolecules *11*, 3652–3659.

Barazzouk, S. and Daneault, C. (2011). Spectroscopic characterization of oxidized nanocellulose grafted with fluorescent amino acids. Cellulose *18*, 643–653.

Ben Azouz, K., Ramires, E.C., Van den Fonteyne, W., El Kissi, N. and Dufresne A. (2012). A simple method for the melt extrusion of cellulose nanocrystal reinforced hydrophobic polymer. ACS Macro Lett. *1*, 236–240.

Bergenstråhle, M., Mazeau, K. and Berglund, L.A. (2008). Molecular modeling of interfaces between cellulose crystals and surrounding molecules: Effects of caprolactone surface grafting. Eur. Polym. J. *44*, 3662–3669.

Berlioz, S., Molina-Boisseau, S., Nishiyama, Y. and Heux, L. (2009). Gas-phase surface esterification of cellulose microfibrils and whiskers. Biomacromolecules *10*, 2144–2151.

Besbes, I., Alila, S. and Boufi, S. (2011a). Nanofibrillated cellulose from TEMPO-oxidized eucalyptus fibres: Effect of the carboxyl content. Carbohydr. Polym. *84*, 975–983.

Besbes, I., Rei Vilar, M. and Boufi, S. (2011b). Nanofibrillated cellulose from alfa, eucalyptus and pine fibres: preparation, characteristics and reinforcing potential. Carbohydr. Polym. *86*, 1198–1206.

Bondeson, D. and Oksman, K. (2007). Dispersion and characteristics of surfactant modified cellulose whiskers nanocomposites. Compos. Interfaces *14*, 617–630.

Bonini, C., Heux, L. Cavaillé, J.Y., Lindner, P., Dewhurst, C. and Terech, P. (2002). Rodlike cellulose whiskers coated with surfactant: A small-angle neutron scattering characterization. Langmuir *18*, 3311–3314.

Braun, B. and Dorgan, J.R. (2009). Single-step method for the isolation and surface functionalization of cellulosic nanowhiskers. Biomacromolecules *10*, 334–341.

Cao, X., Habibi, Y. and Lucia, L.A. (2009). One-pot polymerization, surface grafting, and processing of waterborne polyurethane-cellulose nanocrystal nanocomposites. J. Mater. Chem. *19*, 7137–7145.

Cateto, C.A. and Ragauskas, A. (2011). Amino acid modified cellulose whiskers. RSC Advances *1*, 1695–1697.

Çetin, N.S., Tingaut, P., Ozmen, N., Henry, N., Harper, D., Dadmun, M. and Sebe, G. (2009). Acetylation of cellulose nanowhiskers with vinyl acetate under moderate conditions. Macromol. Biosci. *9*, 997–1003.

Dankovich, T.A. and Gray, D.G. (2011). Contact angle measurements on smooth nanocrystalline cellulose (I) thin films, J. Adhesion Sci. Technol. *25*, 699–708.

de Menezes, A.J., Siqueira, G., Curvelo, A.A.S. and Dufresne, A. (2009). Extrusion and characterization of functionalized cellulose whisker reinforced polyethylene nanocomposites. Polymer *50*, 4552–4563.

De Nooy, A.E.J., Besemer, A.C. and Van Bekkum, H. (1995). Highly selective nitroxyl radical-mediated oxidation of primary alcohol groups in water-soluble glucans. Carbohydr. Res. 269, 89–98.

Dong, S. and Roman, M. (2007). Fluorescently labeled cellulose nanocrystals for bioimaging applications. J. Am. Chem. Soc. *129*, 13810–13811.

Dufresne, A. (2010). Processing of polymer nanocomposites reinforced with polysaccharide nanocrystals. Molecules *15*, 4111–4128.

Elazzouzi-Hafraoui, S., Putaux, J.L. and Heux, L. (2009). Self-assembling and chiral nematic properties of organophilic cellulose nanocrystals. J. Phys. Chem. B *113*, 11069–11075.

Eriksson, M., Notley, S.M. and Wågberg, L. (2007). Cellulose thin films: Degree of cellulose ordering and its influence on adhesion. Biomacromolecules *8*, 912–919.

Eyholzer, C., Bordeanu, N., Lopez-Suevos, F., Rentsch, D., Zimmermann, T. and Oksman, K. (2010). Preparation and characterization of water-redispersible nanofibrillated cellulose in powder form. Cellulose, *17*, 19–30.

Eyley, S. and Thielemans, W. (2011). Imidazolium grafted cellulose nanocrystals for ion exchange applications. Chem. Comm. *47*, 4177–4179.

Favier, V., Canova, G.R., Shrivastava, S.C. and Cavaillé, J.Y. (1997). Mechanical percolation in cellulose whiskers nanocomposites. Polym. Eng. Sci. *37*, 1732–1739.

Filpponen, I. and Argyropoulos, D.S. (2010). Regular linking of cellulose nanocrystals via click chemistry: Synthesis and formation of cellulose nanoplatelet gels. Biomacromolecules *11*, 1060–1066.

Follain, N., Marais, M.F., Montanari, S. and Vignon, M.R. (2010). Coupling onto surface carboxylated cellulose nanocrystals. Polymer *51*, 5332–5344.

Fortunati, E., Armentano, I., Zhou, Q., Iannoni, A., Saino, E., Visai, L., Berglund, L.A. and Kenny, J.M. (2012). Multifunctional bionanocomposite films of poly(lactic acid), cellulose nanocrystals and silver nanoparticles. Carbohydr. Polym. *87*, 1596–1605.

Gardner, K.H. and Blackwell, J. (1974). The structure of native cellulose. Biopolymers *13*, 1975–2001.

Goffin, A.-L. (2010). Polymer bionanocomposites reinforced by functionalized nanoparticles: Impact of nanofiller size, nature and composition. PhD Thesis, Mons University, Belgium.

Goffin, A.-L., Raquez, J.-M., Duquesne, E., Siqueira, G., Habibi, Y., Dufresne, A. and Dubois, Ph. (2011a). Poly(e-caprolactone) based nanocomposites reinforced by surface-grafted cellulose nanowhiskers via extrusion processing: Morphology, rheology, and thermo-mechanical properties. Polymer *52*, 1532–1538.

Goffin, A.-L., Raquez, J.-M., Duquesne, E., Siqueira, G., Habibi, Y., Dufresne, A. and Dubois, Ph. (2011b). From interfacial ring-opening polymerization to melt processing of cellulose nanowhisker-filled polylactide-based nanocomposites. Biomacromolecules *12*, 2456–2465.

Gopalan Nair, K., Dufresne, A., Gandini, A. and Belgacem, M.N. (2003). Crab shell chitin whiskers reinforced natural rubber nanocomposites. 3. Effect of chemical modification of chitin whiskers. Biomacromolecules *4*, 1835–1842.

Goussé, C., Chanzy, H., Excoffier, G., Soubeyrand, L. and Fleury, E. (2002). Stable suspensions of partially silylated cellulose whiskers dispersed in organic solvents. Polymer *43*, 2645–2651.

Goussé, C., Chanzy, H., Cerrada, M.L. and Fleury, E. (2004). Surface silylation of cellulose microfibrils: Preparation and rheological properties. Polymer *45*, 1569–1575.

Grunert, M. and Winter, W. T. (2002a). Nanocomposites of cellulose acetate butyrate reinforced with cellulose nanocrystals. J. Polym. Envir. *10*, 27–30.

Grunert, M. and Winter, W. T. (2002b). Cellulose nanocrystal reinforced acetate butyrate nanocomposites. Polym. Mater. Sci. Eng. *86*, 367–368.

Habibi, Y., Chanzy, H. and Vignon, M.R. (2006). TEMPO-mediated surface oxidation of cellulose whiskers. Cellulose *13*, 679–687.

Habibi, Y. and Dufresne, A. (2008). Highly filled bionanocomposites from functionalized polysaccharide nanocrystals. Biomacromolecules *9*, 1974–1980.

Habibi, Y., Goffin, A.L., Schiltz, N., Duquesne, E., Dubois, P. and Dufresne A. (2008). Bionanocomposites based on poly(ε-caprolactone)-grafted cellulose nanocrystals by ring opening polymerization. J. Mat. Chem. *18*, 5002–5010.

Habibi, Y., Hoegert, I., Kelley, S.S., and Rojas, O.J. (2010). Development of Langmuir–Schaeffer cellulose nanocrystal monolayers and their interfacial behaviours. Langmuir *26*, 990–1001.

Harrisson, S., Drisko, G.L., Malmström, E., Hult, A. and Wooley, K.L. (2011). Hybrid rigid/soft and biologic/synthetic materials: Polymers grafted onto cellulose microcrystals. Biomacromolecules *12*, 1214–1223.

Hasani, M., Cranston, E.D., Westman, G. and Gray, D.G. (2008). Cationic surface functionalisation of cellulose nanocrystals. Soft Matter *4*, 2238–2244.

Hebeish, A. and Guthrie, J.T. (1981). The chemistry and technology of cellulosic copolymers (Springer-Verlag, Berlin).

Helbert, W., Cavaillé, J.Y. and Dufresne, A. (1996). Thermoplastic nanocomposites filled with wheat straw cellulose whiskers. Part I: Processing and mechanical behavior. Polym. Comp. *17*, 604–611.

Heux, L., Chauve, G. and Bonini, C. (2000). Nonflocculating and chiral-nematic self-ordering of cellulose microcrystals suspensions in nonpolar solvents. Langmuir *16*, 8210–8212.

Ifuku, S., Nogi, M., Abe, K., Handa, K., Nakatsubo, F. and Yano, H. (2007). Surface modification of bacterial cellulose nanofibers for property enhancement of optically transparent composites: Dependence on acetyl-group DS. Biomacromolecules *8*, 1973–1978.

Isogai, A., Saito, T. and Fukuzumi, H. (2011). TEMPO-oxidized cellulose nanofibers. Nanoscale *3*, 71–85.

Johansson, L.S., Tammelin, T., Campbell, J.M., Setälä, H. and Österberg, M. (2011). Experimental evidence on medium driven cellulose surface adaptation demonstrated using nanofibrillated cellulose. Soft Matter *7*, 10917–10924.

Jonoobi, M., Harun, J., Mathew, A.P., Hussein, M.Z.B. and Oksman, K. (2010). Preparation of cellulose nanofibers with hydrophobic surface characteristics. Cellulose *17*, 299–307.

Kim, J., Montero, G. Habibi, Y., Hinestroza, J.P., Genzer, J., Argyropoulos, D.S. and Rojas, O.J. (2009). Dispersion of cellulose crystallites by nonionic surfactants in a hydrophobic polymer matrix. Polym. Eng. Sci. *49*, 2054–2061.

Klemm, D., Philipp, B., Heinze, T. Heinze, U., and Wagenknecht, W. (1998). Comprehensive cellulose chemistry. Vol. 1: Fundamentals and analytical methods (Wiley-VCH Verlag GmbH & Co, Weinheim, Germany).

Kloser, E. and Gray, D.G. (2010). Surface grafting of cellulose nanocrystals with poly(ethylene oxide) in aqueous media. Langmuir *26*, 13450–13456.

Kowalski, A., Libiszowski, J., Biela, T., Cypryk, M., Duda, A. and Penczek, S. (2005). Kinetics and mechanism of cyclic esters polymerization initiated with Tin(II) octoate. Polymerization of e-caprolactone and L,L-lactide co-initiated with primary amines. Macromolecules *38*, 8170–8176.

Krassig, H.A. (1985). Structure of cellulose and its relation to properties of cellulose fibers. In: Cellulose and its derivatives: Chemistry, biochemistry and applications, J.F. Kennedy, G.O. Phillips, D.J. Wedlock and P.A. Williams, eds. (Ellis Horwood Limited, Chichester, UK), pp. 3–25.

Krassig, H.A. (1993). Cellulose-structure, accessibility and reactivity (Gordon and Breach Science Publisher, Yverdon, Switzerland).

Kvien, I., Tanem, B.S. and Oksman, K. (2005). Characterization of cellulose whiskers and their nanocomposites by atomic force and electron microscopy. Biomacromolecules *6*, 3160–3165.

Labet, M., Thielemans, W. and Dufresne A. (2007). Polymer grafting onto starch nanocrystals. Biomacromolecules *8*, 2916–2927.

Labet, M. and Thielemans, W. (2011). Improving the reproducibility of chemical reactions on the surface of cellulose nanocrystals: ROP of e-caprolactone as a case study. Cellulose *18*, 607–617.

Lee, K.Y., Quero, F., Blaker, J.J., Hill, C.A.S., Eichhorn, S.J. and Bismarck, A. (2011). Surface only modification of bacterial cellulose nanofibres with organic acids. Cellulose *18*, 595–605.

Lin, N., Chen, G., Huang, J., Dufresne, A. and Chang, P.R. (2009). Effects of polymer-grafted natural nanocrystals on the structure and mechanical properties of poly(lactic acid): a case of cellulose whisker-graft-polycaprolactone. J. Appl. Polym. Sci. *113*, 3417–3425.

Lin, N., Huang, J., Chang, P.R., Feng, J. and Yu, J. (2011). Surface acetylation of cellulose nanocrystal and its reinforcing function in poly(lactic acid). Carbohydr. Polym. *83*, 1834–1842.

Lin, N., Huang, J. and Dufresne A. (2012). Preparations, properties and applications of polysaccharide nanocrystals in advanced functional nanomaterials: A review. Nanoscale *4*, 3274–3294.

Littunen, K., Hippi, U., Johansson, L.S., Österberg, M., Tammelin, T., Laine, J. and Seppälä, J. (2011). Free radical graft polymerization of nanofibrillated cellulose with acrylic monomers. Carbohydr. Polym. *84*, 1039–1047.

Lönnberg, H., Fogelström, L., Azizi Samir, M.A.S., Berglund, L., Malmström, E. and Hult, A. (2008). Surface grafting of microfibrillated cellulose with poly(e-caprolactone) – Synthesis and characterization. Eur. Polym. J. *44*, 2991–2997.

Lu, J., Askeland, P. and Drzal, L.T. (2008). Surface modification of microfibrillated cellulose for epoxy composite applications. Polymer *49*, 1285–1296.

Mahmoud, K.A., Mena, J.A., Male, K.B., Hrapovic, S., Kamen, A. and Luong, J.H.T. (2010). Effect of surface charge on the cellular uptake and cytotoxicity of fluorescent labeled cellulose nanocrystals. ACS Appl. Mater. Interfaces *2*, 2924–2932.

Majoinen, J., Walther, A., McKee, J.R., Kontturi, E., Aseyev, V., Malho, J.M., Ruokolainen, J. and Ikkala, O. (2011). Polyelectrolyte brushes grafted from cellulose nanocrystals using Cu-mediated surface-initiated controlled radical polymerization. Biomacromolecules *12*, 2997–3006.

Mangalam, A.P., Simonsen, J. and Benight, A.S. (2009). Cellulose/DNA hybrid nanomaterials. Biomacromolecules *10*, 497–504.

Montanari, S., Roumani, M., Heux, L. and Vignon, M.R. (2005). Topochemistry of carboxymethylated nanocrystals resulting from TEMPO-mediated oxidation. Macromolecules *38*, 1665–1671.

Moon, R.J., Martini, A., Nairn, J., Simonsen, J. and Youngblood, J. (2011). Cellulose nanomaterials review: Structure, properties and nanocomposites. Chem. Soc. Rev. *40*, 3941–3994.

Morandi, G., Heath, L. and Thielemans, W. (2009). Cellulose nanocrystals grafted with polystyrene chains through surface-initiated atom transfer radical polymerization (SI-ATRP). Langmuir *25*, 8280–8286.

Nielsen, L.J., Eyley, S., Thielemans, W. and Aylott, J.W. (2010). Dual fluorescent labelling of cellulose nanocrystals for pH sensing. Chem. Commun. *46*, 8929–8931.

Nogi, M., Abe, K., Handa, K., Nakatsubo, F., Ifuku, S. and Yano, H. (2006). Property enhancement of optically transparent bionanofiber composites by acetylation. Appl. Phys. Lett. *89*, 233123-1-233123-3.

Owens, D.K. and Wendt, R.C. (1969). Estimation of the surface free energy of polymers. J. Appl. Polym. Sci. *13*, 1741–1747.

Pei, A., Zhou, Q. and Berglund, L. (2010). Functionalized cellulose nanocrystals as biobased nucleation agents in poly(L-lactide) (PLLA) – Crystallization and mechanical property effects. Compos. Sci. Technol. *70*, 815–821.

Raquez, J.M., Murena, Y., Goffin, A.L., Habibi, Y., Ruelle, B., DeBuyl, F. and Dubois, P. (2012). Surface-modification of cellulose nanowhiskers and their use as nanoreinforcers into polylactide: A sustainably-integrated approach. Compos. Sci. Technol. *72*, 544–549.

Rodionova, G., Lenes, M., Eriksen, Ø. and Gregersen, Ø. (2011). Surface chemical modification of microfibrillated cellulose: Improvement of barrier properties for packaging applications. Cellulose *18*, 127–134.

Rojas, O.J., Montero, G.A. and Habibi, Y. (2009). Electrospun nanocomposites from polystyrene loaded with cellulose nanowhiskers. J. Appl. Polym. Sci. *113*, 927–935.

Roy, D., Semsarilar, M., Guthrie, J.T. and Perrier, S. (2009). Cellulose modification by polymer grafting: A review. Chem. Soc. Rev. *38*, 2046–2064.

Saito, T. and Isogai, A. (2004). TEMPO-mediated oxidation of native cellulose. The effect of oxidation conditions on chemical and crystal structures of the water-insoluble fractions. Biomacromolecules *5*, 1983–1989.

Saito, T. and Isogai, A. (2006). Introduction of aldehyde groups on surfaces of native cellulose fibers by TEMPO-mediated oxidation. Colloid Surface A *289*, 219–225.

Saito, T., Hirota, M., Tamura, N., Kimura, S., Fukuzumi, H., Heux, L. and Isogai, A. (2009). Individualization of nano-sized plant cellulose fibrils by direct surface carboxylation using TEMPO catalyst under neutral conditions. Biomacromolecules *10*, 1992–1996.

Sassi, J.F. and Chanzy, H. (1995). Ultrastructural aspects of the acetylation of cellulose. Cellulose *2*, 111–127.

Shibata, I. and Isogai, A. (2003). Depolymerization of cellouronic acid during TEMPO-mediated oxidation. Cellulose *10*, 151–158.

Siqueira, G., Bras, J. and Dufresne, A. (2009). New process of chemical grafting of cellulose nanoparticles with a long chain isocyanate. Langmuir *26*, 402–411.

Sobkowicz, M.J., Braun, B. and Dorgan, J.R. (2009). Decorating in green: Surface esterification of carbon and cellulosic nanoparticles. Green Chem. *11*, 680–682.

Stenstad, P., Andresen, M., Tanem, B.S. and Stenius, P. (2008). Chemical surface modifications of microfibrillated cellulose. Cellulose *15*, 35–45.

Storey, R.F. and Sherman, J.W. (2002). Kinetics and mechanism of the stannous octoate-catalyzed bulk polymerization of e-caprolactone. Macromolecules *35*, 1504–1512.

Syverud, K., Xhanari, K., Chinga-Carrasco, G., Yu, Y. and Stenius, P. (2011). Films made of cellulose nanofibrils: surface modification by adsorption of a cationic surfactant and characterization by computer-assisted electron microscopy J. Nanoparticle Res. *13*, 773–782.

Tingaut, P., Zimmermann, T. and Lopez-Suevos, F. (2010). Synthesis and characterization of bionanocomposites with tunable properties from poly(lactic acid) and acetylated microfibrillated cellulose. Biomacromolecules *11*, 454–464.

Trejo-O'Reilly, J.A., Cavaillé, J.Y. and Gandini, A. (1997). The surface chemical modification of cellulosic fibres in view of their use in composite materials. Cellulose *4*, 305–320.
Wågberg, L., Decher, G., Norgren, M., Lindström, T., Ankerfors, M. and Axnäs, K. (2008). The build-up of polyelectrolyte multilayers of microfibrillated cellulose and cationic polyelectrolytes. Langmuir *24*, 784–795.
Wakelyn, P.J. (1998). Chemical properties of cotton. In: Handbook of fiber chemistry, M. Levin and E.M. Pearce, eds. (Marcel Dekker, New York), pp. 642–654.
Wang, N., Ding, E. and Cheng, R. (2007). Surface modification of cellulose nanocrystals. Front. Chem. Eng. China *1*, 228–232.
Watts, J.F. and Wolstenholme, J. (2003). An introduction to surface analysis by XPS and AES (John Wiley & Sons Ltd, Chichester).
Xhanari, K., Syverud, K., Chinga-Carrasco, G., Paso, K. and Stenius, P. (2011). Reduction of water wettability of nanofibrillated cellulose by adsorption of cationic surfactants. Cellulose *18*, 257–270.
Xu, Q., Yi, J., Zhang, X. and Zhang, H. (2008). A novel amphotropic polymer based on cellulose nanocrystals grafted with azo polymers. Eur. Polym. J. *44*, 2830–2837.
Yi, J., Xu, Q., Zhang, X. and Zhang, H. (2008). Chiral-nematic self-ordering of rodlike cellulose nanocrystals grafted with poly(styrene) in both thermotropic and lyotropic states. Polymer *49*, 4406–4412.
Yi, J., Xu, Q., Zhang, X. and Zhang, H. (2009). Temperature-induced chiral nematic changes of suspensions of poly(N,N-dimethylaminoethyl methacrylate)-grafted cellulose nanocrystals. Cellulose *16*, 989–997.
Yuan, H., Nishiyama, Y., Wada, M. and Kuga, S. (2006). Surface acylation of cellulose whiskers by drying aqueous emulsion. Biomacromolecules *7*, 696–700.
Zhou, Q., Brumer, H. and Teeri, T.T. (2009). Self-organization of cellulose nanocrystals adsorbed with xyloglucan oligosaccharide-poly(ethylene glycol)-polystyrene triblock copolymer. Macromolecules *42*, 5430–5432.
Zoppe, J.O., Peresin, M.S., Habibi, Y., Venditti, R.A. and Rojas, O.J. (2009). Reinforcing poly(e-caprolactone) nanofibers with cellulose nanocrystals. ACS Appl. Mat. Interfaces *1*, 1996–2004.
Zoppe, J.O., Habibi, Y., Rojas, O.J., Venditti, R.A., Johansson, L.S., Efimenko, K., Österberg, M. and Laine, J. (2010). Poly(N-isopropylacrylamide) brushes grafted from cellulose nanocrystals via surface-initiated single-electron transfer living radical polymerization. Biomacromolecules *11*, 2683–2691.

6 Rheological behavior of nanocellulose suspensions and self-assembly

Cellulosic nanoparticles exhibit both a high specific surface area and high density of surface hydroxyl groups. These properties influence their interactions either in the liquid state as suspension or in the solid state as film. The rheological behavior of suspensions plays an important role in some processing techniques of nanocomposites. Moreover, hydrolyzed rod-like cellulosic nanoparticle suspensions display a crystalline liquid behavior that can lead to some interesting properties and applications.

6.1 Rheological behavior of microfibrillated cellulose suspensions

After disintegration, MFC is typically available as a suspension in liquid, usually water. During homogenization, the suspension changes from a low viscosity to a high viscosity medium. This is partly due to the increase in the Einstein coefficient, which increases with the increasing length to diameter ratio of suspended particles. Moreover, the highly reactive surface of cellulose also plays an important role. Rheological studies can thus give information about the fibrillation state of the particles as reported in Chapter 2. Normally a 2 wt% fiber suspension is used. At higher concentrations, the increased viscosity during processing becomes too high so that the suspension cannot be moved forward by the pumping system. The MFC suspensions display a gel-like behavior as shown in Figure 6.1.

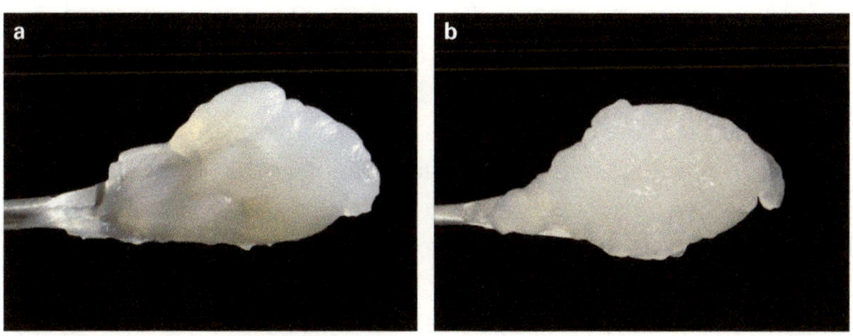

Fig. 6.1: Pictures of microfibrillated cellulose aqueous suspensions: (a) 2 wt% suspension from eucalyptus, enzymatically pretreated, furnished by FCBA, and (b) 10 wt% ARBOCEL NANO MF 40–10, furnished by Arbocel (Lavoine et al., 2012).

The rheological behavior of MFC suspensions was investigated in the pioneering works reported in the 1980's (Herrick et al., 1983; Turbak et al., 1983) and some potential applications were considered from the unusual properties of these suspensions. A gel-like appearance at 2% concentration in water was observed. The authors reported a pseudoplastic, or shear-thinning, behavior and the gel became very fluid when stirred at high shear rate. Thixotropic viscosity properties and stability of the gel on storage, or when subjected to freeze-thaw cycles were also observed.

The investigation of the rheological behavior of aqueous suspensions of MFC obtained from sugar beet pulp has been reported (Lowys et al., 2001). Stress-imposed rheometer was used with two different geometries, viz. cone-plate for flow measurements and plate-plate for dynamic measurements. It was shown that the viscosity of the suspension increased with the concentration of the cellulosic nanoparticles and the existence of a critical concentration ($3\ g \cdot L^{-1}$) above which MFC forms a physical network leading to a gel-like behavior was proposed. Entanglements between the fibrils were assumed to instigate the formation of the network. The influence of the mode of dispersion of MFC was also analyzed in this study. Ultrasonic dispersion was found to be more effective than mechanical stirring. No variation of the storage modulus was reported upon varying the temperature of the suspension in the range 25–60°C.

The rheological behavior of enzymatically pretreated MFC was also reported (Pääkkö et al., 2007). Dynamic tests showed that the aqueous suspensions behaved as gels in the whole investigated concentration range 0.125–5.9 wt%, with storage shear modulus values ranging from 1.5 Pa to 10^5 Pa as shown in Figure 6.2. A controlled strain rheometer was used with plate-plate and cone-plate geometry for low and high MFC concentration suspensions, respectively. The strongly concentration dependent modulus of the gel can therefore be tuned over a wide range around 5 orders of magnitude upon adjusting the concentration within this quite narrow concentration zone, with an extensive shear thinning behavior. Two different fractions of fibrils with lateral dimensions within the range 5–6 nm and 10–20 nm were observed. The thicker particles were assumed to act as the junction zones for the network. The well-controlled networking reflected in the high elastic modulus values roughly 2 orders of magnitude higher than for corresponding gels made by acid-hydrolyzed cellulose nanocrystals. A storage modulus scaling with concentration as the power of ca. 3 was reported. The presence of swollen MFC aggregates was also assumed to account for the much higher modulus values measured for corn cob gels than for well-dispersed MFC gels, although other factors such as cellulose microfibril length and composition could also be involved (Shogren et al., 2011). Rheological studies showed that samples were gel-like from 0.5 to 2.0% and that the storage shear modulus increased following a power law with exponents of 3.7 and 3.2 for blended and homogenized samples, respectively. Both storage and loss moduli were found to be maximal after two passes through the homogenizer and were about ten times larger than typical literature values for microfibrillar cellulose suspensions. SEM and AFM of two-pass sample showed networks of microfibrils and larger expanded fibrillar aggregates while bundles of more separated microfibrils were observed after eight passes.

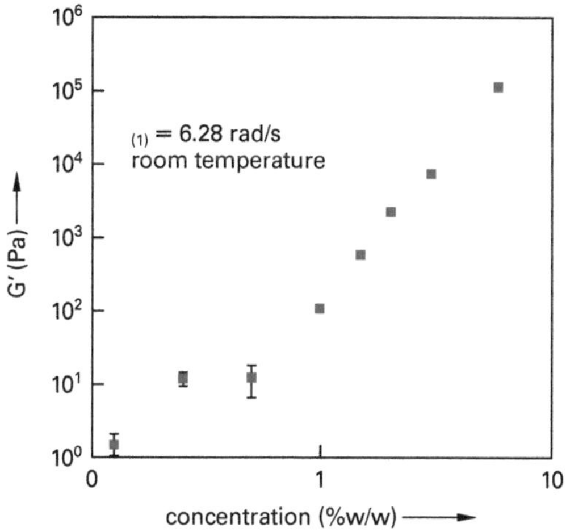

Fig. 6.2: Evolution of the storage modulus as a function of concentration at 25°C and a frequency of 1 Hz (w = 6.28 rad · s^{-1}) for aqueous suspensions of enzymatically pretreated bleached sulfite softwood cellulose pulp MFC (Pääkkö et al., 2007).

The shear-dependent rheological behavior of MFC suspensions over a broad shear rate range has been investigated (Iotti et al., 2011). At low shear rate (0–1,000 s^{-1}) a hysteresis loop was observed in the shear rate–viscosity relationship. The hypothesis for a mechanism of interaction and formation of a fibrils network in slow dynamic studies was suggested. At high shear rates (180,000–330,000 s^{-1}) it was shown that a 1 wt% MFC aqueous suspension has a dilatant behavior with viscosity values increasing from 120 to 300 Pa · s.

The flow properties of a 0.1% TEMPO-oxidized MFC suspension have been investigated in terms of the stirring time (Saito et al., 2007). After stirring for 12 hours with a magnetic stirrer, the suspension exhibited a pseudoplastic behavior. At this stage, a significant swelling of TEMPO-oxidized MFC was observed. However, after three days, the viscosity began to decrease. The partial aggregates of MFC might have disappeared and mostly could have been converted into individual MFC in water by stirring for 10 days. The shear rate dependence of the viscosity of a 1.5 wt% MFC suspension with different oxidation extent was also investigated (Besbes et al., 2011). No significant effect was observed for carboxyl content up to 500 µmol · g^{-1}. However, for higher carboxyl contents, a lower viscosity was reported even if a higher fibrillation yield was obtained. It was ascribed to electrostatic repulsion resulting from the presence of a high level of ionized carboxyl groups on the surface of MFC.

It was also observed that the viscosity of a 1 wt% MFC suspension obtained from sulfite softwood-dissolving pulp using a carboxymethylation pretreatment increased with the number of passes through the homogenization equipment, i.e. with increasing fibrillation of the material (Aulin et al., 2010). It was ascribed to an increased swelling capacity and increased water uptake due to more exposed surface areas of the carboxymethylated MFC.

The influence of various parameters such as concentration, temperature, ionic strength and pH on the rheological behavior of sugar beet pulp MFC suspensions was reported (Agoda-Tandjawa et al., 2010). Different suspensions were compared, namely fresh suspensions or water redispersed MFC obtained after freezing or freeze-drying at two concentrations 0.5 and 1 wt%. No deterioration of the suspension was observed after freezing whereas freeze-drying induced a decrease by more than three times of the storage modulus. Reinforcement of the viscoelastic properties when increasing the ionic strength was reported, but no influence of both the temperature and pH was observed. The influence of low-methoxyl pectin on the rheological and microstructural properties of MFC suspensions was also investigated with the aim to propose controlled "cellulose/pectins" model systems with interesting texturing properties (Agoda-Tandjawa et al., 2012). No significant alteration was reported, but the viscosifier effect of pectins induced an increase of the thixotropic character and shear-thinning behavior of the suspension. A synergistic effect and increase of rheological properties was observed when adding sodium and calcium ions to the suspension. It was ascribed to the gelation of pectins whose networking was assumed to contribute to the formation of a stronger gel.

6.2 Stability of colloidal cellulose nanocrystal suspensions

Cellulose rod-like nanocrystals form colloidal suspensions when suspended in water. In such suspensions, the dispersed-phase particles have a diameter ranging from 1 nm to 1 μm. The stability of these suspensions depends on the dimensions of the dispersed particles, their size polydispersity and surface charge. The use of sulfuric acid for cellulose nanocrystals preparation leads to more stable aqueous suspension than that prepared using hydrochloric acid (Araki et al., 1998). Indeed, the H_2SO_4-prepared nanoparticles present a negatively charged surface while the HCl-prepared nanoparticles are not charged.

During acid hydrolysis of most clean cellulose sources via sulfuric acid, acidic sulfate ester groups are likely to form on the nanoparticle surface. This creates electric double layer repulsion between the nanoparticles in suspension, which plays an important role in their interaction with a polymeric matrix, with the suspension medium and with each other. The density of charges on the polysaccharide nanocrystals surface depends on the hydrolysis conditions and can be determined by elemental analysis or conductimetric titration to exactly know the sulfur content. The sulfate group content increases with acid concentration, acid-to-cellulose ratio and hydrolysis time.

Based on the density and size of the cellulose nanocrystals, the charge density was estimated around $0.155\ e \cdot nm^{-2}$, where e is the elementary charge for a nanocrystal with dimensions of $7 \cdot 7 \cdot 115\ nm^3$ (Araki et al., 1998; Araki et al., 1999). With the following conditions: cellulose concentration of 10 wt% in 60% sulfuric acid at 46°C for 75 min, the charge coverage was estimated at 0.2 negative ester groups per nm (Revol et al., 1992). Other typical values of the sulfur content of cellulose nanocrystals prepared by sulfuric acid hydrolysis have been reported in the literature (Marchessault et al., 1961; Revol et al., 1994). However, it was shown that even at low levels, the sulfate groups caused a significant decrease in degradation temperature and an increase in char fraction confirming that the sulfate groups act as flame retardants (Roman and Winter, 2004). For higher thermostability of the crystals, low acid concentrations, small acid-to-cellulose ratios and short hydrolysis times should be used.

Another way to achieve charged nanocrystals consists in the oxidation of the nanocrystal surface (Araki et al., 2001; Isogai and Kato, 1998) or the post-sulfation of HCl-prepared nanocrystals (Araki et al., 1999).

The overall stability of colloidal suspensions can be determined by examining all of the forces acting on the colloid particles. The random forces involved in Brownian motion and van der Waals forces are both universal forces that act on colloidal particles. Electrostatic forces also play a role if the particles are charged. External forces such as gravity and magnetic and electrical fields may also be present and have an influence on the colloidal stability. The different forces present in a system often have opposing effects. Therefore, it is important to determine which force is dominant for a particular system.

To model the stability of electrostatically stabilized colloids, the Derjaguin, Landau, Verwey and Overbeek (DLVO) theory can be used. This theory was proposed in the 1940s to primarily explain colloidal stability (Derjaguin and Landau, 1941; Verwey and Overbeek, 1947). It combines the effects of both universal van der

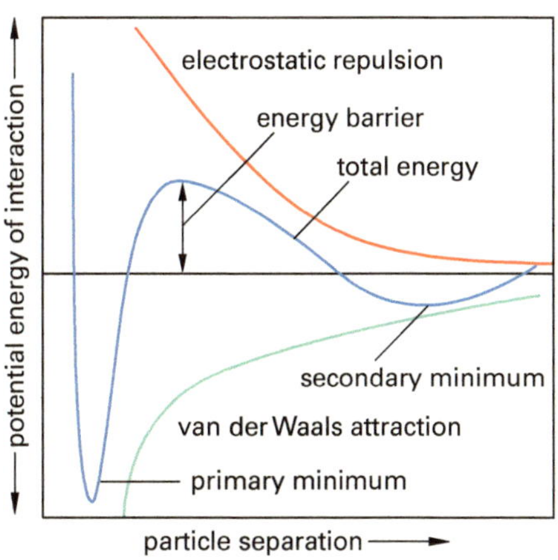

Fig. 6.3: Schematic representation of the total energy of interaction between two colloidal particles, according to DLVO theory.

Waals interactions and the effects of electrostatic repulsions. The balance between these interactions determines the dominant force as a function of particle separation. Electrostatic forces start at a finite value and the repulsion decreases as particle separation increases. Van der Waals forces are attractive and become larger as particles get closer until strong hard-core repulsion occurs at extremely small separations. The total potential energy of interaction between the colloids is the sum of the repulsion energy and the attraction energy, as shown in Figure 6.3. At very small interparticle separations, steep hard-core repulsion is experienced. The net interaction gives rise to an energy barrier, which needs to be overcome before the two surfaces can come into primary adhesive contact.

The blue line in Figure 6.3 depicts a deep well, which is called the primary minimum, and a smaller well to the right, which is the secondary minimum. The hill between both minima corresponds to the energy barrier. This positive energy barrier between the secondary and primary minima is the main parameter in determining the colloid stability since it prevents the colloidal particles from coming into contact and aggregating. If the energy barrier is large, the suspension will be stable. For a small energy barrier, the particles may reach the primary minimum and coagulate. Particles may come into contact in the secondary minimum as well, but if it is not very deep Brownian motion can break up the flocs. The magnitude of the energy barrier depends on the surface potential of the particles and on the ionic strength of the liquid medium. A higher energy barrier arises when the particle surface potentials are larger, leading to more stable suspensions. In contrast, an increase in electrolyte concentrations leads to a decrease in the energy barrier. Thus, adding salt to a suspension will cause coagulation at a critical concentration. If the surface charge is negligible, as for cellulose nanocrystals obtained by hydrolysis with hydrochloric acid (Araki et al., 1998), or if enough ionic species are present in solution to screen the electrostatic double layer repulsions, the suspension becomes unstable and the particles tend to precipitate or flocculate.

Even if they are electrostatically stabilized, the random behavior of the colloidal particles is altered as their concentration in suspension increases. For a given number N of hard particles in a volume V, the free volume V_{free} available to the particles is only a fraction of the total volume because each particle excludes the other particles from a certain volume surrounding it. This reduces the total number of configurations available to the set of N particles and therefore lowers the entropy of the system. For a given density of colloidal particles, it can be demonstrated that if the excluded volume regions overlap, the free volume available to each particle increases, thereby increasing the entropy of the system and decreasing its free energy. In order to achieve this more thermodynamically favorable state, the system must phase shift to a different arrangement of particles in order to maximize the available free volume.

The plot in Figure 6.4 shows qualitatively the calculated free energy for both the fluid phase and solid phase for a suspension of colloidal spheres as a function of the suspension density. Due to their relatively large size (as compared with atoms), the phase behavior of these colloidal spheres can often be studied by optical means. It can be seen that at low densities, the random fluid phase is favored, i.e. the spheres are in a "fluid phase" with all the particles distributed randomly. As the density of the spherical particles within the host liquid is increased, a disorder-order transition may be observed. At high sphere densities, the close-packed solid phase is more stable and the spheres form a "solid phase" or "colloid crystal phase" with the spheres adopting regular positions in a structured lattice. The system thus undergoes a phase shift in order to maximize its free volume and hence its entropy.

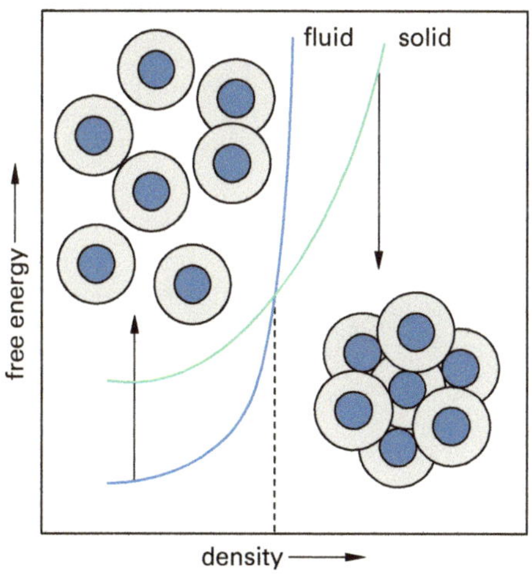

Fig. 6.4: Qualitative representation of the free energy of a dispersion of colloidal spheres as a function of the sphere density for the fluid (random) and solid (ordered) phases. At low particle densities, the fluid phase is energetically favored, while above a critical density (green line), the solid phase becomes more stable and a phase transition occurs (adapted from Baus et al., 1996).

6.3 Birefringence properties of cellulose nanocrystal suspensions

Polarized light is generally used for the observation and the identification of oriented structures, in particular crystalline structures. This technique gives a quick indication of the dispersion state of cellulose nanocrystals obtained by sulfuric acid hydrolysis in the suspension. Theoretically, a beam of unpolarized light entering certain anisotropic material, such as a crystal, will be divided into two beams, vibrating in different planes perpendicular to each other (polarized rays). The simplest instance of the effect arises in materials with uniaxial anisotropy. Cellulose nanocrystals display this property with a strong absorption of one of the polarized rays, because they display two refractive indices. The light emerging from such a material will be plane-polarized light. This is the property of birefringence, or double refraction. This effect was also described in Chapter 3 (Section 3.7.1).

In order to observe the birefringence of a nanocrystal suspension, two linear polarizers crossed at 90° are used. Figure 6.5 shows photographs of suspensions of tunicin nanocrystals observed in polarized light between crossed polars and showing the formation of birefringent domains. The nanocrystal content was 0.5 wt% in both suspensions, and the suspending medium was either water (Figure 6.5(a)) or N,N-dimethylformamide (DMF) (Figure 6.5(b)). The presence of negative charges on the cellulose nanocrystal surface induces an electrostatic repulsion that can be directly observed. This birefringent character of cellulose nanocrystal suspensions was first observed in 1959 (Marchessault et al., 1959).

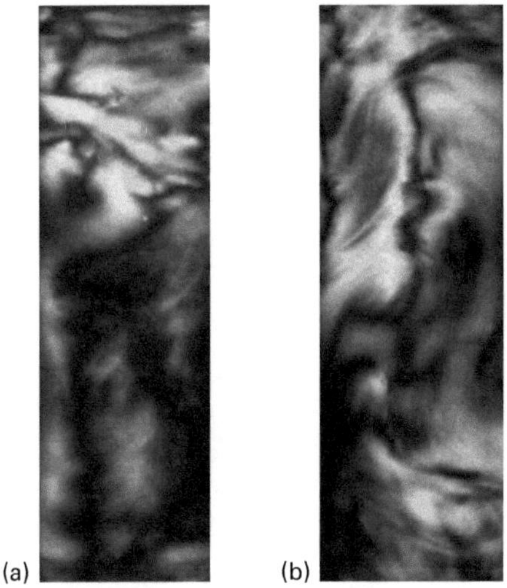

Fig. 6.5: Photographs of suspensions of tunicin nanocrystals (0.5 wt%) in (a) water and (b) N,N-dimethylformamide (DMF) observed between cross nicols, showing the formation of birefringent domains (Azizi Samir et al., 2004).

6.4 Liquid crystalline behavior

6.4.1 Liquid crystalline state

Liquid crystals are an intermediate state of matter with the characteristics of both conventional liquids (fluidity) and solid crystals (some long-range order and anisotropy) (Chandrasekhar, 1992). The first liquid crystal observed was composed of rigid molecules, and still today typical liquid crystals are molecular in nature. They may flow like a liquid, but the molecules may be oriented in a crystal-like way and are typically formed by anisotropic molecules or particles that are rigid and rod-like (Radel and Navidi, 1994). Different types of liquid crystal states can be found, depending on the amount of order in the material, which have different organizational patterns. Figure 6.6 illustrates the classification of liquid crystals according to the arrangement of the molecules or particles. Just as cellulose varies in its degree of order from crystalline to amorphous regions, liquid crystals may vary in their degree of order, undergoing phase transitions between more or less ordered states, similar to solid-liquid-gas transitions.

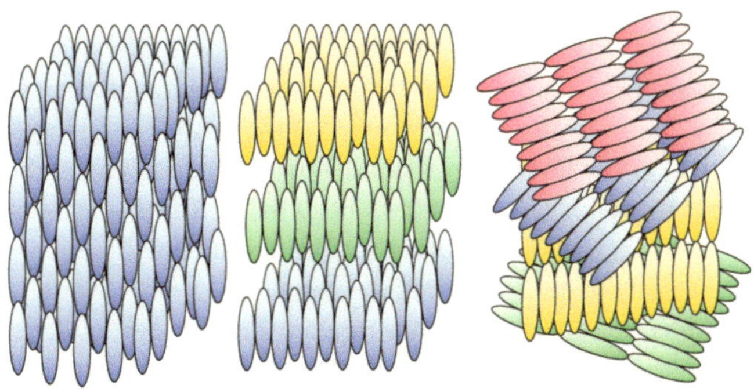

Fig. 6.6: Schematic diagram of the different types of liquid crystal structure.

Although ordered structures can be formed with spherical particles, liquid crystals are typically formed by anisotropic particles, such as a rod, sheet, or disc (DuPré, 1981). This anisotropic shape allows the liquid crystal to have both positional and orientational order, whereas spherical particles can only adopt positional ordering. Most liquid crystals possess orientational order, meaning that the particles align with their long axes oriented in a particular direction, and some liquid crystals possess positional order, meaning that the particles occupy a specific place relative to the other particles (Collings and Hird, 1998).

The nematic liquid crystal phase is characterized by molecules that have no positional order but tend to point in the same direction (along the director). Therefore, nematic liquid crystals possess orientational order, but no positional order, as the rods are free to slide or roll past each other in all directions. The situation is similar to a large number of toothpicks put into a rectangular or cylindrical box and shaken. When the box is opened, the toothpicks will be facing in about the same direction, but will have no definite spatial organization. They are free to move, but like to line up in about the same direction. The smectic state is another distinct mesophase of liquid crystal substances and shows one more degree of orientational order than do the nematics. In the smectic state, the molecules maintain the general orientational order of nematics, but also tend to align themselves in layers or planes, leading to a loss of positional order. The molecules or rods are therefore arranged in layers with their long axes perpendicular to the plane of the layers. Motion is restricted to within these planes, and separate planes are observed to flow with respect to each other. Because of this increased order, the smectic state is more "solid-like" than the nematic. Chiral nematic liquid crystals possess a further degree of order compared to smectic and nematic textures. Chiral refers to the unique ability to selectively reflect one component of circularly polarized light. The term chiral nematic is used interchangeably with cholesteric. As in smectic textures, the molecules or rods are arranged in planes or layers but are free to move perpendicularly to the plane of those layers. The lack of positional order in the perpendicular direction makes the term "pseudo-layer" or "quasi-nematic" layer more appropriate. Within a pseudo-layer, the molecules are aligned on average with a vector or director (see Figure 6.6), the director of each layer is rotated slightly with respect to the layers above and below it, thereby giving the liquid crystal long-range helical order. The pitch of the helix is defined as the distance required for the director to make one full rotation about the chiral nematic or cholesteric axis, which lies perpendicular to the plane of the layers.

The conventional liquid phase is characterized by random and isotropic molecular ordering (little to no long-range order), and fluid-like flow behavior. An isotropic liquid can be made to undergo the phase transition to an ordered liquid crystal by a change in temperature (thermotropic mesomorphism) or by a change in concentration of the mesogen (lyotropic mesomorphism) (de Gennes and Proust, 1993).

Since the liquid crystalline state is intermediate between the solid and liquid states, slow heating of a thermotropic solid induces a transition through the liquid crystal phase, which will appear cloudy, before progressing into the clear liquid phase. Sometimes the material may pass through more than one liquid crystal phase, such as from a highly ordered smectic phase to a less ordered nematic phase, as it melts into a liquid. As the material is heated, the rigid crystal structure of the solid is lost, but the particles maintain some order by generally orienting with their long axes along a given direction. Many molecular mesogens form thermotropic systems.

Lyotropic colloidal liquid crystals were first recognized in the 1920s by Zocher, who investigated nematic textures in solutions of colloidal rod-like V_2O_5 particles (Zocher, 1925; Watson et al., 1949). Similar lyotropic isotropic–nematic transitions of colloidal particles were later reported for clay platelets (Langmuir, 1938) and for tobacco mosaic virus (TMV) rods (Bawden et al., 1936; Bawden and Pirie, 1937). More recently, the isotropic–nematic transition has been observed in systems of rod-shaped particles, such as polytetrafluoroethylene (PTFE) whiskers (Folda et al., 1988), other stiff viruses such as fd bacteriophage and M13 virus (Guo and Gray, 1989; Siekmeyer and Zugenmaier, 1987; Kimura et al., 2005; Baus et al., 1996), rod-like micelles of amphiphilic surfactants (Yu and Saupe, 1980), and colloidal boehmite rods (Buining and Lekkerkerker, 1993; Buining et al., 1994).

In contrast to rigid rod-like particles, somewhat flexible whiskers have been polymerized from tetrafluoroethylene, resulting in long rods with a uniform width of 20 nm and a length that varies from 1 μm to over 20 μm (Folda et al., 1988). Upon standing for about four hours, the suspension of PTFE separated into an upper isotropic phase and a lower phase that is highly birefringent, characteristic of liquid crystalline nematic order. Although the longer PTFE particles (20 μm) fall outside the colloidal range, they still have enough mobility to form an ordered phase without flocculating. In addition to their high length, these PTFE whiskers are also flexible, showing that the isotropic-nematic phase transition is not restricted to rigid rods.

Increasing the concentration of rod-like particles suspended in a liquid will bring about the phase transition for a lyotropic liquid crystal. At dilute concentrations, the rods are all tumbling randomly, but as the concentration increases the particles tend to align parallel to neighboring rods. It is important to note that this should not be viewed as a rigid, static alignment, but more as a preferred direction as the rods continue to move over time. This preferred direction is called the director of the liquid crystal. By averaging the directions of the rod-like particles or molecules with respect to the director, an order parameter, S, can be determined (Haller, 1975):

$$S = \frac{1}{2}(3\cos^2\theta - 1) \tag{6.1}$$

where θ corresponds to the angle between the particle and the director. Therefore, a value of 1 for the order parameter indicates perfect alignment, and 0 indicates complete randomness, as found in a normal liquid. For thermotropic liquid crystals, the order parameter decreases as the temperature increases.

6.4.2 Liquid crystalline behavior of cellulose derivatives

Both lyotropic and thermotropic liquid crystals have been obtained from cellulose derivatives (Tseng et al., 1981). Evidence for these phases can be obtained by optical analysis since chiral nematic liquid crystals reflect light in the visible spectrum if their pitch is of the order of the wavelength of visible light. Liquid crystalline properties of cellulose derivatives were discovered in 1976 (Werbowyj and Gray, 1976). It was found that concentrated aqueous solutions of hydroxypropyl cellulose (HPC) appeared to form the helicoidal structure of a cholesteric liquid crystal. The reflected color, which is determined by the pitch of the chiral nematic helix, varied inversely with cellulose concentration. The color was observed to change from a red-orange to green and finally to violet as the cellulose concentration increased. Other cellulose derivatives changed color as the temperature changed and are therefore thermotropic. However, in contrast to the lyotropic HPC (Werbowyj and Gray, 1984), acetoxypropyl cellulose (APC) was subsequently found to exhibit thermotropic cholesteric liquid crystal phases and appeared violet at 85°C (Tseng et al., 1981). This thin layer was placed on a hot stage and as the temperature increased, the color systematically progressed through the rainbow, becoming red at 125°C, showing that the distance between nematic layers of APC increases with temperature.

Evidence for the chiral nematic helix can also be obtained by measuring the optical rotatory dispersion (ORD) or circular dichroism (CD) of the sample. These measurements also determine whether the helix is right- or left-handed. The orientation of nematic layers in a right-handed chiral nematic changes in a clockwise direction on moving away from the observer along the chiral nematic axis. The layers rotate counterclockwise in a left-handed chiral nematic.

Other cellulose derivatives including ethyl cellulose and ether and ester derivatives of HPC in various liquids including water, methanol and acetic acid were studied. It was believed that the handedness of various cellulose derivatives would be the same since they share a common cellulose backbone. However, both right- and left-handed helices have been observed. While HPC in water, acetic acid, and methanol as well as ether and ester derivatives of HPC resulted in right-handed helices (Werbowyj and Gray, 1984), left-handed structures were observed for ethyl cellulose in acetic acid (Vogt and Zugenmaier, 1985).

The handedness of the chiral nematic helices was found to vary with the nature of the substituents on the cellulose backbone (Guo and Gray, 1989; Gray, 1996) as well as with the nature of the liquid medium (Siekmeyer and Zugenmaier, 1987). For example, the acetylation of ethyl cellulose changes its helical orientation (Guo and Gray, 1989). Changes in liquid also have an effect on the chiral nematic twist sense: in methylpropyl ketone cellulose tricarbanilate forms a right-handed twist, but in diethyleneglycol monoethyl ether the twist sense is left-handed (Siekmeyer and Zugenmaier, 1987). Thus, some cellulose derivatives form right-handed helices and some left-handed, while still others can change from one form to the other. Interactions between the cellulose backbone, chain substituents, and the solvent all influence chiral forces and the resultant helical superstructure.

The electro-optical behavior and selective light reflection of liquid crystal suspensions of cellulose derivatives such as HPC, as well as that of free-standing films and networks retaining the cholesteric organization of the original solutions have been studied. The color and turbidity of iridescently colored polymer composite films of HPC when immersed and swollen in water can be controlled by the addition of different inorganic salts and/or by the application of an electric potential (Nishio et al., 1998; Chiba et al., 2003; Chiba et al., 2006). Ethyl-cyanoethyl cellulose, a cellulose ether, when copolymerized with acrylic acid forms films which reflect visible light and whose cholesteric pitch dependence on polymerization temperature varied with the concentration of cellulose ether (Zhou and Huang, 2005).

6.4.3 Liquid crystalline behavior of cellulose nanocrystal suspensions

Cellulose nanocrystal suspensions also form lyotropic liquid crystals. When viewed using a polarized light source, liquid crystal phases appear to have distinct textures. It was shown that, owing to their rod-like nature, cellulose nanocrystals can form birefringent gels and liquid crystalline structures (Marchessault et al., 1959). Figure 6.7 shows a suspension of cotton cellulose nanocrystals which has formed a chiral nematic liquid crystal phase. Phase separation is observed with the isotropic phase in the upper part of the suspension and chiral nematic liquid crystalline phase in the lower part. The contrasting areas in the textures correspond to domains where the liquid crystalline nanoparticles are oriented in different directions. Within a domain, however, the nanoparticles are well ordered.

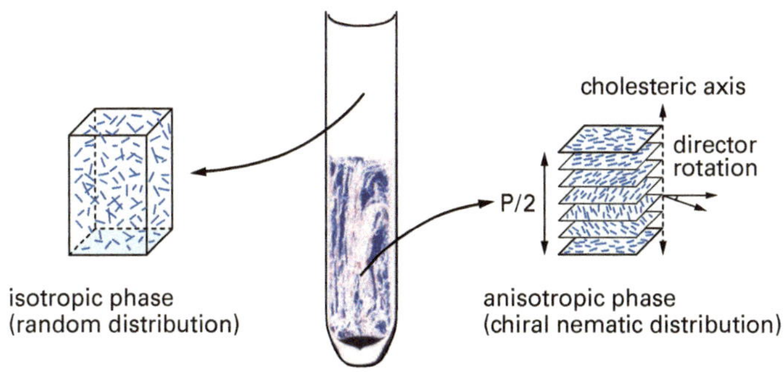

Fig. 6.7: A typical biphasic suspension of cotton cellulose nanocrystals in water, viewed between partially crossed polarizers. The concentration in the upper isotropic phase is about 6.9% (w/v) cellulose in water. In the lower chiral nematic liquid crystalline phase, the concentration is 8.7% (w/v) cellulose. The organization of each phase is schematically illustrated. Half a chiral nematic pitch, $P/2$, is shown (adapted from Beck-Candanedo et al., 2008; photograph from Dong et al., 1998).

Induced circular dichroism (ICD) measurements were used to determine both the handedness and the degree of order of a chiral nematic suspension of cellulose nanocrystal suspensions (Dong and Gray, 1997a). ICD is defined as the difference in adsorption of left- and right-handed circularly polarized light. It can be measured using a spectropolarimeter. The differences in absorption, or apparent absorption, of left- and right-handed circularly polarized light can be determined and this difference is expressed in terms of ellipticity.

At low concentrations the cellulose suspensions are isotropic with the rods tumbling randomly, while at high cellulose concentrations the rods tend to align parallel to neighboring rods forming an anisotropic phase. As water evaporates from a dilute sample, tactoids, small domains of ordered particles, begin to form and as the concentration continues to increase the tactoids coalesce to form the continuous fingerprint texture characteristic of a chiral nematic (Revol et al., 1994). Upon reaching the critical concentration required for phase separation, most cellulose suspensions exhibit the classic fingerprint texture. However, a "birefringent glassy phase" (Araki et al., 2000) and a nematic phase (Araki and Kuga, 2001) have also been observed. The observed chiral nematic order can be enhanced by placing the suspension in a magnetic field (Revol et al., 1994), as cellulose has a negative diamagnetic susceptibility (Sugiyama et al., 1992), meaning that the rods tend to align perpendicular to the magnetic field.

Therefore, as the concentration of rod-like colloids in a suspension increases, the particles adopt a close-packed arrangement in order to maximize the free space. This entropy-driven phase transition is similar to the behavior of colloidal spheres (Figure 6.4). However, the anisotropic shape of the rod-like particles allows for the formation of more intricate phases. In addition to the positional order, observed in a spherical colloid crystal, the rod-like colloids can have orientational order. This extra degree of order enables the formation of several unique liquid crystalline phases, including nematic, smectic, and chiral nematic.

Chiral nematic cellulose nanocrystal liquid crystals form two optical "textures" which are identifiable when a sample is placed on a light microscope stage between two linear polarizers crossed at 90°. The birefringence (anisotropic refractive index, $\Delta n \approx 0.05$) of the cellulose nanocrystals and the order of the liquid crystal allow light to pass through areas of the sample. The planar texture is observed when the cholesteric axis is parallel to the viewing direction and disclinations show up between crossed polars as dark lines against a colored background (Chilaya and Lisetski, 1986). When the cholesteric axis is perpendicular to the viewing direction, a characteristic "fingerprint" texture is seen, in which dark and light lines alternate, the width of each line corresponding to one quarter of the chiral nematic pitch, P. This is illustrated in Figure 6.8. Environmental scanning electron microscopy (ESEM) was also used to examine the effect of concentration on anisotropic phase behavior of cellulose nanocrystal suspensions (Miller and Donald, 2003). The advantage of this technique over conventional SEM was that the sample chamber did not need to be kept under high vacuum conditions and when water vapor is selected as the gas, hydrated and insulating samples can be analyzed. It is therefore ideal for the high-resolution study of biological samples in their natural environment.

Liquid crystal phases of cellulose nanocrystals are easier to work with than molecular cellulose derivatives for several reasons. First, because molecular cellulose derivatives are somewhat flexible, they do not behave as predictably as the rigid rod-like nanocrystals. Moreover, highly concentrated solutions (e.g. 55–70 % for HPC (Werbowyj and Gray, 1976)) of these polymers are required to obtain the liquid crystal phase, which makes handling the viscous solutions difficult. Cellulose nanocrystals, in contrast, begin to form chiral nematic phases at concentrations of only 1–7 wt%, depending on their aspect ratio (Dong et al., 1996; Kimura et al., 2005).

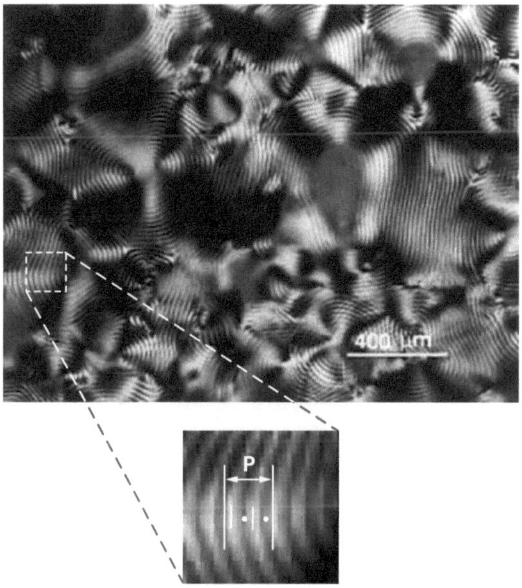

Fig. 6.8: Chiral nematic liquid crystal texture of the anisotropic phase of a cotton cellulose nanocrystal suspension viewed in a polarizing microscope. The nanocrystals lie parallel to the plane of the page in the light regions and perpendicular in the dark regions (adapted from Dong et al., 1996).

In order to understand the liquid crystal phase behavior of cellulose nanocrystals, it is necessary to understand how and why the phases form. Indeed, it seems counterintuitive that randomly-distributed particles should spontaneously align to give an ordered phase, and even more counterintuitive that entropy should be the driving force behind this phenomenon. Nevertheless, this is the case, as explained below.

6.5 Onsager theory for neutral rod-like particles

Onsager was the first to show theoretically that spontaneous ordering, or nematic phase transition, occurred with rod-like particles (Onsager, 1949). His theory predicts that rod-like particles can undergo an orientational disorder-order transition from a disordered isotropic phase to an orientationally ordered phase. This effect was earlier observed for the tobacco mosaic virus where an isotropic low density phase and a nematic higher density phase were found to co-exist in suspension (Bawden and Pirie, 1937).

Phase shifts are observed in suspensions of rod-like particles for entropic reasons. Onsager's theory is a cornerstone in the theory of liquid crystal phase formation for systems of long monodisperse rod-like particles. It accurately predicts an isotropic to nematic transition for these systems, and although the predicted critical concentrations of the phase transition do not coincide with typical experimental values, it provides a simple explanation for the phase separation of dispersions of repulsive rod-like particles. The critical concentration for phase separation predicted theoretically always tends to be greater than experimental concentrations (Dong et al., 1996).

The theory is based on the free energy of rigid rods in the limit of the second virial coefficient. The second virial coefficient can be obtained by averaging the excluded volume over the equilibrium distribution of orientations at the minimum free energy. The excluded volume corresponds to the volume occupied by one rod in which the mass centers of other rods are not allowed to enter. For a pair of spheres of radius r, the excluded volume is obviously a sphere of radius 2r. For rod-like particles with length L and diameter D, the excluded volume is schematically shown in Figure 6.9. In this figure, two particular cases are depicted, i.e. the two particles are parallel (Figure 6.9(a)) or perpendicular (Figure 6.9(b)) to each other. For a pair of long cylinders or rods with a high aspect ratio ($L_1 >> D_1$ and $L_2 >> D_2$), the excluded volume V_{excl} is given by (Onsager, 1949):

$$V_{excl} = L_1 L_2 (D_1 + D_2) \, |\sin\gamma| \tag{6.2}$$

where γ is the angle between the rods.

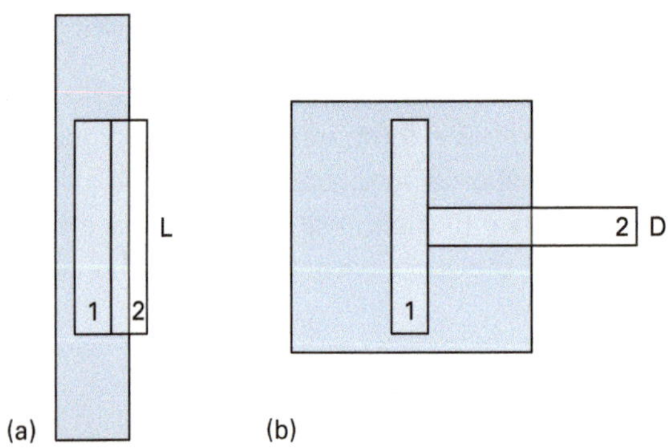

Fig. 6.9: Schematic for the excluded volume (grey area) of the rod 1: (a) parallel to the neighbor rod 2, and (b) perpendicular to the neighbor rod 2.

The excluded volume represented in grey color in Figure 6.9 depends therefore on the angle between the rods. The angular dependence function of the excluded volume for the pair of long rods gives a maximum value when the rods are oriented perpendicularly (Figure 6.9(b)) to each other and a minimum when they are parallel (Figure 6.9(a)). Conversely, for charged rods, the potential of the average force between two cylindrical particles based on the Poisson–Boltzmann equation for electric potential is a minimum when the rods are oriented perpendicularly to each other and a maximum when they are parallel. The potential of average force between the rod-like particles at the equilibrium state must be calculated by summing over the total number of particles in the system. The free energy, osmotic pressure and chemical potential can then be found in terms of the average force (in order for the isotropic and anisotropic phases to coexist, their chemical potential and osmotic pressure must be equal). Assuming the forces to be pairwise additive, two-particle and three-particle interactions can be evaluated and used to correct the average force as first- and second-order correction terms (virial coefficients). Onsager estimated the more complex three-particle interactions, and limited the theory to dilute suspensions in which two-particle interactions predominate (Onsager, 1949). The free energy of the system can then be calculated using the virial coefficients.

The Onsager theory is somewhat restricted as it is accurate only in the limit of very long rods (i.e. $L/D \rightarrow \infty$) and low particle concentration. In this limit, however, excellent agreement between theory and experiment has been observed for suspensions of monodisperse rods.

Onsager's theory may also be understood by looking at the entropy of the system of particles. Their anisotropy implies that in addition to positional or translational entropy, the rod-like particles also possess orientational entropy. With the formation of the ordered phase, the density of particles is no longer uniform and the system's orientational entropy is decreased. However, the overall increase in free volume available for the particles to move about in leads to a gain in the system's translational entropy which more than offsets the loss of orientational entropy. Thus, the phase transition is a purely entropic one, based on particle shape. Attractive forces are therefore not required for ordered phase formation in these systems. For hard rods, the critical concentration for phase separation is determined only by the aspect ratio L/D of the particles (Onsager, 1949):

$$\begin{aligned} \phi_i &= 3.3399 D/L \\ \phi_a &= 4.4858 D/L \end{aligned} \tag{6.3}$$

where ϕ_i and ϕ_a correspond to the volume fractions of the rods in the isotropic and anisotropic phases, respectively. This means that ϕ_i represents the critical concentration at which the nematic phase initially forms from an isotropic suspension and ϕ_a represents the (higher) concentration at which the suspension becomes completely anisotropic. Between these two concentrations, the suspension will separate into two coexisting phases (Onsager, 1949).

For nanocrystals prepared using sulfuric acid, and therefore charge stabilized by sulfate groups, this region has a lower boundary of approximately 1–7 wt% and an upper boundary of approximately 5–13 wt%, depending on the cellulose source. The anisotropic phase typically forms around 5 wt%. Then a fairly large biphasic region exists, due to the polydispersity of the sample, and a pure anisotropic phase is obtained around 10–12 wt%. The precise values for phase separation are influenced by the ionic strength of the suspension (Dong et al., 1996) and the nature of the counterions (Dong and Gray, 1997b). As the ionic strength increases, the effective diameter of the rods decreases and therefore a higher rod concentration is required for phase separation to occur. It can be seen from the expression of the volume fraction in both phases that longer rods decrease the critical concentration for anisotropic phase formation. This means that longer rods experience a larger excluded volume effect. For example, suspensions of boehmite rods with an aspect ratio of 20 were found to phase separate, while rods of aspect ratio of 8 did not (Buining et al., 1994). Fractionation of longer rods into the anisotropic phase was also predicted by Onsager (Onsager, 1949), and has been observed for cellulose nanocrystals (Dong et al., 1998).

Onsager's phase equilibrium theory for rod-like particles validates the experimental observations for cellulose nanocrystal phase separation above a critical concentration. Other theories such as the lattice-based theories of Flory and co-workers (Flory, 1956; Flory and Abe, 1978; Abe and Flory, 1978; Flory and Frost, 1978; Frost and Flory, 1978; Flory, 1978a; Flory, 1978b) and DiMarzio (DiMarzio, 1961), also model the phase separation behavior of liquid crystalline polymers, for example at higher densities and in length-bidisperse systems. However, these theories are qualitative and do not lead to the exact Onsager results for infinitely thin hard rods.

Experimentally, systems of sterically-stabilized rods are used to approximate hard-core rod–rod repulsions, for example polyisobutylene-grafted boehmite particles (Buining and Lekkerkerker, 1993; van Bruggen et al., 1999). The slender bacteriophage fd virus, although it is highly charged, also leads to results in agreement with Onsager's theory when the double layer is taken into account (Maret and Dransfeld., 1985). In some systems, the particle length can be tailored to obtain bidisperse or polydisperse systems to examine the effect on the phase separation behavior (Buining and Lekkerkerker, 1993). For example, broadening the particle length distribution introduces curvature into (and increases the range of) the biphasic coexistence region of the experimental phase diagram and increases the coexisting phase concentrations ϕ_i and ϕ_a with increasing particle concentration according to a lattice theory of the phase separation of rod-like particles (Moscicki and Williams, 1982). Such curvature has been seen for sterically-stabilized boehmite rods (van Bruggen et al., 1999; van Bruggen and Lekkerkerker, 2000).

It is worth noting that the free energy difference between a chiral nematic phase and a nematic phase is much smaller than the free energy difference between an isotropic phase and an anisotropic phase of rod-like particles (de Gennes and Proust, 1993). Experimental data for chiral nematic liquid crystals can therefore be compared with theories developed for nematic phases.

6.6 Theoretical treatment for charged rod-like particles

Onsager's theory accounted for the electrostatic repulsion of charged particles by treating the double layer as part of the particle, with the effect of increasing the effective particle diameter (Onsager, 1949). However, although the electrostatic double layer was accounted for up to the second virial coefficient, phase separation still could not be accurately predicted for typical experimental systems.

Other theories have been developed to predict the phase separation of rod-like polyelectrolytes (Stroobants et al., 1986; Lee, 1987; Semenov and Kokhlov, 1988; Sato and Teramoto, 1991). Stroobants, Lekkerkerker and Odjik's theory (SLO theory) for the phase equilibrium of charged rods modifies Onsager's equations to account for the increased effective rod diameter (D_{eff}) as well as a twisting effect h characterizing the electrostatic interactions between the charged rods. Twisting of rods is due to the preferential perpendicular orientation. A perpendicular orientation between two rods minimizes the repulsion between them, while the larger effective diameter favors a parallel orientation to minimize the effect of the amplified excluded volume and consequent increase in free energy. The magnitude of the twisting effect is given by h, which represents a balance between the electrostatic and entropic factors. The coexisting number densities C_i and C_a of the rods in the isotropic and anisotropic phases are given in the SLO theory by the following equations (Odijk, 1986):

$$
\begin{aligned}
C_i &= 3.290\,[(1-0.675h)\,b]^{-1} \\
C_a &= 4.191\,[(1-0.730h)\,b]^{-1}
\end{aligned}
\tag{6.4}
$$

where h is the twisting factor and b is the second virial coefficient of the system. The values of h and b are given by:

$$
\begin{aligned}
h &= (\kappa D_{eff})^{-1} \\
b &= \frac{\pi}{4} L^2 D_{eff}
\end{aligned}
\tag{6.5}
$$

where κ^{-1} is the Debye length (electrostatic double layer "thickness"), and L and D_{eff} are the length and effective diameter, respectively, of the rod. The effect of the increase in effective diameter due to the electrostatic double layer is to decrease C_i and C_a compared to analogous neutral rods.

6.7 Chiral nematic behavior of cellulose nanocrystal suspensions

6.7.1 Isotropic-chiral nematic phase separation of cellulose nanocrystal suspensions

Nematic liquid crystalline order in cellulose nanocrystal suspensions was demonstrated more than 50 years ago (Marchessault et al., 1959; Hermans, 1963). Macroscopic birefringence of such suspensions observed through crossed polarizers, due to the birefringence of the nanocrystals as well as to a flow anisotropy resulting from the alignment of the nanocrystals under shear, was reported (Marchessault et al., 1959). It was latter shown that the suspensions in fact formed a cholesteric, or chiral nematic, liquid crystalline phase (Revol et al., 1992).

As the concentration of cellulose nanocrystals in the suspension increases, phase separation proceeds. In dilute (isotropic) suspensions, spherical "droplets" of ordered nanocrystals called tactoids form and are visible by polarizing microscopy (Revol et al., 1994). The order in dilute cellulose nanocrystal suspensions has also been observed using static and dynamic light scattering (de Souza Lima and Borsali, 2002) and ultra-small-angle X-ray scattering, by which almost identical scattering profiles were found for the anisotropic and isotropic phases (Furuta et al., 1996). As the cellulose concentration increases and when a critical concentration is reached, the tactoids coalesce to eventually give an ordered phase, which, as has been stated, displays the optical characteristics of a chiral nematic liquid crystal (Marchessault et al., 1959; Marchessault et al., 1961; Revol et al., 1992; Revol et al., 1994; Dong et al., 1996). Remarkably, the nanocrystal concentrations in the two phases do not differ greatly from each other, in contrast to similar coexisting phases of, e.g., ionic polymer latex particle dispersions (Arora and Tata, 1996). A typical chiral nematic pitch for cotton cellulose nanocrystals in suspension is between 10–25 μm (Revol et al., 1992). The addition of salt to the suspension, the concentration of cellulose nanocrystals, temperature and application of a magnetic field were found to affect the pitch of the chiral nematic structure (Pan et al., 2010). Ultrasound treatment was also found to increase the chiral nematic pitch of the suspension (Beck et al., 2011). A much lower value (2 μm) was observed for cellulose nanocrystals obtained from cotton linters and dispersed in apolar solvent (cyclohexane) (Elazzouzi-Hafraoui et al., 2009). Dispersion was carried out using a surfactant. The lowering of the chiral nematic pitch was ascribed to stronger chiral interactions in low dielectric constant media.

The origin of the chirality in the liquid crystal phase formed by cellulose nanocrystals has been the subject of several investigations. It is interesting to note that rod-like cellulose nanocrystals self-order to form a chiral nematic phase, as opposed to a plain nematic phase, whereas there is no enthalpic advantage in so doing. The molecular chirality of the cellulose chain cannot be transmitted between the rods because the distances separating them are too high, being of the order of 20–40 nm (Orts et al., 1998; Beck-Candanedo, 2006a). The arrangement of charged groups in a spiral on the nanocrystal surface has also been ruled out (Araki et al., 2001). Packing by chiral interaction of twisted rods in order to minimize excluded volume has been proposed (Orts et al., 1998; Straley, 1976). In situ small angle neutron scattering (SANS) experiments conducted on cellulose nanocrystal suspensions under magnetic field and shear alignment confirmed that the nanoparticles are helically twisted rods as shown in Figure 6.10 (Orts et al., 1998). The morphological change of the nanocrystal from plain cylindrical configuration to a twisted rod was ascribed to the screening of surface charge (Araki and Kuga, 2001). As shown in Figure 6.10, threaded rods can be packed more tightly when their main axes are offset such that their threads fit within the each other's grooves (Orts et al., 1998) as suggested previously (Straley, 1976). Although the nanocrystal separation prevents actual physical contact, the electrostatic double layer surrounding the rods may be thick enough to transmit the chiral twist if it is of the same order as the lateral dimensions of the rod (Revol and Marchessault, 1993). A helicoidal organization of opposite handedness to the twist along the nanocrystals would be generated from this interaction. The formation of chiral nematic phases with similar properties by PEG-grafted nanocrystals (Araki et al., 2001) and rod-like fd virus (Grelet and Fraden, 2003), as well as surfactant-stabilized cellulose nanocrystals in nonpolar solvents (Heux et al., 2000) supports this hypothesis.

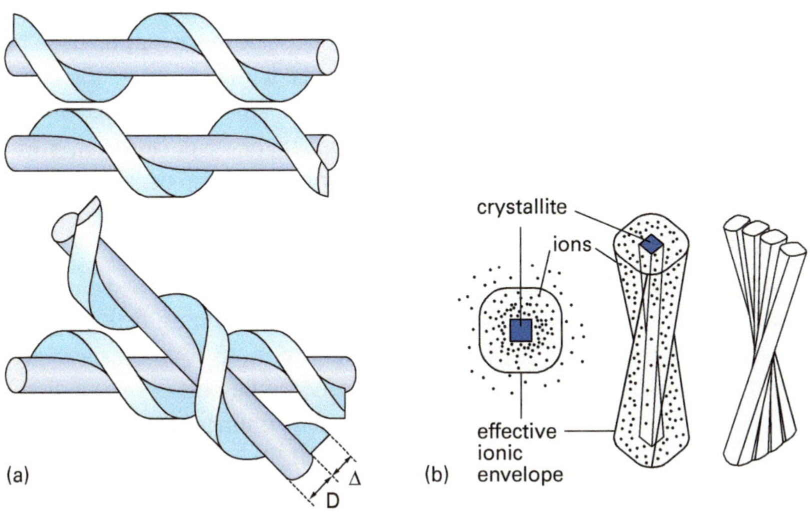

Fig. 6.10: Representations of the tighter packing achievable by the chiral interactions of twisted rods. In (a), the distance between rods (D+2) is reduced to ~ D if, instead of rods packing with axes parallel, they pack with the "thread" of one rod fitting into the "groove" of its neighbor. For nanocrystals with an electrostatic double layer (b), a threaded rod would alter the surrounding electric double layer and affect packing over relatively large distances (Orts et al., 1998).

Owing to their anisometric rod-like shape, cellulose nanocrystals in suspension will align in a shear flow to produce a high degree of orientation (Orts et al., 1995). A film composed of highly oriented cellulose nanocrystals has also been produced by drying a gel-like layer of crystalline cellulose formed under shear flow conditions (Nishiyama et al., 1997).

Cellulose nanocrystals possess negative diamagnetic anisotropy. When dilute aqueous suspensions of tunicin nanocrystals were placed in magnetic fields of up to 7 T, orientation of the nanoparticles with their long axes perpendicular to the field direction was observed (Sugiyama et al., 1992). Thus, the degree of order of liquid crystalline phases of cellulose nanocrystals can be enhanced by placing them in a magnetic field. The cholesteric axis of chiral nematic suspensions of cellulose nanocrystals aligns parallel to the field direction, and it increases the size of the chiral nematic domains (Revol et al., 1994). The direction of the cholesteric axis may therefore be controlled by applying a strong magnetic field. Placing a flat tube containing a chiral nematic suspension of cellulose nanocrystals parallel and perpendicular to the field direction can yield more uniform fingerprint and planar textures, respectively (Dong and Gray, 1997a). Induced circular dichroism of Congo red, an adsorbed dye oriented by the ordered cellulose nanocrystals, provided another method to measure the reorientation of the chiral nematic phase in a strong magnetic field (Dong and Gray, 1997a) and in solid films cast from chiral nematic suspensions oriented in magnetic fields (Sugiyama et al., 1992; Dong and Gray, 1997a). The reorientation of cellulose nanocrystals to a more uniform texture is in contrast with chiral nematic liquid crystals composed of species having positive diamagnetic anisotropy, which simply untwist to give nematic textures when exposed to a magnetic field. However, it was observed that a slowly rotating magnetic field (5 T) applied to chiral nematic suspensions of tunicin nanocrystals caused unwinding of the helical axes to form a nematic-like arrangement (Kimura et al., 2005).

Cellulose nanocrystals have been used as the anionic component in self-assembled thin films of electrostatically adsorbed multilayers containing cationic polymers. The use of magnetic fields to optimize the order of these structures was also investigated (Cranston and Gray, 2006). Cellulose nanocrystal suspensions ordered by placing them in a magnetic field have been used as a medium in which to measure residual dipolar coupling of proteins by C′ decoupled ^{1}H-^{15}N inphase antiphase-heteronuclear single quantum coherence (IPAP-HSQC)[13] NMR (Fleming et al., 2000). Dispersions of cellulose nanocrystals in apolar solvents have also been oriented by an electric field to prepare textured materials (Bordel et al., 2006).

6.7.2 Effect of the polyelectrolyte nature

Interparticle electrostatic interactions have a strong effect on the free energy of the system. They provide stability and promote order. The addition of salts screens the electrostatic charges, destroying the order of anisotropic cellulose nanocrystal suspensions (Furuta et al., 1996) and causing them to flocculate (Dong et al., 1996).

According to Onsager's theory for neutral rods, the aspect ratio determines the critical concentration for phase separation. Cellulose nanocrystals, while rod-like, are far from being infinitely long rods (typical aspect ratios lie in the range 10–67 as seen in Chapter 3), have quite broad length distributions, and moreover are not electrostatically neutral owing to the charged sulfate ester groups on their surfaces when prepared by H_2SO_4 hydrolysis. It is therefore not surprising that the phase behavior of this system agrees only qualitatively with Onsager's theory; the experimental critical concentration tending to be higher than the theoretical value (Dong et al., 1996). The critical concentration and the width of the biphasic coexistence region of suspensions of polyelectrolytic rod-like particles are also very sensitive to variables such as length polydispersity, aspect ratio, surface charge density, ionic strength of the surrounding medium and the nature of the counterions associated with the charged particles (Fraden et al., 1989; Dong and Gray, 1997b; Araki et al., 1998; Araki et al., 1999). The first two variables naturally apply equally well to neutral rods.

Particle dimensions and ionic strength have been shown to govern the phase separation of rod-like polyelectrolytic cellulose nanocrystal suspensions (Dong et al., 1996). Ionic species alter the phase separation behavior of such suspensions by screening out the electrostatic repulsions of the surface sulfate ester groups, thereby reducing the effective rod diameter of the cellulose nanocrystals. For example, increasing the sodium chloride concentration from 0.13 to 1.95 mM increased the critical concentration for phase separation of cellulose nanocrystals from 6.5 to 9.0 wt%. The same effect can be observed by examining the volume fraction of chiral nematic phase which decreases as salt is added. Figure 6.11 summarizes the effects of ionic strength on the phase separation behavior of cellulose nanocrystal suspensions.

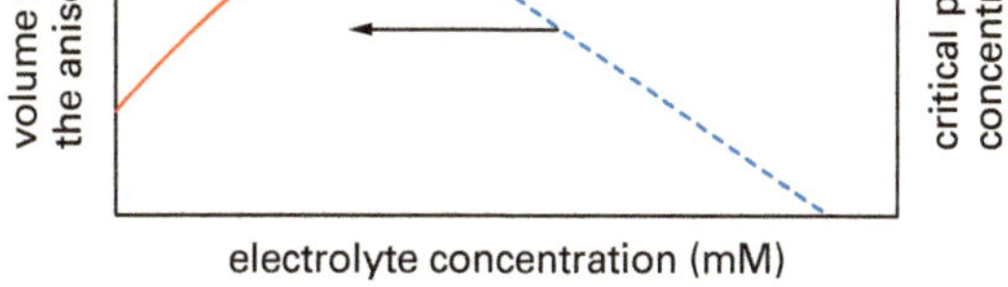

Fig. 6.11: Schematic evolution of the effect of added electrolyte on the phase separation behavior of a suspension of cellulose nanocrystals at constant concentration: volume fraction of rods in the anisotropic phase and critical phase separation concentration.

The phase transition diagram of a pure cotton cellulose nanocrystal suspension in deionized water is shown in Figure 6.12(a). When the total cellulose nanoparticle concentration was below 4.55 wt%, the suspension displayed a single isotropic phase (Dong et al., 1996). When the cellulose concentration increased above this critical value a second phase appeared, giving an upper isotropic phase and a lower anisotropic phase. The volume fraction of the latter increased when increasing the total concentration of cellulose and the complete disappearance of the isotropic phase was observed from 13.13 wt%. The same suspension was characterized at a constant acidic pH (1.61) (Figure 6.12(b)). The reason for considering constant pH and therefore constant hydrogen ion concentration was to maintain a constant ionic strength. A higher suspension concentration was required to form an ordered phase in these suspensions than in deionized water. From isotropic to biphasic and from biphasic to anisotropic, the critical concentration was 10.7 wt% and 15.8 wt%, respectively.

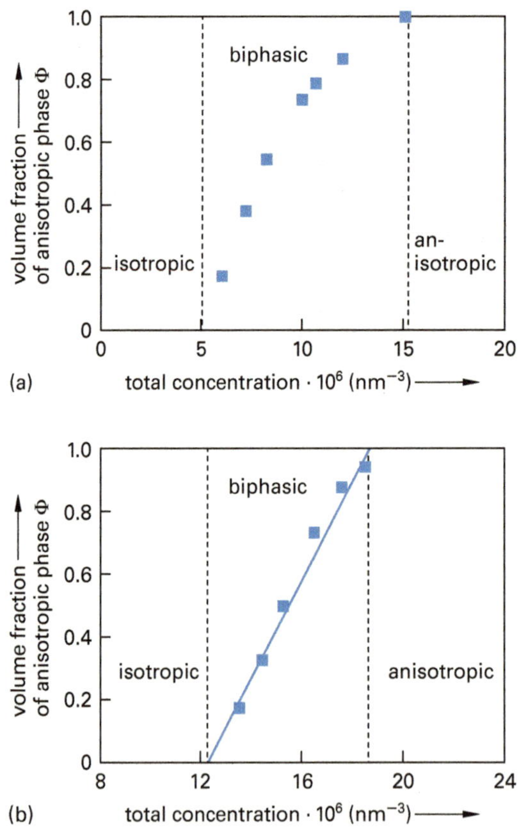

Fig. 6.12: Phase transition diagram of a cotton cellulose nanocrystal suspension (a) in pure water and at constant pH (pH = 1.61) (Dong et al., 1996).

As shown in Figure 6.13 the chiral nematic pitch of the anisotropic phase also decreases as the solvent ionic strength increases, indicating a strengthening of chiral interactions, which are thought to be screened by the electrostatic double layer (Dong et al., 1996). Experimental results for cellulose nanocrystals and other systems are not in complete agreement with SLO theory, possibly because of factors such as difficulty in estimating the contribution of the polyionic particles to the ionic strength of the system (Fraden et al., 1989; Strzelecka and Rill, 1991).

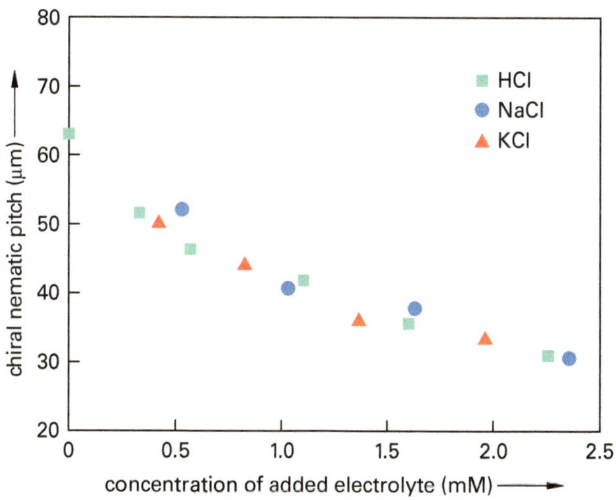

Fig. 6.13: Effect of electrolyte concentration on chiral nematic pitch of the anisotropic phase of a suspension of cotton cellulose nanocrystals. The cellulose nanocrystal concentration was fixed at $9.2 \cdot 10^{-6}$ nm^{-3} (Dong et al., 1996).

Both the stability and phase separation behavior of polyelectrolytic cellulose nanocrystal suspensions are very sensitive to changes in inter-rod interactions such as electrostatic repulsion, steric interactions, hydrophobic and hydration forces (Dong and Gray, 1997b). The nature (e.g. size, hydrophobicity) of the counterions associated with the sulfate ester groups is therefore also an important consideration.

Inorganic counterions tend to increase the critical concentration – in other words, to decrease the tendency for ordered phase formation – in the order $Na^+ < K^+ < Cs^+$, most likely due to decreasing repulsive hydration forces as the hydration number and hydrated ion size decrease. For bulky organic counterions, the phase equilibrium is governed by a balance between hydrophobic attractions and steric repulsions (Araki et al., 2001; Dong and Gray, 1997b).

Investigation of the phase separation behavior in aqueous suspensions of bacterial cellulose (BC) nanocrystals showed a surprising behavior when adding an electrolyte (Hirai et al., 2009). Without the addition of NaCl, the suspensions separated into typical isotropic and chiral nematic phases (Figure 6.14 top). For a 3.0 wt% suspension of BC nanocrystals, when NaCl was added up to a concentration of 1.0 mM, the volume fraction of the chiral nematic phase and the chiral nematic pitch both decreased as a result of the screening of surface charge on the cellulose particles, as observed in previous studies (Gray, 1996). However, cholesteric tactoids become visible in the upper layer upon the addition of NaCl. With further addition of NaCl, the volume fraction of the chiral nematic phase was found to increase along with the chiral nematic pitch. In the upper layer, the cholesteric tactoids become larger and aggregate. Above 2 mM NaCl, phase separation no longer occurred, but domains of the chiral nematic phase were still visible. Finally, at 2.75 mM NaCl and above, only small tactoids are visible (Figure 6.14 bottom). Clearly, the phase separation behavior summarized in Figure 6.14 is not being followed above 1.0 mM added NaCl.

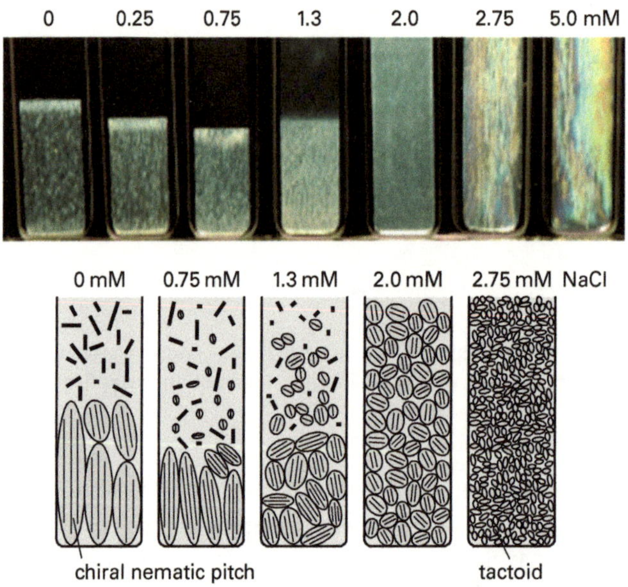

Fig. 6.14: Top: Effect of added NaCl on the phase separation behavior of suspension containing 3 wt% BC nanocrystals after 25 days of standing. Bottom: Schematic illustration of the effects of added NaCl on the phase separation behavior of the BC suspensions for a fixed cellulose concentration (Hirai et al., 2009).

6.7.3 Effect of the presence of macromolecules

Depending on their concentration, cellulose nanocrystal suspensions exhibit three phases: isotropic phase at low concentrations, anisotropic phase at high concentrations and biphasic region at intermediate concentrations, where the isotropic and the anisotropic phases coexist. Thus, addition of high-molecular weight blue dextrans to these solutions would lead to preferential partitioning of dextran into one or the other phase (Edgar and Gray, 2002). Blue dextran is an easily quantifiable non-adsorbing macromolecule consisting of sulfonated triazine dye, Cibacron blue 3G-A, covalently bonded to the high-molecular weight dextran chains. Cibatron blue 3G-A is used as the dye molecule and when it is conjugated with dextran, then the dextran-dye conjugate is referred to as ligand. It was shown that adding dextran to an isotropic suspension caused no demixing of the two components. For a biphasic sample, absorption or preferential partitioning of dextran into the isotropic phase was observed and this effect increased with increasing cellulose concentration (Figure 6.15). Dextran absorbance into the anisotropic phase was considerably lower due to the mutual exclusion of the cellulose nanocrystal ordered domains and dextran molecules.

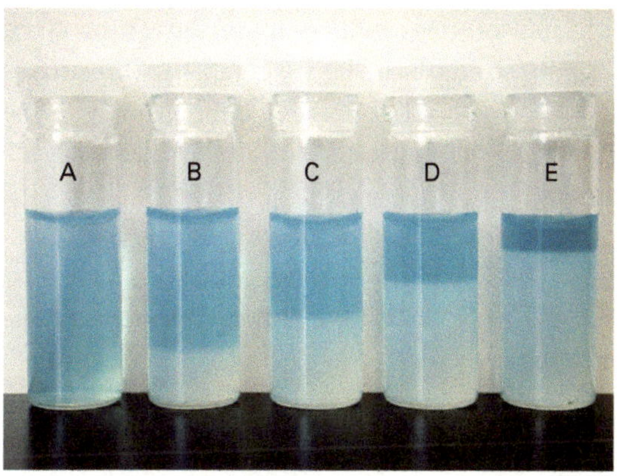

Fig. 6.15: Photographs of vials of biphasic cellulose nanocrystal suspensions with the same blue dextran concentration containing increasing concentrations from left to right of cellulose nanocrystal (A: 6.5 wt%, B: 8.8 wt%, C: 9.4 wt%, D: 11.0 wt%, E: 13.3 wt%), and showing the preferential partitioning of blue dextran into the upper isotropic phase (Edgar and Gray, 2002).

Adding dextran to a completely anisotropic phase caused a phase separation into a dextran-rich isotropic phase and a dextran-poor anisotropic phase (Edgar and Gray, 2002). This phase separation occurred over a period of several days when the blue dextran macromolecules were added to the cellulose nanocrystal suspension. Higher concentrations of blue dextran led to faster separation and larger volume fractions of the isotropic phase. Further investigation of the phase separation behavior induced in highly anisotropic cellulose nanocrystal suspensions (13.8 wt%) by blue dextrans of varying dye content and molecular weight was reported (Beck-Candanedo et al., 2006a). It was shown that the phase separation was associated with the charged dye molecules bonded to the dextran. At increasing ionic strength, depletion attractions due to the blue dextran increasingly contribute to the phase separation. Moreover, addition of dextran induced distortions in the liquid crystal structure of anisotropic suspensions (Edgar and Gray, 2002). Similar results were obtained for the biphasic region. Due to mutual exclusion, dextran molecules migrated into the isotropic phase thereby increasing the osmotic pressure. The cellulose nanocrystals moved into the anisotropic phase (chiral nematic) in order to balance the osmotic pressure. This preferential migration in turn increased the width of the biphasic region.

Dye induced phase separation in anisotropic cellulose nanocrystal suspensions was also investigated using three different kinds of dyes, viz. anionic, cationic and neutral (Beck-Candanedo et al., 2006b). It was observed that anionic dyes of varying charge induced phase separation of an isotropic phase into a biphasic region at much lower ionic strengths than simple 1:1 electrolytes (sodium chlorite). Polyvalence and larger hydration radius were suspected due to their higher effectiveness in inducing phase separation. Cationic dyes did not induce phase separation in anisotropic cellulose nanocrystal suspensions since they interact with the negative surface charges on the nanoparticle surface. Nor could neutral dyes induce phase separation since they were incapable of changing the ionic strength of the suspensions. Direct dyes like Congo red were unable to induce phase separation as they bind strongly to the cellulose nanocrystals and cannot compress the electrical double layer around the nanoparticles and induce phase separation.

The partitioning of neutral and charged dextrans in biphasic cellulose nanocrystal suspensions (Beck-Candanedo et al., 2008) and the triphase equilibrium in mixtures of blue dextrans, undyed dextrans and cellulose nanocrystal suspensions (Beck-Candanedo et al., 2007) were also investigated.

6.8 Liquid crystalline phases of spherical cellulose nanocrystal suspensions

Spherical cellulose nanocrystals have been prepared by hydrolyzing microcrystalline cellulose (MCC) with a mixture of hydrochloric and sulfuric acids under prolonged ultrasonic treatment (Wang et al., 2008). Spheres with a size polydispersity of up to 49% were produced (Figure 6.16(a)). According to Onsager's theory, it should be difficult for a suspension of spherical particles to form a liquid crystalline phase due to the lack of thermodynamic driving force arising from the symmetrical shape of the particles. However, liquid crystalline phase formation was observed (Figure 6.16(b)) in suspensions when the concentration of nanoparticles was higher than 3.9 wt%. The high polydispersity and the charged surface sulfate groups were considered to play an important role in forming the liquid crystalline phase in these suspensions. However, the nature of the liquid crystalline phase is currently unknown.

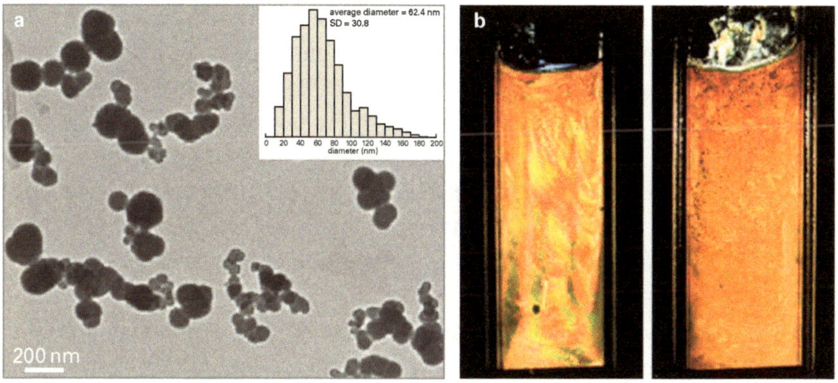

Fig. 6.16: (a) Transmission electron micrograph of the hydrolysis product of microcrystalline cellulose by mixed acid under ultrasonic treatment after 10 h. Inset corresponds to the size distribution of the spherical cellulose nanocrystals from the TEM images, and (b) birefringent patterns of spherical cellulose nanocrystal suspensions observed between crossed nicols in a sample cell with optical path of 1.0 mm: 4.5 wt% (left) and 7.1 wt% (right), both suspensions at rest 3 days after injection (Wang et al., 2008).

6.9 Rheological behavior of cellulose nanocrystal suspensions

Properties of cellulose nanocrystal suspensions have been much more investigated than for MFC suspensions because of their liquid crystalline behavior. Indeed, as previously discussed, rod-like nanoparticles resulting from acid hydrolysis cellulosic fibers can give spontaneously ordered phases. Obviously, the alignment of cellulose nanocrystals is strongly influenced under flow by applying shear rate and liquid crystal transitions can be evidenced. In the pioneering work reported by Marchessault et al., the hydrodynamics properties of nanocrystal suspensions were found to be directly related to the size and length distribution of the nanoparticles (Marchessault et al., 1961).

In the dilute regime, shear thinning behavior is observed and this phenomenon is emphasized as the concentration of nanoparticles increases. This means that the suspension shows concentration dependence at low shear rates exhibiting higher viscosities as the concentration increases and very little concentration dependence at higher rates. At higher concentrations, where the suspensions become lyotropic, a typical behavior of liquid crystalline polymers in solutions is generally reported. Three distinct regions are observed for the shear dependence of the viscosity. At lower shear rates the viscosity continuously decreases. It was supposed to correspond to a shear thinning behavior where the domains formed by the nanocrystals begin to flow and disclination lines of the polydomain texture vary. As the shear rate is increased, the domains start to break up and a semi-plateau region, where the shear thinning is less pronounced as the rate is raised, is observed in the flow curve. Director wagging (Larson, 1999) and alignment in the vorticity direction (Montesi et al., 2004) have been also reported. At a critical rate, a further shear thinning occurred corresponding to the alignment of individual nanocrystals in the flow direction. However, the slope is lower than for the shear thinning observed at lower shear rates. The chirality of the suspension breaks down in favor of a simple nematic structure. Observations with polarized filters reveal that the "fingerprint" patterns indicative of the chiral nematic phase were deformed and disappeared with increasing the shear rate (Orts et al., 1998).

Such behavior was observed from diffraction/scattering experiments (Orts et al., 1998; Ebeling et al., 1999), scanning electron microscopy (SEM) observations (Nishiyama et al., 1997), and rheometrical measurements (Bercea and Navard, 2000; de Souza Lima and Borsali, 2004). It was found that the aspect ratio is a key parameter in determining the degree of shear-induced order, as well as the relaxation behavior after shear ceased (Orts et al., 1998). The viscous suspensions derived from long nanocrystals (~ 280 nm) remained aligned for hours and even days after shearing, while samples with shorter nanocrystals (~ 180 nm) were found to relax quickly.

The shear induced alignment of cotton nanocrystals with an aspect ratio around 10 in colloidal aqueous suspensions was also investigated by small angle X-ray scattering (SAXS), and it was demonstrated that the nanoparticles horizontally aligned along the shear direction when the shear rate exceeded 5 s^{-1} and that this alignment was completely reversible (Ebeling et al., 1999). The experiments were conducted with 6.9 wt% suspensions and a couette cell was positioned in the X-ray beam. Measurements were performed in two stages, i.e. radial scattering and tangential scattering patterns. At low shear rates (0.05 s^{-1}), the radial and tangential patterns possessed isotropic rings and anisotropic peaks, respectively, indicating that cellulose nanocrystals aligned themselves preferentially in the vertical direction. At high shear rates (500 s^{-1}), the radial pattern showed anisotropic rings while the tangential pattern showed an isotropic ring, indicating that the cellulose nanocrystals changed alignment from vertical direction to the direction of shear.

Investigating the rheological behavior of aqueous suspensions of tunicin nanocrystals a critical concentration of 0.8 wt% was reported (Bercea and Navard, 2000). The rheological properties of isotropic and anisotropic suspensions were studied. In the isotropic phase, i.e. below this critical concentration, where the rod-like particles are randomly oriented, their alignment induced a continuous decrease of the viscosity upon increasing the shear rate. Two plateaus were reported at low and high shear rate. At low shear rates, the nanocrystals were randomly oriented and no change in viscosity was observed. At high shear rates, the rods became well oriented and viscosity became independent of shear rate. For higher concentrations, i.e. in the anisotropic-at-rest phase (1–3.5 wt%), the behavior was similar to that of liquid crystal liquid polymers with a weak shear thinning region surrounded by two other shear thinning regions. These different behaviors are represented in Figure 6.17. An attempt to correlate the intrinsic viscosity of cellulose nanocrystal suspensions to the aspect ratio of constituting nanoparticles was reported (Boluk et al., 2011). The intrinsic viscosity was determined for suspensions containing different concentrations of NaCl and extrapolated to 1 nm Debye length to calculate the intrinsic viscosity of electroviscous effect-free hard rods.

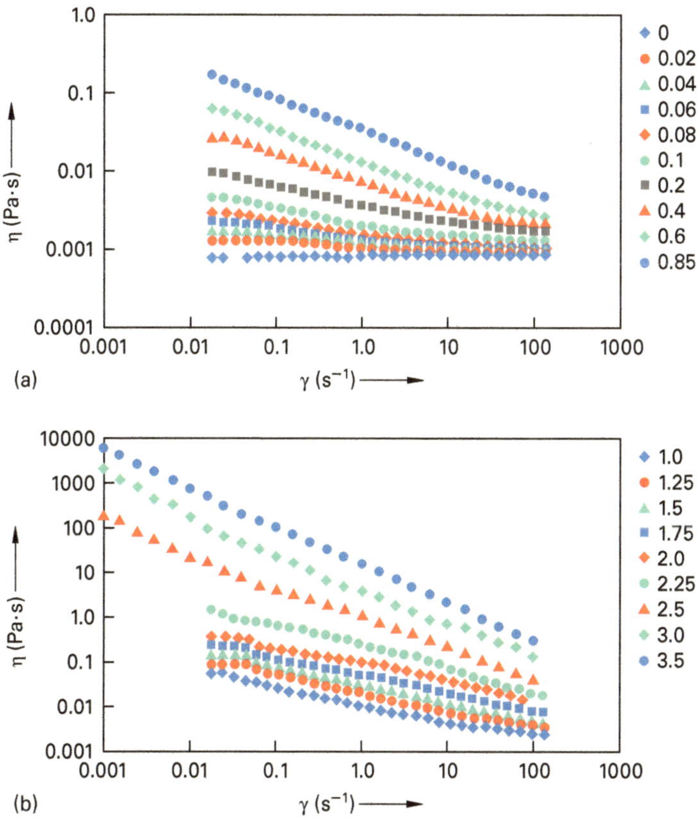

Fig. 6.17: Viscosity of tunicin nanocrystal suspensions as a function of shear rate for different nanocrystal concentrations; (a) in the isotropic at-rest regime up to the lyotropic transition, and (b) above the lyotropic transition in the anisotropic at-rest regime (Bercea and Navard, 2000).

The effects of both the concentration and temperature on the microstructure and shear response of aqueous cotton cellulose nanocrystals have been reported (Ureña-Benavides, 2011). The phase behavior was evaluated using a combination of low-magnification imaging and hot-stage cross-polarized optical microscopy. It was shown that with increasing concentration up to 17.3 vol% the microstructure of the suspensions changed from dispersed liquid crystalline domains in a continuous isotropic matrix to a seemingly co-continuous blend, then to dispersed droplets of isotropic domains in a continuous liquid crystalline phase, followed by a totally liquid crystalline material and finally to a gel. At high concentrations, a marked increase of the consistency coefficient and decrease of the rate index from the power law, as well as loss of fingerprint texture occurred between 12.6 and 14.5 vol% at the liquid crystalline to gel transition.

The rheological properties of a dispersion of cellulose nanocrystals in a 1 wt% aqueous solution of high molecular weight polyoxyethylene (PEO) have been investigated (Ben Azouz, 2012). Upon adding nanoparticles, the viscosity of the suspension surprisingly first decreased and then increased, showing a minimum value around 6 wt% on PEO basis (Figure 6.18). Adsorption of PEO chains on the surface of the nanoparticles has been suspected. This was attributed to strong affinity between PEO chains and the cellulosic surface through interactions between the oxygen groups of PEO and hydroxyl groups of cellulose. This wrapping PEO layer was further used as a "buffer" for melt extrusion of cellulose nanocrystals with a low density polyethylene matrix.

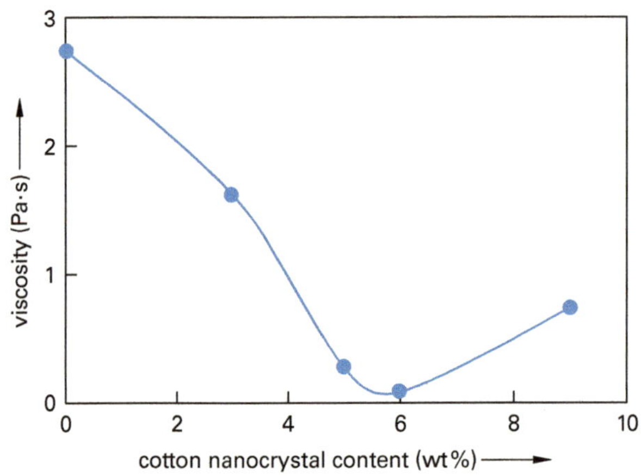

Fig. 6.18: Steady shear viscosity measured for a shear rate of 0.3 s^{-1} for 1 wt % PEO (Mw = $5 \cdot 10^6$ $g \cdot mol^{-1}$) solution as a function of cellulose nanocrystal content (Ben Azouz et al., 2012).

6.10 Light scattering studies

Few studies on light scattering experiments conducted with aqueous cellulose nanocrystal suspensions can be found in the literature. The dimensions and structure of the nanoparticles can be determined from this technique. Both static and dynamic light scattering experiments on tunicin nanocrystal suspensions have been reported (de Souza Lima and Borsali, 2002). In this study, the sample was fractionated by ultracentrifugation aided by a saccharose gradient as reported in Chapter 3 (Section

3.5.2). Before ultracentrifugation two phases existed, viz. a clear phase corresponding to cellulose nanocrystals accumulated on the sucrose gradient and a black phase representing the nanocrystals that did not accumulate on the sucrose gradient. After ultracentrifugation, the sample was fractioned into different bands and TEM analysis on each fraction confirmed that the fractionated samples had narrow polydispersity and reduced nanocrystal aggregation.

For static light scattering studies, reduced elastic scattering I(q)/kC was measured and plotted as a function of the wave vector q (de Souza Lima and Borsali, 2002). Plotting was performed in a normalized form to obtain normalized scattering curves at different concentrations. Normalized scattering curves in the absence of salt showed the presence of angular scattering peaks at q_{max} depending on the concentration of cellulose nanocrystal suspensions. On conducting the same experiment before fractionation, the peaks disappear due to formation of nanoparticle aggregates and increased polydispersity.

For spherical particles, the scattered intensity I(q) can be correlated with the structure factor S(q) and form factor P(q) by the relation:

$$I(q) = S(q) \cdot P(q) \tag{6.6}$$

This relation may be valid for rod-like particles if two conditions are fulfilled:

1. The concentration of the rods is sufficiently high, so that the contributions from larger aggregates supersede the contribution from smaller/pair aggregates.
2. Most rods are aligned perpendicular to the scattering vector q, so that these perpendicular rods have a maximum contribution to scattering making $qL >> 1$. Therefore, P(q) becomes more or less constant and the above relation can be used to calculate the structure factor of the rods.

For elongated tunicin nanocrystals, $qL >> 1$ and the above relation can be used in combination with the equation for the form factor for rigid rod-like particles to derive the structure factor S(q). Broad scattering peaks of S(q) with second and third maxima peaks indicated the presence of strong and long-range electrostatic interactions between the nanoparticles. When adding a salt these scattering peaks disappeared because of the screening of the electrostatic interactions between the cellulose nanocrystals.

The translational and rotational dynamics of cellulose nanocrystals extracted from cotton and tunicate, and therefore having different dimensions, was also investigated (de Souza Lima et al., 2003). Both polarized and depolarized dynamic light scattering, as well as transient electric birefringence decay (TEB) were used to study their dynamical behavior. When the frequency was plotted as a function of the square of scattering wave-vector q^2, slow and fast relaxation modes for both cotton and tunicin nanocrystals were observed with the slower mode contributing 90% of the scattered intensity. Moreover, the frequency varied with q^2, and all relaxation modes were said to be diffusive. The values of translational (D) and diffusion (Θ) coefficients in dilute suspensions were deduced. The obtained values were used in Broersma's equations to determine the dimensions of cotton and tunicin nanocrystals.

The dynamics of cotton cellulose nanocrystals trapped in agarose hydrogels of varying concentrations was studied by polarized dynamic light scattering (Bica et al., 2001). No rotational modes were detected and the observed fast and slow modes were purely translational. For low agarose gel concentrations, i.e. when the mesh size of the agarose network was larger than the length of the rod-like nanoparticles, the dynamics did not depend on the nanocrystal content indicating that the network did not hinder the nanocrystals. On the contrary, when the mesh size of the network was smaller than the nanocrystal length, i.e. for higher agarose gel concentrations, frictional effects arose and the nanoparticle content influenced the dynamics. Moreover, in this condition domains with different nanocrystal contents develop in the gel since the nanoparticles may be excluded from regions of small pore size and concentrate into large ones. Depolarized dynamic light scattering measurements were conducted on the same system (Bica et al., 2006). The rotational relaxation rate of the nanoparticles was found to increase when increasing the concentration of the gel, reflecting reduced amplitude of rotational fluctuations of the nanocrystals due to repulsive interactions with their surroundings.

6.11 Preserving the chiral nematic order in solid films

Above a critical concentration of nanoparticle, chiral nematic liquid crystalline order forms in aqueous suspensions of cellulose nanocrystals. The ratio of this anisotropic phase increases as the concentration increases, i.e. as the water evaporates. Under proper conditions, the suspensions can be slowly evaporated to produce solid semi-translucent films that retain the self-assembled chiral nematic liquid crystalline order formed in the suspension. These films exhibit iridescence by reflecting left-handed circularly polarized light in a narrow wavelength band determined by the chiral nematic pitch and the refractive index of the film. The reflected wavelength, λ, is given by:

$$\lambda = n\,P\,\sin\theta \tag{6.7}$$

where n is the refractive index (n = 1.55 for crystalline cellulose), P is the chiral nematic pitch, and θ is the angle of reflection relative to the surface of the film. This phenomenon of reflectance was explained on the basis of a helicoidal arrangement of birefringent layers, as is the case for cellulose nanocrystals in a chiral nematic liquid crystal (de Vries, 1951).

Therefore, the reflected wavelength becomes shorter at oblique viewing angles, giving rise to visible iridescence colors when the pitch of the helix is of the order of the wavelengths of visible light (between 400 and 700 nm). Increasing the concentration of the electrolyte (e.g., NaCl, KCl) in the cellulose nanocrystal suspensions prior to film casting partially screens the negative charges of the sulfate ester groups on the nanocrystal surfaces. It allows the particles to approach each other more closely and reduces the chiral nematic pitch. The iridescence is consequently shifted towards shorter wavelengths (the ultraviolet region) (Revol et al., 1998). However, this method of "blue-shifting" cellulose nanocrystal film iridescence is limited by the amount of salt that can be added before suspension is destabilized by too much screening and gelation occurs (Dong et al., 1996; Revol et al., 1998). The cellulose nanocrystal film iridescence colors also depend somewhat on the cellulose source and the hydrolysis conditions. Desulfation was also found to reduce the chiral nematic pitch (Revol et al., 1998).

The optical properties of solid cellulose nanocrystal films were characterized by induced circular dichroism (ICD) measurements (Edgar and Gray, 2001). Salt-free suspensions form films that reflect in the infrared region. To produce films that reflect in the visible and ultraviolet regions an appropriate amount of NaCl solution was added prior to evaporation in order to decrease the pitch of the films. If the solid films have a helicoidal structure an ICD peak should be observed on addition of a suitable dye, such as Congo red or Trypan blue. Indeed, since cellulose has no chromophores that absorb in regions easily accessible to most spectropolarimeters, doping with an achiral dye that forms a close association with the cellulose backbone is a necessary step. However, films cast from suspensions with high electrolyte concentrations did not show any ICD signal, suggesting the formation of an isotropic film. Moreover, films prepared in the presence of a magnetic field showed an increase in intensity of the ICD signal, indicating an increase in the degree of order in the film.

The chiral nematic textures are not the only liquid crystalline textures that can be preserved in solid cellulose nanocrystal films. The microstructure of cellulose nanocrystal films strongly depends on the drying conditions. Generally, films obtained by casting under ambient conditions have a polydomain structure in which the cholesteric axis of the chiral nematic domains point in random directions. Therefore, planar and fingerprint textures can be observed by optical microscopy studies of these films. However, such a study of the microstructure of solid cellulose nanocrystal films revealed areas with parabolic focal conic (PFC) defect structure, a symmetric form of focal conic defects in which the line defects form a pair of perpendicular, antiparallel, and confocal parabolas (Roman and Gray, 2005). The cellulose films were characterized by polarized-light (Figure 6.19) and atomic force microscopy. The film surface showed a regular array of large and small elevations resulting from the displacement of the structural layers. This was the first indication that a rod-like colloidal particle can self-assemble into the complex, symmetrical structure of the PFC texture.

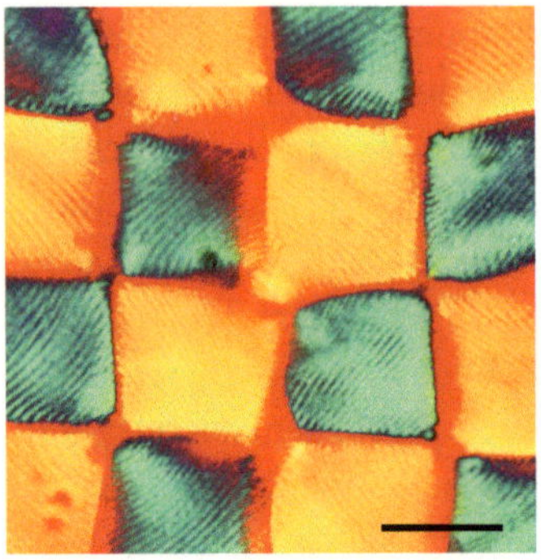

Fig. 6.19: Square lattice in a solid film of cellulose nanocrystals viewed with crossed polars and full-wave retardation plate inserted into the microscope. Scale bar 20 μm (Roman and Gray, 2005).

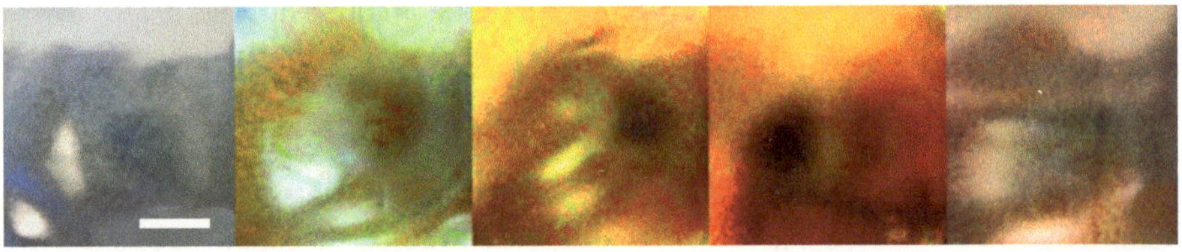

Fig. 6.20: Cellulose nanocrystal films produced from suspensions treated with increasing applied ultrasonic energy (0, 250, 700, 1800, and 7200 $J \cdot g^{-1}$ of cellulose nanocrystal) from left to right. Viewing is normal to the film surface under diffuse lighting. Scale marker 1 cm (Beck et al., 2011).

The influence of ionic strength, temperature, suspension concentration, and exposure to magnetic field on chiral nematic phase was investigated at a macroscopic level using circular dichroism and polarized microscopy (Pan et al., 2010). It was shown that drying a cellulose nanocrystal film in the presence of a 0.2 T external magnetic field increased the pitch to a value that was found to depend on drying time. Ultrasound treatment was also found to increase the chiral nematic pitch in suspension and red-shift the reflection wavelength of cellulose nanocrystal films as the applied energy increased (Beck et al., 2011). Figure 6.20 shows solid films cast from aliquots of 2.8 wt% cellulose nanocrystal suspensions prepared by sulfuric acid hydrolysis from bleached softwood kraft pulp and sonicated with increasing (left to right) energy inputs. The energy was measured in $J \cdot g^{-1}$ of cellulose nanocrystals. The films exhibit reflected iridescence with colors ranging from blue-violet to red. By combining sonication and electrolyte addition, the reflective properties of the film can be predictably tuned. The effects of sonicating a cellulose nanocrystal suspension were shown to be cumulative and permanent. Moreover, suspensions sonicated with different energy inputs can be mixed to prepare films having a reflection band intermediate between those obtained from the individual suspensions. It was suggested that the ultrasound-induced red-shift is electrostatic in nature.

6.12 Conclusions

The unique rheological behavior of nanocellulose dispersions was recognized by the early investigators. The high viscosity at low concentrations makes MFC very interesting as a non-calorie stabilizer and gellant in food applications. However, it imparts some severe challenging issues for the processing of nanocomposites. Due to their unique rod-like structure, colloidal suspensions of cellulose nanocrystals display liquid crystalline behavior that can be tuned and captured in solid films by slow evaporation of the liquid. Ensuing films have novel optical properties generating interesting potential applications.

6.13 References

Abe, A. and Flory, P.J. (1978). Statistical thermodynamics of mixtures of rodlike particles. 2. Ternary mixtures. Macromolecules *11*, 1122–1126.

Agoda-Tandjawa, G., Durand, S., Berot, S., Blassel, C., Gaillard, C., Garnier, C. and Doublier, J.L. (2010). Rheological characterization of microfibrillated cellulose suspensions after freezing. Carbohydr. Polym. *80*, 677–686.

Agoda-Tandjawa, G., Durand, S., Gaillard, C., Garnier, C. and Doublier, J.L. (2012). Rheological behaviour and microstructure of microfibrillated cellulose suspensions/low-methoxyl pectin mixed systems. Effect of calcium ions. Carbohydr. Polym. *87*, 1045–1057.

Araki, J., Wada, M., Kuga, S. and Okano, T. (1998). Flow properties of microcrystalline cellulose prepared by acid treatment of native cellulose. Colloid Surface A *142*, 75–82.

Araki, J., Wada, M., Kuga, S. and Okano, T. (1999). Influence of surface charge on viscosity behaviour of cellulose microcrystal suspension. J. Wood Sci. *45*, 258–261.

Araki, J., Wada, M., Kuga, S. and Okano, T. (2000). Birefringent glassy phase of a cellulose microcrystal suspension. Langmuir *16*, 2413–2415.

Araki, J., Wada, M. and Kuga, S. (2001). Steric stabilization of a cellulose microcrystal suspension by poly(ethylene glycol) grafting. Langmuir *17*, 21–27.

Araki, J. and Kuga, S. (2001). Effect of trace electrolyte on liquid crystal type of cellulose microcrystals. Langmuir *17*, 4493–4496.

Arora, A.K. and Tata, B.V.R. (1996). Ordering and phase transitions in charged colloids (VCH Publishers, New York). Chap. 6.

Aulin, C., Gällstedt, M. and Lindström, T. (2010). Oxygen and oil barrier properties of microfibrillated cellulose films and coatings. Cellulose *17*, 559–574.

Azizi Samir, M.A.S., Alloin, F., Sanchez, J.Y., El Kissi, N. and Dufresne, A. (2004). Preparation of cellulose whiskers reinforced nanocomposites from an organic medium suspension. Macromolecules *37*, 1386–1393.

Baus, M., Coussaert, T. and Achrayah, R. (1996). Colloidal crystals: A van der Waals approach. Physica A *232*, 575–584.

Bawden, F.C., Pirie, N.W. and Bernal, J.D. (1936). Liquid crystalline substances from virus-infected plants. Nature *138*, 1051–1052.

Bawden, F.C. and Pirie, N.W. (1937). The isolation and some properties of liquid crystalline substances from solanaceous plants infected with three strains of tobacco mosaic virus. Proc. R. Soc. London Ser. B *123*, 274–320.

Beck-Candanedo, S. Viet, D. and Gray, D.G. (2006a). Induced phase separation in low ionic strength cellulose nanocrystal suspensions containing high molecular weight blue dextrans. Langmuir *22*, 8690–8695.

Beck-Candanedo, S., Viet, D. and Gray, D.G. (2006b). Induced phase separation in cellulose nanocrystal suspensions containing ionic dye species. Cellulose *13*, 629–635.

Beck-Candanedo, S., Viet, D. and Gray, D.G. (2007). Triphase equilibria in cellulose nanocrystal suspensions containing neutral and charged macromolecules. Macromolecules *40*, 3429–3436.

Beck-Candanedo, S., Viet, D. and Gray, D.G. (2008). Partitioning of charged and neutral dextran-dye derivatives in biphasic cellulose nanocrystal suspensions. Can. J. Chem. *86*, 503–511.

Beck, S., Bouchard, J. and Berry, R. (2011). Controlling the reflection wavelength of iridescent solid films of nanocrystalline cellulose. Biomacromolecules *12*, 167–172.

Ben Azouz, K., Ramires, E.C., Van den Fonteyne, W., El Kissi, N. and Dufresne A. (2012). A simple method for the melt extrusion of cellulose nanocrystal reinforced hydrophobic polymer. ACS Macro Lett. *1*, 236–240.

Bercea, M. and Navard, P. (2000). Shear dynamics of aqueous suspensions of cellulose whiskers. Macromolecules *33*, 6011–6016.

Besbes, I., Alila, S. and Boufi, S. (2011). Nanofibrillated cellulose from TEMPO-oxidized eucalyptus fibres: Effect of the carboxyl content. Carbohydr. Polym. *84*, 975–983.

Bica, C.I.D., Borsali, R., Geissler, E. and Rochas, C. (2001). Dynamics of cellulose whiskers in agarose gels. 1. Polarized dynamic light scattering. Macromolecules *34*, 5275–5279.

Bica, C.I.D., Borsali, R., Rochas, C. and Geissler, E. (2006). Dynamics of cellulose whiskers spatially trapped in agarose hydrogels. Macromolecules *39*, 3622–3627.

Boluk, Y., Lahiji, R., Zhao, L. and McDermott, M.T. (2011). Suspension viscosities and shape parameter of cellulose nanocrystals (CNC). Colloid Surface A *377*, 297–303.

Bordel, D., Putaux, J.L. and Heux, L. (2006). Orientation of native cellulose in an electric field. Langmuir *22*, 4899–4901.

Buining, P.A. and Lekkerkerker, H.N.W. (1993). Isotropic-nematic phase separation of a dispersion of organophilic boehmite rods. J. Phys. Chem. *97*, 11510–11516.

Buining, P.A., Philipse, A.P. and Lekkerkerker, H.N.W. (1994). Phase behavior of aqueous dispersions of colloidal boehmite rods. Langmuir *10*, 2106–2114.

Chandrasekhar, S. (1992). Liquid crystals. 2nd ed. (Cambridge University Press, New York).

Chiba, R., Nishio, Y. and Miyashita, Y. (2003). Electrooptical behavior of liquid-crystalline (hydroxypropyl)cellulose/inorganic salt aqueous solutions. Macromolecules *36*, 1706–1712.

Chiba, R., Nishio, Y., Sato, Y., Ohtaki, M. and Miyashita, Y. (2006). Preparation of cholesteric (hydroxypropyl)cellulose/polymer networks and ion-mMediated control of their optical properties. Biomacromolecules *7*, 3076–3082.

Chilaya, G.S. and Lisetski, L.N. (1986). Cholesteric liquid crystals: Physical properties and molecular – statistical theories. Mol. Cryst. Liq. Cryst. *140*, 243–386.

Collings, P.J. and Hird, M. (1998). Introduction to liquid crystals (Taylor & Francis, London).

Cranston, E.D. and Gray, D.G. (2006). Formation of cellulose-based electrostatic layer-by-layer films in a magnetic field. Sci. Technol. Adv. Mater. *7*, 319–321.

de Gennes, P.G. and Proust, J. (1993). The physics of liquid crystals. 2nd ed. (Oxford University Press, Oxford, UK). p. 110.

de Souza Lima, M.M. and Borsali, R. (2002). Static and dynamic light scattering from polyelectrolyte microcrystal cellulose. Langmuir *18*, 992–996.

de Souza Lima, M.M., Wong, J.T., Paillet, M., Borsali, R. and Pecora, R. (2003). Translational and rotational dynamics of rodlike cellulose whiskers. Langmuir *19*, 24–29.
de Souza Lima, M.M. and Borsali, R. (2004). Rodlike cellulose microcrystals: Structure, properties and applications. Macromol. Rapid Commun. *25*, 771–787.
de Vries, H.I. (1951). Rotatory power and other optical properties of certain liquid crystals. Acta Cryst. *4*, 219–226.
Derjaguin, B.V. and Landau, L. (1941). Theory of the stability of strongly charged lyophobics sols and of the adhesion of strongly charged particles in solutions of electrolytes. USSR Acta Physiochim. *14*, 633–662.
DiMarzio, E.A. (1961). Statistics of orientation effects in linear polymer molecules. J. Chem. Phys. *35*, 658–669.
Dong, X.M., Kimura, T., Revol, J.F. and Gray, D.G. (1996). Effects of ionic strength on the isotropic-chiral nematic phase transition of suspensions of cellulose crystallites. Langmuir *12*, 2076–2082.
Dong, X.M. and Gray, D.G. (1997a). Induced circular dichroism of isotropic and magnetically-oriented chiral nematic suspensions of cellulose crystallites. Langmuir *13*, 3029–3034.
Dong, X.M. and Gray, D.G. (1997b). Effect of counterions on ordered phase formation in suspensions of charged rodlike cellulose crystallites. Langmuir *13*, 2404–2409.
Dong, X.M., Revol, J.F. and Gray, D.G. (1998). Effects of microcrystallite preparation conditions on the formation of colloid crystals of cellulose. Cellulose *5*, 19–32.
DuPré, D.B. (1981). Liquid crystals. In: Kirk-Othmer encyclopedia of chemical technology, Vol. 14 (Wiley-Interscience, New York), pp. 395–427.
Ebeling, T. Paillet, M., Borsali, R., Diat, O., Dufresne, A., Cavaillé, J.Y. and Chanzy, H. (1999). Shear-induced orientation phenomena in suspensions of cellulose microcrystals, revealed by small angle X-ray scattering, Langmuir *15*, 6123–6126.
Edgar, C.D. and Gray, D.G. (2001). Induced circular dichroism of chiral nematic cellulose films. Cellulose *8*, 5–12.
Edgar, C.D. and Gray, D.G. (2002). Influence of dextran on the phase behavior of suspensions of cellulose nanocrystals. Macromolecules *35*, 7400–7406.
Elazzouzi-Hafraoui, S., Putaux, J.L. and Heux, L. (2009). Self-assembling and chiral nematic properties of organophilic cellulose nanocrystals. J. Phys. Chem. B *113*, 11069–11075.
Fleming, K., Gray, D., Prasannan, S. and Matthews, S. (2000). Cellulose crystallites: A new and robust liquid crystalline medium for the measurement of residual dipolar couplings. J. Am. Chem. Soc. *122*, 5224–5225.
Flory, P.J. (1956). Phase equilibria in solutions of rod-like particles. Proc. R. Soc. London Ser. A *234*, 73–89.
Flory, P.J. and Abe, A. (1978). Statistical thermodynamics of mixtures of rodlike particles. 1. Theory for polydisperse systems. Macromolecules *11*, 1119–1122.
Flory, P.J. and Frost, R.S. (1978). Statistical thermodynamics of mixtures of rodlike particles. 3. The most probable distribution. Macromolecules *11*, 1126–1133.
Flory, P.J. (1978a). Statistical thermodynamics of mixtures of rodlike particles. 5. Mixtures with random coils. Macromolecules *11*, 1138–1141.
Flory, P.J. (1978b). Statistical thermodynamics of mixtures of rodlike particles. 6. Rods connected by flexible joints. Macromolecules *11*, 1141–1144.
Folda, T., Hoffmann, H., Chanzy, H. and Smith, P. (1988). Liquid crystalline suspensions of poly(tetrafluoroethylene). Nature *333*, 55–56.
Fraden, S., Maret, G., Caspar, D.L.D. and Meyer, R.B. (1989). The isotropic-nematic phase transition and angular correlations in isotropic suspensions of tobacco mosaic virus. Phys. Rev. Lett. *63*, 2068–2071.

Frost, R.S. and Flory, P.J. (1978). Statistical thermodynamics of mixtures of rodlike particles. 4. The Poisson distribution. Macromolecules *11*, 1134–1138.

Furuta, T., Yamahara, E., Konishi, T. and Ise, N. (1996). Ordering in aqueous cellulose hydrolysate dispersions – An ultra-small-angle X-ray-scattering study. Macromolecules *29*, 8994–8995.

Gray, D.G. (1996). Chirality in cellulose and cellulose-based materials. Polymer Prepr. *37*, 485–486.

Grelet, E. and Fraden, S. (2003). What is the origin of chirality in the cholesteric phase of virus suspensions? Phys. Rev. Lett. *90*, 198302-1–198302-4.

Guo, J.X. and Gray, D.G. (1989). Chiroptical behavior of (acetyl)(ethyl) cellulose liquid crystalline solutions in chloroform. Macromolecules *22*, 2086–2090.

Haller, I. (1975). Thermotropic and static properties of liquid crystals. Prog. Solid State Chem. *10*, 103–118.

Hermans, J. (1963). Flow of gels of cellulose microcrystals. I. Random and liquid crystalline gels. J. Polym. Sci. C *2*, 129–144.

Herrick, F.W., Casebier, R.L., Hamilton, J.K. and Sandberg, K.R. (1983). Microfibrillated cellulose: morphology and accessibility. J. Appl. Polym. Sci. Polym. Symp. *37*, 797–813.

Heux, L., Chauve, G. and Bonini, C. (2000). Nonflocculating and chiral-nematic self-ordering of cellulose microcrystals suspensions in nonpolar solvents. Langmuir *16*, 8210–8212.

Hirai, A., Inui, O., Horii, F. and Tsuji, M. (2009). Phase separation behavior in aqueous suspensions of bacterial cellulose nanocrystals prepared by sulfulic acid treatment. Langmuir *25*, 497–502.

Iotti, M., Gregersen, Ø.W., Moe, S. and Lenes, M. (2011). Rheological studies of microfibrillar cellulose water dispersions. J. Polym. Environ. *19*, 137–145.

Isogai, A. and Kato, Y. (1998). Preparation of polyglucuronic acid from cellulose by TEMPO-mediated oxidation. Cellulose *5*, 153–164.

Kimura, F., Kimura, T., Tamura, M., Hirai, A., Ikuno, M. and Horii, F. (2005). Magnetic alignment of chiral nematic phase of cellulose microfibril suspension. Langmuir *21*, 2034–2037.

Langmuir, I. (1938). The role of attractive and repulsive forces in the formation of tactoids, thixotropic gels, protein crystals and coacervates. J. Chem. Phys. *6*, 873–896.

Larson, R.G. (1999). The structure and rheology of complex fluids (Oxford University Press, New York).

Lavoine, N., Desloges, I. Dufresne, A. and Bras, J. (2012). Microfibrillated cellulose – Its barrier properties and applications in cellulosic materials: A review. Carbohydr. Polym. *90*, 735–764.

Lee, S.D. (1987). A numerical investigation of nematic ordering based on a simple hard-rod model. J. Chem. Phys. *87*, 4972–4974.

Lowys, M.P., Desbrières, J. and Rinaudo, M. (2001). Rheological characterization of cellulosic micro®bril suspensions. Role of polymeric additives. Food hydrocolloid *15*, 25–32.

Marchessault, R.H., Morehead, F.F. and Walter, N.M. (1959). Liquid crystal systems from fibrillar polysaccharides. Nature *184*, 632–633.

Marchessault, R.H., Morehead, F.F. and Joan Koch, M. (1961). Hydrodynamics properties of neutral suspensions of cellulose crystallites as related to size and shape. J. Colloid Sci. *16*, 327–344.

Maret G. and Dransfeld, K. (1985). Biomolecules and polymers in high steady magnetic fields. In: Strong and ultrastrong magnetic fields and their applications – Topics in applied physics, Vol. 57, F. Herlach, ed. (Springer-Verlag, Berlin), pp. 143–204.

Miller, A.F. and Donald, A.M. (2003). Imaging of anisotropic cellulose suspensions using environmental scanning electron microscopy. Biomacromolecules *4*, 510–517.

Montesi, A., Peña, A. and Pasquali, M. (2004). Vorticity alignment and negative normal stresses in sheared attractive emulsions. Phys. Rev. Lett. *92*, 058303-1–058303-4.

Moscicki, J.K. and Williams, G. (1982). The effect of a Gaussian distribution of chain-lengths on the phase behaviour of a model system of rod-like macromolecules in solution. Polymer *23*, 558–568.

Nishio, Y., Kai, T., Kimura, N., Oshima, K. and Susuki, H. (1998). Controlling the selective light reflection of a cholesteric liquid crystal of (hydroxypropyl)cellulose by electrical stimulation. Macromolecules *31*, 2384–2386.

Nishiyama, Y., Kuga, S., Wada, M. and Okano, T. (1997). Cellulose microcrystal film of high uniaxial orientation. Macromolecules, *30*, 6395–6397.

Odijk, T. (1986). Theory of lyotropic liquid crystals. Macromolecules *19*, 2313–2329.

Onsager, L. (1949). The effects of shape on the interaction of colloidal particles. Ann. N. Y. Acad. Sci. *51*, 627–659.

Orts, W.J., Revol, J.F., Godbout, L. and Marchessault, R.H. (1995). SANS study of chirality and order in liquid crystalline cellulose suspensions. Mater. Res. Soc. Symp. Proc. *376*, 317–322.

Orts, W.J., Godbout, L., Marchessault, R.H. and Revol, J.F. (1998). Enhanced ordering of liquid crystalline suspensions of cellulose microfibrils: A small angle neutron scattering study. Macromolecules *31*, 5717–5725.

Pääkkö, M., Ankerfors, M., Kosonen, H., Nykänen, A., Ahola, S., Österberg, M., Ruokolainen, J., Laine, J., Larsson, P.T., Ikkala, O. and Lindström, T. (2007). Enzymatic hydrolysis combined with mechanical shearing and high-pressure homogenization for nanoscale cellulose fibrils and strong gels. Biomacromolecules *8*, 1934–1941.

Pan, J., Hamad, W. and Straus, S.K. (2010). Parameters affecting the chiral nematic phase of nanocrystalline cellulose films. Macromolecules *43*, 3851–3858.

Radel, S.R. and Navidi, M.H. (1994). Chemistry. 2nd ed. (West Publishing Company, New York). p. 563.

Revol, J.F., Bradford, H., Giasson, J., Marchessault, R.H. and Gray, D.G. (1992). Helicoidal self-ordering of cellulose microfibrils in aqueous suspension. Int. J. Biol. Macromol. *14*, 170–172.

Revol, J.F. and Marchessault, R.H. (1993). *In vitro* chiral nematic ordering of chitin crystallites. Int. J. Biol. Macromol. *15*, 329–335.

Revol, J.F., Godbout, L., Dong, X.M., Gray, D.G., Chanzy, H. and Maret, G. (1994). Chiral nematic suspensions of cellulose crystallites: Phase separation and magnetic field orientation. Liq. Cryst. *16*, 127–134.

Revol, J.F., Godbout, L. and Gray, D.G (1998). Solid films of cellulose with chiral nematic order and optically variable properties. J. Pulp Pap. Sci. *24*, 146–149.

Roman, M. and Winter, W.T. (2004). Effect of sulphate groups from sulphuric acid hydrolysis on the thermal degradation behaviour of bacterial cellulose. Biomacromolecules *5*, 1671–1677.

Roman, M. and Gray, D.G. (2005). Parabolic focal conics in self-assembled solid films of cellulose nanocrystals. Langmuir *21*, 5555–5561.

Saito, T., Kimura, S., Nishiyama, Y. and Isogai, A. (2007). Cellulose nanofibers prepared by TEMPO-mediated oxidation of native cellulose. Biomacromolecules *8*, 2485–2491.

Sato, T. and Teramoto, A. (1991). Pertubation theory of isotropic-liquid-crystal phase equilibria in polyelectrolyte solutions. Physica A *176*, 72–86.

Semenov, A.N. and Kokhlov, A.R. (1988). Statistical physics of liquid crystalline polymers. Sov. Phys. Usp. *31*, 988–1014.

Shogren, R.L., Peterson, S.C., Evans, K.O. and Kenar, J.A. (2011). Preparation and characterization of cellulose gels from corn cobs. Carbohydr. Polym. *86*, 1351–1357.

Siekmeyer, M. Zugenmaier, P. (1987). Investigations of molar mass dependence of the lyotropic liquid-crystal-line system: cellulose tricarbanilate/diethylene glycol monoethyl ether. Macromol. Chem. Rapid Comm. *8*, 511–517.

Straley, J.P. (1976). Theory of piezoelectricity in nematic liquid crystals, and of the cholesteric ordering. Phys. Rev. A *14*, 1835–1841.

Stroobants, A., Lekkerkerker, H.N.W. and Odijk, T. (1986). Effect of electrostatic interaction on the liquid crystal phase separation in solutions of rodlike polyelectrolytes. Macromolecules *19*, 2232–2238.

Strzelecka, T.E. and Rill, R.L. (1991). Phase transitions in concentrated DNA solutions: Ionic strength dependence. Macromolecules *24*, 5124–5133.

Sugiyama, J., Chanzy, H. and Maret, G. (1992). Orientation of cellulose microcrystals by strong magnetic fields. Macromolecule, *25*, 4232–4234.

Tseng, S.L., Valente, A. and Gray, D.G. (1981). Cholesteric liquid crystalline phases based on acetoxypropyl cellulose. Macromolecules *14*, 715–719.

Turbak, A.F., Snyder, F.W. and Sandberg, K.R. (1983). Microfibrillated cellulose: A new cellulose product: Properties, uses, and commercial potential. J. Appl. Polym. Sci. Polym. Symp. *37*, 815–827.

Ureña-benavides, E.E., Ao, G., Davis, V.A. and Kitchens, C.L. (2011). Rheology and phase behavior of lyotropic cellulose nanocrystal suspensions. Macromolecules *44*, 8990–8998.

van Bruggen, M.P.B., Dhont, J.K.G. and Lekkerkerker, H.N.W. (1999). Morphology and kinetics of the isotropic-nematic phase transition in dispersions of hard rods. Macromolecules *32*, 2256–2264.

van Bruggen, M.P.B. and Lekkerkerker, H.N.W. (2000). Tunable attractions directing nonequilibrium states in dispersions of hard rods. Macromolecules *33*, 5532–5535.

Verwey, E.J., and Overbeek, J.Th.G. (1947). Theory of the stability of lyophobic colloids (Elsevier, New York).

Vogt,U. and Zugenmaier, P. (1985). Structural models for some liquid crystalline cellulose derivatives. Ber. Bunsenges. Phys. Chem. *89*, 1217–1224.

Wang, N., Ding, E. and Cheng, R. (2008). Preparation and liquid crystalline properties of spherical cellulose nanocrystals. Langmuir *24*, 5–8.

Watson, J.H.L., Heller, W. and Wojtowicz, W. (1949). Comparative electron and light microscopic investigations of tactoid structures in V2O5-sols. Science *109*, 274–278.

Werbowyj, R.S. and Gray, D.G. (1976). Liquid crystalline structure in aqueous hydroxypropyl cellulose solutions. Mol. Cryst. Liq. Cryst. *34*, 97–103.

Werbowyj, R.S. and Gray, D.G. (1984). Optical properties of (hydroxypropyl)cellulose; cholesteric pitch and polymer concentration. Macromolecules *17*, 1512–1520.

Yu, L.J. and Saupe, A. (1980). Observation of a biaxial nematic phase in potassium laurate-1-decanol-water mixtures. Phys. Rev. Lett. *45*, 1000–1003.

Zhou, X. and Huang, Y. (2005). Cellulose derivative-based cholesteric networks. J. Appl. Polym. Sci. *96*, 1648–1653.

Zocher, H. (1925). Über freiwillige Strukturbildung in Solen. Z. Anorg. Allg. Chem. *147*, 91–110.

7 Processing of nanocellulose-based materials

This chapter describes the different processing ways leading to materials involving cellulose nanocrystals of microfibrillated cellulose. It mainly consists of polymer nanocomposites, but other materials have also been reported in the literature. Cellulose nanoparticles have a strong tendency for self-association because of the omnipresence of interacting surface hydroxyl groups. This property, which is the basis of the strength of paper sheets, is a desirable feature for the formation of load-bearing percolating architectures within the host polymer matrix. However, these inter-particle interactions can cause aggregation during the preparation of the nanocomposite, thus inducing the loss of the nanoscale and limit the potential of mechanical reinforcement. This aggregation phenomenon is magnified when the size of the particle decreases.

7.1 Polymer latexes

Latex may be natural or synthetic and is the stable dispersion (emulsion) of polymer microparticles in an aqueous medium. As found in nature, it is a complex milky fluid found in 10% of all flowering plants (angiosperms) (Agrawal and Konno, 2009) consisting of proteins, alkaloids, starches, sugars, oils, tannins, resins, and gums that coagulates on exposure to air. It is usually exuded after tissue injury and mainly serves as defense against herbivorous insects. The word is also used to refer to natural latex rubber, particularly non-vulcanized rubber. Latex can also be made synthetically by polymerizing a monomer that has been emulsified with surfactants

Processing of nanocomposite materials from cellulosic nanoparticles and polymer latex was historically the first one reported in the literature (Favier et al., 1995a). The nanocomposite system reported in this study consisted of tunicin nanocrystals and the polymeric matrix was obtained by the statistical copolymerization of styrene and butyl acrylate (poly(S-co-BuA)). The aqueous dispersion of polymer contained spherical particles 150 nm in average diameter and the solid content was 50 wt%. This mode of processing allows preserving the individualization state of the nanoparticles resulting from their colloidal dispersion in water. For this earlier work, the nature of the polymer was dictated by the fact that it was a fully amorphous polymer because of the random sequence of monomeric units, making it a model polymeric matrix. Moreover, the chemical composition of the copolymer consisting of 35 wt% and 65 wt% of styrene and butyl acrylate, respectively, resulted in a glass transition temperature around 0°C facilitating film processing by casting/evaporation at room temperature. This pioneering report was shortly followed by a second paper (Favier et al., 1995b).

From this kind of polymer, solid nanocomposite films were obtained by mixing and casting the two aqueous suspensions followed by water evaporation performed above the glass transition temperature of the polymer as shown in Figure 7.1. During water evaporation, the solid content in the medium continuously increases and the latex particles get closer and closer. The polymeric particles act as impenetrable domains to nanocrystals during evaporation due to their high viscosity. Therefore, the size of these particles influences the final dispersion of the rod-like nanoparticles (Dubief et al., 1999). When they come into contact with each other, these soft polymeric particles deform and adopt a polyhedral form, trapping the cellulosic nanoparticles in their dispersion state. The boundary between the former latex particles disappears by chain diffusion leading to a continuous polymer film containing the dispersed polysaccharide nanoparticles. The evaporation temperature needs to be above the glass transition temperature of the matrix to allow self-diffusion of the chains but not too high to avoid the formation of a dry skin on the surface of the film that could hinder water evaporation from the core.

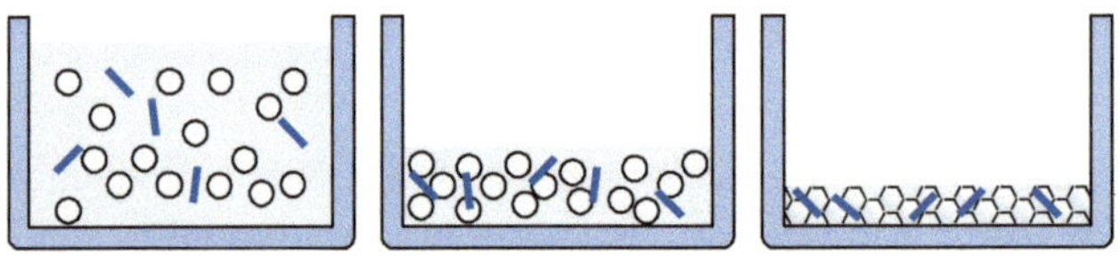

Fig. 7.1: Processing of polysaccharide nanocrystals reinforced polymer nanocomposite films using polymer latex.

The same copolymer was used in association with wheat straw (Helbert et al., 1996; Dufresne et al., 1997), tunicin (Hajji et al., 1996) or sugar beet (Azizi Samir et al., 2004a) cellulose nanocrystals, but also with MFC prepared from sugar beet pulp (Dalmas et al., 2006; Dalmas et al., 2007) or *Opuntia ficus-indica* parenchyma cell (Malainine et al., 2005).

Other latexes such as poly (β-hydroxyoctanoate) (PHO), polyvinylchloride (PVC), waterborne epoxy, waterborne polyurethane (WPU), natural rubber (NR), polyvinyl acetate (PVAc), and acrylic were also used as matrix. For PVC-based nanocomposites di-ethyl-hexyl phthalate (DOP) was used as plasticizer (30 per hundred ratio of PVC).

These different systems, the nature of the nanocellulosic reinforcing phase and the processing technique used are summarized in Table 7.1.

Except casting/evaporation, alternative methods consist in freeze-drying and hot-pressing or freeze-drying, extruding and hot-pressing the mixture. The dispersion of nanoparticles in the nanocomposite film strongly depends on the processing technique and conditions. Scanning electron microscopy (SEM) comparison between either cast and evaporated or freeze-dried and subsequently hot-pressed composites based on poly(S-co-BuA) reinforced with wheat straw nanocrystals, demonstrated that the former were less homogeneous and displayed a gradient of nanoparticle concentration between the upper and lower faces of the composite film (Helbert et al.,

Polymer	Nano-particle	Source of Cellulose	Processing Technique	Reference
Acrylic (UCAR™)	CNC	Acacia Pulp	Casting/Evaporation	(Pu et al., 2007)
Acrylic copolymer	CNC	Bacterial Cellulose	Casting/Evaporation/ Hot Pressing	(Trovatti et al., 2010)
NR	CNC/MFC	Date Palm Tree	Casting/Evaporation	(Bendahou et al., 2009; 2010; 2011)
	CNC	Cassava Bagasse	Casting/Evaporation	(Pasquini et al., 2010)
	CNC	Capim Dourado	Casting/Evaporation	(Siqueira et al., 2010)
	CNC	Sugarcane Bagasse	Casting/Evaporation	(Bras et al., 2010)
	CNC/MFC	Sisal	Casting/Evaporation	(Siqueira et al., 2011)
PVAc	CNC	Sisal	Casting/Evaporation	(Garcia de Rodriguez et al., 2006)
	MFC	Bleached Beech Pulp	Casting/Evaporation	(López-Suevos et al., 2010)
PHO	CNC	Tunicin	Casting/Evaporation	(Dubief et al., 1999; Dufresne et al., 1999; 2000)
Poly (S-co-BuA)	CNC	Tunicin	Casting/Evaporation	(Favier et al., 1995a; 1995b; Hajji et al. 1996)
	CNC	Wheat Straw	Casting/Evaporation	(Helbert et al., 1996; Dufresne et al., 1997)
	CNC/MFC	Sugar Beet Pulp	Casting/Evaporation	(Azizi Samir et al., 2004a)
	MFC	*Opuntia ficus indica* Cladodes	Casting/Evaporation	(Malainine et al., 2005)
	MFC	Sugar Beet Pulp	Casting/Evaporation and Freeze-Drying/Hot Pressing	(Dalmas et al., 2006; 2007)
	MFC	Eucalyptus	Casting/Evaporation	(Besbes et al., 2011a)
	MFC	Alfa, Eucalyptus, Pine	Casting/Evaporation	(Besbes et al., 2011b)
PVC/DOP	CNC	Tunicin	Freeze-Drying/Hot-Mixing/Hot-Pressing	(Chazeau et al., 1999a; 1999b; 1999c; 2000)
Waterborne Epoxy	CNC	Tunicin	Casting/Evaporation/ Curing	(Matos Ruiz et al., 2000; 2001)
WPU	CNC	Flax	Casting/Evaporation	(Cao et al., 2007)
	CNC	Cotton Linter	Casting/Evaporation	(Wang et al., 2010a)

Table 7.1: Polymer nanocomposites obtained from nanocellulose and polymeric matrix in the latex form.

1996; Dufresne et al., 1997). This sedimentation phenomenon was confirmed using wide angle X-ray scattering (WAXS) by comparing the diffracted X-ray beams by the two faces. It was suggested that this observation was most probably induced by the processing technique itself and that casting/evaporation technique results in the less homogenous films, where the nanocrystals have a tendency to orient randomly into horizontal plans.

The polymeric particle size seems to play a predominant role (Dubief et al., 1999). Larger latex particle size resulted in higher mechanical properties of the ensuing nanocomposite film. Indeed, the polymeric particles act as impenetrable domains to polysaccharide nanoparticles during the film formation due to their high viscosity. Increasing latex particle size leads to an increase of the excluded volume and these geometrical constraints seem to affect the nanocrystal network formation.

Stable aqueous nanocomposite dispersions containing cellulose nanocrystals and a poly(styrene-co-hexyl-acrylate) matrix were prepared via miniemulsion polymerization (Ben Elmabrouk et al., 2009). Addition of a reactive silane was used to stabilize the dispersion. Dispersions obtained by this method could be used in the form of a thin film after water evaporation and particle coalescence, or in agglomerated form to obtain granules that could be further transformed by other techniques such as extrusion or injection. This innovative approach was expected to enable the dispersion of cellulose nanoparticles with a relatively high solid content combined with a homogenous distribution in a polymeric matrix without the necessity to isolate them from water in which they form a stable suspension. It also allows for the generation of a premixed dry cellulose nanoparticles/polymer mixtures that can be fed to an extruder. This was expected to result in extruded nanocomposites with improved nanoparticle dispersion, and thus improved mechanical performance, over the use of two separate dry feeds.

7.2 Hydrosoluble or hydrodispersible polymers

Hydrosoluble or water-soluble polymers bear functional groups that can impart water solubility. They therefore form solutions in water. The degree of solubility is dependent on the number, position, and frequency of these moieties. Hydration relies on interactions at polar (ionic and hydrogen bonding) sites. The solubility of hydrosoluble polymers originates from the competition between polymer-water and water-water interactions, balanced by hydrophobic interactions induced by apolar components from the polymer.

After dissolution of the hydrosoluble or at least hydrodispersible polymer, the aqueous solution can be mixed with the aqueous suspension of cellulose nanoparticles. The ensuing mixture is generally evaporated to obtain a solid nanocomposite film. It can also be freeze-dried and hot-pressed. Filtration method can also be used, whose advantage lies in the possibility of preparing high cellulose nanoparticle content composites, which is usually challenging for other preparation routes. The dissolution process ensures the homogeneity of the composite at the molecular level provided that both the suspension and the solution have a sufficient dispersion level. The solubility of the polymer depends on temperature, polymer concentration and molecular weight. Stirring of the solution is generally performed to promote the dissolution of the polymer. However, aggregation and chain scission can be induced by mechanical stirring and special attention should be paid to the dissolution process. Chain scission induced by high-speed stirring is likely to occur near the center of the macromolecule, leading to a decrease of the polydispersity index. For instance, it was shown for polyethylene oxide (PEO) that the ability of polymer scission induced by the dispersion procedure increases with increasing its molecular weight (Bossard et al., 2010). In the concentrated regime, it was shown that aggregation of PEO chains was favored by increasing the concentration. Hydrodynamic forces combined to additional stresses due to entanglement enhanced the breakup of polymeric chains and aggregates.

The preparation of cellulosic nanoparticles reinforced starch, silk fibroin, polyoxyethylene oxide (PEO), polyvinyl alcohol (PVA), hydroxypropyl cellulose (HPC), carboxymethyl cellulose (CMC), soy protein isolate (SPI), phenol-formaldehyde resin, hemicelluloses or hydroxyethylcellulose (HEC) has been reported in the literature. Chitosan was also used as matrix, but the liquid medium contained a low acetic acid content to allow dissolution of the polymer (Nordqvist et al., 2007). Polyelectrolyte-macroion complexes (PMC) between chitosan and cellulose nanocrystals were prepared for potential applications in drug delivery (Wang and Roman, 2011). The concentration of cellulose nanocrystals was found to have a strong effect on PMC formation, with higher cellulose nanoparticle concentrations resulting in larger or more aggregate PMC particles. Hydrogels based on carboxymethylated MFC were prepared by UV polymerization of N-vinyl-2-pyrrolidone with Tween 20 trimethacrylate as a cross-linking agent (Eyholzer et al., 2011).

These different systems, the nature of the nanocellulosic reinforcing phase and the processing technique used are summarized in Table 7.2.

Polymer	Nano-particle	Source of Cellulose	Processing Technique	Reference
Alginate-Acerola Puree/ Corn Syrup	CNC	Cotton	Casting/Evaporation	(Azeredo et al., 2012)
Carrageenan	CNC	α-cellulose microfiber	Casting/Evaporation	(Sanchez-Garcia et al., 2010)
Chitosan	MFC	Bleached Sulfite Softwood Pulp	Casting/Evaporation	(Nordqvist et al., 2007; Fernandes et al., 2010)

Polymer	Nano-particle	Source of Cellulose	Processing Technique	Reference
CMC/Glycerin	CNC	Cotton	Casting/Evaporation	(Choi and Simonsen, 2006)
Glucomannan/ Glycerol	CNC	MCC	Casting/Evaporation	(Mikkonen et al., 2010)
	MFC	Sulfite Softwood Pulp	Casting/Evaporation	(Mikkonen et al., 2011)
Glucuronoxylan		Bacterial Cellulose	Casting/Evaporation	(Dammström et al., 2005)
HEC	MFC	Sulfite Softwood Pulp	Vacuum Filtration/ Drying	(Sehaqui et al., 2011a)
HPC	MFC	Sulfite Pulp	Casting/Evaporation	(Zimmermann et al., 2004; 2005)
	MFC	Wheat Straw, Beech Wood Pulp	Casting/Evaporation	(Zimmermann et al., 2010)
PEO	CNC	Tunicin	Casting/Evaporation	(Azizi Samir et al., 2004b; 2004c; 2004d; 2005; 2006)
Phenol-Formaldehyde	MFC	Sugar Beet Pulp	Casting/Evaporation	(Leitner et al., 2007)
	CNC	MCC	Casting/Curing	(Liu and Laborie, 2011)
PMVEMA/PEG	CNC	MCC	Casting/Evaporation	(Goetz et al., 2009; 2010)
PVA	MFC	Sulfite Pulp	Casting/Evaporation	(Zimmermann et al., 2004; 2005)
	MFC	Sugar Beet Pulp	Casting/Evaporation	(Leitner et al., 2007)
	MFC	Soybean	Casting/Evaporation	(Wang and Sain, 2007a; 2007b)
	CNC	Cotton	Casting/Evaporation	(Roohani et al., 2008)
	CNC	Cotton	Casting/Evaporation	(Paralikar et al., 2008)
	MFC	Daicel	Casting/Evaporation	(Lu et al., 2008)
	CNC	MCC	Casting/Evaporation	(Lee et al., 2009)
	MFC	Regenerated, Pure Cellulose, MCC	Casting/Evaporation	(Cheng et al., 2009)
	MFC	MCC	Casting/Evaporation	(Frone et al., 2011)
Silk Fibroin	CNC	Tunicin	Casting/Evaporation	(Noishiki et al., 2002)
SPI	CNC	Cotton Linter	Casting/Evaporation	(Wang et al., 2006)

Polymer	Nano-particle	Source of Cellulose	Processing Technique	Reference
Starch/ Glycerol	MFC	Sugar Beet Pulp	Casting/Evaporation	(Dufresne and Vignon, 1998)
	MFC	Potato Pulp	Casting/Evaporation	(Dufresne et al., 2000)
	CNC	Tunicin	Casting/Evaporation	(Anglès and Dufresne, 2000; 2001)
	CNC	Cottonseed Linter	Casting/Evaporation	(Lu et al., 2005)
	CNC	Ramie	Casting/Evaporation	(Lu et al., 2006)
	MFC	Bleached Sulfite Softwood Pulp	Casting/Evaporation	(López-Rubio et al., 2007)
	CNC	Hemp	Casting/Evaporation	(Cao et al., 2008a)
	CNC	Flax	Casting/Evaporation	(Cao et al., 2008b)
	CNC	Pea Hull Fiber	Casting/Evaporation	(Chen et al., 2009)
	MFC	Bleached Sulfite Softwood Pulp	Casting/Evaporation	(Svagan et al., 2007; Svagan et al., 2009)
	CNC	Bamboo	Casting/Evaporation	(Liu et al., 2010)
	CNC	Bacterial Cellulose	Casting/Evaporation	(Woehl et al., 2010)
	MFC	Wheat Straw	Casting/Evaporation	(Kaushik et al., 2010)
	MFC	Bleached Sulfite Softwood Pulp	Casting/Evaporation	(Plackett et al., 2010)
Starch/Glycerol/ Sorbitol	CNC	Cassava Bagasse	Casting/Evaporation	(Teixeira et al., 2009)
Starch/Sorbitol	CNC	Tunicin	Casting/Evaporation	(Mathew and Dufresne, 2002; Mathew et al., 2008)
Oxidized Starch	MFC	Bleached Softwood Kraft Pulp	Molding	(Yano and Nakahara, 2004)
Hydroxypropylated and Oxidized Starch/ Sorbitol	CNC	MCC	Casting/Evaporation	(Kvien et al., 2007)
Xylan/Sorbitol	CNC	Bleached Softwood Kraft Pulp	Casting/Evaporation	(Saxena et al., 2009)

Table 7.2: Polymer nanocomposites obtained from nanocellulose and hydrosoluble polymeric matrix.

Starch has been extensively used as hydrosoluble polymer matrix because of its renewable nature and abundance of the material. Moreover, favorable filler/matrix interactions are expected. For these materials, the use of a plasticizer, generally a polyol (glycerol or sorbitol), is essential to reduce the brittleness of ensuing nanocomposites. Heat treatment-induced cross-linking of PVA-based nanocomposite membranes has been reported by adding poly(acrylic acid) (PAA) as a cross-linking agent (Paralikar et al., 2008). The carboxyl group in PAA and the hydroxyl groups of PVA and cellulose nanocrystals formed ester linkages along with presumed hydrogen bonding between PVA and nanoparticles. This allows the overall nanocomposite to have effective cross-linking and barrier resistance. The goal was for these barrier membranes to have the ability to prevent permeation of harmful chemicals and to be chemically inert, mechanically strong, tough, flexible and water resistant. Cellulose nanocomposites in which the cellulose nanocrystals are cross-linked with poly(methyl vinyl ether-co-maleic acid) and poly(ethylene glycol) (PMVEMA/PEG) have been prepared (Goetz et al., 2009; Goetz et al., 2010).

Water-redispersible MFC in powder form was recently prepared from refined bleached beech pulp by carboxymethylation and mechanical disintegration (Eyholzer et al., 2010). However, the carboxymethylated sample displayed a loss of crystallinity and a strong decrease in thermal stability, limiting its use for nanocomposite processing. Bacterial cellulose/PEO nanocomposites of controlled composition have been produced by modifying the culture medium with various concentrations of PEO (Brown and Laborie, 2007). It was shown that as the PEO content increased, cellulose crystallized into smaller nanofibers. Epoxy resin systems consisting of glycerol polyglycidyl ether and sorbitol polyglycidyl ether cured with tannic acid were reinforced with MFC (Shibata and Nakai, 2010). The prepolymer and curing agent were solubilized in water and this solution was mixed with the MFC suspension. This mixture was freeze-dried, pre-cured at 50°C and compression-molded at 160°C.

7.3 Non-aqueous systems

Except the use of an aqueous polymer dispersion, or latex, an alternative way to process non-polar polymer nanocomposites reinforced with polysaccharide nanoparticles consists in their dispersion in an adequate (with regard to matrix) organic medium. This processing technique is therefore similar to the one consisting of using a hydrosoluble polymer except that the processing liquid medium is non-aqueous. The preparation of stable dispersions of cellulose nanoparticles in any non-aqueous media can be achieved either by coating their surface with a surfactant or by chemically modifying it. The global objective is to reduce their surface energy in order to improve their dispersibility/compatibility with non-polar media. For some specific liquid media, the dispersion of cellulose nanoparticles can be obtained without any surfactant or chemical modification step.

7.3.1 Non-aqueous polar medium

Stable suspensions of cellulose nanocrystals with negatively charged sulfate groups, commonly produced by hydrolysis of the native cellulose with sulfuric acid, can be obtained in various polar liquid media. Both the high value of the dielectric constant of the liquid and the medium wettability of cellulose nanocrystals were supposed to control the stability of the suspension (Azizi Samir et al., 2004e). This approach generally involves freeze-drying of the initial aqueous suspension followed by the further redispersion in the non-aqueous polar liquid. For instance, stable cellulose nanocrystal suspensions have been prepared in N,N-dimethyl formamide (DMF) (Azizi Samir et al., 2004e; Marcovich et al., 2006), dimethyl sulfoxide (DMSO), N-methyl pyrrolidine (NMP), formic acid and m-cresol (van den Berg et al., 2007a). Figure 7.2 shows optical photographs of suspensions of tunicin nanocrystals in these different liquid media observed in polarized light between cross polarizers. In each case, a strong birefringence was observed, indicating a good dispersion level of the nanoparticles in the liquid medium. Similar dispersions in polar aprotic organic solvents (DMF and DMSO) using nanocrystals obtained from cotton were reported (Viet et al., 2007). The dispersions showed flow birefringence. However, the redispersion was found to be incomplete, and there was evidence for aggregation in the suspensions. A small amount of water appeared to be critical to suspension stability. Cellulose nanocrystals without surface charge prepared by hydrolysis with hydrochloric acid do not disperse as well in aprotic solvents (DMSO, DMF, NMP) (van den Berg et al., 2007a). However, formic acid and m-cresol have been shown to also disperse non-charged nanocrystals properly.

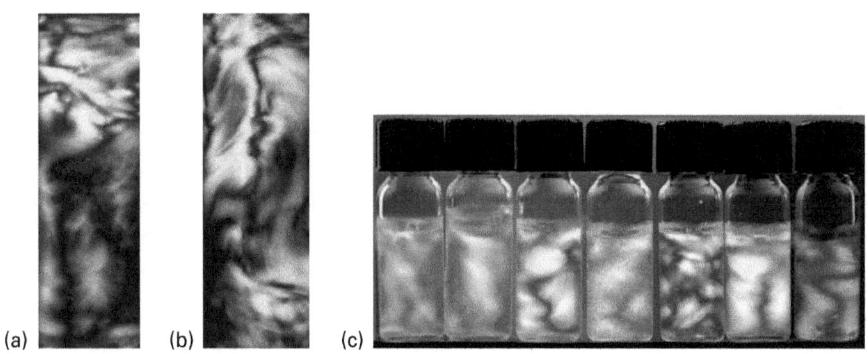

Fig. 7.2: Photographs of suspensions of sulfuric acid-prepared tunicin nanocrystals viewed through cross polarizers: in (a) water and (b) DMF (0.5 wt%) (Azizi Samir et al., 2004e), and (c) from left to right: as prepared in water, freeze-dried and redispersed in water, DMF, DMSO, NMP, formic acid and m-cresol (5.0 mg · mL^{-1}) (van den Berg et al., 2007a).

From DMF suspensions, tunicin nanocrystal reinforced POE plasticized with tetraethylene glycol dimethyl ether (TEGDME) was prepared by casting and evaporation of DMF (Azizi Samir et al., 2004d). Cross-linked nanocomposites were also prepared by dispersing cellulose nanocrystals in a solution of an unsaturated linear polycondensate, addition of a photo-initiator, casting, evaporating the solvent and UV-curing (Azizi Samir et al., 2004f). Stable cellulose nanocrystal suspensions in DMF were incorporated in different polyurethane (PU) matrices (Marcovich et al., 2006; Auad et al., 2008), ethylene oxide-epichlorohydrin (EO-EPI) copolymer (Capadona et al., 2007; Capadona et al., 2008), poly(3-hydroxybutyrate-co-3-hydroxyvalerate) (PHBV) (Jiang et al., 2008), and polymethylmethacrylate (PMMA) solution (Liu et al., 2010). The effect of reinforcing shape memory PUs with cellulose nanocrystals was also investigated (Auad et al., 2008; Auad et al., 2011; Auad et al., 2012). Polyaniline (PANI) films reinforced with tunicin nanocrystals were prepared by mixing a solution of PANI with a dispersion of freeze-dried cellulose nanocrystals in formic acid (van den Berg et al., 2007b).

7.3.2 Solvent mixture and solvent exchange

Casting from a mixture of solvents can also be used to prepare cellulose nanoparticle-reinforced nanocomposites. In this method, the aqueous suspension of nanoparticles is mixed with a polymer solution involving a solvent miscible with water. Homogeneous nanocomposite films based on $LiClO_4$-doped ethylene oxide-epichlorohydrin (EO-EPI) copolymers and tunicin nanocrystals were prepared by solution-casting of tetrahydrofuran (THF)/water mixtures and subsequent compression-molding (Schroers et al., 2004). All-cellulose films have also been prepared from cellulose nanocrystals and a cellulosic matrix regenerated from aqueous NaOH-urea solvent system (cellulose II) on the basis of their temperature-dependent solubility (Qi et al., 2009).

Solvent exchange procedure can be used to suspend cellulosic nanoparticles in the proper liquid medium for further surface chemical modification. It can also be used for mixing the suspension with a polymer solution. Cotton cellulose nanocrystals were solvent exchanged from water to NMP and polysulfone (PSf) was dissolved in this dispersion, which was then used to make films by the phase inversion process (Noorani et al., 2007). A slow solvent exchange procedure from water to acetone and then to DMF was also used to prepare cellulose nanocrystal reinforced epoxy nanocomposites (Tang and Weder, 2010). The ensuing materials were compared with those prepared by directly redispersing freeze-dried cellulose nanocrystals in DMF.

MFC-reinforced polylactic acid (PLA) nanocomposites were prepared by stirring MFC in a mixture of water and acetone (Suryanegara et al., 2009; Suryanegara et al., 2011). Successive centrifugations were performed to replace water by acetone, and then acetone was replaced by dichloromethane using the same procedure. The precipitate of MFC was suspended in dichloromethane and treated by a homogenizer to obtain a well-dispersed suspension, in which PLA was added and dissolved upon stirring. Cellulose nanocrystal suspensions were solvent exchanged from water to chloroform to prepare nanocomposites with a PLA matrix in presence of a compatibilizer (Pandey et al., 2009). A dispersion of cellulose nanocrystals was also solvent exchanged from water to chloroform and PLA was added to this dispersion (Sanchez-Garcia and Laragon, 2010). However, microscopic observations showed that the dispersion of the nanoparticles in PLA was poorer than when using freeze-dried nanocrystals redispersed in chloroform.

The preparation of all-cellulose nanocomposite films consisting of cellulose nanocrystals and regenerated cellulose (cellulose II) using ionic liquid was also reported (Ma et al., 2011). Cellulose nanocrystal suspensions in 1-(2-hydroxylethyl)-3-methyl imidazolium chloride ([HeMIM]Cl) were prepared by solvent exchange from water. It was shown that the native crystalline structure and needle-like morphology of the nanoparticles was retained at room temperature. This suspension was added to a viscous cellulose ionic liquid solution at 50°C. After stirring, the nanocomposite films were obtained by casting/evaporation and the ionic liquid was removed by washing with running deionized water. All-cellulose composite films were also obtained by carrying out a partial dissolution of MFC films (Duchemin et al., 2009). In this study, the ionic liquid used was 1-butyl-3-methylimidazolium chloride ([C4MIM]Cl). Effective dissolution was obtained at 80°C for different times. Lower temperatures did not promote any dissolution, whereas higher temperatures led to fast disintegration of the cellulosic substrate.

Hand lay-up of epoxy/MFC mixture with plain woven carbon fiber was reported (Gabr et al., 2010). The epoxy/MFC mixture was prepared by adding epoxy into MFC ethanol suspension obtained by a solvent exchange procedure. Aqueous suspension of MCC cellulose nanocrystals was homogeneously mixed with a phenolic resin and solvent exchanged to DMF to avoid bubble formation during cure and obtain a more transparent and stable dispersion (Liu and Laborie, 2011). After casting the mixture, pre-curing was performed at 80°C and post-curing was carried out under vacuum stepwise at 140°C and then 185°C.

Another approach involving a solvent exchange process consists in forming a network of nanofibers that will serve as a template that can be filled with a polymer (Capadona et al., 2007; Capadona et al., 2008). This method is schematically represented in Figure 7.3. The first step is the formation of the nanocrystal template through a sol-gel process involving the formation of a homogeneous aqueous dispersion. It is followed by gelation through solvent exchange with a water miscible solvent. Acetone is routinely used, but a range of solvents such as methanol, THF, ethanol, acetonitrile or isopropanol is suitable. The solvent is gently added by letting it run down the walls of the beaker so as to avoid mixing and form an organic layer on top of the aqueous nanocrystal dispersion. The solvent gradually replaces water and the top organic

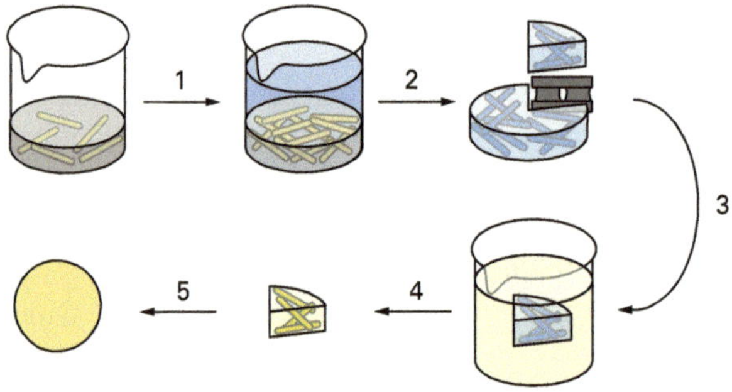

Fig. 7.3: Schematic of the nanocomposite preparation by a template approach: (1) A non-solvent is added to a nanocrystal dispersion in the absence of any polymer. (2) Solvent exchange promotes the self-assembly of a nanocrystal gel. (3) The gelled nanocrystal scaffold is imbibed with a polymer by immersion in a polymer solution, before the nanocomposite is dried (4) and compacted (5) (Capadona et al., 2007).

layer is exchanged with fresh solvent daily until the bottom portion has assembled into a mechanically coherent nanocrystal-solvent gel (typically several days).

In the second step, the nanocrystal template is filled with a polymer matrix by immersing the solvent gel into a polymer solution (typically several hours). The polymer solvent must be miscible with the gel solvent and must not redisperse the cellulose nanocrystals. The solvent is then evaporated. A reference material was prepared by immersion of an acetone gel in neat toluene (without polymer) as a representative polymer solvent. Supercritical extraction with CO_2 and microscopic inspection of this material show that the network structure remained intact throughout this step (Capadona et al., 2007). A broad range of host polymers was used, namely polystyrene (PS), polyvinyl acetate (PVAc), poly(ethylene oxide-co-epichlorohydrin) (poly(EO-co-EPI), polybutadiene (PBU), styrene-butadiene-styrene (SBS) copolymers and polyurethane (PU) (Capadona et al., 2007; Capadona et al., 2008; Capadona et al., 2009; Shanmuganathan et al., 2010a; Shanmuganathan et al., 2010b; Shanmuganathan et al., 2010c; Mendez et al., 2011).

7.3.3 In situ polymerization

In situ polymerization is an effective way to process a nanocomposite material. In this method the nanoparticles are impregnated with a monomer and the monomer subsequently polymerized. Usually the monomer is very fluid, allowing impregnation without unduly disturbing the nanoparticle arrangement. In this approach, the monomer has a dual function, serving as both an effective dispersant for the nanoparticles and the matrix precursor for the in situ polymerization. After polymerization, the material is processed with little necessary finishing.

The in situ polymerization approach has been used to prepare polyfurfuryl alcohol nanocomposites without the use of solvents or surfactants (Pranger and Tannenbaum, 2008). The polymerization of furfuryl alcohol was catalyzed by sulfate groups from the surface of nanocrystals prepared from MCC. Polystyrene-based nanocomposites were obtained by immersing for 3 h freeze-dried bacterial cellulose membranes in a styrene solution containing 1 wt% azodiisobutyronitrile (AIBN) as initiator (Peng et al., 2011). The immersed samples were subsequently pressed between two glass plates at 85°C for 5 h to allow polymerization of styrene. A similar strategy was use to prepare conducting porous composite membranes of bacterial cellulose and polypyrrole through in situ oxidative chemical polymerization of pyrrole (Müller et al., 2011).

MFC-coated polypyrrole was also prepared using in situ polymerization to obtain electrically conducting continuous high-surface-area materials (Nyström et al., 2010). The reaction between an oxidant solution of pyrrole and the MFC dispersion led to a layer of polypyrrole on the cellulose surface. It was found that the high surface area of MFC was maintained upon normal drying without the use of any solvent exchange. The dry material had a surface area and conductivity around 90 $m^2 \cdot g^{-1}$ and 1.5 $S \cdot cm^{-1}$, respectively.

7.3.4 Surfactant

Surfactants or surface active agents are compounds that lower the surface tension of a liquid, the interfacial tension between two liquids, or that between a liquid and a solid. Structurally, they are organic amphiphilic compounds, meaning that they contain hydrophobic groups (tail) and hydrophilic groups (head). Therefore, a surfactant molecule contains both a water insoluble (or oil soluble) component and a water soluble component. They can be classified according to the composition of their tail, head or composition of their counter-ion in the case of ionic surfactants.

Coating of cotton and tunicin nanocrystals by a surfactant such as a phosphoric ester of polyoxyethylene (9) nonyl phenyl ether was found to lead to stable suspensions in toluene and cyclohexane (Heux et al., 2000) or chloroform (Kvien et al., 2005). The excess surfactant can be removed by centrifugation and the freeze-dried material can be redispersed in apolar solvent. Coated tunicin nanocrystals reinforced atactic polypropylene (aPP) (Ljungberg et al., 2005), isotactic polypropylene (iPP) (Ljungberg et al., 2006), or poly(ethylene-co-vinyl acetate) (EVA) (Chauve et al., 2005) were obtained by solvent casting using toluene. For aPP-based nanocomposites, the amount of surfactant adsorbed on the cellulosic nanoparticles was found to be 1.6 times the cellulose weight. The same procedure was used to disperse cellulosic nanoparticles in chloroform and process composites with poly lactic acid (PLA) (Kvien et al., 2005; Petersson and Oksman, 2006).

7.3.5 Surface chemical modification

Surface chemical modification of cellulose nanoparticles is another way to decrease their surface energy and disperse them in organic liquids of low polarity. It generally involves reactive hydroxyl groups from the surface. Experimental conditions should avoid swelling media and peeling effect of surface-grafted chains inducing their dissolution in the reaction medium. The chemical grafting process has to be mild in order to preserve the integrity of the nanoparticle. The different surface chemical modification strategies described in Chapter 5 can be used depending on the nature of the processing liquid medium.

7.4 Foams and aerogels

A foam is a substance that is formed by trapping many gaseous bubbles in a liquid or solid. It is normally an extremely complex system consisting of polydisperse gas bubbles separated by draining films. A polymeric foam is a solid cellular material formed by a solid polymer and a gas. The gas occupies the major volume of the foam making it much lighter than the solid polymer of which it is made. Aerogels are manufactured materials derived from a gel in which the liquid component has been replaced with a gas. The resulting material has a very low density with several remarkable properties including the ability to thermal insulation. Aerogels are highly porous solids made up to 99.8% air with a density as low as 4 $mg \cdot cm^{-3}$, making it the lightest known solid. Production of aerogels is done by the sol-gel process. First a gel is created in solution and then the liquid is carefully removed to leave the aerogel intact.

Nanocellulose can be used to make aerogels/foams, either by itself or in composite formulations. Aerogels and foams can be used as porous templates, potentially useful in various nanoapplications.

MFC-based starch foams have been studied for packaging applications in order to replace polystyrene-based foams (Svagan et al., 2008; Svagan et al., 2010; Svagan et al., 2011). The advantage of using MFC instead of micrometric fibers is that the nanofibrils can reinforce the thin cells in the foam. Starch-based foams can be prepared using a number of different techniques, including backing in hot mold, extrusion, compression/explosion processing, freeze-drying and microwave heating. However, MFC are delivered as diluted aqueous suspensions and high water content presents a major challenge when foaming. Preliminary results using foaming by microwave heating show that high water contents produce foams with coarse cells (Svagan et al., 2008). Reduction of the water content of the MFC suspension results in nanofibers aggregation and difficulties in achieving a homogeneous dispersion. Therefore, only foams with low filler content (≤ 10 wt%) can be successfully prepared by the microwave heating technique. Nevertheless, even at these low contents, significant reinforcing effects can be observed. In the freeze-drying technique, the high water content does not present a problem, although the energy involved in the process to eliminate water increases as its content increases. The water is frozen into ice crystal and upon sublimation, a porous structured material can be obtained. Heating and mixing of starch granules and MFC suspension were performed using a microwave oven and kitchen aid (Svagan et al., 2008) or a water bath at 95°C whilst stirring with an overhead mixer (Svagan et al., 2010; Svagan et al., 2011). The porosity of the foam can be controlled through the water content. Foams with high MFC contents (up to 70 wt%) were obtained using this technique. Figure 7.4 presents the cell structure of starch-based foam with increasing MFC content. Up to 40 wt% MFC content anisotropic irregular shaped cells were obtained, but at 70 wt% no well-defined cells were observed in the major part of the foam. Moreover, at 0 and 10 wt% MFC content the cells were closed, whereas at 40 wt% MFC both closed and open cells were attained. Even if the freeze-drying technique is time-consuming and not ideal from an industrial manufacturer perspective, the potential of MFC-reinforced microcellular foams was demonstrated (Svagan et al., 2008). Improved dimensional stability, compression strength and modulus were reported by adding cellulose nanocrystals to starch/PVA foams (Wang et al., 2010b). The aqueous mixture was frozen at –20°C, and subsequently thawed to 25°C. This freezing/thawing cycle was repeated up to 7 times to obtain white sponges and investigate the effect of repeated cycles on their structure and properties.

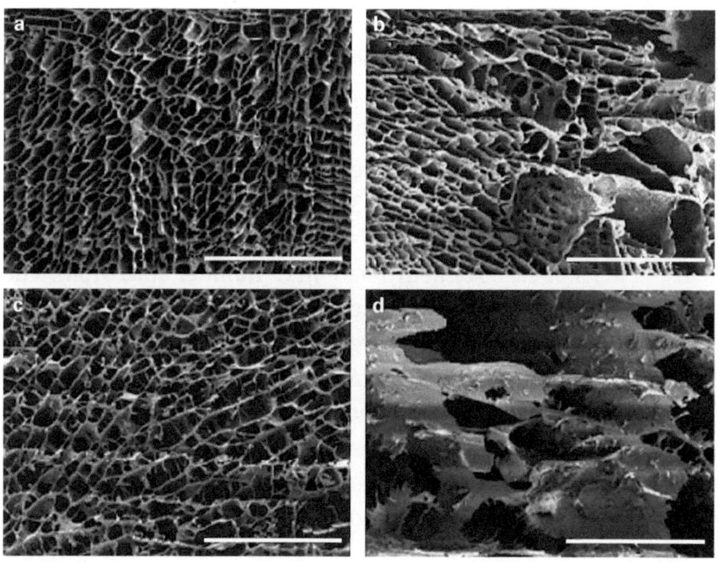

Fig. 7.4: The cell structure of starch-based foams with 0 (a), 10 (b), 40 (c) and 70 wt% MFC (d). The sections are normal to the cylinder axis and at ca. half the height of the original foam. The scale bars are 300 μm (Svagan et al., 2008).

Bacterial cellulose/chitosan composite scaffolds of open pore microstructure with interconnecting pores were also fabricated through freezing and lyophilization method (Nge et al., 2010). Pore size ranged from 120 to 280 μm. Improved mechanical properties were ascribed to closer packing as well as integration of nanofibrils within chitosan matrix forming the pore walls.

Moreover, it is possible to prepare pure nanocellulose aerogels applying various freeze-drying and super critical CO_2 drying techniques. Indeed, during water evaporation from a nanocellulose suspension, capillary forces are exerted on the nanoparticles leading to consolidation into a high density, low specific surface area film. The challenge consists therefore in preserving the surface area of the nanoparticles by drying the hydrogel by means other than water evaporation so that nanoparticle aggregation is avoided.

Ductile low-density aerogels were prepared from MFC hydrogels by two different freeze-drying procedures (Pääkkö et al., 2008). The 3D open nano-scale network structure of the aqueous gel was preserved by using cryogenic freeze-drying with liquid propane, while the nanofibers aggregate to form 2D extended sheet-like structures by using vacuum freeze-drying. Micro-scale pores were found in the aerogels prepared by both procedures and their porosity was 95–98%. Higher porosity (up to 99.5%) cellular structure was obtained by using cryogenic freeze-drying with liquid nitrogen (Sehaqui et al., 2010). A centrifugation method was developed to concentrate the aqueous MFC dispersion, and freeze-drying of dispersions of different concentrations led to MFC foams with densities ranging between 7 $kg \cdot m^{-3}$ and 103 $kg \cdot m^{-3}$ (Figure 7.5). A wide range of mechanical properties including compression was obtained by controlling density and nanofibrill interaction in the foams. MFC aerogels of ultra-high porosity (93%–99%) were also prepared from MFC hydrocolloidal dispersions using solvent exchange from water to tert-butanol followed by tert-butanol freeze-drying (Sehaqui et al., 2011b). The "effective" diameter of the nanofibrils was around 10 nm and specific surface areas were as high as 153–284 $m^2 \cdot g^{-1}$.

In another study, nanofibrillar cellulose aerogels were prepared by vacuum freeze-drying of aqueous dispersions of carboxymethylated MFC (Aulin et al., 2010a). The aerogel network was modified by chemical vapor deposition of a fluorinated silane to achieve a conformal coating of low surface energy and tune its wetting properties towards non-polar liquids/oils. The morphology and porosity of the aerogels were tuned by selecting the concentration of MFC dispersions before freeze-drying. The wettability behavior of the cellulose surfaces can be switched between super-wetting and super-repellent, using different scales of roughness and porosity created by the freeze-drying technique and change of concentration of the nanoparticle dispersion. Aerogels were also prepared from bacterial cellulose and impregnated with metalhydroxide/oxide precursors, which can readily be transformed into grafted magnetic nanoparticles along the cellulose nanofibers (Olsson et al., 2010). Figure 7.6 shows the magnetic nanoparticles precipitated on the cellulose hydrogel. The porosity of

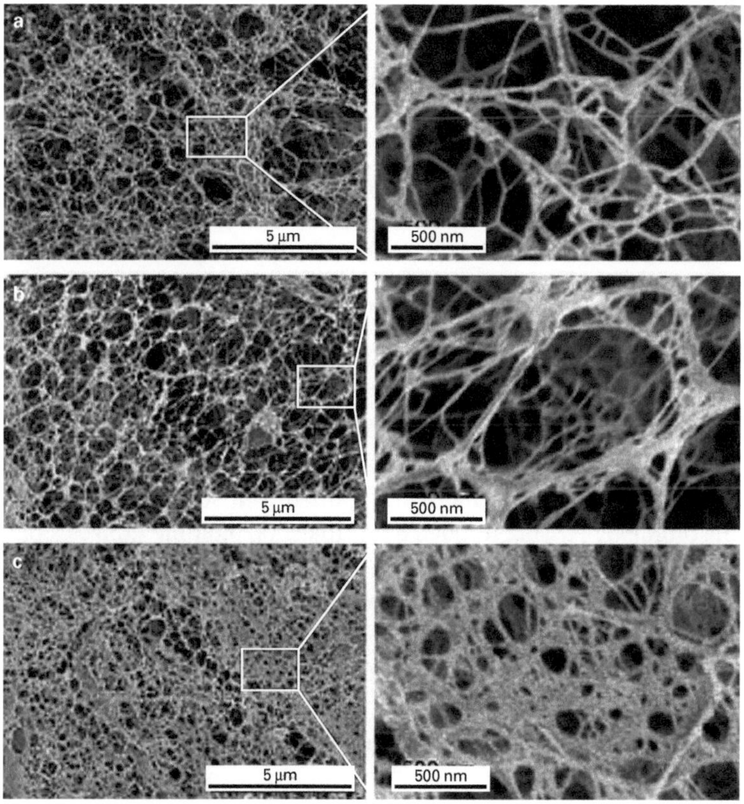

Fig. 7.5: FE-SEM micrographs of the bottom surface of MFC foams with a density of 7 (a), 32 (b), and 79 $kg \cdot m^{-3}$ (c) (Sehaqui et al., 2010).

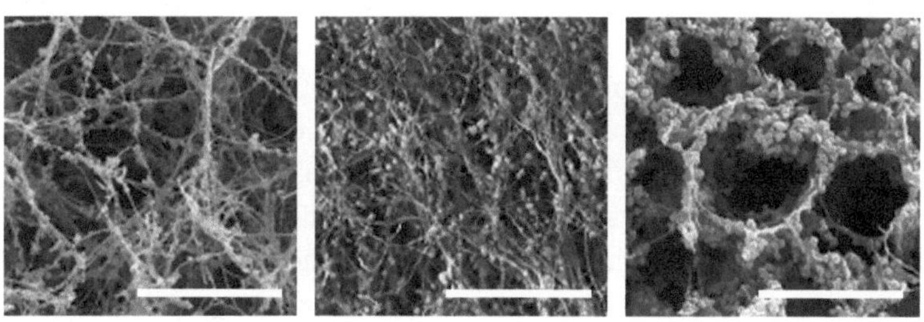

Fig. 7.6: SEM images of magnetic aerogels at different loadings of cobalt ferrite nanoparticles (from left to right): sample C1 (70 wt% of particles), sample C2 (80 wt% of particles) and sample C3 (95 wt% of particles). Scale bars, 4 µm (Olsson et al., 2010).

the resulting dried nanocomposite was controlled from 98% (upon removal of water during freeze-drying) in the aerogel to 10% (upon compaction) in the nanopaper.

Cellulose nanocrystals can also be made to gel in water under low power sonication giving rise to aerogels. Solvent exchange from water to ethanol and supercritical CO_2 drying were used to prepare aerogels with low density (down to 78 $mg \cdot cm^{-3}$) from cotton nanocrystals (Heath and Thielemans, 2010). They present the highest reported surface area (> $600m^2 \cdot g^{-1}$) and lowest shrinkage during drying (6.5%) for cellulose aerogels

7.5 Melt compounding

Melt-compounding techniques, such as extrusion or injection molding, are commonly used to process thermoplastic polymers. Extrusion is a high volume manufacturing process in which solid polymeric material is transported by a screw and melted to form a continuous profile by passing through a die. Injection molding utilizes a ram or screw-type plunger to force molten plastic material into a mold cavity.

These conventional processing techniques are infrequently employed for the preparation of cellulose nanoparticle reinforced polymer nanocomposites. This is ascribed to inherent incompatibility and thermal stability issues. The hydrophilic nature of polysaccharides causes irreversible agglomeration during drying and aggregation in non-polar matrices because of the formation of additional hydrogen bonds between nanoparticles.

7.5.1 Drying of the nanoparticles

Four methods have been examined to dry cellulose nanocrystal and MFC suspensions, viz. oven drying, freeze-drying, supercritical drying and spray-drying (Peng et al., 2012). The particle size and morphology of the nanoparticles were determined by dynamic light scattering, transmission and scanning electron microscopies (TEM, SEM), and morphological analysis. Spray drying was proposed as the technically suitable process to dry nanocellulose suspensions and composite processing because particle size ranging from nano to micron were obtained. The effects of freeze-drying and air-drying were also studied for cellulose nanocrystals prepared from MCC based on cotton linters or wood pulp (Rämänen et al., 2012). It was observed that the crystal structure and crystallinity of nanoparticles remained during the treatments, whereas their nanoscale structure was significantly influenced by drying method, neutralization and source of cellulose. Drying method was found to slightly influence the thermal stability of cellulose nanocrystals, whereas the char residue varied significantly depending on the drying process. It was reported that the weight residue was higher after air drying than freeze-drying and it was explained by more regular packing of air-dried nanocrystals.

Moreover, cellulose nanocrystals present low thermal stability when heated at moderated temperatures, which prevent their processing with methods involving heat. This is ascribed to the introduction of sulfate groups resulting from the acid hydrolysis process involving sulfuric acid. All of these issues have mainly limited the processing of MFC or cellulose nanocrystal-based nanocomposites to wet processing methods such as casting/evaporation, which was extensively studied.

Table 7.3 summarizes the different nanocomposite systems processed from cellulose nanoparticles by melt compounding. A close look to these different systems allows us to discern different strategies. If a polar polymer is used as matrix, there is no a priori need to compatibilize the filler with the matrix material. For apolar polymeric matrices, a solvent exchange procedure can be used to mix the cellulosic nanoparticles with the polymer in solution or functionalization of the filler can be conducted. Adsorption of small molecules or macromolecules on the surface of the nanoparticle can also be used as processing aids to tune the surface of the nanoparticles and compatibilize with matrix. Finally, other physical methods have been investigated.

Polymer	**Nano-particle**	**Source of Cellulose**	**Processing Aid/ Surface Functionalization**	**Processing Technique**	**Reference**
BIOPLAST GF 106/02	CNC/ MFC	Ramie/ *Luffa cylindrica*	DOPE Process	Extrusion	(Lemahieu et al., 2011)
NR	CNC	MCC	Silane	Milling/Hot-Pressing	(Xu et al., 2012)
PCL	CNC	Ramie	ROP PCL Grafting	Extrusion/Injection	(Goffin et al., 2011a)
PE	MFC	Soybean	Ethylene-acrylic Oligomer Coating	Mixing/Hot-Pressing	(Wang and Sain, 2007a)
	CNC	Ramie	Aliphatic Chains Grafting	Extrusion	(de Menezes et al., 2009)
	CNC	Cotton	PEO	Extrusion	(Ben Azouz et al., 2012)
PEO	CNC	Ramie	–	Extrusion	(Alloin et al., 2011)
PHB	MFC	Hemp	Different Coatings	Extrusion/Injection Molding	(Wang and Sain, 2007c)
PHBV	CNC	MCC	PEG	Extrusion/Injection Molding	(Jiang et al., 2008)

Polymer	Nano-particle	Source of Cellulose	Processing Aid/ Surface Functionalization	Processing Technique	Reference
PLA	CNC	MCC	PEG/Maleated PLA	Extrusion	(Oksman et al., 2006)
	CNC	MCC	PVA	Extrusion	(Bondeson and Oksman, 2007)
	MFC	Daicel	–	Mixing/Hot-Pressing	(Iwatake et al., 2008)
	MFC	Daicel	–	Mixing/Hot-Pressing	(Suryanegara et al., 2009)
	MFC	Kenaf	–	Extrusion/Injection	(Jonoobi et al., 2010)
	MFC	Daicel	–	Injection	(Suryanegara et al., 2011)
	CNC	Ramie	ROP PLA Grafting	Extrusion/Injection	(Goffin et al., 2011b)
	CNC	Ramie	Silane	Extrusion/Injection	(Raquez et al., 2012)
	CNC	MCC	Surfactant	Extrusion	(Fortunati et al., 2012)
PP	MFC	Soybean	Ethylene-acrylic Oligomer Coating	Mixing/Hot-Pressing	(Wang and Sain, 2007b)
Starch	CNC	Softwood Dissolving Pulp, Cotton, Bacterial Cellulose	–	Extrusion	(Orts et al., 2005)

Table 7.3: Polymer nanocomposites obtained from nanocellulose by melt compounding.

7.5.2 Melt compounding with a polar matrix

Starch reinforced cellulose nanocrystals composites were obtained by extrusion (Orts et al., 2005). Pre-gelatinized starch was first prepared by heating starch in water at 95°C. It was then mixed with nanocrystals, and the mixture was introduced into the extruder and formed into monofilaments and films under low and high shear mode. The data suggested that shear alignment significantly improved tensile strength of the nanocomposites.

High molecular weight PEO reinforced cellulose nanocrystal nanocomposites have been obtained by extrusion (Alloin et al., 2011). The aqueous mixture was freeze-dried and allowed to melt in the extruder under nitrogen flow at 180°C. In order to evaluate the influence of the extrusion process, the nanocrystal length and diameter after extrusion were determined through microscopic observations. The nanoparticles were extracted from the extruded nanocomposite by dissolving the PEO matrix in water. A decrease of the nanoparticle dimensions was observed upon extrusion. The length and cross section of the nanocrystals was found to decrease from about 200 nm to 120 nm, and from 7 nm to 5 nm, respectively (Figure 7.7). No significant change of the aspect ratio, from 28 to 24, was observed after extrusion. Moreover, a significant narrowing of the length distribution was reported. Compared to cast/evaporated samples, absence of a percolating nanocrystal network was reported for extruded samples. This was verified from creep measurements. The strain of the cast/evaporated nanocomposite reached a plateau value, corresponding to the mechanical response of a solid with a delayed elasticity, while the one of the extruded nanocomposite gradually increased with time, which is characteristic of a fluid. A poor dispersion of the filler within the polymeric matrix was observed despite the hydrophilic nature of the host material.

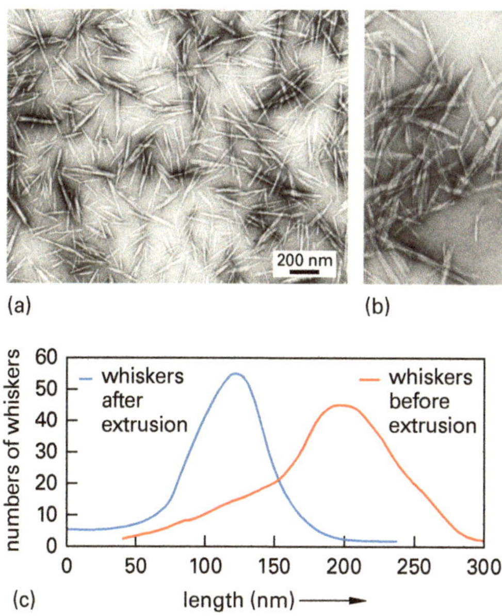

Fig. 7.7: TEM of ramie nanocrystals before extrusion (a) and after extrusion and dissolution of the PEO matrix in water (b), and their length distribution (c) (Alloin et al., 2011).

7.5.3 Melt compounding using solvent exchange

MFC water dispersion was mixed with acetone and PLA was added gradually under stirring (Iwatake et al., 2008). After evaporation of the liquid (acetone and water), the mixture was kneaded with a twin rotary roller mixture. As a comparison, MFC dispersion was added directly to the melted PLA, and the mixture was kneaded. The compounds were crushed into small pieces and compressed at 160°C. A similar method was reported for MFC, in which the water phase was removed and replaced by acetone by successive centrifugations (Suryanegara et al., 2009). Next, acetone was replaced by dichloromethane using the same process, PLA was added to this suspension and the solvent was evaporated. The mixture was then kneaded with a twin rotary roller mixture, crushed into small pieces and hot pressed. It was found that addition of 10 wt% MFC increased the stiffness of PLA at high temperatures and allowed ejection of the injected samples without distortion at a holding time of just 10 s (Suryanegara et al., 2011). No change of the crystallinity was observed and this effect was ascribed to the reinforcement and increase of the rubbery modulus provided by MFC.

PLA-based nanocomposites were also processed using MFC prepared from kenaf pulp as the reinforcing phase (Jonoobi et al., 2010). The first step consisted in preparing master batches with high contents of MFC in PLA using a mixture of acetone and chloroform as the liquid medium. After drying, the crushed master batches were extruded and injection molded with PLA. However, from these methods an organic solvent is still used.

7.5.4 Melt compounding with processing aids

The preparation of MCC cellulose nanocrystal reinforced PLA nanocomposites by melt extrusion was carried out by pumping the suspension of nanocrystals into the polymer melt during the extrusion process, inducing a probable strong hydrolytic degradation of the polymer (Oksman et al., 2006). Prior to extrusion, MCC was swelled with N,N-dimethylacetamide (DMAc) containing lithium chloride (LiCl) and partly separated to nanocrystals by ultra-sonication. This suspension was concentrated to 17 wt% by vacuum drying before pumping it into the co-rotating twin screw extruder to minimize the amount of chemicals introduced. Polyethylene glycol (PEG), used as a processing aid to decrease the viscosity of the system, and maleated PLA with a maleic anhydride content of 2.2%, used as a processing aid and coupling agent, were premixed with PLA. Extrusion was carried out in the temperature range 170–200°C and the liquid phase was removed by atmospheric venting and vacuum venting. The extruded material was compression molded. Microscopic analysis showed partly dispersed nanocrystals when using both PEG and maleated PLA, whereas no single nanoparticle was observed when using only maleated PLA, showing the need of a processing aid to disperse the nanocrystals in the matrix.

An attempt to use PVA as a compatibilizer to promote the dispersion of cellulose nanocrystals within the PLA matrix was reported (Bondeson and Oksman, 2007). Two feeding methods of PVA and cellulose nanocrystals were used, dry-mixing with PLA prior to extrusion or pumping as suspension directly into the extruder. However, due to immiscibility of the polymers, phase separation occurred with a continuous PLA phase and a discontinuous PVA phase. The cellulose nanocrystals were primarily located in the PVA phase and only a negligible amount was located in the PLA phase, leading to poor performance of the nanocomposites. This specific localization of the cellulosic nanoparticles was obviously ascribed to the more hydrophilic nature of PVA compared to PLA.

Conversely, cellulose nanocrystals prepared from MCC and stabilized with a surfactant, an acid phosphate ester of ethoxylated nonylphenol, were used as nucleating agent and found to significantly increase the crystallization rate and degree of crystallinity of the extruded PLA matrix (Fortunati et al., 2012).

Soybean MFC was used to prepare nanocomposites with a polyethylene (PE) (Wang and Sain, 2007a) and polypropylene (PP) (Wang and Sain, 2007b) matrix in a laboratory Brabender internal mixer. Ethylene-acrylic oligomer aqueous emulsion was used as a dispersant to improve the compatibility with the PE or PP matrix. Coated MFC was compounded with the polymeric matrix at 170°C and the mixture was hot-pressed at 180°C into sheet form.

Melt processing (extrusion and injection molding) of cellulose nanocrystals reinforced PHBV was also attempted (Jiang et al., 2008). Despite using polyethylene glycol (PEG) as a compatibilizer, the nanoparticles agglomerates formed during Freeze-drying could not be broken and well dispersed by the extrusion process. PEG is miscible with PHBV and a lack of strong interaction between PEG and cellulose nanocrystals was suspected. Therefore, during the high shear twin-screw compounding PEG could be removed from nanoparticle surface and blended with the PHBV matrix. Without the shielding of the PEG coating, the nanocrystals could not be well dispersed as evidenced from microscopic observations.

A similar strategy was adopted but using a much higher molecular weight compatibilizing polymer (Ben Azouz et al., 2012). PEO with molecular weight as high as 5 million was used. First, the rheological properties of a dispersion of cellulose nanocrystals in an aqueous solution of PEO were investigated. A peculiar behavior was reported. Upon adding cellulosic nanoparticles, the viscosity of the suspension first decreased and then increased. Adsorption of PEO chains on the surface of the nanoparticles was suspected. This may be attributed to the strong affinity between PEO chains and the cellulosic surface through interactions between the oxygen groups of PEO and hydroxyl groups of cellulose. However, heat flow microcalorimetry experiments have suggested a stronger affinity of the cellulose nanocrystal surface with hydroxyl en groups of PEO than its ether oxygen (Azizi Samir et al., 2004b). In the experimental conditions investigated (1 wt% PEO in water), it appeared that a cotton nanocrystal concentration around 6 wt% corresponded to a critical concentration, sufficient to adsorb all the PEO chains available in the suspension. Above this critical value, the viscosity increased with filler content and the suspension displays a typical suspension behavior with a viscosity increasing with the suspension concentration. Freeze-drying of this PEO-adsorbed cellulose nanocrystal dispersion was performed and the ensuing lyophilisate was successfully extruded with low density polyethylene. Compared to neat CNC-based nanocomposites, both improved dispersibility and thermal stability were observed (Figure 7.8). The improved thermal stability of PEO-adsorbed nanoparticles was confirmed by thermogravimetric analysis. It was ascribed to a protection role of adsorbed PEO chains that hidden the surface sulfate groups of cellulose nanocrystals.

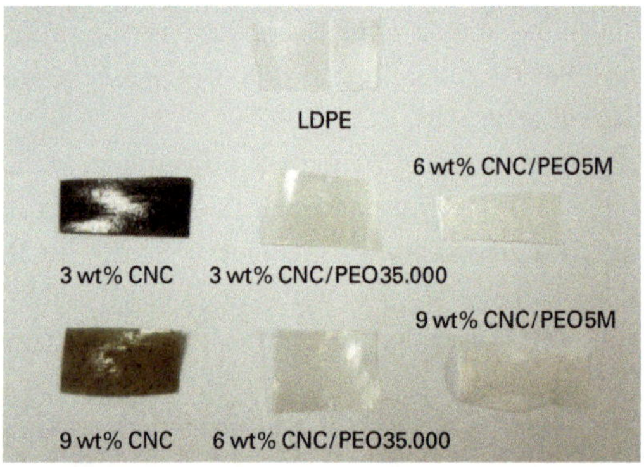

Fig. 7.8: Pictures of the extruded films: unfilled low density polyethylene (LDPE) matrix, and LDPE reinforced with neat cellulose nanocrystals (CNC) and PEO-adsorbed cellulose nanocrystals (Ben Azouz et al., 2012). Two different molecular weights PEO (35,000 and 5 million) have been used.

7.5.5 Melt compounding with chemically grafted nanoparticles

Functionalization of the surface of the nanoparticles is most of the time a necessary step to avoid irreversible agglomeration during drying and aggregation in non-polar matrices because of the formation of additional hydrogen bonds. However, this strategy, that induces an additional functionalization step, is time-consuming and hardly compatible with an industrial application of cellulose-based nanocomposites.

Organic acid chlorides-grafted cellulose nanocrystals extracted from ramie fibers were extruded with low density polyethylene (LDPE) (de Menezes et al., 2009). Aliphatic chains presenting different lengths were used and grafted by an esterification reaction. The homogeneity of the ensuing nanocomposite was found to increase with the length of the grafted chains (Figure 7.9). A significant improvement in terms of elongation at break was observed when sufficiently long chains were grafted on the surface of the nanoparticles. It was ascribed to improved dispersion of the nanoparticles within the LDPE matrix.

Fig. 7.9: Photographs of the neat LDPE film and extruded nanocomposite films reinforced with 10 wt% of unmodified and C18 acid chloride-grafted cellulose whiskers (de Menezes et al., 2009).

Poly(ε-caprolactone) (PCL)-grafted cellulose nanocrystals extracted from ramie fibers and synthesized by ring-opening polymerization of the corresponding lactone were studied as master batches by melt-blending within its commercial PCL matrix (Goffin et al., 2011a). The relative cellulose content of PCL-grafted cellulose nanocrystals was estimated by gravimetry at 14 wt%. An excellent dispersion of the nanoparticles within the hydrophobic matrix was reported. Interestingly, a solid-like behavior was observed by rheological analyses most likely due to the formation of a polymer physical network. In order to attest the formation of this physical chains network between the PCL-grafted chains and the free PCL chains of the matrix, a nanocomposite was prepared with a PCL-grafted nanohybrid characterized by shorter PCL chains. The entanglement of the surface-grafted polyester chains with the PCL matrix contributed to increase thermomechanical properties and decrease the chain relaxation phenomenon.

A similar strategy involving ring-opening polymerization of L-lactide initiated from the hydroxyl groups available at the nanocrystal surface was also used to prepared PLA-based nanocomposites by extrusion and injection-molding at 165–180°C (Goffin et al., 2011b). The relative cellulose content of PLA-grafted cellulose nanocrystals was estimated by gravimetry at 14 wt%. It was clearly evidenced that the chemical grafting of nanocrystals enhanced their compatibility with the polymeric matrix and thus improved the final properties of the nanocomposites. Large modification of the crystalline properties such as the crystallization half-time was evidenced according to the nature of the PLA matrix and the content of nanofiller. Moreover, PLA grafting was shown to avoid the degradation of the cellulosic nanoparticles (Figure 7.10). It seems that the presence of grafted PLA chains at the surface of the nanocrystals acted as a protective shell around the nanoparticles and therefore prevented their thermal degradation, which allows their processing at high temperature with very limited degradation of cellulose during the melt-processing. The removal of sulfate moieties from the surface of the nanoparticles during the chemical grafting of PLA chains was also suspected, which limits their thermal degradation. Functional triakoxysilanes

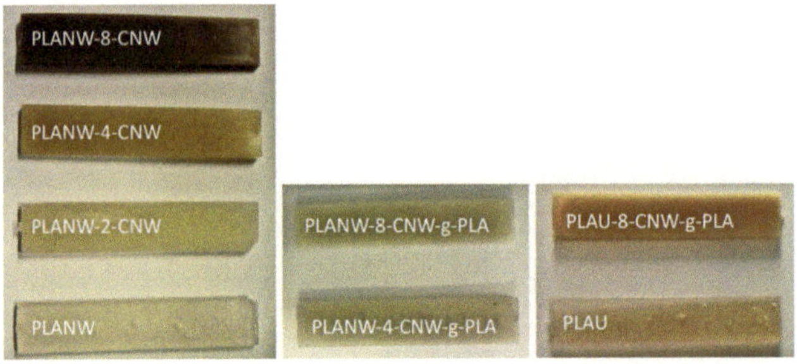

Fig. 7.10: Pictures of the injection-molded PLA-based samples (PLANW and PLAU refer to two different commercial PLA) filled or not filled with unmodified ramie cellulose nanocrystals (CNW) and PLA-grafted ramie cellulose nanocrystals (CNW-g-PLA) (Goffin et al., 2011b).

bearing various organic moieties (alkyl, amino, and (meth)acryloxy) were also used to modify the surface of cellulose nanocrystals (Raquez et al., 2012).

7.5.6 Melt compounding using physical process

An attempt to use a recently patented concept (Dispersed Nano-Objects Protective Encapsulation – DOPE process) intended to disperse carbon nanotubes in polymeric matrices was reported. Physically cross-linked alginate capsules were successfully formed in the presence of either cellulose nanocrystals or MFC (Lemahieu et al., 2011). The ensuing capsules have been extruded with a thermoplastic material.

7.6 Filtration and impregnation

Another possible processing technique of nanocomposites using cellulosic nanoparticles in the dry state consists in the filtration of the aqueous suspension to obtain a film or dried mat of particles followed by immersion in a polymer solution. The impregnation of the dried mat is performed under vacuum to promote wetting of the mat. Composites were processed by filling the cavities with transparent thermosetting resins such as phenol formaldehyde (Nakagaito and Yano, 2004; Nakagaito et al., 2005; Nakagaito and Yano, 2008a; Nakagaito and Yano, 2008b), epoxy (Shimazaki et al., 2007), acrylic (Nakagaito et al., 2005; Nogi et al., 2005; Iwamoto et al., 2005; Iwamoto et al., 2008) and melamine formaldehyde (Henriksson and Berglund, 2007). Nonwoven mats of cellulose microfibrils were also used to prepare polyurethane composite materials using film stacking method (Seydibeyoğlu and Oksman, 2008). The final fiber content of the composite can be controlled by swelling the vacuum filtered MFC mat in water just before freeze-drying to obtain samples with lower density (higher void volume) (Nakagaito and Yano, 2008b).

A manufacturing process similar to papermaking, which enables the production of thin sheets made of uniformly dispersed MFC with PLA fibers, was proposed (Nakagaito et al., 2009). The mixture was carried out in water and sieved, dried under load and finally compression molded.

7.7 Spinning and electrospinning

Spinning is a manufacturing process used for producing polymer fibers. The operating principle of electrospinning has been described in Chapter 2 (Section 2.1.6.) and an application for the preparation of cellulose nanofibers has been reported. Nanocomposite electrospun fibers can also be obtained by using a suspension of cellulose nanoparticles in a polymer solution. It has emerged as an alternative processing method for cellulose nanocrystals in polymer matrices, but as for the casting-evaporation technique, their dispersion can be challenging. Moreover, control and correlation between operational conditions and morphology of the ensuing electrospun micro- or nanofibers is difficult because of the numerous concurrent phenomena involved during spinning. Processing parameters such as voltage and distance between the spinning tip and the collector, properties of the spinning solution (conductivity, viscosity, density, surface tension, etc.), and its flow rate can drastically affect the outcome of the spinning process.

Electrospinning of water soluble polymers reinforced nanocomposites has been reported in the literature. Bacterial cellulose nanocrystals were incorporated into PEO nanofibers with a diameter of less than 1 μm by the electrospinning process to enhance the mechanical properties of the electrospun fibers (Park et al., 2007). The nanoparticles were found to be globally well embedded and aligned inside the fibers, even though they were partially aggregated. Likewise, electrospun PVA fiber mats reinforced with cellulose nanocrystals (Figure 7.11), with diameter in the nanoscale

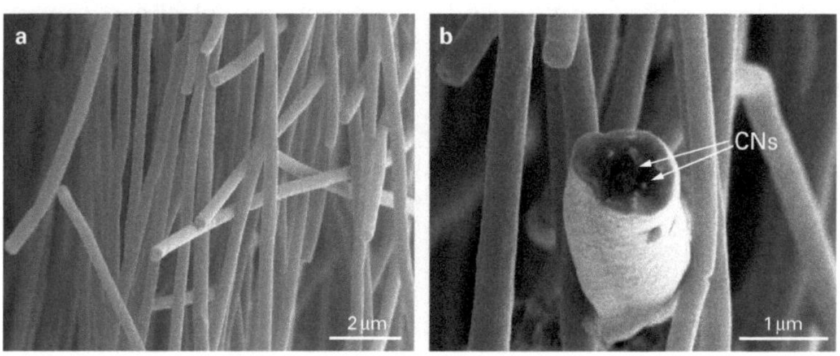

Fig. 7.11: Cryo-scanning electron micrograph (a) and variable pressure, ultrahigh resolution field emission scanning electron micrograph (b) of electrospun polyvinyl alcohol loaded with 15% of cellulose nanocrystals (CNs) (Peresin et al., 2010).

range and enhanced mechanical properties, were successfully produced (Medeiros et al., 2008; Peresin et al., 2010).

MFC-based macrofibers have been prepared by wet-extrusion of a MFC hydrogel (1 wt%) into a coagulation bath of an organic solvent and drying (Walther et al., 2011). The prerequisites for the coagulant were (i) miscibility with water, and (ii) moderate polarity and hydrogen bonding capability. Good mechanical properties were obtained that could be increased by a further alignment of the constituent nanofibrils by post-drawing processes. PVA-cotton nanocrystals suspensions were spun using a syringe pump and a syringe with a needle and spinning dopes were injected into cooled methanol maintained at a temperature between –15 and –20°C (Uddin et al., 2011). The spun gel fibers were kept immersed in the cooled methanol bath for 24 h, wound into a bobbin, kept again in methanol at room temperature for 4 h and dried in air for 24 h. Moreover, as-spun fibers were drawn in a hot oven (210°C) with a hand-operated drawing apparatus. The maximal draw ratio defined as the ratio of the diameter of the as-spun fiber to the one of the drawn fiber ranged between 20 and 38. The drawn fibers exhibited extremely high orientation of nanocrystals and excellent mechanical properties, with a Young's modulus of 56 GPa for PVA fibers reinforced with 30 wt% cotton nanocrystals.

Different strategies have been proposed to electrospin polymer solutions with cellulose nanocrystal in organic media. For instance, surfactant-coated cellulose nanocrystals have been dispersed in polystyrene (PS) dissolved in THF (Rojas et al., 2009). Nonionic surfactant sorbitan monostearate was used. Nanoparticle surface polymer grafting was proved to be efficient to obtain PCL reinforced with cellulose nanocrystals from DMF-dichloromethane (Zoppe et al., 2009). The use of polar solvents, such as DMF, has also been studied to electrospin PLA with cellulose nanocrystals (Xiang et al., 2009). An indirect, sequestered "core-in-shell" electrospinning technique has been reported (Magalhães et al., 2009) in which an aqueous dispersion of sulfated cellulose nanocrystals constitutes the discrete "core" component surrounded by a cellulose "shell". Aqueous suspensions of cellulose nanocrystals prepared by TEMPO-mediated oxidation of wood pulp and tunicin were also electrospun in an acetone coagulation bath (Iwamoto et al., 2011). The wood fibers obtained at high spinning rate formed hollow structures, whereas the higher viscosity tunicin nanocrystal suspension produced spun fibers with a porous structure. Increased spinning rate also increased the nanocrystal alignment along the fiber axis.

7.8 Multilayer films

The layer-by-layer assembly (LBL) is a method by which thin films particularly of oppositely charged layers are deposited. Thin film LBL assembly technique can also be utilized for nanoparticles. In general the LBL assembly is described as sequential adsorption of positive or negative charged species by alternatively dipping into the solutions. The excess or remaining solution after each adsorption step is rinsed with solvent and thus a thin layer of charged species on the surface ready for next adsorption step is obtained. This process is schematically outlined in Figure 7.12.

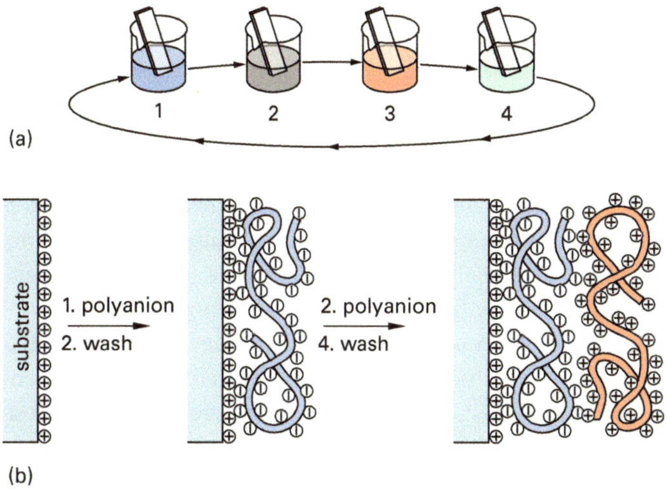

Fig. 7.12: (a) Schematic of the film deposition process using slides and beakers. Steps 1 and 3 represent the adsorption of a polyanion and polycation, respectively, and steps 2 and 4 are washing steps. The four steps are the basic build-up sequence for the simplest film architecture, $(A/B)_n$. The construction of more complex film architectures requires only additional beakers and a different deposition sequence. (b) Simplified molecular picture of the first two adsorption steps, depicting film deposition starting with a positively charged substrate. Counterions are omitted for clarity. The polyion conformation and layer interpenetration are an idealization of the surface charge reversal with each adsorption step (Decher, 1997).

There are many advantages of LBL assembly technique and these include simplicity, universality and thickness control in nanoscale. Further, the LBL assembly process does not require highly pure components and sophisticated hardware. For almost all aqueous dispersions of even high molecular weight species, it is easy to find an LBL pair that can be useful for building thin layers. In each adsorption step, either a monolayer or a sub-monolayer of the species is obtained and therefore the number of adsorption steps needed for a particular nanoscale layer can be founded. The use of the LBL technique is expected to maximize the interaction between cellulose nanocrystals and a polar polymeric matrix, such as chitosan (de Mesquita et al., 2010). It also allows the incorporation of high amounts of cellulose nanocrystals, presenting a dense and homogeneous distribution in each layer.

The preparation of cotton nanocrystal multilayer composites with a polycation, poly(diallyldimethylamonium chloride) (PDDA) using LBL technique has been reported (Podsiadlo et al., 2005). The authors concluded that the multilayer films presented high uniformity and dense packing of nanocrystals. Higher aspect ratio tunicin nanocrystals were also used with polyethyleneimine (PEI) (Podsiadlo et al., 2007). Their LBL assembled films show strong antireflection properties having an origin in a novel highly porous architecture reminiscent of a "flattened matchsticks pile", with film-thickness-dependent porosity and optical properties created by randomly oriented and overlapping nanocrystals. At an optimum number of LBL deposition cycles, light transmittance reached nearly 100%.

Orientated self-assembled films were also prepared using a strong 7 T magnetic field (Cranston and Gray, 2006a). Orientation of the deposited nanocrystals was observed, but only after long exposure to the field. The same authors prepared multilayered films containing cellulose nanocrystals and poly(allylamine hydrochloride) (Cranston and Gray, 2006b). Both solution-dipping and spin-coating assembly methods were used. Relatively few deposition cycles were needed to give full surface coverage, with film thicknesses ranging from 10 to 500 nm. Films prepared by spin-coating were substantially thicker than solution-dipped films and displayed radial orientation of the rod-shaped nanoparticles. The films prepared with these techniques had good optical properties, displaying thin-film interference colors. These polyelectrolyte multilayer films were found to be ideally suited for surface force measurements using the colloid-probe atomic force microscopy (CP-AFM) technique (Cranston et al., 2010).

Polyelectrolyte multilayers were prepared from MFC obtained from carboxymethylated cellulose fibers and different cationic polyelectrolytes (Wågberg et al., 2008). Silicon oxide surfaces were first treated with cationic polyelectrolytes before exposing the surfaces to MFC. Very-well defined layers were formed. A polyelectrolyte with a 3D structure led to the build-up of thick layers of MFC, whereas the use of a highly charged linear polyelectrolyte led to the formation of thinner layers of MFC. The films of polyelectrolytes and MFC were so smooth and well-defined that they showed clearly different interference colors, depending on the film thickness. A good correlation was found between the thickness of the films measured with ellipsometry and the thickness estimated from their colors. It was also shown that the amounts of polyethyleneimine (PEI) and MFC adsorbed can be adjusted by changing the pH and the electrolyte and polyelectrolyte concentrations (Aulin et al., 2008). High pH and high electrolyte concentration of PEI solution were found to promote the adsorption of MFC during the build-up of the multilayer, while an increase in the electrolyte concentration of the MFC dispersion was found to have the opposite effect. The Young's modulus of multilayer films of MFC and PEI was determined using the strain-induced elastic buckling instability for mechanical measurements (SIEBIMM) technique (Cranston et al., 2011). At 50% relative humidity, a value of 1.5 GPa was reported for 35–75 nm thick films, whereas it was 10 times larger, around 17.2 GPa in vacuum.

The preparation of thin films composed of alternating layers of orientated rigid cellulose nanocrystals and flexible polycation chains was reported (Jean et al., 2008). Alignment of the rod-like nanoparticles was achieved using anisotropic suspensions of cellulose nanocrystals. Green composites based on cellulose nanocrystals/xyloglucan multilayers have been prepared using the non-electrostatic cellulose-hemicellulose interaction (Jean et al., 2009). The thin films were characterized using neutron reflectivity experiments and AFM observations. More recently, biodegradable nanocomposites were obtained with the LBL technique using highly deacetylated chitosan and cellulose nanocrystals (de Mesquita et al., 2010). Hydrogen bonds and electrostatic interactions between the negatively charged sulfate groups on the nanoparticles surface and the ammonium groups of chitosan were the driving forces for the growth of the multilayered films. A high density and homogeneous distribution of cellulose nanocrystals adsorbed on each chitosan layer, each bilayer being around 7 nm thick, were reported. Self-organized films were also obtained using only charge-stabilized dispersions of celluloses nanoparticles with opposite charges from the LBL technique (Aulin et al., 2010b).

7.9 Conclusions

The processing of nanocellulose-based materials is crucial since it determines their usage properties. As for any nanoparticle, the main challenge is related to their homogeneous dispersion within a polymeric matrix. In most studies, nanocellulose reinforced nanocomposites are prepared in liquid medium, using polymer solution or polymer dispersion (latex). Its main advantage is that it allows preserving the dispersion state of the nanoparticles in the liquid medium. However, the number of polymeric matrices can be restricted and this processing technique is both non-industrial and non-economic. The polymer melt approach is most probably the most convenient processing technique because it is a green process and it is industrially and economically viable. However, it is challenging to find the suitable conditions because of the inherent incompatibility of cellulose with most polymeric matrices and thermal stability issues.

7.10 References

Agrawal, A.A. and Konno, K. (2009). Latex: A Model for understanding mechanisms, ecology, and evolution of plant defense against herbivory. Annu. Rev. Ecol. Evol. Syst. *40*, 311–331.

Alloin, F., D'Aprea, A., Dufresne, A., El Kissi, N. and Bossard, F. (2011). Poly(oxyethylene) and ramie whiskers based nanocomposites: Influence of processing: Extrusion and casting/evaporation. Cellulose *18*, 957–973.

Anglès, M.N. and Dufresne, A. (2000). Plasticized starch/tunicin whiskers nanocomposites: 1. Structural analysis. Macromolecules *33*, 8344–8353.

Anglès, M.N. and Dufresne, A. (2001). Plasticized starch/tunicin whiskers nanocomposites: 2. Mechanical behavior. Macromolecules *34*, 2921–2931.

Auad, M.L., Contos, V.S., Nutt, S., Aranguren, M.I. and Marcovich, N.E. (2008). Characterization of nanocellulose-reinforced shape memory polyurethanes. Polym. Inter. *57*, 651–659.

Auad, M.L., Richardson, T., Orts, W.J., Medeiros, E.S., Mattoso, L.H.C., Mosiewicki, M.A., Marcovich, N.E. and Aranguren, M.I. (2011). Polyaniline-modified cellulose nanofibrils as reinforcement of a smart polyurethane. Polym. Inter. *60*, 743–750.

Auad, M.L., Richardson, T., Hicks, M., Mosiewicki, M.A., Aranguren, M.I. and Marcovich, N.E. (2012). Shape memory segmented polyurethanes: Dependence of behaviour on nanocellulose addition and testing conditions. Polym. Inter. *61*, 321–327.

Aulin, C., Varga, I., Claesson, P.M., Wågberg, L. and Lindström, T. (2008). Buildup of polyelectrolyte multilayers of polyethyleneimine and microfibrillated cellulose studied by *in situ* dual-polarization interferometry and quartz crystal microbalance with dissipation. Langmuir *24*, 2509–2518.

Aulin, C. Netrval, J., Wågberg, L. and Lindström, T. (2010a). Aerogels from nanofibrillated cellulose with tunable oleophobicity. Soft Matter *6*, 3298–3305.

Aulin, C., Johansson, E., Wågberg, L. and Lindström, T. (2010b). Self-organized films from cellulose I nanofibrils using the layer-by-layer technique. Biomacromolecules *11*, 872–882.

Azeredo, H.M.C., Miranda, K.W.E., Rosa, M.F., Nascimento, D.M. and de Moura, M.R. (2012). Edible films from alginate-acerola puree reinforced with cellulose whiskers. Food Sci. Technol. *46*, 294–297.

Azizi Samir, M.A.S., Alloin, F., Paillet, M. and Dufresne, A. (2004a). Tangling effect in fibrillated cellulose reinforced nanocomposites. Macromolecules *37*, 4313–4316.

Azizi Samir, M.A.S., Alloin, F., Sanchez, J.-Y. and Dufresne, A. (2004b). Cellulose nanocrystals reinforced poly(oxyethylene). Polymer *45*, 4149–4157.

Azizi Samir, M.A.S., Alloin, F., Gorecki, W., Sanchez, J.-Y. and Dufresne, A. (2004c). Nanocomposite polymer electrolytes based on poly(oxyethylene) and cellulose nanocrystals. J. Phys. Chem. B *108*, 10845–10852.

Azizi Samir, M.A.S., Montero Mateos, A., Alloin, F., Sanchez, J.-Y. and Dufresne, A. (2004d). Plasticized nanocomposite polymer electrolytes based on poly(oxyethylene) and cellulose whiskers. Electrochim. Acta *49*, 4667–4677.

Azizi Samir, M.A.S., Alloin, F., Sanchez, J.Y., El Kissi, N. and Dufresne, A. (2004e). Preparation of cellulose whiskers reinforced nanocomposites from an organic medium suspension. Macromolecules *37*, 1386–1393.

Azizi Samir, M.A.S., Alloin, F., Sanchez, J.Y. and Dufresne, A. (2004f). Cross-linked nanocomposite polymer electrolytes reinforced with cellulose whiskers. Macromolecules *37*, 4839–4844

Azizi Samir, M.A.S., Chazeau, L., Alloin, F., Cavaillé, J.-Y., Dufresne, A. and Sanchez, J.-Y. (2005). POE-based nanocomposite polymer electrolytes reinforced with cellulose whiskers. Electrochim. Acta *50*, 3897–3903.

Azizi Samir, M.A.S., Alloin, F. and Dufresne, A. (2006). High performance nanocomposite polymer electrolytes. Compos. Interfaces *13*, 545–559.

Ben Azouz, K., Ramires, E.C., Van den Fonteyne, W., El Kissi, N. and Dufresne A. (2012). Simple method for the melt extrusion of a cellulose nanocrystal reinforced hydrophobic polymer. ACS Macro Lett. *1*, 236–240.

Ben Elmabrouk, A., Thielemans, W., Dufresne, A. and Boufi, S. (2009). Preparation of poly(styrene-co-hexylacrylate)/cellulose whiskers nanocomposites via miniemulsion polymerization. J. Appl. Polym. Sci. *114*, 2946–2955.

Bendahou, A., Habibi, Y., Kaddami, H. and Dufresne, A. (2009). Physico-chemical characterization of palm from *Phoenix dactylifera* – L, Preparation of cellulose whiskers and natural rubber–based nanocomposites. J. Biobased Mat. Bioenergy *3*, 81–90.

Bendahou, A., Kaddami, H. and Dufresne, A. (2010). Investigation on the effect of cellulosic nanoparticles' morphology on the properties of natural rubber based nanocomposites. Eur. Polym. J. *46*, 609–620.

Bendahou, A., Kaddami, H., Espuche, E., Gouanvé, F. and Dufresne, A. (2011). Synergism effect of montmorillonite and cellulose whiskers on the mechanical and barrier properties of natural rubber. Macromol. Mat. Eng. *296*, 760–769.

Besbes, I., Alila, S. and Boufi, S. (2011a). Nanofibrillated cellulose from TEMPO-oxidized eucalyptus fibres: Effect of the carboxyl content. Carbohydr. Polym. *84*, 975–983.

Besbes, I., Rei Vilar, M. and Boufi, S. (2011b). Nanofibrillated cellulose from alfa, eucalyptus and pine fibres: preparation, characteristics and reinforcing potential. Carbohydr. Polym. *86*, 1198–1206.

Bondeson, D. and Oksman, K. (2007). Polylactic acid/cellulose whisker nanocomposites modified by polyvinyl alcohol. Composites Part A *38*, 2486–2492.

Bossard, F., El Kissi, N., D'Aprea, A., Alloin, F., Sanchez, J.-Y. and Dufresne, A. (2010). Influence of dispersion procedure on rheological properties of aqueous solutions of high molecular weight PEO. Rheol. Acta *49*, 529–540.

Bras, J., Hassan, M.L., Bruzesse, C., Hassan, E.A., El-Wakil, N.A. and Dufresne, A. (2010). Mechanical, barrier, and biodegradability properties of bagasse cellulose whiskers reinforced natural rubber nanocomposites. Ind. Crops Prod. *32*, 627–633.

Brown, E.E. and Laborie, M.-P. (2007). Bioengineering bacterial cellulose/poly(ethylene oxide) nanocomposites. Biomacromolecules *8*, 3074–3081.

Cao, X., Dong, H and Li, C.M. (2007). New nanocomposite materials reinforced with flax cellulose nanocrystals in waterborne polyurethane. Biomacromolecules *8*, 899–904.

Cao, X., Chen, Y., Chang, P.R., Stumborg, M. and Huneault, M.A. (2008a). Green composites reinforced with hemp nanocrystals in plasticized starch. J. Appl. Polym. Sci. *109*, 3804–3810.

Cao, X., Chen, Y., Chang, P.R., Muir, A.D. and Falk G. (2008b). Starch-based nanocomposites reinforced with flax cellulose nanocrystals. eXPRESS Polym. Let. *2*, 502–510.

Capadona, J.R., Van den Berg, O., Capadona, L., Schroeter, M., Tyler, D., Rowan, S.J. and Weder, C. (2007). A versatile approach for the processing of polymer nanocomposites with self assembled nanofibre templates. Nature Nanotechnology *2*, 765–769.

Capadona, J.R., Shanmuganathan, K., Tyler, D.J., Rowan, S.J. and Weder, C. (2008). Stimuli-responsive polymer nanocomposites inspired by the sea cucumber dermis. Science *319*, 1370–1374.

Capadona, J.R., Shanmuganathan, K., Trittschuh, S., Seidel, S., Rowan, S.J. and Weder, C. (2009). Polymer nanocomposites with nanowhiskers isolated from microcrystalline cellulose. Biomacromolecules *10*, 712–716.

Chauve, G., Heux, L., Arouini, R. and Mazeau, K. (2005). Cellulose poly(ethylene-co-vinyl acetate) nanocomposites studied by molecular modeling and mechanical spectroscopy. Biomacromolecules *6*, 2025–2031.

Chazeau, L., Cavaillé, J.Y., Canova, G.R., Dendievel, R. and Boutherin, B. (1999a). Viscoelastic properties of plasticized PVC reinforced with cellulose whiskers. J. Appl. Polym. Sci. *71*, 1797–1808.

Chazeau, L., Cavaillé, J.Y. and Terech, P. (1999b). Mechanical behaviour above Tg of a plasticized PVC reinforced with cellulose whiskers, a SANS structural study. Polymer *40*, 5333–5344.

Chazeau, L., Paillet, M. and Cavaillé, J.Y. (1999c). Plasticized PVC reinforced with cellulose whiskers: 1-linear viscoelastic behavior analyzed through the quasi point defect theory. J. Polym. Sci. Polym. Phys. *37*, 2151–2164.

Chazeau, L., Cavaillé, J.Y. and Perez, J. (2000). Plasticized PVC reinforced with cellulose whiskers. II. Plastic behavior. J. Polym. Sci. B *38*, 383–392.

Chen, Y., Liu, C., Chang, P.R., Cao, X. and Anderson, D.P. (2009). Bionanocomposites based on pea starch and cellulose nanowhiskers hydrolyzed from pea hull fibre: Effect of hydrolysis time. Carbohydr. Polym. *76*, 607–615.

Cheng, Q., Wang, S. and Rials T.G. (2009). Poly(vinyl alcohol) nanocomposites reinforced with cellulose fibrils isolated by high intensity ultrasonication. Composites Part A *40*, 218–224.

Choi, Y.J. and Simonsen, J. (2006). Cellulose nanocrystal-filled carboxymethyl cellulose nanocomposites. J. Nanosci. Nanotechnol. *6*, 633–639.

Cranston, E.D. and Gray, D.G. (2006a). Formation of cellulose-based electrostatic layer-by-layer films in a magnetic field. Sci. Technol. Adv. Mater. *7*, 319–321.

Cranston, E.D. and Gray, D.G. (2006b). Morphological and optical characterization of polyelectrolyte multilayers incorporating nanocrystalline cellulose. Biomacromolecules *7*, 2522–2530.

Cranston, E.D., Gray, D.G. and Rutland, M.W. (2010). Direct surface force measurements of polyelectrolyte multilayer films containing nanocrystalline cellulose. Langmuir *26*, 17190–17197.

Cranston, E.D., Eita, M., Johansson, E., Netrval, J., salajková, M., Arwin, H. and Wågberg, L. (2011). Determination of Young's modulus for nanofibrillated cellulose multilayer thin films using buckling mechanics. Biomacromolecules *12*, 961–969.

Dalmas, F., Chazeau, L., Gauthier, C., Cavaillé, J.Y., and Dendievel, R. (2006). Large deformation mechanical behaviour of flexible nanofiber filled polymer nanocomposites. Polymer *47*, 2802–2812.

Dalmas, F., Cavaillé, J.Y., Gauthier, C., Chazeau, L. and Dendievel, R. (2007). Viscoelastic behaviour and electrical properties of flexible nanofiber filled polymer nanocomposites. Influence of processing conditions. Compos. Sci. Technol. *67*, 829–839.

Dammström, S., Salmén, L. and Gatenholm, P. (2005). The effect of moisture on the dynamic mechanical properties of bacterial cellulose/glucuronoxylan nanocomposites. Polymer *46*, 10364–10371.

de Menezes, A.J., Siqueira, G., Curvelo, A.A.S. and Dufresne, A. (2009). Extrusion and characterization of functionalized cellulose whisker reinforced polyethylene nanocomposites. Polymer *50*, 4552–4563.

de Mesquita, J.P., Donnici, C.L. and Pereira, F.V. (2010). Biobased nanocomposites from layer-by-layer assembly of cellulose nanowhiskers with chitosan. Biomacromolecules *11*, 473–480.

Decher, G. (1997). Fuzzy nanoassemblies: Toward layered polymeric multicomposites. Science *277*, 1232–1237.

Dubief, D., Samain, E. and Dufresne, A. (1999). Polysaccharide microcrystals reinforced amorphous poly(b-hydroxyoctanoate) nanocomposite materials. Macromolecules *32*, 5765–5771.

Duchemin, B.J.C., Mathew, A.P. and Oksman, K. (2009). All-cellulose composites by partial dissolution in the ionic liquid 1-butyl-3-methylimidazolium chloride. Composites A *40*, 2031–2037.

Dufresne, A., Cavaillé, J.Y. and Helbert, W. (1997). Thermoplastic nanocomposites filled with wheat straw cellulose whiskers. Part II: Effect of processing and modeling, Polym Compos. *18*, 198–210.

Dufresne, A. and Vignon, M.R. (1998). Improvement of starch films performances using cellulose microfibrils. Macromolecules *31*, 2693–2696.

Dufresne, A., Kellerhals, M.B. and Witholt, B. (1999). Transcrystallization in mcl-PHAs/ cellulose whiskers composites. Macromolecules *32*, 7396–7401.

Dufresne, A., Dupeyre, D. and Vignon, M.R. (2000). Cellulose microfibrils from potato cells: processing and characterization of starch/cellulose microfibrils composites. J. Appl. Polym. Sci. *76*, 2080–2092.

Dufresne, A. (2000). Dynamic mechanical analysis of the interphase in bacterial polyester/ cellulose whiskers natural composites. Compos. Interfaces *7*, 53–67.

Eyholzer, Ch., Bordeanu, N., López-Suevos, F., Rentsch, D., Zimmermann, T. and Oksman, K. (2010). Preparation and characterization of water-redispersible nanofibrillated cellulose in powder form. Cellulose *17*, 19–30.

Eyholzer, C., Borges de Couraça, A., Duc, F., Bourban, P.E., Tingaut, P., Zimmermann, T., Månson, J.A.E. and Oksman, K. (2011). Biocomposite hydrogels with carboxymethylated, nanofibrillated cellulose powder for replacement of the nucleus pulposus. Biomacromolecules *12*, 1419–1427.

Favier, V., Canova, G.R., Cavaillé, J.Y., Chanzy, H., Dufresne, A. and Gauthier, C. (1995a). Nanocomposites materials from latex and cellulose whiskers. Polym. Adv. Technol. *6*, 351–355.

Favier, V., Chanzy, H. and Cavaillé, J.Y. (1995b). Polymer nanocomposites reinforced by cellulose whiskers. Macromolecules *28*, 6365–6367.

Fernandes, S.C.M., Freire, C.S.R., Silvestre, A.J.D., Pascoal Neto, C., Gandini, A., Berglund, L.A. and Salmén, L. (2010). Transparent chitosan films reinforced with a high content of nanofibrillated cellulose. Carbohydr. Polym. *81*, 394–401.

Fortunati, E., Armentano, I., Zhou, Q., Iannoni, A., Saino, E., Visai, L., Berglund, L.A. and Kenny, J.M. (2012). Multifunctional bionanocomposite films of poly(lactic acid), cellulose nanocrystals and silver nanoparticles. Carbohydr. Polym. *87*, 1596–1605.

Frone, A.N., Panaitescu, D.M., Donescu, D., Spataru, C.I., Radovici, C., Trusca, R. and Somoghi, R. (2011). Preparation and characterization of PVA composites with cellulose nanofibers obtained by ultrasonication. BioResources *6*, 487–512.

Gabr, M.H., Elrahman, M.A., Okubo, K. and Fujii, T. (2010). Effect of microfibrillated cellulose on mechanical properties of plain-woven CFRP reinforced epoxy. Compos. Struct. *92*, 1999–2006.

Garcia de Rodriguez, N.L., Thielemans, W. and Dufresne, A. (2006). Sisal cellulose whiskers reinforced polyvinyl acetate nanocomposites. Cellulose *13*, 261–270.

Goetz, L., Mathew, A., Oksman, K., Gatenholm, P. and Ragauskas, A.J. (2009). A novel nanocomposite film prepared from crosslinked cellulosic whiskers. Carbohydr. Polym. *75*, 85–89.

Goetz, L., Foston, M., Mathew, A.P., Oksman, K. and Ragauskas, A.J. (2010). Poly(methyl vinyl ether-co-maleic acid)-polyethylene glycol nanocomposites cross-linked *in situ* with cellulose nanowhiskers. Biomacromolecules *11*, 2660–2666.

Goffin, A.-L., Raquez, J.-M., Duquesne, E., Siqueira, G., Habibi, Y., Dufresne, A. and Dubois, Ph. (2011a). Poly(e-caprolactone) based nanocomposites reinforced by surface-grafted cellulose nanowhiskers via extrusion processing: Morphology, rheology, and thermo-mechanical properties. Polymer *52*, 1532–1538.

Goffin, A.-L., Raquez, J.-M., Duquesne, E., Siqueira, G., Habibi, Y., Dufresne, A. and Dubois, Ph. (2011b). From interfacial ring-opening polymerization to melt processing of cellulose nanowhisker-filled polylactide-based nanocomposites. Biomacromolecules *12*, 2456–2465.

Hajji, P., Cavaillé, J.Y., Favier, V., Gauthier, C. and Vigier, G. (1996). Tensile behavior of nanocomposites from latex and cellulose whiskers. Polym. Compos. *17*, 612–619.

Heath, L. and Thielemans, W. (2010). Cellulose nanowhisker aerogels. Green Chemistry *12*, 1448–1453.

Helbert, W., Cavaillé, J.Y. and Dufresne, A. (1996). Thermoplastic nanocomposites filled with wheat straw cellulose whiskers. Part I: Processing and mechanical behavior. Polym. Compos. *17*, 604–611.

Henriksson, M. and Berglund, L.A. (2007). Structure and properties of cellulose nanocomposite films containing melamine formaldehyde. J. Appl. Polym. Sci. *106*, 2817–2824.

Heux, L., Chauve, G. and Bonini, C. (2000). Nonflocculating and chiral-nematic self-ordering of cellulose microcrystals suspensions in nonpolar solvents. Langmuir *16*, 8210–8212.

Iwamoto, S., Nakagaito, A.N., Yano, H. and Nogi, M. (2005). Optically transparent composites reinforced with plant fiber-based nanofibers. Appl. Phys. A *81*, 1109–1112.

Iwamoto, S., Abe, K. and Yano, H. (2008). The effect of hemicelluloses on wood pulp nanofibrillation and nanofiber network characteristics. Biomacromolecules *9*, 1022–1026.

Iwamoto, S., Isogai, A. and Iwata, T. (2011). Structure and mechanical properties of wet-spun fibers made from natural cellulose nanofibers. Biomacromolecules *12*, 831–836.

Iwatake, A., Nogi, M. and Yano, H. (2008). Cellulose nanofiber-reinforced polylactic acid. Compos. Sci. Technol. *68*, 2103–2106.

Jean, B., Dubreuil, F., Heux, L. and Cousin, F. (2008). Structural details of cellulose nanocrystals/ polyelectrolytes multilayers probed by neutron reflectivity and AFM. Langmuir *24*, 3452–3458.

Jean, B., Heux, L., Dubreuil, F., Chambat, G. and Cousin, F. (2009). Non-electrostatic building of biomimetic cellulose-xyloglucan multilayers. Langmuir *25*, 3920–3923.

Jiang, L., Morelius, E., Zhang, J., Wolcott, M. and Holbery, J. (2008). Study of the poly(3-hydroxybutyrate-co-3-hydroxyvalerate)/cellulose nanowhiskers composites prepared by solution casting and melt processing. J. Compos. Mat. *42*, 2629–2645.

Jonoobi, M., Harun, J., Mathew, A.P. and Oksman, K. (2010). Mechanical properties of cellulose nanofiber (CNF) reinforced polylactic acid (PLA) prepared by twin screw extrusion. Compos. Sci. technol. *70*, 1742–1747.

Kaushik, A., Singh, M. and Verma, G. (2010). Green nanocomposites based on thermoplastic starch and steam exploded cellulose nanofibrils from wheat straw. Carbohydr. Polym. *82*, 337–345.

Kvien, I., Tanem, B.S. and Oksman, K. (2005). Characterization of cellulose whiskers and their nanocomposites by atomic force and electron microscopy. Biomacromolecules *6*, 3160–3165.

Kvien, I., Sugiyama, J., Votrubec, M. and Oksman, K. (2007). Characterization of starch based nanocomposites. J. Mater. Sci. *42*, 8163–8171.

Lee, S.Y., Mohan, D.J., Kang, I.A., Doh, G.H., Lee, S. and Han, N.O. (2009). Nanocellulose reinforced PVA composite films: Effects of acid treatment and filler loading. Fibers and Polymers *10*, 77–82.

Leitner, J., Hinterstoisser, B., Wastyn, M., Keckes, J. and Gindl, W. (2007). Sugar beet cellulose nanofibril-reinforced composites. Cellulose *14*, 419–425.

Lemahieu, L., Bras, J., Tiquet, P., Augier, S. and Dufresne, A. (2011). Extrusion of nanocellulose reinforced nanocomposites using the dispersed nano-objects protective encapsulation (DOPE) process. Macromol. Mat. Eng. *296*, 984–991.

Liu, H., Liu, D., Yao, F. and Wu, Q. (2010). Fabrication and properties of transparent polymethylmethacrylate/ cellulose nanocrystals composites. Bioresource Technol. *101*, 5685–5692.

Liu H. and Laborie, M.-P. (2011). Bio-based nanocomposites by *in situ* cure of phenolic prepolymers with cellulose whiskers. Cellulose *18*, 619–630.

Ljungberg, N., Bonini, C., Bortolussi, F., Boisson, C., Heux, L. and Cavaillé, J.Y. (2005). New nanocomposite materials reinforced with cellulose whiskers in atactic polypropylene: effect of surface and dispersion characteristics. Biomacromolecules *6*, 2732–2739.

Ljungberg, N., Cavaillé, J.Y. and Heux, L. (2006). Nanocomposites of isotactic polypropylene reinforced with rod-like cellulose whiskers. Polymer *47*, 6285–6292.

López-Rubio, A., Lagaron, J.M., Ankerfors, M., Lindström, T., Nordqvist, D., Mattozzi, A. and Hedenqvist, M.S. (2007). Enhanced film forming and film properties of amylopectin using micro-fibrillated cellulose. Carbohydr. Polym. *68*, 718–727.

López-Suevos, F., Eyholzer, C., Bordeanu, N. and Richter K. (2010). DMA analysis and wood bonding of PVAc latex reinforced with cellulose nanofibrils. Cellulose *17*, 387–398.

Lu, J., Wang, T. and Drzal, L.T. (2008). Preparation and properties of microfibrillated cellulose polyvinyl alcohol composite materials. Composites Part A *39*, 738–746.

Lu, Y., Weng, L. and Cao, X. (2005). Biocomposites of plasticized starch reinforced with cellulose crystallites from cottonseed linter. Macromol. Biosci. *5*, 1101–1107.

Lu, Y., Weng, L. and Cao, X. (2006). Morphological, thermal and mechanical properties of ramie crystallites-reinforced plasticized starch biocomposites. Carbohydr. Polym. *63*, 198–204.

Ma, H., Zhou, B., Li, H.S., Li, Y.Q. and Ou, S.Y. (2011). Green composite films composed of nanocrystalline cellulose and a cellulose matrix regenerated from functionalized ionic liquid solution. Carbohydr. Polym. *84*, 383–389.

Malainine, M.E., Mahrouz, M. and Dufresne, A. (2005). Thermoplastic nanocomposites based on cellulose microfibrils from Opuntia ficus-indica parenchyma cell. Compos. Sci. Technol. *65*, 1520–1526.

Magalhães, W.L.E., Cao, X. and Lucia, L.A. (2009). Cellulose nanocrystals/cellulose core-in-shell nanocomposite assemblies. Langmuir *25*, 13250–13257.

Marcovich, N.E., Auad, M.L., Belessi, N.E., Nutt, S.R. and Aranguren, M.I. (2006). Cellulose micro/nanocrystals reinforced polyurethane. J. Mater. Res. *21*, 870–881.

Mathew, A.P. and Dufresne, A. (2002). Morphological investigation of nanocomposites from sorbitol plasticized starch and tunicin whiskers. Biomacromolecules *3*, 609–617.

Mathew, A.P., Thielemans, W. and Dufresne, A. (2008). Mechanical properties of nanocomposites from sorbitol plasticized starch and tunicin whiskers. J. Appl. Polym. Sci. *109*, 4065–4074.

Matos Ruiz, M., Cavaillé, J.Y., Dufresne, A., Gérard, J.F. and Graillat, C. (2000). Processing and characterization of new thermoset nanocomposites based on cellulose whiskers. Compos. Interfaces *7*, 117–131.

Matos Ruiz, M., Cavaillé, J.Y., Dufresne, A., Graillat, C. and Gérard, J.F. (2001). New waterborne epoxy coatings based on cellulose nanofillers. Macromol. Symp. *169*, 211–222.

Medeiros, E.S., Mattoso, L.H.C., Ito, E.N., Gregorski, K.S., Robertson, G.H., Offeman, R.D., Wood, D.F., Orts, W.J. and Imam S.H. (2008). Electrospun nanofibers of poly(vinyl alcohol) reinforced with cellulose nanofibrils. J. Biobased Mater. Bioenergy *2*, 231–242.

Mendez, J., Annamalai, P.K., Eichhorn, S.J., Rusli, R., Rowan, S.J., Foster, E.J. and Weder, C. (2011). Bioinspired mechanically adaptive polymer nanocomposites with water-activated shape-memory effect. Macromolecules *44*, 6827–6835.

Mikkonen, K.S., Mathew, A.P., Pirkkalainen, K., Serimaa, R., Xu, C., Willför, S., Oksman, K. and Tenkanen, M. (2010). Glucomannan composite films with cellulose nanowhiskers. Cellulose *17*, 69–81.

Mikkonen, K.S., Stevanic, J.S., Joly, C., Dole, P., Pirkkalainen, K., Serimaa, R., Salmén, L. and Tenkanen, M. (2011). Composite films from spruce galactoglucomannans with microfibrillated spruce wood cellulose. Cellulose *18*, 713–726.

Müller, D., Rambo, C.R., Recouvreux, D.O.S., Porto, L.M. and Barra, G.M.O. (2011). Chemical *in situ* polymerization of polypyrrole on bacterial cellulose nanofibers. Synth. Met. *161*, 106–111.

Nakagaito, A.N. and Yano, H. (2004). The effect of morphological changes from pulp fiber towards nano-scale fibrillated cellulose on the mechanical properties of high-strength plant fiber based composites. Appl. Phys. A *78*, 547–552.

Nakagaito, A.N., Iwamoto, S. and Yano, H. (2005). Bacterial cellulose: the ultimate nano-scalar cellulose morphology for the production of high-strength composites. Appl. Phys. A *80*, 93–97.

Nakagaito, A.N. and Yano, H. (2008a). Toughness enhancement of cellulose nanocomposites by alkali treatment of the reinforcing cellulose nanofibres. Cellulose *15*, 323–331.

Nakagaito, A.N. and Yano, H. (2008b). The effect of fiber content on the mechanical and thermal expansion properties of biocomposites based on microfibrillated cellulose. Cellulose *15*, 555–559.

Nakagaito, A.N., Fujimura, A., Sakai, T., Hama, Y. and Yano, H. (2009). Production of microfibrillated cellulose (MFC)-reinforced polylactic acid (PLA) nanocomposites from sheets obtained by a papermaking-like process. Compos. Sci. Technol. *69*, 1293–1297.

Nge, T.T., Nogi, M., Yano, H. and Sugiyama, J. (2010). Microstructure and mechanical properties of bacterial cellulose/chitosan porous scaffold. Cellulose *17*, 349–363.

Nogi, M., Handa, K., Nakagaito, A.N. and Yano, H. (2005). Optically transparent bionanofiber composites with low sensitivity to refractive index of the polymer matrix. Appl. Phys. Lett. *87*, 243110–243112.

Noishiki, Y., Nishiyama, Y., Wada, M., Kuga, S. and Magoshi, J. (2002). Mechanical properties of silk fibroin-microcrystalline cellulose composite films. J. Appl. Polym. Sci. *86*, 3425–3429.

Noorani, S., Simonsen, J. and Atre, S. (2007). Nano-enabled microtechnology: Polysulfone nanocomposites incorporating cellulose nanocrystals. Cellulose *14*, 577–584.

Nordqvist, D., Idermark, J., Hedenqvist, M.S., Gällstedt, M., Ankerfors, M. and Lindström, T. (2007). Enhancement of the wet properties of transparent chitosan-acetic acid-salt films using microfibrillated cellulose. Biomacromolecules *8*, 2398–2403.

Nyström, G., Mihranyan, A., Razaq, A., Lindström, T., Nyholm, L. and Strømme, M. (2010). A nanocellulose polypyrrole composite based on microfibrillated cellulose from wood. J. Phys. Chem. B *114*, 4178–4182.

Oksman, K., Mathew, A.P., Bondeson, D. and Kvien, I. (2006). Manufacturing process of cellulose whiskers/polylactic acid nanocomposites. Compos. Sci. Technol. *66*, 2776–2784.

Olsson, R.T., Azizi Samir, M.A.S., Salazar-Alvarez, G., Belova, L.; Ström, V., Berglund, L.A., Ikkala, O., Nogués, J. and Gedde, U.W. (2010). Making flexible magnetic aerogels and stiff magnetic nanopaper using cellulose nanofibrils as templates. Nature Nanotechnology *5*, 584–588.

Orts, W.J., Shey, J., Imam, S.H., Glenn, G.M., Guttman, M.E. and Revol, J.F. (2005). Application of cellulose microfibrils in polymer nanocomposites. J. Polym. Env. *13*, 301–306.

Pääkkö, M., Vapaavuori, J., Silvennoinen, R., Kosonen, H., Ankerfors, M. Lindström, T., Berglund, L.A. and Ikkala, O. (2008). Long and entangled native cellulose I nanofibers allow flexible aerogels and hierarchically porous templates for functionalities. Soft Matter *4*, 2492–2499.

Pandey, J.K., Chu, W.S., Kim, C.S., Lee, C.S. and Ahn, S.H. (2009). Bio-nano reinforcement of environmentally degradable polymer matrix by cellulose whiskers from grass. Composites: Part B *40*, 676–680.

Paralikar, S.A., Simonsen, J. and Lombardi, J. (2008). Poly(vinyl alcohol)/cellulose nanocrystals barrier membranes. J. Membrane Sci. *320*, 248–258.

Park, I., Kang, M., Kim, H.S. and Jin, H.J. (2007). Electrospinning of poly(ethylene oxide) with bacterial cellulose whiskers. Macromol. Symp. *249–250*, 289–294.

Pasquini, D., Teixeira, E.M., Curvelo, A.A.S., Belgacem, M.N. and Dufresne, A. (2010). Extraction of starch nanocrystals and cellulose whiskers from cassava bagasse and their applications as reinforcing agent in natural rubber. Ind. Crops Prod. *32*, 486–490.

Peng, K., Wang, B., Chen, S., Zhong, C. and Wang, H. (2011). Preparation and properties of polystyrene/bacteria cellulose nanocomposites by *in situ* polymerization. J. Macromol. Sci. B *50*, 1921–1927.

Peng, Y., Gardner, D.J. and Han, Y. (2012). Drying cellulose nanofibrils: In search of a suitable method. Cellulose *19*, 91–102.

Peresin, M.S., Habibi, Y., Zoppe, J.O., Pawlak, J.J. and Rojas, O.J. (2010). Nanofiber composites of polyvinyl alcohol and cellulose nanocrystals: Manufacture and characterization, Biomacromolecules *11*, 674–681.

Petersson, L. and Oksman, K. (2006). Biopolymer based nanocomposites: comparing layered silicates and microcrystalline cellulose as nanoreinforcement. Compos. Sci. Technol. *66*, 2187–2196.

Plackett, D., Anturi, H., Hedenqvist, M., Ankerfors, M., Gällstedt, M., Lindström, T. and Siró, I. (2010). Physical properties and morphology of films prepared from microfibrillated cellulose and microfibrillated cellulose in combination with amylopectin. J. Appl. Polym. Sci. *117*, 3601–3609.

Podsiadlo, P., Choi, S.Y., Shim, B., Lee, J., Cuddihy, M. and Kotov, N.A. (2005). Molecularly engineered nanocomposites: Layer-by-layer assembly of cellulose nanocrystals. Biomacromolecules *6*, 2914–2918.

Podsiadlo, P., Sui, L., Elkasabi, Y., Burgardt, P., Lee, J., Miryala, A., Kusumaatmaja, W., Carman, M.R., Shtein, M., Kieffer, J. Lahann, J. and Kotov, N.A. (2007). Layer-by-layer assembled films of cellulose nanowires with antireflective properties. Langmuir *23*, 7901–7906.

Pranger, L. and Tannenbaum, R. (2008). Biobased nanocomposites prepared by *in situ* polymerization of furfuryl alcohol with cellulose whiskers or montmorillonite clay. Macromolecules *41*, 8682–8687.

Pu, Y., Zhang, J., Elder, T., Deng, Y., Gatenholm, P. and Ragauskas, A.J. (2007). Investigation into nanocellulosics versus acacia reinforced acrylic films. Composites: Part B *38*, 360–366.

Qi, H., Cai, J., Zhang, L. and Kuga, S. (2009). Properties of films composed of cellulose nanowhiskers and a cellulose matrix regenerated from alkali/urea solution. Biomacromolecules *10*, 1597–1602.

Rämänen, P., Penttilä, P.A., Svedström, K., Maunu, S.L. and Serimaa, R. (2012). The effect of drying method on the properties and nanoscale structure of cellulose whiskers. Cellulose *19*, 901–912.

Raquez, J.M., Murena, Y., Goffin, A.L., Habibi, Y., Ruelle, B., DeBuyl, F. and Dubois, P. (2012). Surface-modification of cellulose nanowhiskers and their use as nanoreinforcers into polylactide: A sustainably-integrated approach. Compos. Sci. Technol. *72*, 544–549.

Rojas, O.J., Montero, G.A. and Habibi, Y. (2009). Electrospun nanocomposites from polystyrene loaded with cellulose nanowhiskers. J. Appl. Polym. Sci. *113*, 927–935.

Roohani, M., Habibi, Y., Belgacem, N.M., Ebrahim, G., Karimi, A.N. and Dufresne, A. (2008). Cellulose whiskers reinforced polyvinyl alcohol copolymers nanocomposites. Eur. Polym. J. *44*, 2489–2498.

Sanchez-Garcia, M.D. and Laragon, J.M. (2010). On the use of plant cellulose nanowhiskers to enhance the barrier properties of polylactic acid. Cellulose *17*, 987–1004.

Sanchez-Garcia, M.D., Hilliou, L. and Laragon, J.M. (2010). Morphology and water barrier properties of nanobiocomposites of k/i-hybrid carrageenan and cellulose nanowhiskers. J. Agric. Food Chem. *58*, 12847–12857.

Saxena, A., Elder, T.J., Pan, S. and Ragauskas, A.J. (2009). Novel nanocellulosic xylan composite film. Composites Part B *40*, 727–730.

Schroers, M., Kokil, A. and Weder, C. (2004). Solid polymer electrolytes based on nanocomposites of ethylene oxide-epichlorohydrin copolymers and cellulose whiskers. J. Appl. Polym. Sci. *93*, 2883–2888.

Sehaqui, H., Salajková, M., Zhou, Q. and Berglund, L.A. (2010). Mechanical performance tailoring of tough ultra-high porosity foams prepared from cellulose I nanofiber suspensions. Soft Matter *6*, 1824–1832.

Sehaqui, H., Zhou, Q. and Berglund, L.A. (2011a). Nanostructured biocomposites of high toughness – A wood cellulose nanofiber network in ductile hydroxyethylcellulose matrix. Soft Matter *7*, 7342–7350.

Sehaqui, H., Zhou, Q. and Berglund, L.A. (2011b). High-porosity aerogels of high specific surface area prepared by new route using nanofibrillated cellulose (NFC) disintegrated from the wood cell wall. Compos. Sci. Technol. *71*, 1593–1599.

Seydibeyoğlu, M.Ö. and Oksman, K. (2008). Novel nanocomposites based on polyurethane and micro fibrilatted cellulose. Compos. Sci. Technol. *68*, 908–914.

Shanmuganathan, K., Capadona, J.R., Rowan, S.J. and Weder, C. (2010a). Biomimetic mechanically adaptive nanocomposites. Prog. Polym. Sci. *35*, 212–222.

Shanmuganathan, K., Capadona, J.R., Rowan, S.J. and Weder, C. (2010b). Bio-inspired mechanically-adaptive nanocomposites derived from cotton cellulose whiskers. J. Mater. Chem. *20*, 180–186.

Shanmuganathan, K., Capadona, J.R., Rowan, S.J. and Weder, C. (2010c). Stimuli-responsive mechanically adaptive polymer nanocomposites. ACS Appl. Mater. Interfaces *2*, 165–174.

Shibata, M. and Nakai, K. (2010). Preparation and properties of biocomposites composed of bio-based epoxy resin, tannic acid, and microfibrillated cellulose. J. Polym. Sci. B *48*, 425–433.

Shimazaki, Y., Miyazaki, Y., Takezawa, Y., Nogi, M., Abe, K., Ifuku, S. and Yano, H. (2007). Excellent thermal conductivity of transparent cellulose nanofiber/epoxy resin nanocomposites. Biomacromolecules *8*, 2976–2978.

Siqueira, G., Abdillahi, H., Bras, J. and Dufresne, A. (2010). High reinforcing capability cellulose nanocrystals extracted from *Syngonanthus nitens* (capim dourado). Cellulose *17*, 289–298.
Siqueira, G., Tapin-Lingua, S. Bras, J., da Silva Perez, D. and Dufresne, A. (2011). Mechanical properties of natural rubber nanocomposites reinforced with cellulosic nanoparticles obtained from combined mechanical shearing, and enzymatic and acid hydrolysis of sisal fibers. Cellulose *18*, 57–65.
Suryanegara, L., Nakagaito, A.N. and Yano, H. (2009). The effect of crystallization of PLA on the thermal and mechanical properties of microfibrillated cellulose-reinforced PLA composites. Compos. Sci. Technol. *69*, 1187–1192.
Suryanegara, L., Okumura, H., Nakagaito, A.N. and Yano, H. (2011). The synergetic effect of phenylphosphonic acid zinc and microfibrillated cellulose on the injection molding cycle time of PLA composites. Cellulose *18*, 689–698.
Svagan, A.J., Azizi Samir, M.A.S. and Berglund, L.A. (2007). Biomimetic polysaccharide nanocomposites of high cellulose content and high toughness. Biomacromolecules *8*, 2556–2563.
Svagan, A.J., Azizi Samir, M.A.S. and Berglund, L.A. (2008). Biomimetic foams of high mechanical performance based on nanostructured cell walls reinforced by native cellulose nanofibrils. Adv. Mat. *20*, 1263–1269.
Svagan, A.J., Hedenqvist, M.S. and Berglund, L.A. (2009). Reduced water vapour sorption in cellulose nanocomposites with starch matrix. Compos. Sci.. Technol. *69*, 500–506.
Svagan, A.J., Jensen, P., Dvinskikh, S.V., Furó, I. and Berglund, L.A. (2010). Towards tailored hierarchical structures in cellulose nanocomposite biofoams prepared by freezing/freeze-drying. J. Mat. Chem. *20*, 6646–6654.
Svagan, A.J., Berglund, L.A. and Jensen, P. (2011). A cellulose nanocomposite biopolymer foam – Hierarchical structure effects on energy absorption. ACS Appl. Mater. Interfaces *3*, 1411–1417.
Tang and Weder (2010). Cellulose whisker/epoxy resin nanocomposites. ACS Appl. Mater. Interfaces *2*, 1073–1080.
Teixeira, E.M., Pasquini, D., Curvelo, A.A.S., Corradini, E., Belgacem, M.N. and Dufresne, A. (2009). Cassava bagasse cellulose nanofibrils reinforced thermoplastic cassava starch. Carbohydr. Polym. *78*, 422–431.
Trovatti, E., Oliveira, L., Freire, C.S.R., Silvestre, A.J.D., Neto, C.P., Cruz Pinto, J.J.C. and Gandini, A. (2010). Novel bacterial cellulose-acrylic resin nanocomposites. Compos. Sci. Technol. *70*, 1148–1153.
Uddin, A.J., Araki, J. and Gotoh, Y. (2011). Toward "strong" green nanocomposites: Polyvinyl alcohol reinforced with extremely oriented cellulose whiskers. Biomacromolecules *12*, 617–624.
van den Berg, O., Capadona, J.R. and Weder, C. (2007a). Preparation of homogeneous dispersions of tunicate cellulose whiskers in organic solvents. Biomacromolecules *8*, 1353–1357.
van den Berg, O., Schroeter, M., Capadona, J.R. and Weder, C. (2007b). Nanocomposites based on cellulose whiskers and (semi)conducting conjugated polymers. J. Mater. Chem. *17*, 2746–2753.
Viet, D., Beck-Candanedo, S. and Gray, D.G. (2007). Dispersion of cellulose nanocrystals in polar organic solvent. Cellulose *14*, 109–113.
Wågberg, L. Decher, G., Norgren, M., Lindström, T., Ankerfors, M. and Axnäs, K. (2008). The build-up of polyelectrolyte multilayers of microfibrillated cellulose and cationic polyelectrolytes. Langmuir *24*, 784–795.
Walther, A., Timonen, J.V.I., Diez, I., Laukkanen, A. and Ikkala, O. (2011). Multifunctional high-performance biofibers based on wet-extrusion of renewable native cellulose nanofibrils. Adv. Mat. *23*, 2924–2928.
Wang, B. and Sain, M. (2007a). Dispersion of soybean stock-based nanofiber in a plastic matrix. Polym. Inter. *56*, 538–546.
Wang, B. and Sain, M. (2007b). Isolation of nanofibers from soybean source and their reinforcing capability on synthetic polymers. Compos. Sci. Technol. *67*, 2521–2527.

Wang, B. and Sain, M. (2007c). The effect of chemically coated nanofiber reinforcement on biopolymer based nanocomposites. BioResources *2*, 371–388.
Wang, H. and Roman, M. (2011). Formation and properties of chitosan-cellulose nanocrystal polyelectrolyte-macroion complexes for drug delivery applications. Biomacromolecules *12*, 1585–1593.
Wang, Y., Cao, X. and Zhang, L. (2006). Effects of cellulose whiskers on properties of soy protein thermoplastics. Macromol. Biosci. *6*, 524–531.
Wang, Y., Tian, H. and Zhang, L. (2010a). Role of starch nanocrystals and cellulose whiskers in synergistic reinforcement of waterborne polyurethane. Carbohydr. Polym. *80*, 665–671.
Wang, Y., Chang, C. and Zhang, L. (2010b). Effects of freeze/thawing cycles and cellulose nanowhiskers on structure and properties of biocompatible starch/PVA sponges. Macromol. Mat. Eng. *295*, 137–145.
Woehl, M.A., Canestraro, C.D., Mikowski, A., Sierakowski, M.R., Ramos, L.P. and Wypych, F. (2010). Bionanocomposites of thermoplastic starch reinforced with bacterial cellulose nanofibres: Effect of enzymatic treatment on mechanical properties. Carbohydr. Polym. *80*, 866–873.
Xiang, C., Joo, Y.L. and Frey, M.W. (2009). Nanocomposite fibers electrospun from poly(lactic acid)/ cellulose nanocrystals. J. Biobased Mater. Bioenergy *3*, 147–155.
Xu S.H., Gu J., Luo Y.F. and Jia D.M. (2012). Effects of partial replacement of silica with surface modified nanocrystalline cellulose on properties of natural rubber nanocomposites. eXPRESS Polym. Lett. *6*, 14–25.
Yano, H. and Nakahara, S. (2004). Bio-composites produced from plant microfiber bundles with a nanometer unit web-like network. J. Mat. Sci. *39*, 1635–1638.
Zimmermann, T., Pöhler, E. and Geiger, T. (2004). Cellulose fibrils for polymer reinforcement. Adv. Eng. Mat. *6*, 754–761.
Zimmermann, T., Pöhler, E. and Schwaller, P. (2005). Mechanical and morphological properties of cellulose fibril reinforced nanocomposites. Adv. Eng. Mat. *7*, 1156–1161.
Zimmermann, T., Bordeanu, N. and Strub E. (2010). Properties of nanofibrillated cellulose from different raw materials and its reinforcement potential. Carbohydr. Polym. *79*, 1086–1093.
Zoppe, J.O., Peresin, M.S., Habibi, Y., Venditti, R.A. and Rojas, O.J. (2009). Reinforcing poly(e-caprolactone) nanofibers with cellulose nanocrystals. ACS Appl. Mater. Interfaces *1*, 1996–2004.

8 Thermal properties

Thermal properties of materials are of importance for processing issues and some practical uses. In this chapter the thermal properties of both cellulose nanoparticles and related polymer nanocomposites are reported. The main thermal characteristics of polymeric systems are the glass-rubber transition, melting point and thermal stability. The effect of cellulosic nanoparticles on these parameters as well as crystallization of the polymeric matrix are also reported.

8.1 Thermal expansion of cellulose

When a material is heated, the motion of its constituent atoms increases and a higher average separation results. The rate of expansion as a function of temperature corresponds to the thermal expansion coefficient (TEC) which generally varies with temperature. Therefore, it measures the relative volumetric, area or linear change per degree.

8.1.1 Thermal expansion coefficient of cellulose crystal

X-ray diffraction patterns obtained at different temperatures can be used to determine the TEC of cellulose crystal. From the positions of the diffraction peaks, the d-spacing and unit cell parameters can be calculated. The linear and volume TECs, α and β, are then determined as follow:

$$\alpha = \frac{1}{\ell}\frac{\Delta\ell}{\Delta T}$$

$$\beta = \frac{1}{V}\frac{\Delta V}{\Delta T} \tag{8.1}$$

where ℓ is either the d-spacing or unit cell parameter, V is the low-temperature volume of the unit cell, and T is the temperature (in °C or K). This method has been applied to different cellulose sources (Wakelin et al., 1960; Seitsonen and Mikkonen, 1973; Takahashi and Takenaka, 1982; Wada, 2002; Hori and Wada, 2005; Langan et al., 2005; Hori and Wada, 2006).

The thermal behavior of oriented films of highly crystalline cellulose Iβ and III_I were investigated using X-ray diffraction between room temperature and 250°C (Wada, 2002). It was shown that cellulose Iβ underwent a transition into the high-temperature phase when increasing the temperature above 220–230°C, while cellulose III_I was transformed into cellulose Iβ when the sample was heated above 200°C.

For cellulose Iβ, the TEC of the a axis increased linearly from $\alpha_a = 4.3 \cdot 10^{-5}\,K^{-1}$ at room temperature to $\alpha_a = 17.0 \cdot 10^{-5}\,K^{-1}$ at 200°C, while the TEC of the b axis remained constant at $\alpha_b = 0.5 \cdot 10^{-5}\,K^{-1}$. Cellulose III_I also showed an anisotropic thermal expansion in the lateral direction, and the TECs of the a and b axes were $\alpha_a = 7.6 \cdot 10^{-5}\,K^{-1}$ and $\alpha_b = 0.8 \cdot 10^{-5}\,K^{-1}$, respectively. The anisotropic thermal expansion behaviors in the lateral direction for Iβ and III_I were found to be closely related to the intermolecular hydrogen-bonding systems.

Tension wood cellulose obtained from *Populus maximowiczii* was also investigated at temperatures ranging from room temperature to 250°C (Hori and Wada, 2005). Three equatorial and one meridional d-spacings showed a gradual linear increase with increasing temperature. For temperatures above 180°C, however, the equatorial *d*-spacing increased dramatically. The linear TECs below 180°C of the a, b, and c axes were: $\alpha_a = 13.6 \cdot 10^{-5}\,K^{-1}$, $\alpha_b = -3.0 \cdot 10^{-5}\,K^{-1}$, and $\alpha_c = 0.6 \cdot 10^{-5}\,K^{-1}$, respectively, and the volume TEC was $\beta = 11.1 \cdot 10^{-5}\,K^{-1}$. The anisotropic thermal expansion in the three coordinate directions was closely related to the crystal structure of the wood cellulose, and it governed the macroscopic thermal behavior of solid wood.

Highly crystalline samples of cellulose II and III_{II} have been prepared from repeated mercerization of ramie fibers and supercritical ammonia treatment of the mercerized ramie fibers, respectively (Hori and Wada, 2006). The thermal expansion behavior of these samples was investigated using X-ray diffraction at temperatures ranging from room temperature to 250 °C. With increasing temperature, the unit cell of cellulose II expanded in the lateral direction and contracted in the longitudinal direction, with the a and b axes increasing by 0.54 and 3.4%, respectively, and the c axis decreasing by 0.09%. The anisotropic thermal expansion in these three directions was closely related to the crystal structure and the hydrogen bonding in cellulose II. A similar anisotropic thermal expansion was also observed for cellulose III_{II}. Cellulose III_{II} expanded in the lateral direction but contracted in the longitudinal direction.

Synchrotron X-ray data have been collected to 1.4 Å for fibers of cellulose Iβ and regenerated cellulose II (Fortisan) at ambient temperature and at 100 K (Langan et al., 2005). It was found that the unit cell of regenerated cellulose II contracted, with decreasing temperature, by 0.25%, 0.22% and 0.1% along the a, b, and c axes, respectively, whereas that of cellulose Iβ contracted only in the direction of the a axis, by 0.9%. The value of $4.6 \cdot 10^{-5}\,K^{-1}$ for the TEC of cellulose Iβ in the a axis direction was explained by simple harmonic molecular oscillations and the lack of hydrogen bonding in this direction. The molecular conformations of each allomorph were essentially unchanged by cooling to 100 K.

The influence of temperature on structure and properties of the cellulose Iβ crystal was also studied by molecular dynamics simulations with the GROMOS 45a4 force-field (Bergenstråhle et al., 2007). When the temperature increased from 300 K to 500 K, it was found that the density decreased while the a and b parameter cells expanded from 7.4% and 6%, respectively, and the c parameter (chain axis) slightly contracted by 0.5%.

8.1.2 Thermal expansion coefficient of nanocellulose films

At a different scale of observation, the TEC of cellulose nanoparticle films can be investigated. The experimental determination of the TEC can be obtained by subjecting the film to a constant uniaxial stress and following the resulting strain as a function of temperature using a thermomechanical analyzer.

Influence of acetylation on the TEC of bacterial cellulose (BC) sheets in the temperature range 20–150°C has been investigated (Ifuku et al., 2007). It was found that the TEC decreased to values lower than 10^{-6} K^{-1} when increasing the acetyl group content up to a DS value of 0.6. It was hypothesized that the surface and amorphous region of BC became preferentially acetylated and therefore less mobile by the introduction of bulky acetyl groups, while the crystallinity remained unchanged. It induced a restriction of the thermal expansion. For higher DS values, the TEC increased because of the reduction of the degree of crystallinity of BC nanofibers upon extensive acetylation. TEC of pre-dried TEMPO-oxidized MFC films was determined at 0.03 N load in a nitrogen atmosphere from 28 to 102°C (Fukuzumi et al., 2009). A value of $2.7 \cdot 10^{-6}$ K^{-1} was reported, which is much lower than that of glass (about $9 \cdot 10^{-6}$ K^{-1}). The TEC of MFC sheets was found to increase as the degree of fibrillation increases (Iwamoto et al., 2007). No significant different was found when using never-dried or once-dried pulp for the preparation of MFC (Iwamoto et al., 2008).

8.1.3 Thermal expansion coefficient of nanocellulose-based composites

The low TEC of cellulose nanoparticles can be used to reduce the TEC of polymer sheets. The linear thermal expansion was studied for all-cellulose composites obtained by a wet process by controlling the solubility of cellulose (Nishino et al., 2004). Compared to the cellulose matrix, which showed a linear TEC of $1.4 \cdot 10^{-5}$ K^{-1}, the all-cellulose composite exhibited almost no thermal expansion or contraction. The TEC value of the composite was about 10^{-7} K^{-1} (0.1 ppm $\cdot$ K^{-1}), which is much lower than those of metals, e.g. $11.8 \cdot 10^{-6}$ K^{-1} for Fe or $2.49 \cdot 10^{-6}$ K^{-1} for Si, and comparable to quartz. The TEC of a BC sheet impregnated with epoxy resin or phenol formaldehyde between 50°C and 150°C was reported to be $6 \cdot 10^{-6}$ K^{-1} and $3 \cdot 10^{-6}$ K^{-1}, respectively (Yano et al., 2005). It is much lower than the value reported for the neat epoxy resin sheet ($1.2 \cdot 10^{-4}$ K^{-1}).

The TEC of MFC mats immersed in acrylic resin was measured between 20°C and 150°C during elongation with a heating rate of 5°C $\cdot$ min^{-1} in a nitrogen atmosphere at a load of 3 g (Iwamoto et al., 2005). A value of $1.7 \cdot 10^{-5}$ K^{-1}, two times lower than for BC-based composites (Yano et al., 2005), and much lower than for the neat acrylic resin sheet ($8.6 \cdot 10^{-5}\ K^{-1}$) was reported. The TEC of MFC reinforced phenol formaldehyde resin was found to decrease significantly from $6 \cdot 10^{-5}\ K^{-1}$ for the neat matrix to $10^{-5}\ K^{-1}$ when adding 40 wt% MFC and to stabilize at this value for higher filler content (Nakagaito and Yano, 2008). Similar results were obtained for acrylic resin reinforced with BC (Nogi et al., 2006). Moreover, acetylation of the BC sheet still reduced the TEC of the composite. This decrease of the TEC value originates from the much higher modulus of MFC compared to the neat matrix, low intrinsic value of the TEC of cellulose and filler/matrix interactions (Nakagaito and Yano, 2008). However, no influence of the size of the reinforcing cellulosic phase was observed. A good correlation between the TEC and inverse of the Young's modulus was reported for MFC reinforced phenol formaldehyde resin as shown in Figure 8.1. Moreover, it was shown that whereas an acrylic resin sheet expanded by 3.1% over a temperature range from 20°C to 150°C, introduction of only 5 wt% BC reduced the planar thermal expansion to 0.05% over the same temperature range (Nogi and Yano, 2008). The corresponding TEC were $2.45 \cdot 10^{-4}\ K^{-1}$ and $4 \cdot 10^{-6}\ K^{-1}$, respectively. However, the TEC of the nanocomposite in the thickness direction was found to be $5.55 \cdot 10^{-4}\ K^{-1}$ (expansion of 7.2% from 20°C to 150°C). This anisotropic behavior was ascribed to the formation of a dense and finely branched nanofiber network in the planar direction and weakly interacting multiple layers of planar nanofiber networks in the thickness direction. However, in a study involving nine types of matrix resin, it was shown that at the same MFC content the nanocomposites using lower Young's modulus matrix resin exhibited lower CTE values than using higher Young's modulus matrix resins (Okahisa et al., 2009).

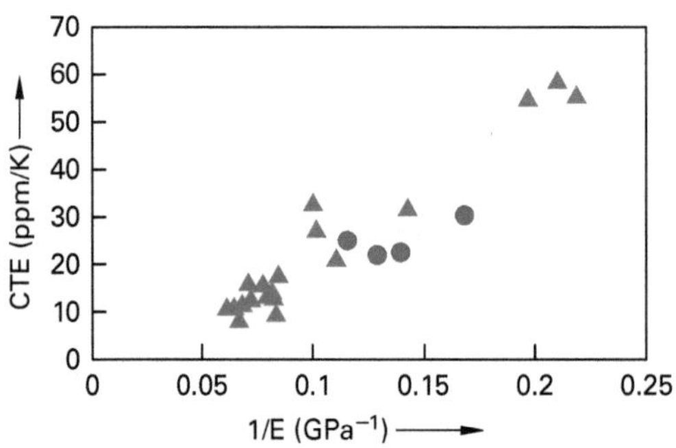

Fig. 8.1: Linear relationship between thermal expansion coefficient and inverse of Young's modulus for MFC (▲) and pulp (●) reinforced phenol formaldehyde resin (Nakagaito and Yano, 2008).

An important parameter to restrict the TEC of composites is the strength of the cellulose nanofiber network interactions. These interactions need to be strong enough to restrict the thermal expansion of the matrix. It was confirmed for acetylated BC impregnated with an acrylic resin (Ifuku et al., 2007). The crystallinity of the cellulose nanofibers is also an important parameter. The thermal expansion of sheets prepared from fibrillated pulp fibers with different crystallinity and impregnated with acrylic resin was investigated (Iwamoto et al., 2007). The crystallinity of MFC was increased by treating the nanofibers with concentrated acid to hydrolyze the amorphous cellulose domains. Compared to untreated fibrillated pulp, reduced thermal expansion was reported for composite sheets prepared from hydrolyzed cellulose.

8.2 Thermal conductivity of nanocellulose-based nanocomposites

Literature is scarce in data regarding the thermal conductivity of nanocellulose. Data have been reported for commercial nanofibers (Daicel Chemical Industries) reinforced epoxy resin nanocomposites (Shimazaki et al., 2007). A sheet of nanofibers was obtained by filtration and immersed in the epoxy resin containing a curing agent. After curing, the thermal diffusibility α and the thermal conductivity ($\lambda = \alpha \cdot \rho \cdot Cp$, where ρ and Cp are the density and specific heat capacity of the sample at constant pressure, respectively) of the nanocomposite sheet were measured. The in-plane thermal conductivity λ_{xy} of the nanocomposite sheet was 1.1 and 0.7 $W \cdot m^{-1} \cdot K^{-1}$ for a nanofiber content of 58 wt% and 39 wt%, respectively, whereas it was only 0.15 $W \cdot m^{-1} \cdot K^{-1}$ for the neat epoxy resin. The transverse thermal conductivity λ_z was almost the same for the nanocomposite sheet and neat epoxy resin because of the in-plane orientation of nanofibers that occurred during the preparation of the nanocomposite sheet. The enhanced thermal conductivity was attributed to the better phonon-transporting efficiency of cellulose nanofibers in the resin.

The thermal conductivity of polypropylene (PP) reinforced with cellulose nanocrystals coated with maleic anhydride grafted PP (MAPP) was also reported (Bahar et al., 2012). The presence of both MAPP and cellulose nanoparticles increased the thermal conductivity of the sample.

8.3 Thermal transitions of cellulose nanoparticles

Thermal transitions in polymers are generally investigated using differential scanning calorimetry (DSC). During DSC experiments the difference in the heat flow required to increase or decrease the temperature of a sample and reference is measured as a function of time or temperature. The temperature is varied linearly as a function of time and both the sample and reference are maintained at the same temperature. Exothermic and endothermic transitions in the sample can be detected. Differential thermal analysis (DTA) is an alternative technique in which the heat flow is maintained constant and differences in temperature between the sample and reference are measured as a function of temperature.

Hydrophilic polymers like cellulose always contain some amount of water. Three different types of water are generally distinguished, viz. free (or not bonded) water, freezing-bonded water and bonded (non-freezing) water. Free water behaves like normal water in terms of freezing and melting, whereas the freezing-bonded water exhibits considerable supercooling effect. The strongly bonded water molecules do not freeze, but change the physical and chemical properties of cellulose. An endothermic peak can be observed around 80–120°C in the DSC thermogram of cellulose nanoparticles (de Menezes et al., 2009). It is ascribed to the vaporization of water. The magnitude of this endothermic peak is much lower when performing a second temperature scan after heating and cooling the sample. Moreover, it shifts towards higher temperatures (Szcześniak et al., 2008).

It is well known that the temperature of thermal transitions depends on the degree of crystallinity of the polymer and heating rate used for the experiment. When cellulose nanoparticles are subjected to a heating rate, no thermal event is observed in the DSC trace before their degradation. The degradation of cellulose occurs around 220°C but its color already starts to become yellow-brown at 180°C under ambient conditions because of a loss of chemical stability. The glass transition and melting point of dry cellulose usually observed for semicrystalline polymers occur at temperatures higher than the degradation temperature because of the highly cohesive nature of this biopolymer. This is ascribed to strong H-bonding interactions between cellulose chains. Extrapolation of the glass transition temperature (Tg) dependence to zero concentration of solvent, yields values for dry cellulose of 220–250°C (Goring, 1963; Akim, 1977; Back and Salmén, 1982; Kalaschnik et al., 1991).

Furthermore, the melting point (Tm) of cellulose has been estimated from its glass transition temperature (Tg = 230°C) to be roughly 450°C using the empirical correlation Tg = 0.7 · Tm(K) (Nordin et al., 1973). However, melting effects were produced on the surface of cellulose fibers by applying a rapid heating of the sample with a continuous CO_2 laser allowing a short heating time of 0.1 ms to 500°C. Formation of heating bubbles and disappearance of the fibrillar structure were observed.

For polymer-grafted cellulose nanoparticles melting endotherms can be observed (Habibi and Dufresne, 2008; Habibi et al., 2008; Lönnberg et al., 2008; de Menezes et al., 2009; Cao et al., 2009; Lin et al., 2009). This thermal transition is not attributed to cellulose but to the grafted polymeric chains that form a crystalline brush-like structure at the surface of the nanoparticles. Compared to the typical values of the melting point reported for the free polymer, the temperature position of this melting endotherm is lower because of the restricted size of these crystallites. As the molecular weight of the grafted chains increased, their melting temperature increased as shown for polycaprolactone (PCL)-grafted MFC (Lönnberg et al., 2008). Because of the restricted mobility of grafted chains compared to free polymer chains, a lower degree of crystallinity and longer crystallization time were observed (Lönnberg et al., 2008). For cellulose nanocrystals grafted with thermoresponsive poly(N-isopropylacrylamide), the glass transition of the grafted polymer was observed (Zoppe et al., 2010). It occurred at higher temperature (143°C) than for the bulk polymer (130°C) because of restricted mobility.

8.4 Thermal stability of cellulose nanoparticles

Thermal stability of cellulose nanoparticles is a crucial factor in order to gauge their applicability for biocomposite processing, as melt processing requires elevated temperatures. For thermoplastic polymers, it rises above 200°C. Indeed, compared to inorganic fillers, many biomaterials suffer from inferior thermal properties. It is therefore important to verify that the preparation conditions and potential alterations of the surface chemistry do not change the onset temperature of thermal degradation. Thermogravimetric analysis (TGA) is typically performed by heating the sample from room temperature to 600°C at a given heating rate (typically $10°C \cdot min^{-1}$) in a nitrogen, air or helium atmosphere and recording the weight loss as a function of temperature. Use of an inert atmosphere avoids thermo-oxidation degradation. The derivative traces (DTG) are best suited to determine the decomposition temperatures while the weight traces can be used to measure the weight loss associated with the different degradation processes.

Upon chemical purification, cellulose fibers exhibit an increased thermal stability because of the removal of hemicelluloses, lignin and pectins, which degrade at lower temperatures. TGA curves of cellulose generally show an initial small weight drop between 50°C and 150°C, which corresponds to absorbed water and residual moisture. This weight loss can be used to determine the water content of the material.

8.4.1 Thermal degradation of cellulose

The first step in the thermal degradation of cellulose is a cleavage of the macromolecules or depolymerization producing alkali-soluble products (Fengel and Wegener, 1984). This takes place when the cellulose structure has absorbed enough energy to activate the cleavage of the glycosidic linkage to produce glucose. Moreover, cellulose is principally responsible for the production of flammable volatiles during the thermal degradation of wood. Indeed, by raising the temperature above 200°C the thermal degradation of cellulose and the formation of volatile products proceed rapidly. Degradation of cellulose also occurs through dehydration, oxidation reactions, decarboxylation and transglycosylation. It can be accelerated in the presence of water, acids and oxygen. Chain cleavage and dehydration follow a zero-order reaction, whereas oxidation is a first-order reaction (Hernádi, 1976). Heating in air causes oxidation of the hydroxyl groups resulting in an increase of carbonyl and subsequently of carboxyl groups. As the temperature increases, the degree of polymerization of cellulose decreases further, free radicals appear and carbonyl, carboxyl and hydroperoxide groups are formed. Thermal degradation rates increase as heating continues. Glucose resulting from depolymerization is mainly dehydrated to levoglucosan (1,6-anhydro-β-D-glucopyranose), but other anhydroglucoses (1,2-, 1,4-anhydroglucose, 1,6-anhydroglucofuranose, enones), furan and furan derivatives are also produced. During pyrolysis, water and acids are produced. As temperature increases to around 450°C, the production of volatile compounds is complete and the continuing weight loss is due to degradation of the remaining char.

The reactions taking place during the thermal degradation of cellulose and other polysaccharides have been classified into the following categories (Shafizadeh and DeGroot, 1976):

- Depolymerization of the polysaccharides by transglycosylation at about 300°C to provide a mixture of levoglucosan, other monosaccharide derivatives, and a variety of randomly linked oligosaccharides. This mixture is generally referred to as the tar fraction.
- The above reactions are accompanied by dehydration of sugar units in cellulose. These give unsaturated compounds, including 3-deoxyglucosenone, levoglucosenone, furfural and a variety of furan derivatives which are found partly in the tar fraction and partly among the volatiles.
- At somewhat higher temperatures, fission of sugar units provides a variety of carbonyl compounds, such as acetaldehyde, glyoxal and acrolein, which readily evaporate.
- Condensation of the unsaturated products and cleavage of the side chains by means of free radical mechanisms leave a highly reactive carbonaceous residue containing trapped free radicals.

Decomposition of cellulose crystallites in highly crystalline tension wood was studied by X-ray diffraction (Kim et al., 2001a). With one-hour isothermal treatments, the cellulose crystallites did not decompose at 300°C, but completely decomposed at 340°C.

8.4.2 Thermal stability of microfibrillated cellulose

Degradation temperatures reported in the literature for MFC prepared from different cellulosic sources are collected in Table 8.1. When available, the degradation temperature of the starting purified cellulose fibers used for the production of MFC is also reported. This list is not exhaustive.

The thermal stability of MFC produced from wheat straw and soy hulls fibers was investigated using TGA in a nitrogen environment (Alemdar and Sain, 2008). Compared to untreated and chemically treated fibers, the degradation temperature increased and reached 290°C. Moreover, the char content was found to decrease upon fibrillation. It was ascribed to the mechanical treatment that induced removal of calcium oxalate crystals present in the starting raw fiber and to a lesser extent in the

Source of Cellulose	Atmosphere	Before Fibrillation	Chemical Modification	After Fibrillation	Method	Reference
Bleached Sulfite Wood Pulp	Nitrogen	–	–	355	T_p	(Lönnberg et al., 2008)
	Helium	–	–	290	$T_{5wt\%}$	(Tingaut et al., 2010)
			Acetylated 1.5%	309		
			Acetylated 4.5%	324		
			Acetylated 8.5%	338		
			Acetylated 10.5%	345		
			Acetylated 17%	330		
	Air	–	–	250	T_o	(Nyström et al., 2010)
Kraft Wood Pulp	Nitrogen	–	–	280/350	T_o/T_p	(Lu et al., 2008a)
			Titanate Treatment	250	T_o	
			–	365	T_p	(Lu et al., 2008b)
Refined Bleached Beech Pulp	Nitrogen	250/345	–	250/345	T_o/T_p	(Eyholzer et al., 2010)
			Carboxy-methylation	200/300	T_o/T_p	
Soy Hulls Fibers	Nitrogen	–	–	290	T_o	(Alemdar and Sain, 2008)
Wheat Straw	Nitrogen	232	–	296	T_o	
		276/346	–	283/338	T_o/T_p	(Kaushik et al., 2010)
Wood Bleached Kraft Pulp	Nitrogen	300	TEMPO oxidation	200	T_o	(Fukuzumi et al., 2009)

$T_{5wt\%}$: 5 wt% weight loss temperature
T_o : onset temperature of degradation
T_p : DTG peak temperature

Table 8.1: Thermal degradation temperature of microfibrillated cellulose.

purified cellulose fibers. The increase of the thermal stability after the defibrillation treatment was not so clear for wheat straw (Kaushik et al., 2010) and refined bleached beech pulp (Eyholzer et al., 2010). Also, the fibrillation treatment induced a reduction of the char content because of the removal of non-organic components (Kaushik et al., 2010).

The thermal stability can also be affected by the surface chemistry of MFC. It was shown that the thermal degradation of TEMPO-oxidized MFC started around 200°C in a nitrogen atmosphere, while degradation began at 300°C for the original cellulose (Fukuzumi et al., 2009). It was concluded that the formation of sodium carboxylate groups from the C6 primary hydroxyls of MFC surface by TEMPO-mediated oxidation significantly decreased the thermal degradation temperature. Similar results were obtained for carboxymethylated MFC prepared from refined bleached beech pulp (Eyholzer et al., 2010). The opposite effect was reported for acetylated MFC (Tingaut et al., 2010). The degradation temperature was systematically shifted towards higher temperatures for acetylated samples and a gradual increase of the thermal stability was observed when increasing the acetyl group content, up to 10.5%. For the most acetylated sample (17%), a lower degradation temperature was observed and associated with a significant decrease in crystallinity. Preparation of hybrid clay-MFC films reduces the thermal degradation rate of the cellulosic material (Liu et al., 2011). This system showed self-extinguishing characteristics when submitted to open flames and also considerably delayed the thermal degradation of cellulose, primarily because of the favorable gas barrier properties of ordered clay platelets.

8.4.3 Thermal stability of cellulose nanocrystals

Degradation temperatures reported in the literature for nanocrystals prepared from different cellulosic sources are collected in Table 8.2. When available, the degradation temperature of the starting purified cellulose fibers used for the production of nanocrystals is also reported. These values are only indicative since the thermal degradation of cellulose nanocrystals depends on experimental details such as heating scanning rate and hydrolysis conditions used to prepare the nanoparticles. This list is not exhaustive.

However, regardless of the source of cellulose it is clearly seen in Table 8.2 that compared to the raw starting material, cellulose nanocrystals display a significantly reduced thermal stability. This effect is ascribed to the acid hydrolysis step which generally involves sulfuric acid for suspension stability issues. During the hydrolysis reaction, sulfate groups are introduced on the surface of the nanoparticles promoting at the same time improved stability of the aqueous suspension and lowering of the thermal stability. Carbonization of cellulose impregnated with sulfuric acid by immersion of the sample in dilute acid for a few minutes was reported (Kim et al., 2001b). It was shown that cellulose pyrolysis in the presence of sulfate was divided into a low-temperature process between 110°C and 200°C and a high-temperature process between 300°C and 600°C. It was observed that the mass yield of carbon after 800°C treatment in nitrogen increased 2–3 times by addition of small amounts of sul-

Source of Cellulose	Atmosphere	Before Hydrolysis	Chemical Modification	After Hydrolysis	Method	Reference
Banana Fibers	Nitrogen	–	–	230	T_0	(Elanthikkal et al., 2010)
Bio-Residue from Wood Bioethanol Production	Air	248/290	–	122/133	T_0/T_p	(Oksman et al., 2011)
Cassava Bagasse	Air	280	–	220	T_0	(Teixeira et al., 2009)
Coconut Husk	Nitrogen	200	–	120	T_0	(Rosa et al., 2010)
		200–215		135–145		(Fahma et al., 2011)
Cotton	Nitrogen	–	$NaHCO_3$ Neutralization	230	T_0	(Choi and Simonsen, 2006)
			–	250		(Noorani et al., 2007)
				260*	T_p	(Xu et al., 2008)
			PMMAZO Grafting	320*		
			–	150*	T_0	(Yi et al., 2008)
			PS Grafting	270*		
			–	150*		(Yi et al., 2009)
			PDMAEMA Grafting	240*		
			TEMPO Oxidation	225		(Filpponen and Argyropoulos, 2010)
			–	180*		(Ben Azouz et al., 2012)
			PEO Coating	300–340*		
Cotton Linter**	Air	–	–	300	T_0	(Braun and Dorgan, 2009)
			Esterified	300		
Cottonseed Linter	Nitrogen	–	–	200	T_0	(Cao et al., 2009)
Kenaf	Nitrogen	353	–	198–233	T_p	(Kargarzadeh et al., 2012)

Source of Cellulose	Atmos-phere	Before Hydrolysis	Chemical Modification	After Hydrolysis	Method	Reference
MCC	Nitrogen	320*	–	240*	T_0	(Petersson et al., 2007)
			Surfactant Coating	300*		
	–	–	–	200*		(Jiang et al., 2008)
	Nitrogen	316/364	NaOH Neutrali-zation	285/331	T_0/T_p	(Sanchez and Lagaron, 2010)
		310/364		313/331		(Sanchez-Garcia et al., 2010)
		–	–	307	T_p	(Ten et al., 2010)
		–	–	200*	T_0	(Liu and Laborie, 2011)
		300	–	230		(Fortunati et al., 2012)
			Surfactant Coating	200		
Mulberry	Nitrogen	397	–	335	T_p	(Li et al., 2009)
Ramie	Helium	–	–	290*	T_p	(Habibi et al., 2008)
			NaOH Neutrali-zation	340*		
	Nitrogen		–	210/368	T_0/T_p	(de Menezes et al., 2009)
			Aliphatic Chains Grafting	200–242/ 305–321		
	–		–	220/300*	T_0/T_p	(Zoppe et al., 2009)
			PCL Grafting	250/375*		
			–	270	T_0	(Zoppe et al., 2010)
			Poly(NiPAAm) Grafting	270*		
	Air		–	270		(Alloin et al., 2011)
	Helium			265		
Rice Husk	Nitrogen	350*	–	260*	T_p	(Johar et al., 2012)
Sugarcane Bagasse	Nitrogen	–	–	170	T_0	(Bras et al., 2010)
	Air	270		210–255		(Teixeira et al., 2011)

* : estimated from graphical data ** : Extracted with HCl $T_{5wt\%}$: 5 wt% weight loss temperature
T_0 : onset temperature of degradation T_p : DTG peak temperature

Table 8.2: Thermal degradation temperature of cellulose nanocrystals.

furic acid that was considered to act as a dehydration catalyst, thus suppressing the release of volatile organic substances. The shrinkage of the sample during carbonization was also significantly reduced by the addition of sulfuric acid.

A detailed investigation on the thermostability of cellulose nanocrystals acid-hydrolyzed from bacterial cellulose has been reported (Roman and Winter, 2004). Different acid hydrolysis conditions with H_2SO_4 were used to elucidate the relationship between the number of sulfate groups introduced and the thermal degradation behavior. The effect of sulfuric acid concentration, acid-to-pulp ratio, and hydrolysis temperature and time were investigated. The amount of introduced sulfate groups was determined by potentiometric titration with an aqueous NaOH solution and the surface charge density of the nanocrystals was calculated for a specific surface area of 189 $m^2 \cdot g^{-1}$ using an assumed average rectangular-solid particle with dimensions $8 \cdot 40 \cdot 1000$ nm^3. The authors observed that during hydrolysis, the sulfate group content could be increased by increasing the reaction time, reaction temperature, acid concentration and acid-to-pulp ratio. Increasing sulfate groups led to degradation at lower temperatures and a broader temperature range was observed as compared to unhydrolyzed samples as shown in Figure 8.2. Also increasing the number of sulfate groups on nanocrystals increased the amount of charred residue at 350°C, indicating that sulfate groups are flame-retardant in nature.

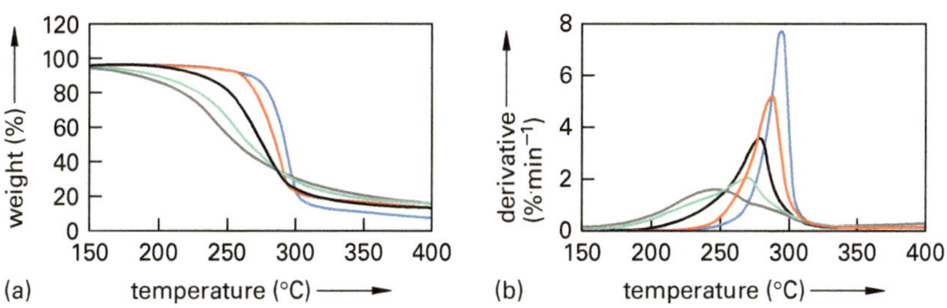

(c)

hydrolysis conditions						
sample	C (% w/v)	R (mmol.g^{-1})	T (°C)	t (h)	n_{SO_3H} (mmol.kg^{-1})	σ (e.nm^{-2})
B	12	188	104	2	2.1	0.007
F	65	66	40	3	10.3	0.033
G	61	618	40	3	50.8	0.162
H	65	66	60	2	73.0	0.233

C: sulfuric acid concentration, R: acid-to-pulp ratio, T: hydrolysis temperature, t: hydrolysis time, n_{SO_3H}: amount of introduced sulfate groups, σ: surface charge density.

Fig. 8.2: (a) Thermogravimetric (TG) and (b) derivative TG curves of unhydrolyzed (—) and sulfuric acid hydrolyzed bacterial cellulose under different conditions specified in panel (c): B (—), F (—), G (—) and H (—) (Roman and Winter, 2004).

The degradation was described as a two-step process, viz. low temperature process and high temperature process (Roman and Winter, 2004). The low temperature process involves the degradation of most accessible amorphous regions which are also highly sulfated. The high temperature process involves the degradation of less accessible interior crystalline regions that are comparatively less sulfated. For highly sulfated samples, an extra step involving the degradation of sulfated ends of crystalline regions was observed. Using the Broido method (Broido, 1969), the authors calculated the activation energies for both degradation processes and concluded that more sulfate groups led to lower activation energies for degradation. It was also suggested that sulfuric acid may be instrumental in catalyzing the degradation process. Direct catalysis by the acidic molecules could be a possible mechanism.

The thermal degradation behavior of spherical cellulose nanocrystals prepared by acid hydrolysis of MCC with mixed acid was investigated (Wang et al., 2007). Profile analysis of derivative TGA traces showed that the two pyrolysis processes consist of multi-step reactions. The influence of cellulose particle size on degradation was also studied. It was shown that the degradation of cellulose for small size particles occurs at lower temperatures and facilitates the char residue formation.

Cellulose nanocrystals prepared by treating MCC with 1-butyl-3-methylimidazolium hydrogen sulfate ionic liquid also displayed a reduced thermal stability compared to MCC (Man et al., 2011). It was ascribed to adhesion of the sulfate group from the ionic liquid onto the surface of the produced nanoparticles. Similarly, the thermal stability of cellulose nanocrystals prepared by swelling MCC in a N,N-dimethylacetamide (DMAc)/lithium chloride (LiCl) solution was found to be lower than for MCC before swelling (Oksman et al., 2006). The thermal stability was accessed by submitting the materials to an isothermal TGA test performed at 180°C for 40 min.

To increase the thermal stability of H_2SO_4-hydrolyzed nanocrystals, neutralization of the nanoparticles by NaOH can be carried out (Wang et al., 2007). A marked increase by 50°C of the degradation temperature for neutralized samples has been observed compared to untreated samples (Habibi et al., 2008). An isothermal TGA test performed at 185°C for 40 min also provided evidence of the higher thermal stability of NaOH-neutralized cellulose nanocrystals (Fortunati et al., 2012).

Improved thermal stability of cellulose nanocrystals was also observed when adsorbing a surfactant (Petersson et al., 2007; Fortunati et al., 2012) or polyoxyethylene (PEO) macromolecules (Ben Azouz et al., 2012) on the surface of the nanoparticles. For PEO-adsorbed cellulose nanocrystals the degradation process was shifted toward higher temperatures and occurred in a narrower temperature range as shown in Figure 8.3. Moreover, this effect was enhanced when increasing the molecular weight of PEO. This phenomenon was ascribed to a protective role of adsorbed macromolecules that hid the surface sulfate groups of cellulose nanocrystals. The cellulose nanocrystals modified with surfactant display a lower weight loss between 150°C and 300°C compared to neat nanoparticles (Fortunati et al., 2012).

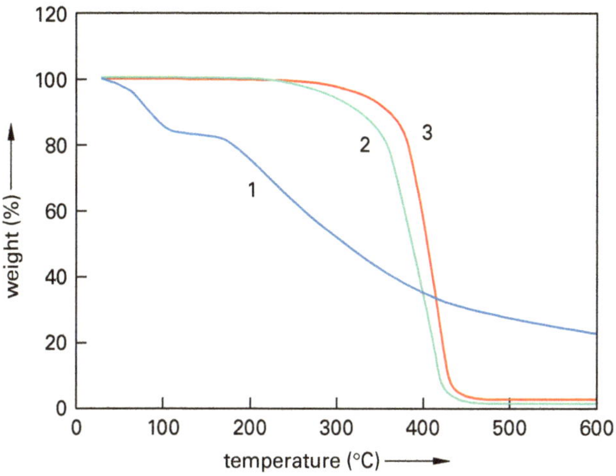

Fig. 8.3: Thermogravimetric curves of freeze-dried cotton cellulose nanocrystals: (1) neat nanocrystals, and PEO-adsorbed nanocrystals with average molecular weight (2) $\overline{M}_w = 3.5 \cdot 10^4\ g \cdot mol^{-1}$ and (3) $\overline{M}_w = 5 \cdot 10^6\ g \cdot mol^{-1}$. The cellulose nanocrystal-to-PEO ratio is 80:20 (Ben Azouz et al., 2012).

Polymer surface chemical grafting also generally improves the thermal stability of cellulose nanocrystals as shown in Table 8.2 probably for similar reasons. This effect has been reported for poly[6-(4-(4-methoxyphenylazo)phenoxyl)hexyl methacrylate] (PMMAZO)- (Xu et al., 2008), polystyrene-(Yi et al., 2008), poly(N,N-dimethylaminoethyl methacrylate) (PDMAEMA)-(Yi et al., 2009), and PCL-grafted nanocrystals (Zoppe et al., 2009).

It was shown by TGA that the presence of acetate or butyrate ester groups on the nanocrystal surface did not affect the thermal stability of cellulose (Braun and Dorgan, 2009). The onset of degradation under air atmosphere remained around 300°C for functionalized nanocrystals. However, in this study cellulose nanocrystals were produced by a one-step acid-catalyzed esterification using a mixture of hydrochloric and acetic (or butyric) acid (and not sulfuric acid like in most studies) to simultaneously promote the hydrolysis of amorphous cellulosic chains and esterification of accessible hydroxyl groups to produce surface functionalized nanocrystals. Similarly, amidation of TEMPO-oxidized cellulose nanocrystals did not change the thermal degradation temperature (Filpponen and Argyropoulos, 2010). However, significant changes were observed in the decomposition profile after click reaction. This material displayed a two-step degradation process, one around 225°C similar to the one observed for the precursors of the click reaction and a second thermal decomposition event at 325°C.

8.4.4 Thermal stability of bacterial cellulose and electrospun fibers

For bacterial cellulose (BC), TGA traces usually show three weight loss stages attributed to the loss of bound water, proteinaceous material from the bacterial cells, and cellulose for increasing temperatures (George et al., 2005; Brown and Laborie, 2007). As for other forms of cellulose, the degradation temperature is around 300°C under nitrogen atmosphere (George et al., 2005; Brown and Laborie, 2007; Peng et al., 2011). The degradation temperature of BC was also found to significantly shift to lower temperatures when organic acid was used to modify its surface (Lee et al., 2011). This shift increased with increasing carbon chain length of the organic acid used. This was thought to be due to the reduction of effective hydrogen bonds between BC nanofibers.

BC, as well as cellulose nanocrystals prepared from tunicin and cotton, was pyrolyzed up to above 2000°C to prepare nanofibrillar carbon structures (Kuga et al., 2002). The surface area of this carbon structure depended on the drying method and was similar to the starting cellulose material.

The thermal stability of cellulose electrospun from an ionic liquid, 1-butyl-3-methylinmidazolium (BMIMCI) was determined and compared to original cellulose and regenerated cellulose (Quan et al., 2010). It was found to be lower for electrospun and regenerated cellulose than for the original cellulose because of reduced degree of crystallinity.

8.5 Glass transition of nanocellulose-based nanocomposites

The glass transition temperature (Tg) is the temperature at which a polymer changes from hard and brittle (glassy state) to soft and ductile (rubbery state). It is probably one of the most important characteristics of polymers since its value conditions different properties of the material, such as its mechanical behavior, matrix chains dynamics and swelling behavior, as well as processing conditions. It can be observed for any polymeric material because even if the polymer can crystallize, some macromolecular chains remain in the amorphous state leading to a semicrystalline state. The glass transition appears as a second order transition since there is no transfer of heat. The heat capacity and volume change to accommodate the increased motion of the wiggling chains, but it does not change discontinuously and hence it is not a thermodynamic transition.

Differential scanning calorimetry (DSC) and dynamic mechanical analysis (DMA) can be used to evaluate Tg of polymers and composites. However, when giving a Tg value determined from DSC experiments attention should be paid to the following points:

- The Tg value depends on the heating/cooling rate used for the experiment. It is shifted by 3–4 K when changing the rate by a factor 10 (for instance when changing the rate from 1 $K \cdot min^{-1}$ to 10 $K \cdot min^{-1}$). The heating/cooling rate used for the experiment should therefore be specified.

- The glass transition occurs over a temperature range. It is therefore important to specify the method which was used to determine the Tg value. For instance, Figure 8.4 shows the different temperatures that could be used. Tg_1, corresponds to the onset of the specific heat increment ascribed to Tg, in others words, the temperature at which some polymer chains start to undergo the transition. Tg_2 is associated with the inflection point of the DSC trace, and Tg_3 corresponds to the offset of the glass-rubber transition, i.e., the temperature at which the DSC curve joins the baseline again.
- The thermal history of the polymer is also an important factor and can influence the Tg value of the polymer. It is ascribed to the well-known structural relaxation or physical aging effect. When a polymer is stored at a temperature lower than its Tg, its structure can evolve to a more thermodynamically stable state. This evolution affects the value of its Tg and gives rise to a characteristic peak which is superimposed to the specific heat increment. This effect can be erased by heating the sample above its Tg and characterizing the heated-treated sample.

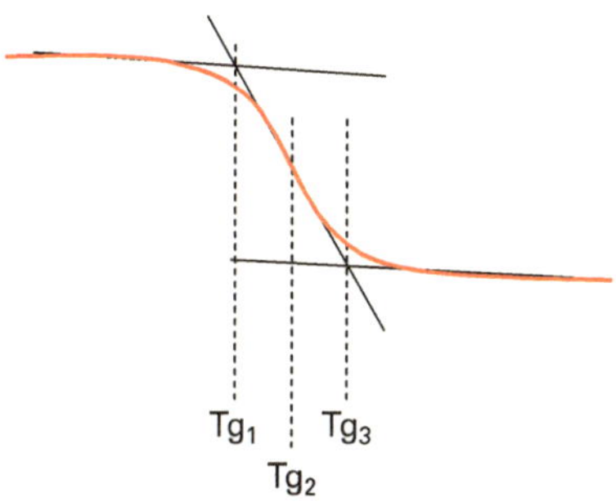

Fig. 8.4: Manifestation of the glass-rubber transition of a polymer in DSC experiments and position of the characteristic temperatures.

In DSC experiments, Tg is generally taken as the inflection point of the specific heat increment at the glass-rubber transition (Tg_2, Figure 8.4). From DMA tests, not a transition but a relaxation process is evidenced in this temperature range. The temperature position of this relaxation process (Tα) depends on Tg, frequency of the measurement and magnitude of the modulus drop associated with Tg (mechanical coupling effect). Its value can be taken as the temperature at the maximum peak of the internal friction factor ($\tan \delta = E''/E'$) or the loss modulus (E"), where E' corresponds to the storage tensile modulus. In shear solicitation, E' and E" are replaced by the storage shear modulus (G') and the loss shear modulus (G"), respectively. The maximum of the loss modulus is close to Tg measured by DSC (with scanning rate around 10 $K \cdot min^{-1}$) when the measurement is performed at 1 Hz. DMA can be a powerful technique to estimate Tg through Tα values for semicrystalline polymers for which the specific heat increment at the glass-rubber transition measured from DSC is generally ill defined.

When preparing nanocomposite materials from a polymeric matrix and cellulose nanoparticles, it is important to verify if the glass transition temperature of the matrix is affected by the presence of the reinforcing phase. The influence of cellulose nanoparticles on the Tg value determined by DSC for different polymer matrices is reported in Tables 8.3 and 8.4. This list is not exhaustive. The first table refers to systems for

Effect	Polymer	Nano-particle	Chemical Modification	Filler Content (wt%)	Reference
No effect	CAB	CNC	–	0–10	(Grunert and Winter, 2002)
			Silylation		
	PEO	CNC	–	0–30	(Azizi Samir et al., 2004a)
		BC		0–59	(Brown and Laborie, 2007)
	PEO-LiTFSI	CNC	–	0–10	(Azizi Samir et al., 2004b)
	Cross-linked PEO	CNC	–	0–6	(Azizi Samir et al., 2004c)
	PCL	CNC	–	0–30	(Habibi and Dufresne, 2008)
			PCL Grafting	0–50	
			–	0–30	(Habibi et al., 2008)
			PCL Grafting	0–40	
	PLA	MFC	–	0–17.3	(Tingaut et al., 2010)
			Acetylation		
		CNC	–	0–5	(Fortunati et al., 2012)
			Surfactant Coating		
	PS	CNC	Surfactant Coating	0–6	(Kim et al., 2009)
	Poly(S-co-BuA)	CNC	–	0–6	(Hajji et al., 1996)
	Poly(S-co-HA)	CNC	–	0–5	(Ben Elmabrouk et al., 2009)
	PVA	MFC	–	0–15	(Lu et al., 2008b)
	PVC/DOP	CNC	–	0–12.4	(Chazeau et al., 1999)
	NR	CNC	–	0–15	(Bendahou et al., 2009)
		MFC			(Bendahou et al., 2010)
		CNC		0–7.5	(Bras et al., 2010)
	Starch/glycerol	CNC	–	0–3.2	(Anglès and Dufresne, 2000)

Table 8.3: Influence of nanocellulose on the glass transition temperature of polymers.

which no alteration of Tg of the matrix was observed when adding the nanoparticles, whereas the second refers to systems for which an alteration was observed. It is worth noting that the Tg value cannot be determined from DSC with a precision better than 1–2 K.

Surprisingly, it can be seen from Table 8.3 that for many systems the addition of cellulose nanoparticles did not affect appreciably the Tg value of the host polymer. This observation is unexpected if one considers the nature and high specific area of cellulose nanocrystals or MFC. Even polymers that display a good affinity for cellulose, like PEO, did not show any increase of Tg resulting from interfacial mobil-

Effect	Polymer	Nano-particle	Chemical Modification	Filler Content (wt%)	Reference
Increase	PCL	CNC	–	0–12	(Siqueira et al., 2009)
			Carbamination		
		MFC	Carbamination		
	Poly (VA-co-VAc)	CNC	–	0–12	(Roohani et al., 2008)
	PVAc	CNC	–	0–10	(Garcia de Rodriguez et al., 2006)
	Starch/glycerol	CNC	–	6.2–25	(Anglès and Dufresne, 2000)
			NaOH Neutralization	0–30	(Lu et al., 2005)
				0–40	(Lu et al., 2006)
				0–30	(Cao et al., 2008a)
					(Cao et al., 2008b)
	Starch/sorbitol	CNC	–	0–15	(Mathew and Dufresne, 2002)
				0–5	(Kvien et al., 2007)
	PU	CNC	–	0–5	(Marcovich et al., 2006)
	WPU	CNC	–	0–10	(Cao et al., 2009)
Decrease	PLA	CNC	PCL Grafting	0–12	(Lin et al., 2009)
			PLA Grafting	0–8	(Goffin et al., 2011a)
	PMMA	CNC	–	0–10	(Liu et al., 2010)
	PS	CNC	Surfactant Coating	0–9	(Rojas et al., 2009)
	Starch/sorbitol	CNC	–	15–25	(Mathew and Dufresne, 2002)
	WPU	CNC	NaOH Neutralization	0–30	(Cao et al., 2007)

Table 8.4: Influence of nanocellulose on the glass transition temperature of polymers.

ity restriction of the polymer chains. Moreover, for tunicin nanocrystals reinforced poly(S-co-BuA) the Tg was also found to be independent of both filler content and processing conditions (Hajji et al., 1996).

However, for some nanocomposite systems an increase of Tg has been observed (Table 8.4). Most of them consist of a plasticized starch matrix.

For glycerol plasticized starch-based composites, peculiar effects of tunicin nanocrystals on Tg of the matrix were reported (Anglès and Dufresne, 2000). It was observed that the unfilled matrix was a complex heterogeneous system composed of glycerol-rich domains dispersed in a starch-rich continuous phase, and each phase exhibited its own Tg. These two Tg values decrease as the moisture content increases owing to the well-known plasticizing effect of water as shown in Figure 8.5. For low nanocrystal contents (up to 3.2 wt%), the same behavior was reported. However, for higher filler contents (6.2 wt% and up), an antiplasticization phenomenon and an increase of the Tg of starch-rich domains were observed. These observations were discussed according to the possible interactions between hydroxyl groups on the cellulosic surface and starch, the selective partitioning of glycerol and water in the bulk starch matrix or at nanocrystal surface, and the restriction of amorphous starch chains mobility in the vicinity of the starch crystallite coated filler surface. Similarly, for glycerol plasticized starch reinforced with cellulose nanocrystals prepared from cottonseed linter (Lu et al., 2005), ramie fibers (Lu et al., 2006), hemp (Cao et al., 2008a) and flax (Cao et al., 2008b), an increase of the Tg of starch-rich domains with filler content was reported and attributed to cellulose/starch interactions. These interactions resulted in a decrease of the flexibility of amorphous starch chains.

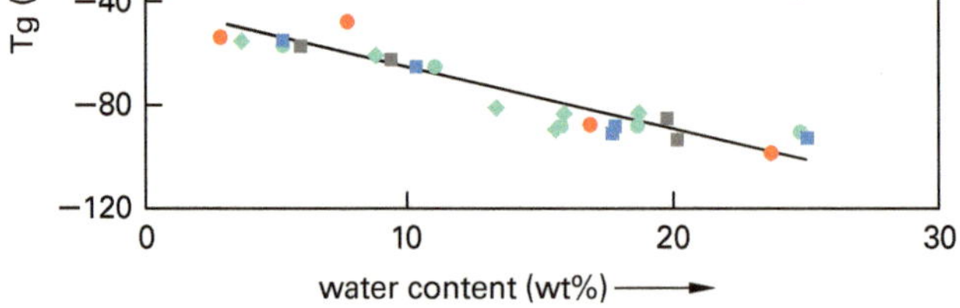

Fig. 8.5: Glass-rubber transition temperatures of glycerol-rich (low temperature) and starch-rich (high temperature) domains versus water content for glycerol plasticized waxy maize starch filled with 0 (●), 3.2 (●), 6.2 (■), 16.7 (■), and 25 wt % (♦) tunicin nanocrystals. Solid lines serve to guide the eye (Anglès and Dufresne, 2000).

For tunicin nanocrystals reinforced sorbitol plasticized starch, a single glass transition was observed (Mathew and Dufresne, 2002). It was associated with the higher molecular weight and lower mobility of sorbitol compared to glycerol. The value of Tg was found to increase slightly up to about 15 wt% nanocrystals and to decrease for higher loading levels. Crystallization of amylopectin chains upon nanocrystal addition and migration of sorbitol molecules to the amorphous domains were proposed to explain the observed modifications. However, for sorbitol plasticized starch reinforced with 5 wt% cellulose nanocrystals, the presence of two glass transitions was reported, whose temperatures increased when adding the nanoparticles (Kvien et al., 2007).

For polyvinyl acetate (PVAc) reinforced with cellulose nanocrystals extracted from sisal fibers, it was shown that the presence of the nanofiller prevented the plasticization of the matrix with water (Garcia de Rodriguez et al., 2006). A much lower depression of Tg was observed for nanocomposites when increasing the moisture content compared to the neat matrix. Therefore, for a given moisture content Tg increased when increasing the nanocrystal content. It was ascribed to limitation of water uptake of the sample because of the formation of a continuous nanoparticle network within the host matrix. Copolymers of PVAc and polyvinyl alcohol (PVA) with different degree of hydrolysis, and then different hydrophilicity, were used as matrix for the processing of cellulose nanocrystal reinforced nanocomposites (Roohani et al., 2008). The samples were conditioned at different relative humidity conditions. In a dry atmosphere, stronger filler/matrix interactions were observed for fully hydrolyzed PVA compared to partially hydrolyzed samples. These stronger interactions induced an increase of the glass temperature. For moist samples, Tg increased significantly upon nanocrystal addition regardless of the degree of hydrolysis of the matrix, because of the formation of a water layer at the interface, the PVA matrix becoming less plasticized by water.

Waterborne polyurethane (WPU)/cellulose nanocrystal composites have been synthesized via in situ polymerization (Cao et al., 2009). The conditions were optimized to induce the grafting of part of the pre-synthesized WPU chains on the surface of cellulose nanocrystals. It was observed that Tg of the soft segments of the WPU matrix increased when increasing the nanoparticle content. The restricted mobility of WPU matrix chains neighboring nanocrystals was attributed to co-crystallization with grafted chains and enhanced interactions between soft and hard segments limiting the microphase separation in WPU. Chemical bonding between cellulose nanocrystals and PU matrix was suggested as the reason for increased Tg in another work (Marcovich et al., 2006).

Both unmodified and chemically modified (nanocrystals and MFC) cellulose nanoparticles were found to increase the Tg of PCL (Siqueira et al., 2009). This effect was more significant with unmodified nanocrystals. Enhanced crystallization of the nanocomposites compared to the neat matrix was suspected to limit the molecular mobility of amorphous PCL chains. For date palm tree MFC reinforced natural rubber (NR), a clear splitting of the main relaxation peak associated with Tg of the NR matrix was observed (Bendahou et al., 2010). It was supposed to be induced by the formation of an interfacial layer surrounding the filler and whose mobility is restricted compared to the bulk matrix.

However, for other systems a decrease of Tg was observed when adding the nanoparticles as shown in Table 8.4. For flax cellulose nanocrystals reinforced WPU, this depression of Tg was assigned to enhanced degrees of freedom of the soft segments in WPU as a result of improved microphase separation between soft and hard segments (Cao et al., 2007). In the case of polystyrene/cellulose nanocrystal composite microfibers prepared by electrospinning, the reduction observed in Tg was explained by the plasticizing effect of the surfactant used to disperse the nanoparticles in tetrahydrofuran (THF) (Rojas et al., 2009). A plasticizing effect of grafted poly(lactic acid) (PLA) chains was also suggested to interpret the decrease in the Tg of the PLA matrix (Goffin et al., 2011a). For PCL-grafted cellulose nanocrystals reinforced PLA the increased mobility of PLA chains was ascribed to the rubbery PCL phase (Lin et al., 2009). The strong reduction observed for Tg of the poly(methyl methacrylate) (PMMA) matrix when adding cellulose nanocrystals was ascribed to an evasive restriction of interactions between the polymer chains from the matrix (Liu et al., 2010).

8.6 Melting/crystallization of nanocellulose-based nanocomposites

In semicrystalline polymeric matrix-based nanocomposites, the melting temperature (Tm), crystallization temperature (Tc) and associated heat of fusion/crystallization (ΔHm/ΔHc) of the thermoplastic matrix can be determined from DSC measurements. Melting/crystallization are first order transitions because a transfer of heat between the polymer and surroundings is involved and the material undergoes an abrupt volume change. Melting and crystallization result in an endothermic and exothermic peak, respectively. X-ray diffraction can also be used as a technique to elucidate the eventual modifications in the crystalline structure of the matrix after cellulose nanoparticle addition.

8.6.1 Melting temperature

Most semicrystalline polymers form lamellae crystals which are 10–30 nm thick and at least one order of magnitude larger in the lateral direction. These lamellae compose larger spheroidal structures named spherulites which correspond to lamellae aggregates when crystallization occurs by cooling the polymer from the melt. The melting point Tm is directly correlated with the thickness d of the lamellae. These two param-

Effect	Polymer	Nano-particle	Chemical Modification	Filler Content (wt%)	Reference
No effect	CAB	CNC	–	0–10	(Grunert and Winter, 2002)
	LDPE	CNC	–	0–15	(de Menezes et al., 2009)
			Aliphatic Chains Grafting		
	PCL	CNC	–	0–12	(Siqueira et al., 2009)
			Carbamination		
		MFC	Carbamination		
		CNC	–	0–30	(Habibi and Dufresne, 2008)
			PCL Grafting	0–50	
			–	0–30	(Habibi et al., 2008)
			–	0–7.5	(Zoppe et al., 2009)
			PCL Grafting		
			–	0–8	(Goffin et al., 2011b)
			PCL Grafting		
	PEO	CNC	–	0–10	(Azizi Samir et al., 2004a)
	PEO-LiTFSI	CNC	–	0–10	(Azizi Samir et al., 2004b)
	PLA	CNC	–	0–2	(Pei et al., 2010)
			Silylation		
	PP	CNC	–	0–6	(Ljungberg et al., 2006)
			MAPP Grafting		
			Surfactant Coating		
			MAPP Grafting	0–15	(Bahar et al., 2012)
	PU	CNC	–	0–2	(Auad et al., 2012)
	PVA	CNC	–	0–30	(Uddin et al., 2011)
	Poly(VA-co-VAc)	CNC	–	0–15	(Peresin et al., 2010)
	Starch/sorbitol	CNC	–	0–25	(Mathew and Dufresne, 2002)
Increase	CAB	CNC	Silylation	0–10	(Grunert and Winter, 2002)
	PU	CNC	PANI Coating	0–15	(Auad et al., 2011)
	Starch/glycerol	CNC	–	0–25	(Anglès and Dufresne, 2000)
	Starch/sorbitol	CNC		0–5	(Kvien et al., 2007)

Effect	Polymer	Nano-particle	Chemical Modification	Filler Content (wt%)	Reference
Decrease	PCL	CNC	PCL Grafting	0–40	(Habibi et al., 2008)
	PEO	CNC	–	10–30	(Azizi Samir et al., 2004a)
		BC		0–59	(Brown and Laborie, 2007)
		CNC		0–30	(Alloin et al., 2011)
	Poly(VA-co-VAc)	CNC	–	0–12	(Roohani et al., 2008)

Table 8.5: Influence of nanocellulose on the melting temperature of polymers.

eters are linked through the Thomson equation which is based on the fact that the melting temperature decreases on changing from infinite crystal sizes to finite ones:

$$\mathrm{Tm} = \mathrm{T}_{\mathrm{m}}^{\mathrm{o}}\left(1 - \frac{2\sigma}{\Delta\mathrm{H}_{\mathrm{m}}^{\mathrm{o}} \cdot \rho \cdot \mathrm{d}}\right) \tag{8.2}$$

where $\mathrm{T}_{\mathrm{m}}^{\mathrm{o}}$ is the melting temperature of infinite (bulk) crystal, σ the specific free energy of the crystal-melt phase boundary, $\Delta\mathrm{H}_{\mathrm{m}}^{\mathrm{o}}$ the heat of fusion for 100% crystalline matrix (bulk crystalline polymer), and ρ the density of the solid. Small crystals are less stable because of the more dominant surface energy that reduces the cohesion energy and shifts the melting point to lower values. $\mathrm{T}_{\mathrm{m}}^{\mathrm{o}}$ and $\mathrm{H}_{\mathrm{m}}^{\mathrm{o}}$ values cannot be experimentally determined but extrapolated by varying the crystallization conditions.

The influence of cellulose nanoparticles on the Tm value of the polymer matrix determined by DSC for different polymer matrices is reported in Table 8.5. This list is not exhaustive. Interpretation of the results can be achieved from the above mentioned concepts. An increase of Tm reveals the formation of thicker crystalline lamellae. However, it can be seen from Table 8.5 that for most systems no variation of the melting point was observed.

8.6.2 Crystallization temperature

When a semicrystalline polymer is cooled from the melt, crystallization occurs at a temperature which is lower than the melting point. For polymer crystallization to start, the primary nucleation first needs to take place. The nucleation itself can be defined as the formation of a small amount of crystalline material due to fluctuations in density or order in the supercooled melt. These crystalline germs can either dissociate, if thermal motion destroys the molecular order, or grow further, if the germ size exceeds a certain critical value. Crystal growth is achieved by the further addition of folded polymer chain segments and only occurs for temperatures above the glass transition temperature. The crystallization temperature (Tc) value is therefore directly correlated with the easiness of polymer chains to form crystalline domains, i.e. its value is higher the more the crystallization is favored.

The influence of cellulose nanoparticles on the Tc value of the polymer matrix determined by DSC for different polymer matrices is reported in Table 8.6. This list is not exhaustive. Interpretation of the results can be achieved from the above mentioned concepts. An increase of Tc is indicative of an easier crystallization process. It is generally associated with a nucleating agent action of the cellulosic nanoparticles.

Effect	Polymer	Nanoparticle	Chemical Modification	Filler Content (wt%)	Reference
No effect	CAB	CNC	–	0–10	(Grunert and Winter, 2002)
	PCL			0–8	(Goffin et al., 2011b)
	PP		MAPP Grafting	0–6	(Ljungberg et al., 2006)
				0–15	(Bahar et al., 2012)
Increase	PCL	CNC	–	0–12	(Siqueira et al., 2009)
			Carbamination		
		MFC	Carbamination		
		CNC	PCL Grafting	0–8	(Goffin et al., 2011b)
	PLA	MFC	–	0–10	(Suryanegara et al., 2009)
	PU	CNC	–	0–2	(Auad et al., 2012)
Decrease	PEO	CNC	–	0–30	(Azizi Samir et al., 2004a)
					(Alloin et al., 2011)
	PHBV	CNC	NaOH Neutralization	0–5	(Jiang et al., 2008)

Table 8.6: Influence of nanocellulose on the crystallization temperature of polymers.

In heating DSC scans an exothermic peak is sometimes observed, attributed to the so-called cold crystallization during heating process. This phenomenon has been observed for some cellulose nanoparticle reinforced polymer nanocomposites. For poly(3-hydroxybutyrate-co-3-hydroxyvalerate) (PHBV), the cold crystallization was observed at 66°C; this temperature was decreased to 37°C when adding 2 or 5 wt% cellulose nanocrystals (Jiang et al., 2008). It was ascribed to an enhanced crystallization ability of PHBV, the nanocrystals acting as nucleation agent and inducing crystallization of the polymer at lower temperatures (Ten et al., 2010). A similar trend was reported for PLA reinforced with MFC (Suryanegara et al., 2009) and MCC nanocrystals (Sanchez-Garcia and Lagaron, 2010; Fortunati et al., 2012). Moreover, the heat of cold crystallization increased for nanocomposites (Sanchez-Garcia and Lagaron, 2010; Fortunati et al., 2012) or the cold crystallization transition appeared while it was absent for the neat matrix (Goffin et al., 2011a).

8.6.3 Degree of crystallinity

The fraction of the ordered molecules in polymer is characterized by the degree of crystallinity, which typically ranges between 10 and 80%. The heat of fusion/crystallization ΔHm/ΔHc can be used to calculate the degree of crystallinity χ_c of the matrix using the following equation:

$$\chi_c = \frac{\Delta Hm}{w \cdot \Delta H_m^o} \tag{8.3}$$

where w is the weight fraction of polymer matrix material in the composite and is the heat of fusion for 100% crystalline matrix (bulk crystalline polymer).

The influence of cellulose nanoparticles on the degree of crystallinity of the polymer matrix determined by DSC for different polymer matrices is reported in Table 8.7. This list is not exhaustive. Any alteration of the crystallinity of the polymer matrix is particularly important because it directly affects important properties of the material such as optical, mechanical and barrier properties which are the main assets of cellulose nanoparticles. For instance, MFC has been added to PLA to improve its stiffness at high temperature and reduce deformation of molded samples during ejection (Suryanegara et al., 2011).

As can be seen from Table 8.7, in most cases adding cellulose nanoparticles induces an increase of the degree of crystallinity of the matrix. It is generally ascribed to a nucleating agent action of the dispersed phase. It is well known that nucleation is strongly affected by impurities, dyes, plasticizers, fillers and other additives in the polymer. However, a simple comparison of the data reported in Table 8.7 is difficult because the nanoparticle-induced crystallization effect is strongly dependent on the nanocomposite processing technique, thermal history of the material, as well as shape, surface area and rigidity of the filler.

Moreover, it seems that the nucleating effect of cellulosic nanocrystals is mainly governed by surface chemistry considerations. Indeed, when untreated or surfactant-coated nanocrystals were reported to be very good nucleating agents for PP, the unmodified nanoparticles having the largest nucleating effect (Ljungberg et al., 2006). Conversely, nanocrystals grafted with maleated polypropylene did not modify the crystallization of PP. It was shown from both X-ray diffraction and DSC analysis that the crystallization behavior of films containing unmodified and surfactant-modified nanocrystals displayed two crystalline forms (α and β), whereas the neat matrix

Effect	Polymer	Nano-particle	Chemical Modification	Filler Content (wt%)	Reference
Increase	CAB	CNC	Silylation	0–10	(Grunert and Winter, 2002)
	LDPE	CNC	–	0–15	(de Menezes et al., 2009)
			Aliphatic Chains Grafting		
	PCL	CNC	–	0–30	(Habibi and Dufresne, 2008)
			PCL Grafting	0–50	
			–	0–30	(Habibi et al., 2008)
			PCL Grafting	0–40	
			–	0–12	(Siqueira et al., 2009)
			Carbamination		
		MFC	Carbamination		
		CNC	–	0–8	(Goffin et al., 2011b)
			PCL Grafting		
	PLA	CNC	PCL Grafting	0–12	(Lin et al., 2009)
			–	0–2	(Pei et al., 2010)
			Silylation		
			NaOH Neutralization	0–5	(Sanchez-Garcia and Lagaron, 2010)
		MFC	–	0–10	(Suryanegara et al., 2009)
				0–10	(Suryanegara et al., 2009)
		CNC	–	0–8	(Goffin et al., 2011a)
			PLA Grafting		
			–	0–5	(Fortunati et al., 2012)
			Surfactant Coating		
	PP	CNC	MAPP Grafting	0–15	(Bahar et al., 2012)
	PU	CNC	–	0–1	(Auad et al., 2008)
			PANI Coating	0–15	(Auad et al., 2011)
			–	0–2	(Auad et al., 2012)
	Starch/glycerol	MFC	–	0–15	(Kaushik et al., 2010)
	Starch/sorbitol	CNC	–	0–25	(Mathew and Dufresne, 2002)
	WPU	CNC	–	0–10	(Cao et al., 2009)

Effect	Polymer	Nano-particle	Chemical Modification	Filler Content (wt%)	Reference
No effect	PCL	CNC	–	0–7.5	(Zoppe et al., 2009)
	PEO	CNC	–	0–10	(Azizi Samir et al., 2004a)
					(Alloin et al., 2011)
	PEO-LiTFSI	CNC	–	0–10	(Azizi Samir et al., 2004b)
	Cross-linked PEO	CNC	–	0–6	(Azizi Samir et al., 2004c)
	PLA	MFC	–	0–17.3	(Tingaut et al., 2010)
			Acetylation		
	PP	CNC	–	0–6	(Ljungberg et al., 2006)
			Surfactant Coating		
	PVA	CNC	–	0–30	(Uddin et al., 2011)
Decrease	CAB	CNC	–	0–10	(Grunert and Winter, 2002)
	PCL	CNC	PCL Grafting	0–7.5	(Zoppe et al., 2009)
	PEO	BC	–	0–59	(Brown and Laborie, 2009)
		CNC	–	10–30	(Azizi Samir et al., 2004a)
					(Alloin et al., 2011)
	PP	CNC	MAPP Grafting	0–6	(Ljungberg et al., 2006)
	Poly(VA-co-VAc)	CNC	–	0–12	(Roohani et al., 2008)
				0–15	(Peresin et al., 2010)

Table 8.7: Influence of nanocellulose on the degree of crystallinity of polymers.

and the nanocomposite reinforced with nanocrystals grafted with maleated polypropylene only crystallized in the α form. It was suspected that the more hydrophilic the nanocrystal surface, the more it appeared to favor the formation of the β phase. It was observed that BC nanocrystals impede the crystallization of the cellulose acetate butyrate (CAB) matrix whereas silylated ones help to nucleate the crystallization. Silk fibroin reinforced with tunicin nanocrystal composites were investigated using infrared spectroscopy (Noishiki et al., 2002). A conformational change of fibroin chains from a random coil to an ordered structure was reported. This change was ascribed to the highly ordered surface of cellulosic nanocrystals.

The nucleating effect of cellulose nanocrystals for polymer with the development of a transcrystalline layer has been reported for glycerol plasticized starch (Anglès and Dufresne, 2000), poly(hydroxyoctanoate) (PHO) (Dufresne et al., 1999), and PP (Gray, 2008). It consists in a preferential nucleation of the amorphous polymeric matrix chains during cooling at the surface of nanoparticles. Several theories have been suggested to account for the development of a transcrystalline layer. These include epitaxial growth based on lattice matching, wettability and surface energy of the substrates, adsorption of small molecules, stress-induced crystallization by local flow, residual stress caused by mismatch in coefficients of thermal expansion, topography of the substrates, and residual crystals at the surface of the foreign particles. For glycerol plasticized starch-based systems (Anglès and Dufresne, 2000), the formation of the transcrystalline zone around the nanocrystals was assumed to be due to the accumulation of plasticizer in the cellulose/amylopectin interfacial zones improving the ability of amylopectin chains to crystallize. These specific crystallization conditions were evidenced at high moisture content and high nanoparticle content (> 16.7 wt%) by DSC and wide-angle X-ray diffraction. It was displayed through a shoulder on the low temperature side of the melting endotherm and the observation of a new peak in the X-ray diffraction pattern. This transcrystalline zone could originate from a glycerol-starch V structure. In addition, the inherent restricted mobility of amylopectin chains was put forward to explain the lower water uptake of cellulose/starch composites for increasing filler content. Transcrystallization of PP at cellulose nanocrystal surfaces was evidenced and it was found to result from enhanced nucleation due to some form of epitaxy (Gray, 2008). For tunicin nanocrystals reinforced PHO, the thickness of the transcrystalline layer was estimated from dynamic mechanical measurements at around 2.7 nm (Dufresne, 2000).

Transcrystallization of PP was observed from the edge of a piece of film cast from cotton nanocrystal suspension using polarized optical microscopy (Gray, 2008). Even at early stages of the crystallization, nucleation along the edge of the cellulose film was very dense, so that the edge was completely covered by a transcrystalline layer well before there was much nucleation in the bulk. After 20 min isothermal crystallization at 136°C some spherulite growth on the top of the film was observed, in addition to the dense transcrystalline layer at the edge of the film. While the film edge enhances nucleation, the growth rate of the transcrystalline layer was about the same as that of the bulk spherulites, with the width of the layer roughly equal to the radii of the bulk spherulites as shown in Figure 8.6. To overcome the hypothesis of thermal or vapor-induced shear in the cooling melt the nucleating ability of cellulose nanocrystals deposited on a glass surface was examined. It was observed that intense nucleation occurred on the part of the surface where the nanocrystals were deposited, showing that they were the source of the preferential nucleation.

The main reasons for the incapacity of cellulose nanoparticles to serve as efficient nucleating agents and a hindrance to growth of polymer spherulites are reported below. It can be due to:

- The affinity of the polymer chains (e.g. PEO or PVA) with the reactive cellulose surface (Azizi Samir et al., 2004a; Azizi Samir et al., 2004b; Brown and Laborie, 2007; Roohani et al., 2008). This effect should result in a restricted molecular mobility of the polymer chains in contact with the nanoparticle surface. Owing

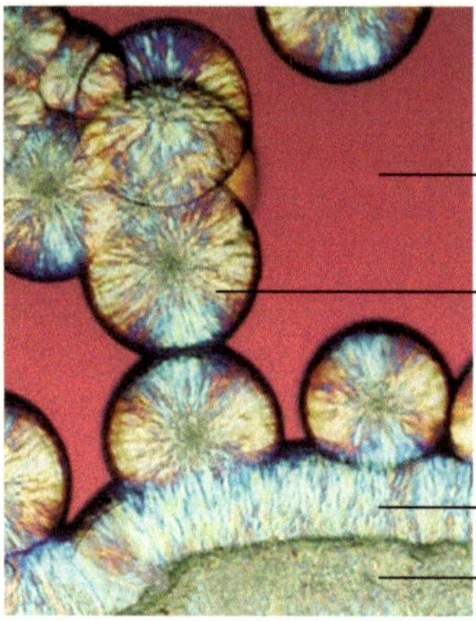

Fig. 8.6: Crystallization of isotactic polypropylene at the edge of cellulose nanocrystal film (initial melt at ~ 220°C). Image after 20 min at 136°C. The red color is the isotropic-PP melt in which spherulites are growing. Crossed polars, first-order red plate. Scale bar 200 μm (Gray, 2008).

to the high specific surface area of these nanoparticles, this hindered mobility could be strong enough to affect the global self-diffusion of the matrix.

- The increase of the viscosity of the polymer melt and confinement of the polymer within a dense cellulosic network ascribed to the presence of the nanoparticles (Azizi Samir et al., 2004a; Azizi Samir et al., 2004c; Brown and Laborie, 2007). This increased viscosity may induce an increase of the activation energy for the diffusion of the chains yielding smaller and less stable crystals. This effect can explain the increased crystallization observed for some systems for low filler loading acting as efficient nuclating agent and decreased crystallization for higher filler contents that hinder the transport of polymer segments delaying the crystallization process.
- A different crystalline form induced by the nanoparticle has also been evoked. For instance, two crystalline forms (α and β) of PP have been observed when adding unmodified and surfactant-coated cellulose nanocrystals, whereas the neat matrix and nanocomposite reinforced with MAPP-grafted nanocrystals only crystallized in the α-form (Ljungberg et al., 2006). However, the nanocrystal-induced crystalline structure can also increase the degree of crystallinity. For quasi amorphous PLA, the development of a double melting peak when adding PLA-grafted cellulose nanocrystals was ascribed to the co-existence of two crystalline structures, i.e. less perfect crystals (α'-form crystals), which have enough time to melt and to reorganize into crystals with higher structural perfection (α-form crystals), before they remelt at higher temperature (Goffin et al., 2011a).

- The processing technique itself, for instance electrospinning that involves very high shear stress and fast solvent evaporation inducing rapid crystallization (Zoppe et al., 2009; Peresin et al., 2010).

8.6.4 Rate of crystallization

The macroscopic study of the crystallization process is based on the evolution of the crystalline fraction of the polymer as a function of time in an isothermal regime. The relative crystallinity (χ_t) of the material as a function of crystallization time (t) can be determined from the crystallization exotherm at a given crystallization temperature:

$$\chi_t = \frac{\int_0^t \left(\frac{dH}{dt}\right) dt}{\int_0^\infty \left(\frac{dH}{dt}\right) dt} \tag{8.4}$$

where dH/dt is the relative heat flow rate. Obviously, χ_t tends to unity for longer times, i.e. after complete crystallization. For example, Figure 8.7(a) shows the isothermal crystallization behavior of PLA reinforced with 1 wt% cotton nanocrystals at different crystallization temperatures (Pei et al., 2010). The shift of exotherms to longer times when increasing the crystallization temperature indicates a decrease of the crystallization rate and that nucleation is the limiting factor of crystallization in this temperature range. Figure 8.7(b) illustrates the effect of cellulose nanoparticles extracted from sisal fibers (both nanocrystals and MFC) on the crystallization isotherm of PCL

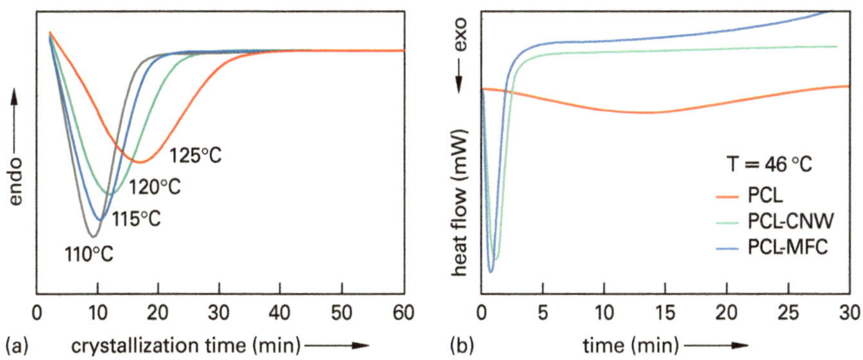

Fig. 8.7: Isothermal crystallization exotherms for (a) PLA reinforced with 1 wt% cotton nanocrystals at different crystallization temperatures (Tc) indicated in the Figure (Pei et al., 2010), and (b) neat PCL and PCL nanocomposites reinforced with 12 wt% chemically modified sisal nanocrystals (CNW) or MFC at Tc = 46°C (Siqueira et al., 2011).

obtained at 46°C (Siqueira et al., 2011). The filler content was 12 wt% and the surface of the nanoparticles was modified with n-octadecyl isocyanate. It can be clearly seen that the exotherm for the nanocomposites occurs much sooner than for the neat PCL. In other words, the nanoparticles significantly accelerate the crystallization of PCL, slightly more with MFC than with cellulose nanocrystals. However, it is worth noting that the degree of substitution (DS) for both type of nanoparticle was different.

This effect can be directly visualized from polarized optical microscopy observations. For these observations, the sample is quenched from the melt down to a temperature lower than the melting point. Examples for tunicin nanocrystals reinforced PEO (Azizi Samir et al., 2004a) and cotton nanocrystals reinforced PLA (Pei et al., 2010) are given in Figures 8.8(a) and 8.8(b), respectively. Birefringent spherulites are clearly identified through the characteristic Maltese cross pattern indicating a spherical symmetry. The heterogeneous nucleating action of the nanocrystals results in an increase of the overall crystallization rate, reduction of the nucleation induction period and increase in the number of primary nucleation sites.

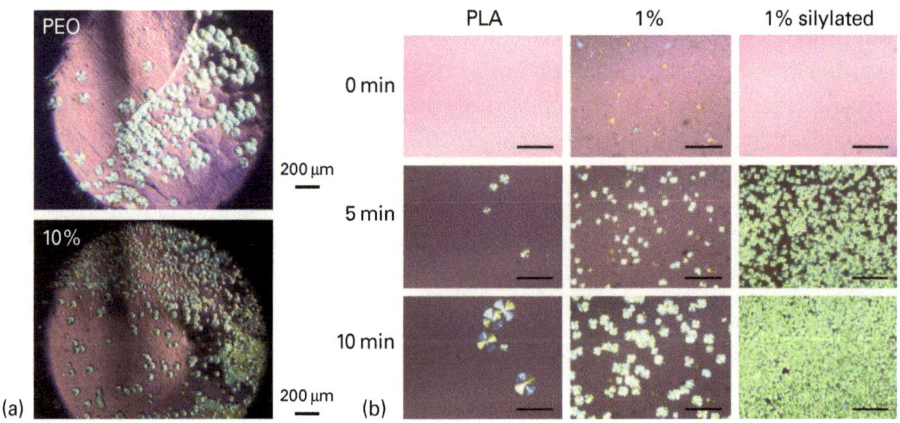

Fig. 8.8: Polarized optical micrographs of (a) neat PEO and 10 wt% tunicin nanocrystals reinforced films obtained after 100 s isothermal crystallization at 53.6°C (Azizi Samir et al., 2004a), and (b) neat PLA and 1 wt% cotton nanocrystals reinforced films (unmodified and silylated) obtained after 0, 5 and 10 min isothermal crystallization at 125°C (Pei et al., 2010). The scale bar in panel (b) is 200 µm.

The development of the crystallization with time can be viewed by plotting the relative crystallinity as a function of time as presented in Figure 8.9 for PCL-based nanocomposites (Siqueira et al., 2011). The typical sigmoid shape of this curve reflects the various phases of crystallization starting with a nucleation step, a radial crystal growth or primary crystallization followed by a secondary crystallization of lamellae thickening. It can be seen from Figure 8.9 that the nanocomposites build up their crystallinity much faster that neat PCL.

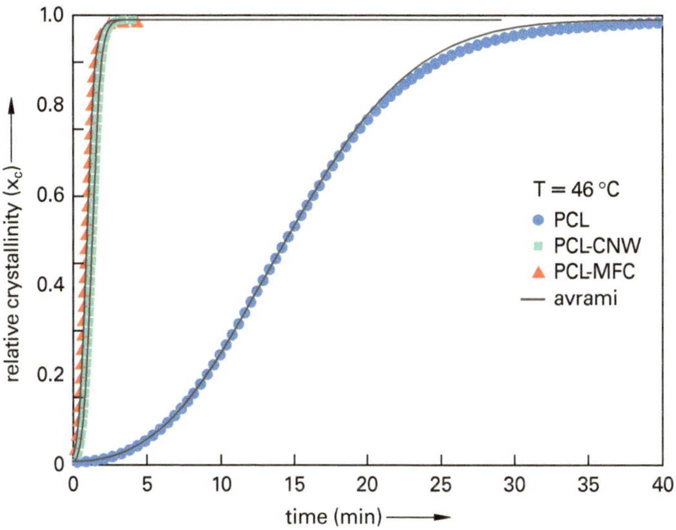

Fig. 8.9: Evolution of the relative crystallinity as a function of time for neat PCL and PCL nanocomposites reinforced with 12 wt% chemically modified sisal nanocrystals (CNW) or MFC at Tc = 46°C (Siqueira et al., 2011).

To describe quantitatively the isothermal crystallization kinetics, the Avrami equation (Avrami, 1939; Avrami, 1940; Avrami, 1941) can be used:

$$1 - \chi_t = \exp(-kt^n) \tag{8.5}$$

where k is the crystallization rate constant (min^{-n}) and n the Avrami exponent, which are assumed to be a diagnostic of the mechanism of crystallization. k depends on the geometry of the growing crystalline phase and n correlated with the nucleation mechanism and crystal growth dimension. These parameters can be assessed for each crystallization temperature by plotting the Avrami plots:

$$\log[-\ln(1 - \chi_t)] = \log(k) + n\log(t) \tag{8.6}$$

From these parameters, it was suggested that a two-dimensional crystallization growth with a heterogeneous nucleation mechanism occurred in cellulose nanocomposites based on PCL (Siqueira et al., 2011), whereas a three-dimensional crystallization growth with a homogeneous nucleation mechanism was suggested in cellulose nanocomposites based on PLA (Pei et al., 2010). The nucleating effect of cellulose nanocrystals was also investigated in isothermal crystallization kinetic studies for PU/cellulose nanocomposites (Han et al., 2012). The study was based on the Avrami kinetics model and it was shown that cellulose nanocrystals increased the rate of crystallization and fostered heterogeneous crystallization and crystal growth in two dimensions.

The half-time of crystallization ($t_{1/2}$) defined as the time when χ_t is equal to 0.5 is often used to quantify the crystallization kinetics:

$$t_{1/2} = \left(\frac{\ln(2)}{k}\right)^{1/n} \tag{8.7}$$

The rate of crystallization of PCL was found to be increased by more than 100-fold with the addition of cellulose nanoparticles (Siqueira et al., 2011). A lower but significant decrease of $t_{1/2}$ was also reported for PLA reinforced with 1 or 2 wt% cellulose nanocrystals (Pei et al., 2010).

Crystallization is assumed to be a thermally activated process such that the crystallization rate parameter k can be described with the Arrhenius equation:

$$\frac{1}{n}\ln(k) = \ln(k_o) - \frac{\Delta E_a}{RT_c} \tag{8.8}$$

where k_o is a temperature-dependent pre-exponential factor, ΔE_a the total activation energy and R the universal gas constant. The faster crystallization of the PCL (Siqueira et al., 2011) or PU (Han et al., 2012) matrix in presence of cellulose nanoparticles correlated with the lower activation energy as assessed from Avrami model.

The Lauritzen–Hoffman nucleation parameter (Hoffman et al., 1975) was also found to decrease for PCL-based nanocomposites, confirming that the cellulose nanoparticles lower the energetic requirements for the polymer nucleation (Siqueira et al., 2011).

8.7 Thermal stability of nanocellulose-based nanocomposites

The thermal stability of materials is a key parameter to determine the temperature range of processing and use. The influence of cellulose nanoparticles on the thermal stability of the polymer matrix determined by TGA for different polymer matrices is reported in Table 8.8. This list is not exhaustive. The effect of the nanoparticles depends obviously on the initial thermal stability of the neat polymer. The char residue can also be altered by the nanofiller.

Enhanced thermal stability of the host polymer results from favorable interactions with the cellulosic filler. Increased thermal stability of commercially available acrylic copolymers was reported when adding BC (Trovatti et al., 2010). It was hypothesized to reflect the good dispersion and compatibility between both components resulting from van der Waals interactions and hydrogen bonds between the hydroxyl groups of cellulose and carboxyl groups of the acrylic polymer. Moreover, the presence of residual surfactant was also suspected to contribute to the interfacial compatibility between BC and the acrylic matrix. In another study performed with PEO and BC, the increased thermal stability was ascribed to mutual thermal stabilization (Brown and Laborie, 2011). For PVA (Lee et al., 2009) and regenerated cellulose (Ma et al., 2011) reinforced with cellulose nanocrystals, strong hydrogen bonding between hydroxyl groups of cellulose and the polar matrix was also suggested to improve the thermal stability of the host polymer. Similarly, the formation of a confined structure in PU matrix was supposed to improve the thermal stability of the matrix upon cellulose

Effect	Polymer	Nano-particle	Atmos-phere	Chemical Modification	Filler Content (wt%)	Reference
No effect	PEO	CNC	Nitrogen	–	0–10	(Azizi Samir et al., 2004a)
	PEO-LiTFSI	CNC	Nitrogen	–	0–10	(Azizi Samir et al., 2004b)
	Phenolic Resin	CNC	Nitrogen	–	0–7.5	(Liu and Laborie, 2011)
	PLA	CNC	Nitrogen	NaOH Neutralization	0–5	(Sanchez-Garcia and Lagaron, 2010)
				–		(Fortunati et al., 2012)
				Surfactant Coating		
			Air	–		
				Surfactant Coating		
Increase	Acrylic Copolymer	BC	Nitrogen	–	0–10	(Trovatti et al., 2010)
	PEO	BC	Nitrogen	–	0–99.5	(Brown and Laborie, 2007)
	PFA	CNC	Nitrogen	–	–	(Pranger and Tannenbaum, 2008)
	PVA	MFC	Nitrogen	–	0–15	(Lu et al., 2008b)
		CNC*			0–5	(Lee et al., 2009)
		CNC			0–7	(Cho and Park, 2011)
	Regenerated Cellulose	CNC	Air	–	0–20	(Qi et al., 2009)
			Nitrogen			(Ma et al., 2011)
	WPU	CNC	Nitrogen	–	0–10	(Cao et al., 2009)

Effect	Polymer	Nano-particle	Atmos-phere	Chemical Modification	Filler Content (wt%)	Reference
Decrease	CMC/glycerin	CNC	Nitrogen	$NaHCO_3$ Neutralization	0–30	(Choi and Simonsen, 2006)
	Epoxy	MFC	Nitrogen	–	0–5	(Lu et al., 2008a)
				Silylation		
				Titanate Treatment		
	NR	CNC	Nitrogen	–	0–10	(Bras et al., 2010)
	PEO	CNC	Nitrogen	–	0–30	(Alloin et al., 2011)
	PHBV	CNC	Nitrogen	–	0–5	(Ten et al., 2010)
	PLA	CNC	Nitrogen	MAPP Grafting	0–10	(Pandey et al., 2009a; Pandey et al., 2009b)
		MFC	Helium	–	0–17.3	(Tingaut et al., 2010)
				Acetylation		
	PMMA	CNC	Nitrogen	–	0–10	(Liu et al., 2010)
	Polysulfone	CNC	Nitrogen	–	0–11	(Noorani et al., 2007)
	Poly(VA-co-VAc)	CNC	–	–	0–15	(Peresin et al., 2010)
	Starch/glycerol	CNC	Nitrogen	–	0–24	(Chen et al., 2009)
		MFC			0–15	(Kaushik et al., 2010)

* : Extracted with HBr

Table 8.8: Influence of nanocellulose on the thermal degradation of polymers.

nanocrystal addition (Cao et al., 2009). However, for higher filler contents a decrease of the thermal stability is sometimes observed because of aggregation (Ma et al., 2011).

The low thermal stability of cellulose and in particular cellulose nanocrystals prepared by sulfuric acid hydrolysis can impact the global behavior of the nanocomposite. Cotton cellulose nanocrystal content appeared to have an effect on the thermal behavior of carboxymethyl cellulose (CMC) plasticized with glycerin (Choi and Simonsen, 2006), suggesting a close association between the filler and the matrix. The thermal degradation of unfilled CMC was observed from its melting point (270°C) and had a very narrow temperature range of degradation. Cellulose nanocrystals were found to degrade at a lower temperature (230°C) than CMC, but showed a very broad degradation temperature range. The degradation of cellulose nanocrystal reinforced CMC was observed between these two limits, but of interest was the lack of steps.

The low onset degradation temperature of H_2SO_4-prepared nanocrystals was supposed to decrease the degradation temperature of polysulfone (Noorani et al., 2007), glycerol plasticized starch (Chen et al., 2009), natural rubber (NR) (Bras et al., 2010),

PMMA (Liu et al., 2010), PVA (Peresin et al., 2010) and PEO (Alloin et al., 2011). For PLA-based nanocomposites the higher water content induced by the presence of the cellulosic nanofiller was expected to decrease the thermal stability of the polymer (Pandey et al., 2009a). Acetylation of the nanoparticles can limit this effect (Tingaut et al., 2010). The increased thermal conductivity of the polymer after cellulose nanocrystal addition was also suggested as the reason for reduced thermal stability of the composite (Ten et al., 2010).

8.8 Conclusions

Cellulose displays a low thermal expansion coefficient that can be advantageously utilized to reduce the thermal expansion of polymeric materials. However, the limited thermal stability of cellulose can be a limiting factor for the processing of nanocomposites at high temperatures and some high-temperature uses of the material. This feature is enhanced when using sulfuric acid-hydrolyzed cellulose nanocrystals or TEMPO-oxidized MFC. For most systems, the thermal characteristics of the polymeric matrix, i.e. glass transition temperature and melting point, are generally slightly affected by the presence of the cellulose nanoparticles. However, many studies report an increase of both the degree of crystallinity and rate of crystallization ascribed to a nucleating effect of nanocellulose.

8.9 References

Akim, E.L. (1977). Changes in cellulose structure during manufacture and converting of paper. In: Cellulose Chemistry and Technology, ACS Symp. Ser. N°48, pp. 153–172.

Alemdar, A. and Sain, M. (2008). Isolation and characterization of nanofibers from agricultural residues – Wheat straw and soy hulls. Biores. Technol. *99*, 1664–1671.

Alloin, F., D'Aprea, A., Dufresne, A., El Kissi, N. and Bossard, F. (2011). Poly(oxyethylene) and ramie whiskers based nanocomposites: Influence of processing: Extrusion and casting/evaporation. Cellulose *18*, 957–973.

Anglès, M.N. and Dufresne, A. (2000). Plasticized starch/tunicin whiskers nanocomposites. 1. Structural analysis. Macromolecules *33*, 8344–8353.

Auad, M.L., Contos, V.S., Nutt, S., Aranguren, M.I. and Marcovich, N.E. (2008). Characterization of nanocellulose-reinforced shape memory polyurethanes. Polym. Inter. *57*, 651–659.

Auad, M.L., Richardson, T., Orts, W.J., Medeiros, E.S., Mattoso, L.H.C., Mosiewicki, M.A., Marcovich, N.E. and Aranguren, M.I. (2011). Polyaniline-modified cellulose nanofibrils as reinforcement of a smart polyurethane. Polym. Inter. *60*, 743–750.

Auad, M.L., Richardson, T., Hicks, M., Mosiewicki, M.A., Aranguren, M.I. and Marcovich, N.E. (2012). Shape memory segmented polyurethanes: Dependence of behaviour on nanocellulose addition and testing conditions. Polym. Inter. *61*, 321–327.

Avrami, M. (1939). Kinetics of phase change. I. General theory. J. Chem. Phys. *7*, 1103–1112.

Avrami, M. (1940). Kinetics of phase change. II. Transformation-time relations for random distribution of nuclei. J. Chem. Phys. *8*, 212–224.

Avrami, M. (1941). Kinetics of phase change. III. Granulation, phase change, and microstructure. J. Chem. Phys. *9*, 177–184.

Azizi Samir, M.A.S., Alloin, F., Sanchez, J.-Y. and Dufresne, A. (2004a). Cellulose nanocrystals reinforced poly(oxyethylene). Polymer *45*, 4149–4157.

Azizi Samir, M.A.S., Alloin, F., Gorecki, W., Sanchez, J.-Y. and Dufresne, A. (2004b). Nanocomposite polymer electrolytes based on poly(oxyethylene) and cellulose nanocrystals. J. Phys. Chem. B *108*, 10845–10852.

Azizi Samir, M.A.S., Alloin, F., Sanchez, J.Y., El Kissi, N. and Dufresne, A. (2004c). Preparation of cellulose whiskers reinforced nanocomposites from an organic medium suspension. Macromolecules *37*, 1386–1393.

Back, E.L. and Salmén, N.L. (1982). Glass transitions of wood components hold implications for molding and pulping processes. Tappi *65*, 107–110.

Bahar, E., Ucar, N., Onen, A., Wang, Y., Oksüz, M., Ayaz, O., Ucar, M. and Demir, A. (2012). Thermal and mechanical properties of polypropylene nanocomposite materials reinforced with cellulose nano whiskers. J. Appl. Polym. Sci. *125*, 2882–2889.

Ben Azouz, K., Ramires, E.C., Van den Fonteyne, W., El Kissi, N. and Dufresne A. (2012). A simple method for the melt extrusion of cellulose nanocrystal reinforced hydrophobic polymer. ACS Macro Lett. *1*, 236–240.

Ben Elmabrouk, A., Thielemans, W., Dufresne, A. and Boufi, S. (2009). Preparation of poly(styrene-co-hexylacrylate)/cellulose whiskers nanocomposites via miniemulsion polymerization. J. Appl. Polym. Sci. *114*, 2946–2955.

Bendahou, A., Habibi, Y., Kaddami, H. and Dufresne, A. (2009). Physico-chemical characterization of palm from *Phoenix dactylifera* – L, preparation of cellulose whiskers and natural rubber–based nanocomposites. J. Biobased Mat. Bioenergy *3*, 81–90.

Bendahou, A., Kaddami, H. and Dufresne, A. (2010). Investigation on the effect of cellulosic nanoparticles' morphology on the properties of natural rubber based nanocomposites. Eur. Polym. J. *46*, 609–620.

Bergenstråhle, M., Berglund, L.A. and Mazeau, K. (2007). Thermal response in crystalline Ib cellulose: A molecular dynamics study. J. Phys. Chem. B *111*, 9138–9145.

Bras, J., Hassan, M.L., Bruzesse, C., Hassan, E.A., El-Wakil, N.A. and Dufresne, A. (2010). Mechanical, barrier, and biodegradability properties of bagasse cellulose whiskers reinforced natural rubber nanocomposites. Ind. Crops Prod. *32*, 627–633.

Braun, B. and Dorgan, J.R. (2009). Single-step method for the isolation and surface functionalization of cellulosic nanowhiskers. Biomacromolecules *10*, 334–341.

Broido, A. (1969). A simple, sensitive graphical method of treating thermogravimetric analysis data. J. Polym. Sci. A-2 Polym. Phys. *7*, 1761–1773.

Brown, E.E. and Laborie, M.P.G. (2007). Bioengineering bacterial cellulose/poly(ethylene oxide) nanocomposites. Biomacromolecules *8*, 3074–3081.

Cao, X., Dong, H and Li, C.M. (2007). New nanocomposite materials reinforced with flax cellulose nanocrystals in waterborne polyurethane. Biomacromolecules *8*, 899–904.

Cao, X., Chen, Y., Chang, P.R., Stumborg, M. and Huneault, M.A. (2008a). Green composites reinforced with hemp nanocrystals in plasticized starch. J. Appl. Polym. Sci. *109*, 3804–3810.

Cao, X., Chen, Y., Chang, P.R., Muir, A.D. and Falk G. (2008b). Starch-based nanocomposites reinforced with flax cellulose nanocrystals. eXPRESS Polym. Let. *2*, 502–510.

Cao, X., Habibi, Y. and Lucia, L.A. (2009). One-pot polymerization, surface grafting, and processing of waterborne polyurethane-cellulose nanocrystal nanocomposites. J. Mater. Chem. *19*, 7137–7145.

Chazeau, L., Cavaillé, J.Y. and Terech, P. (1999). Mechanical behaviour above Tg of a plasticized PVC reinforced with cellulose whiskers, a SANS structural study. Polymer *40*, 5333–5344.

Chen, Y., Liu, C., Chang, P.R., Cao, X. and Anderson, D.P. (2009). Bionanocomposites based on pea starch and cellulose nanowhiskers hydrolyzed from pea hull fibre: Effect of hydrolysis time. Carbohydr. Polym. *76*, 607–615.

Cho, M.J. and Park, B.D. (2011). Tensile and thermal properties of nanocellulose-reinforced poly(vinyl alcohol) nanocomposites. J. Ind. Eng. Chem. *17*, 36–40.

Choi, Y.J. and Simonsen, J. (2006). Cellulose nanocrystal-filled carboxymethyl cellulose nanocomposites. J. Nanosci. Nanotechnol. *6*, 633–639.

de Menezes, A.J., Siqueira, G., Curvelo, A.A.S. and Dufresne, A. (2009). Extrusion and characterization of functionalized cellulose whisker reinforced polyethylene nanocomposites. Polymer *50*, 4552–4563.

Dufresne, A., Kellerhals, M.B. and Witholt, B. (1999). Transcrystallization in mcl-PHAs/ cellulose whiskers composites. Macromolecules *32*, 7396–7401.

Dufresne, A. (2000). Dynamic mechanical analysis of the interphase in bacterial polyester/cellulose whiskers natural composites. Compos. Interfaces *7*, 53–67.

Elanthikkal, S., Gopalakrishnapanicker, U., Varghese, S. and Guthrie, J.T. (2010). Cellulose microfibres produced from banana plant wastes: Isolation and characterization. Carbohydr. Polym. *80*, 852–859.

Eyholzer, Ch., Bordeanu, N., López-Suevos, F., Rentsch, D., Zimmermann, T. and Oksman, K. (2010). Preparation and characterization of water-redispersible nanofibrillated cellulose in powder form. Cellulose *17*, 19–30.

Fahma, F., Iwamoto, S., Hori, N., Iwata, T. and Takemura, A. (2011). Effect of pre-acid-hydrolysis treatment on morphology and properties of cellulose nanowhiskers from coconut husk. Cellulose *18*, 443–450.

Fengel, D. and Wegener, G. (1984). Wood chemistry, ultrastructure, reactions (Walter de Gruyter, Berlin, New York).

Filpponen, I. and Argyropoulos, D.S. (2010). Regular linking of cellulose nanocrystals via click chemistry: Synthesis and formation of cellulose nanoplatelet gels. Biomacromolecules *11*, 1060–1066.

Fortunati, E., Armentano, I., Zhou, Q., Iannoni, A., Saino, E., Visai, L., Berglund, L.A. and Kenny, J.M. (2012). Multifunctional bionanocomposite films of poly(lactic acid), cellulose nanocrystals and silver nanoparticles. Carbohydr. Polym. *87*, 1596–1605.

Fukuzumi, H., Saito, T., Iwata, T., Kumamoto, Y. and Isogai, A. (2009). Transparent and high gas barrier films of cellulose nanofibers prepared by TEMPO-mediated oxidation. Biomacromolecules *10*, 162–165.

Garcia de Rodriguez, N.L., Thielemans, W. and Dufresne, A. (2006). Sisal cellulose whiskers reinforced polyvinyl acetate nanocomposites. Cellulose *13*, 261–270.

George, J., Ramana, K.V., Sabapathy, S.N., Jagannath, J.H. and Bawa, A.S. (2005). Characterization of chemically treated bacterial (*Acetobacter xylinum*) biopolymer: Some thermo-mechanical properties. Int. J. Biol. Macromol. *37*, 189–194.

Goffin, A.-L., Raquez, J.-M., Duquesne, E., Siqueira, G., Habibi, Y., Dufresne, A. and Dubois, Ph. (2011a). From interfacial ring-opening polymerization to melt processing of cellulose nanowhisker-filled polylactide-based nanocomposites. Biomacromolecules *12*, 2456–2465.

Goffin, A.-L., Raquez, J.-M., Duquesne, E., Siqueira, G., Habibi, Y., Dufresne, A. and Dubois, Ph. (2011b). Poly(e-caprolactone) based nanocomposites reinforced by surface-grafted cellulose nanowhiskers via extrusion processing: Morphology, rheology, and thermo-mechanical properties. Polymer *52*, 1532–1538.

Goring, D.A.I. (1963). Thermal softening of lignin, hemicellulose and cellulose. Pulp Paper Mag. Can. *64*, T517–T527.

Gray, D.G. (2008). Transcrystallization of polypropylene at cellulose nanocrystal surfaces. Cellulose *15*, 297–301.

Grunert, M. and Winter, W. T. (2002). Nanocomposites of cellulose acetate butyrate reinforced with cellulose nanocrystals. J. Polym. Envir. *10*, 27–30.

Habibi, Y. and Dufresne, A. (2008). Highly filled bionanocomposites from functionalized polysaccharide nanocrystals. Biomacromolecules *9*, 1974–1980.

Habibi, Y., Goffin, A.L., Schiltz, N., Duquesne, E., Dubois, P. and Dufresne A. (2008). Bionanocomposites based on poly(ε-caprolactone)-grafted cellulose nanocrystals by ring opening polymerization. J. Mat. Chem. *18*, 5002–5010.

Hajji, P., Cavaillé, J.Y., Favier, V., Gauthier, C. and Vigier, G. (1996). Tensile behavior of nanocomposites from latex and cellulose whiskers. Polym. Compos. *17*, 612–619.

Han, J., Zhu, Y., Hu, J., Luo, H., Yeung, L.Y., Li, W., Meng, Q., Ye, G., Zhang, S. and Fan, Y. (2012). Morphology, reversible phase crystallization, and thermal sensitive shape memory effect of cellulose whisker/SMPU nano-composites. J. Appl. Polym. Sci. *123*, 749–762.

Hernádi, S. (1976). Thermal ageing in oxygen of paper made from cellulose at different degrees of beating. *Svensk Papperstidning 79*, 418–423.

Hoffman, J.D., Frolen, L.J., Ross, G.S. and Lauritzen, J.I. (1975). Growth-rate of spherulites and axialites from melt in polyethylene fractions – Regime-1 and regime-2 crystallization. J. Res. Natl. Bur. Stand. Phys. Chem. *79*, 671–699.

Hori, R. and Wada, M. (2005). The thermal expansion of wood cellulose crystals. Cellulose *12*, 479–484.

Hori, R. and Wada, M. (2006). The thermal expansion of cellulose II and III_{II} crystals. Cellulose *13*, 281–290.

Ifuku, S., Nogi, M., Abe, K., Handa, K., Nakatsubo, F. and Yano, H. (2007). Surface modification of bacterial cellulose nanofibers for property enhancement of optically transparent composites: Dependence on acetyl-group DS. Biomacromolecules *8*, 1973–1978.

Iwamoto, S., Nakagaito, A.N., Yano, H. and Nogi, M. (2005). Optically transparent composites reinforced with plant fiber-based nanofibers. Appl. Phys. A *81*, 1109–1112.

Iwamoto, S., Nakagaito, A.N. and Yano, H. (2007). Nano-fibrillation of pulp fibers for the processing of transparent nanocomposites. Appl. Phys. A-Mater. Sci. Process *89*, 461–466.

Iwamoto, S., Abe, K. and Yano, H. (2008). The effect of hemicelluloses on wood pulp nano-fibrillation and nanofiber network characteristics. Biomacromolecules *9*, 1022–1026.

Jiang, L., Morelius, E., Zhang, J., Wolcott, M. and Holbery, J. (2008). Study of the poly(3-hydroxybutyrate-co-3-hydroxyvalerate)/cellulose nanowhiskers composites prepared by solution casting and melt processing. J. Compos. Mat. *42*, 2629–2645.

Johar, N. Ahmad, I. and Dufresne, A. (2012). Extraction, preparation and characterization of cellulose fibres and nanocrystals from rice husk. Ind. Crops *37*, 93–99.

Kalashnik, A.T., Papkov, S.P., Rudinskaya, G.V. and Milkova, L.P. (1991). Liquid crystal state of cellulose. Polym. Sci. USSR *33*, 107–112.

Kargarzadeh, H., Ahmad, I., Abdullah, I., Dufresne, A., Zanariah Zainuddin, S.Y. and Sheltami, R.M. (2012). Effect of hydrolysis conditions on the morphology, crystallinity and thermal stability of cellulose nanocrystals extracted from kenaf bast fibres. Cellulose *19*, 855–866.

Kaushik, A., Singh, M. and Verma, G. (2010). Green nanocomposites based on thermoplastic starch and steam exploded cellulose nanofibrils from wheat straw. Carbohydr. Polym. *82*, 337–345.

Kim, D.Y., Nishiyama, Y., Wada, M., Kuga, S. and Okano, T. (2001a). Thermal decomposition of cellulose crystallites in wood. Holzforschung *55*, 521–524.

Kim, D.Y., Nishiyama, Y., Wada, M. and Kuga, S. (2001b). High-yield carbonization of cellulose by sulfuric acid impregnation. Cellulose *8*, 29–33.

Kim, J., Montero, G. Habibi, Y., Hinestroza, J.P., Genzer, J., Argyropoulos, D.S. and Rojas, O.J. (2009). Dispersion of cellulose crystallites by nonionic surfactants in a hydrophobic polymer matrix. Polym. Eng. Sci. *49*, 2054–2061.

Kuga, S., Kim, D.Y. and Nishiyama, Y. (2002). Nanofibrillar carbon from native cellulose. Mol. Cryst. Liq. Cryst. *387*, [237]/13–[243]19.

Kvien, I., Sugiyama, J., Votrubec, M. and Oksman, K. (2007). Characterization of starch based nanocomposites. J. Mater. Sci. *42*, 8163–8171.

Langan, P., Sukumar, N., Nishiyama, Y. and Chanzy, H. (2005). Synchrotron X-ray structures of cellulose I_b and regenerated cellulose II at ambient temperature and 100 K. Cellulose *12*, 551–562.

Lee, S.Y., Mohan, D.J., Kang, I.A., Doh, G.H., Lee, S. and Han, N.O. (2009). Nanocellulose reinforced PVA composite films: Effects of acid treatment and filler loading. Fibers and Polymers *10*, 77–82.

Lee, K.Y., Quero, F., Blaker, J.J., Hill, C.A.S., Eichhorn, S.J. and Bismarck, A. (2011). Surface only modification of bacterial cellulose nanofibres with organic acids. Cellulose *18*, 595–605.

Li, R., Fei, J., Cai, Y., Li, Y., Feng, J. and Yao, J. (2009). Cellulose whiskers extracted from mulberry: A novel biomass production, Carbohydr. Polym. *76*, 94–99.

Lin, N., Chen, G., Huang, J., Dufresne, A. and Chang, P.R. (2009). Effects of polymer-grafted natural nanocrystals on the structure and mechanical properties of poly(lactic acid): A case of cellulose whisker-graft-polycaprolactone. J. Appl. Polym. Sci. *113*, 3417–3425.

Liu, H., Liu, D., Yao, F. and Wu, Q. (2010). Fabrication and properties of transparent polymethylmethacrylate/ cellulose nanocrystals composites. Bioresource Technol. *101*, 5685–5692.

Liu, A., Walther, A., Ikkala, O., Belova, L., and Berglund, L. A. (2011). Clay Nanopaper with Tough Cellulose Nanofiber Matrix for Fire Retardancy and Gas Barrier Functions. Biomacromolecules *12*(3), 633–641.

Liu H. and Laborie, M.-P. (2011). Bio-based nanocomposites by *in situ* cure of phenolic prepolymers with cellulose whiskers. Cellulose *18*, 619–630.

Ljungberg, N., Cavaillé, J.Y. and Heux, L. (2006). Nanocomposites of isotactic polypropylene reinforced with rod-like cellulose whiskers. Polymer *47*, 6285–6292.

Lönnberg, H., Fogelström, L., Azizi Samir, M.A.S., Berglund, L., Malmström, E. and Hult, A. (2008). Surface grafting of microfibrillated cellulose with poly(e-caprolactone) – Synthesis and characterization. Eur. Polym. J. *44*, 2991–2997.

Lu, Y., Weng, L. and Cao X. (2005). Biocomposites of plasticized starch reinforced with cellulose crystallites from cottonseed linter. Macromol. Biosci. *5*, 1101–1107.

Lu, Y., Weng, L. and Cao, X. (2006). Morphological, thermal and mechanical properties of ramie crystallites – Reinforced plasticized starch biocomposites. Carbohydr. Polym. *63*, 198–204.

Lu, J., Askeland, P. and Drzal, L.T. (2008a). Surface modification of microfibrillated cellulose for epoxy composite applications. Polymer *49*, 1285–1296.

Lu, J., Wang, T. and Drzal, L.T. (2008b). Preparation and properties of microfibrillated cellulose polyvinyl alcohol composite materials. Composites Part A *39*, 738–746.

Ma, H., Zhou, B., Li, H.S., Li, Y.Q. and Ou, S.Y. (2011). Green composite films composed of nanocrystalline cellulose and a cellulose matrix regenerated from functionalized ionic liquid solution. Carbohydr. Polym. *84*, 383–389.

Man, Z., Muhammad, N., Sarwono, A., Bustam, M.A., Kumar, M.V. and Rafiq, S. (2011). Preparation of cellulose nanocrystals using an ionic liquid. J. Polym. Environ. *19*, 726–731.

Marcovich, N.E., Auad, M.L., Belessi, N.E., Nutt, S.R. and Aranguren, M.I. (2006). Cellulose micro/ nanocrystals reinforced polyurethane. J. Mater. Res. *21*, 870–881.

Mathew, A.P. and Dufresne, A. (2002). Morphological investigation of nanocomposites from sorbitol plasticized starch and tunicin whiskers. Biomacromolecules *3*, 609–617.

Nakagaito, A.N. and Yano, H. (2008). The effect of fiber content on the mechanical and thermal expansion properties of biocomposites based on microfibrillated cellulose. Cellulose *15*, 555–559.

Nishino, T., Matsuda, I. and Hirao, K. (2004). All-cellulose composite. Macromolecules *37*, 7683–7687.

Nogi, M., Abe, K., Handa, K., Nakatsubo, F., Ifuku, S. and Yano, H. (2006). Property enhancement of optically transparent bionanofiber composites by acetylation. Appl. Phys. Lett. *89*, 233123-1–233123-3.

Nogi, M. and Yano, H. (2008). Transparent nanocomposites based on cellulose produced by bacteria offer potential innovation in the electronics device industry. Adv. Mater. *20*, 1849–1852.

Noishiki, Y., Nishiyama, Y., Wada, M., Kuga, S. and Magoshi, J. (2002). Mechanical properties of silk fibroin-microcrystalline cellulose composite films. J. Appl. Polym. Sci. *86*, 3425–3429.

Noorani, S., Simonsen, J. and Atre, S. (2007). Nano-enabled microtechnology: Polysulfone nanocomposites incorporating cellulose nanocrystals. Cellulose *14*, 577–584.

Nordin, S.B., Nyrén, J.O. and Back, E.L. (1973). Note on molten cellulose produced by a laser beam. Svensk Papperstidning *76*, 609–610.

Nyström, G., Mihranyan, A., Razaq, A., Lindström, T., Nyholm, L. and Strømme, M. (2010). A nanocellulose polypyrrole composite based on microfibrillated cellulose from wood. J. Phys. Chem. B *114*, 4178–4182.

Okahisa, Y., Yoshida, A., Miyaguchi, S. and Yano, H. (2009). Optically transparent wood-cellulose nanocomposite as a base substrate for flexible organic light-emitting diode displays. Compos. Sci.Technol. *69*, 1958–1961.

Oksman, K., Mathew, A. P., Bondeson, D. and Kvien, I. (2006). Manufacturing process of cellulose whiskers/polylactic acid nanocomposites. Compos. Sci. Technol. *66*, 2776–2784

Oksman, K., Etang, J.A., Mathew, A.P. and Jonoobi, M. (2011). Cellulose nanowhiskers separated from a bio-residue from wood bioethanol production. Biomass & Bioenergy *35*, 146–152.

Pandey, J.K., Chu, W.S., Kim, C.S., Lee, C.S. and Ahn, S.H. (2009a). Bio-nano reinforcement of environementally degradable polymer matrix by cellulose whiskers from grass. Composites: Part B *40*, 676–680.

Pandey, J.K., Lee, C.S and Ahn, S.H. (2009b). Preparation and properties of bio-nanoreinforced composites from biodegradable polymer matrix and cellulose whiskers. J. Appl. Polymer. Sci. *115*, 2493–2501

Pei, A., Zhou, Q. and Berglund, L. (2010). Functionalized cellulose nanocrystals as biobased nucleation agents in poly(L-lactide) (PLLA) – Crystallization and mechanical property effects. Compos. Sci. Technol. *70*, 815–821.

Peng, K., Wang, B., Chen, S., Zhong, C. and Wang, H. (2011). Preparation and properties of polystyrene/bacteria cellulose nanocomposites by *in situ* polymerization. J. Macromol. Sci. B *50*, 1921–1927.

Peresin, M.S., Habibi, Y., Zoppe, J.O., Pawlak, J.J. and Rojas, O.J. (2010). Nanofiber composites of polyvinyl alcohol and cellulose nanocrystals: Manufacture and characterization, Biomacromolecules *11*, 674–681.

Petersson, L., Kvien, I. and Oksman, K. (2007). Structure and thermal properties of poly(lactic acid)/cellulose whiskers nanocomposites materials. Compos. Sci. Technol. *67*, 2535–2544

Pranger, L. and Tannenbaum, R. (2008). Biobased nanocomposites prepared by *in situ* polymerization of furfuryl alcohol with cellulose whiskers or montmorillonite clay. Macromolecules *41*, 8682–8687.

Qi, H., Cai, J., Zhang, L. and Kuga, S. (2009). Properties of films composed of cellulose nanowhiskers and a cellulose matrix regenerated from alkali/urea solution. Biomacromolecules *10*, 1597–1602.

Quan, S.L., Kang, S.G. and Chin, I.J. (2010). Characterization of cellulose fibers electrospun using ionic liquid. Cellulose *17*, 223–230.

Rojas, O.J., Montero, G.A. and Habibi, Y. (2009). Electronspun nancomposites from polystyrene loaded with cellulose nanowhiskers. J. Appl. Polym. Sci. *113*, 927–935.

Roman, M. and Winter, W.T. (2004). Effect of sulphate groups from sulphuric acid hydrolysis on the thermal degradation behaviour of bacterial cellulose. Biomacromolecules *5*, 1671–1677.

Roohani, M., Habibi, Y., Belgacem, N.M., Ebrahim, G., Karimi, A.N. and Dufresne, A. (2008). Cellulose whiskers reinforced polyvinyl alcohol copolymers nanocomposites. Eur. Polym. J. *44*, 2489–2498.

Rosa, M.F., Medeiros, E.S., Malmonge, J.A., Gregorski, K.S., Wood, D.F., Mattoso, L.H.C., Glenn, G., Orts, W.J. and Imam, S.H. (2010). Cellulose nanowhiskers from coconut husk fibers: Effect of preparation conditions on their thermal and morphological behaviour. Carbohydr. Polym. *81*, 83–92.

Sanchez-Garcia, M.D. and Laragon, J.M. (2010). On the use of plant cellulose nanowhiskers to enhance the barrier properties of polylactic acid. Cellulose *17*, 987–1004.

Sanchez-Garcia, M.D., Hilliou, L. and Laragon, J.M. (2010). Morphology and water barrier properties of nanobiocomposites of k/i-hybrid carrageenan and cellulose nanowhiskers. J. Agric. Food Chem. *58*, 12847–12857.

Seitsonen, S. and Mikkonen, I. (1973). X-ray study of the thermal expansion of cotton cellulose. Polymer J. *5*, 263–267.

Shafizadeh, V. and DeGroot, W.F. (1976). Combustion characteristics of cellulosic fuels. In: Thermal uses and properties of carbohydrates and lignins, F. Shafizadeh, K.V. Sarkanen and D.A. Tillman, eds. (Academic Press, New York, San Fransisco, London), pp. 1–17.

Shimazaki, Y., Miyazaki, Y., Takezawa, Y., Nogi, M., Abe, K., Ifuku, S. and Yano, H. (2007). Excellent thermal conductivity of transparent cellulose nanofiber/epoxy resin nanocomposites. Biomacromolecules *8*, 2976–2978.

Siqueira, G., Bras, J. and Dufresne, A. (2009). Cellulose whiskers vs. microfibrils: influence of the nature of the nanoparticle and its surface functionalization on the thermal and mechanical properties of nanocomposites. Biomacromolecules *10*, 425–432.

Siqueira, G., Fraschini, C., Bras, J., Dufresne, A., Prud'homme, R. and Laborie, M.P. (2011). Impact of the nature and shape of cellulosic nanoparticles on the isothermal crystallization kinetics of poly(e-caprolactone). Eur. Polym. J. *47*, 2216–2227.

Suryanegara, L., Nakagaito, A.N. and Yano, H. (2009). The effect of crystallization of PLA on the thermal and mechanical properties of microfibrillated cellulose-reinforced PLA composites. Compos. Sci. Technol. *69*, 1187–1192.

Suryanegara, L., Okumura, H., Nakagaito, A.N. and Yano, H. (2011). The synergetic effect of phenylphosphonic acid zinc and microfibrillated cellulose on the injection molding cycle time of PLA composites. Cellulose *18*, 689–698.

Szcześniak, L., Rachocki, A. and Tritt-Goc, J. (2008). Glass transition temperature and thermal decomposition of cellulose powder. Cellulose *15*, 445–451.

Takahashi, M. and Takenaka, H. (1982). X-ray study of the thermal expansion and transition of crystalline cellulose. Polymer J. *14*, 675–679.

Teixeira, E.M., Pasquini, D., Curvelo, A.A.S., Corradini, E., Belgacem, M.N. and Dufresne, A. (2009). Cassava bagasse cellulose nanofibrils reinforced thermoplastic cassava starch. Carbohydr. Polym. *78*, 422–431.

Teixeira, E.M., Bondancia, T.J., Teodoro, K.B.R., Corrêa, A.C., Marconcini, J.M. and Mattoso, L.H.C. (2011). Sugar cane bagasse whiskers: Extraction and characterizations. Ind. Crops Prod. *33*, 63–66.

Ten, E., Turtle, J., Bahr, D., Jiang, L. and Wolcott, M. (2010). Thermal and mechanical properties of poly(3-hydroxybutyrate-co-3-hydroxyvalerate)/cellulose nanowhiskers composites. Polymer *51*, 2652–2660.

Tingaut, P., Zimmermann, T. and Lopez-Suevos, F. (2010). Synthesis and characterization of bionanocomposites with tunable properties from poly(lactic acid) and acetylated microfibrillated cellulose. Biomacromolecules *11*, 454–464.

Trovatti, E., Oliveira, L., Freire, C.S.R., Silvestre, A.J.D., Neto, C.P., Cruz Pinto, J.J.C. and Gandini, A. (2010). Novel bacterial cellulose-acrylic resin nanocomposites. Compos. Sci. Technol. *70*, 1148–1153.

Uddin, A.J., Araki, J. and Gotoh, Y. (2011). Toward "strong" green nanocomposites: Polyvinyl alcohol reinforced with extremely oriented cellulose whiskers. Biomacromolecules *12*, 617–624.

Wada, M. (2002). Lateral thermal expansion of cellulose $I_β$ and III_I polymorphs. J. Polym. Sci. B *40*, 1095–1102.

Wakelin, J.H., Sutherland, A. and Beck, L.R. (1960). Linear thermal expansion coefficients for the crystalline phase in high polymers. J. Polym. Sci. *42*, 278–280.

Wang, N., Ding, E. and Cheng, R. (2007). Thermal degradation behavior of spherical cellulose nanocrystals with sulfate groups. Polymer *48*, 3486–3493.

Xu, Q., Yi, J., Zhang, X. and Zhang, H. (2008). A novel amphotropic polymer based on cellulose nanocrystals grafted with azo polymers. Eur. Polym. J. *44*, 2830–2837.

Yano, H., Sugiyama, J., Nakagaito, A.N., Nogi, M., Matsuura, T., Hikita, M. and Handa, K. (2005). Optically transparent composites reinforced with networks of bacterial nanofibers. Adv. Mat. *17*, 153–155.

Yi, J., Xu, Q., Zhang, X. and Zhang, H. (2008). Chiral-nematic self-ordering of rodlike cellulose nanocrystals grafted with poly(styrene) in both thermotropic and lyotropic states. Polymer *49*, 4406–4412.

Yi, J., Xu, Q., Zhang, X. and Zhang, H. (2009). Temperature-induced chiral nematic changes of suspensions of poly(N,N-dimethylaminoethyl methacrylate)-grafted cellulose nanocrystals. Cellulose *16*, 989–997.

Zoppe, J.O., Peresin, M.S., Habibi, Y., Venditti, R.A. and Rojas, O.J. (2009). Reinforcing poly(e-caprolactone) nanofibers with cellulose nanocrystals. ACS Appl. Mater. Interfaces *1*, 1996–2004.

Zoppe, J.O., Habibi, Y., Rojas, O.J., Venditti, R.A., Johansson, L.S., Efimenko, K., Österberg, M. and Laine, J. (2010). Poly(N-isopropylacrylamide) brushes grafted from cellulose nanocrystals via surface-initiated single-electron transfer living radical polymerization. Biomacromolecules *11*, 2683–2691.

9 Mechanical properties of nanocellulose-based nanocomposites

The mechanical properties of nanocellulose are its most important and best investigated asset. Indeed, these properties are profitably exploited by Mother Nature since cellulose is a structural polymer that provides the mechanical strength and stiffness to higher plant cells. Impressive mechanical properties, nanoscale dimensions and high aspect ratio of nanocellulose make it therefore an ideal candidate for the reinforcement of polymeric materials. To improve the mechanical properties of the host material and take advantage of this property, special care needs to be paid to the processing of the nanocomposite. Indeed, it is well-known that the macroscopic mechanical properties of heterogeneous materials depend on the specific behavior of each phase and the composition (volume fraction of each phase), but also on the morphology (spatial arrangement of the phases) and the interfacial properties.

By adding conventional high modulus reinforcements into a polymer, usually the modulus and the strength of the composite are improved, while the ductility and the impact strength are decreased. In the case of well dispersed nanoparticles, the modulus and strength can be improved without any significant change in ductility since they do not create high stress concentrations due to their nanosize.

In the recent years, great interest has been focused on investigating the use of cellulose nanocrystals or MFC as a reinforcing phase in a polymeric matrix, evaluating the mechanical properties of the resulting composites and elucidating the origin of the mechanical reinforcing effect. Dynamic mechanical analysis (DMA) is a powerful tool to investigate the linear mechanical behavior of materials in a broad temperature/frequency range and it is strongly sensitive to the morphology of heterogeneous systems. Non-linear mechanical properties are generally accessed through classical tensile or compressive tests.

9.1 Pioneering works

The first demonstration of the reinforcing effect of cellulose nanocrystals was reported in 1995 (Favier et al., 1995a). It was shortly followed by a second paper (Favier et al., 1995b). In these studies, a copolymer of styrene and butyl acrylate (poly(S-co-BuA)) in the latex form was reinforced with cellulose nanocrystals extracted from tunicate. The reinforcing effect of the nanoparticles was accessed by DMA experiments in the shear mode.

Figure 9.1 shows the evolution of the storage shear modulus as a function of temperature for this system. The curve corresponding to the neat matrix is typical of an amorphous thermoplastic. For temperatures below its glass transition temperature (Tg) the copolymer is in the glassy state and the modulus remains roughly constant around 1 GPa. It is well known that the modulus for any glassy polymer, either fully amorphous or semicrystalline, is of the order of GPa (10^9 Pa). It is due to the fact that in the glassy state, molecular motions are largely restricted to vibration and short-range rotational motions. Then, a sharp decrease in the elastic modulus was observed (by more than four decades), associated with the anelastic manifestation of the glass-rubber transition. This modulus drop corresponds to an energy dissipation phenomenon displayed in the concomitant relaxation process where the loss angle tan δ passes through a maximum. This relaxation process (main relaxation) involves cooperative motions of long chain sequences. The temperature position of this relaxation process depends on both Tg and the frequency of the measurement. It also depends on the magnitude of the storage modulus drop (mechanical coupling effect). For an

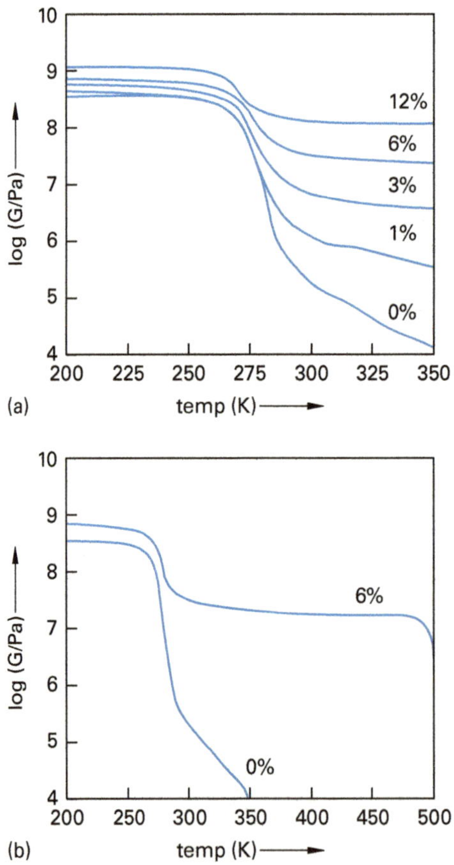

Fig. 9.1: Logarithm of the storage shear modulus as a function of temperature for poly(S-co-BuA) reinforced with tunicin nanocrystals. Panel (a) shows the reinforcing effect obtained for various nanocrystal contents and panel (b) shows the improvement of the thermal stability of the matrix (Favier et al., 1995b).

amorphous polymer, the temperature range of the subsequent rubbery plateau is well known to depend on its molecular weight. In the terminal zone, the modulus became even lower with increasing temperature and the setup failed to measure it. It corresponds to the irreversible flow of the polymer induced by chain disentanglement.

For nanocomposite films, the authors measured a spectacular improvement in the storage modulus after adding tunicin nanocrystals into the host polymer even at low content (Figure 9.1(a)). This increase was especially significant above the glass-rubber transition temperature of the thermoplastic matrix because of its poor mechanical properties in this temperature range. In the rubbery state of the thermoplastic matrix, the modulus of the composite with a loading level as low as 6 wt% was more than two orders of magnitude higher than the one of the unfilled matrix. Moreover, the introduction of 6 wt% or more cellulosic nanocrystals provides an outstanding thermal stability of the matrix modulus (Figure 9.1(b)) up to the temperature at which cellulose starts to degrade (500K).

9.2 Modeling of the mechanical behavior

Modeling of the mechanical behavior of nanocellulose reinforced nanocomposites is much easier to perform with cellulose nanocrystals compared to microfibrillated cellulose (MFC) because of the more defined morphology of the former. In multiphase polymer systems, the mechanical behavior depends on the specific behavior of each phase, the composition (volume fraction of each phase), the morphology (spatial arrangement of the phases) and the interfacial properties. For heterogeneous materials, the relationship between the elastic modulus and these different parameters has been extensively studied. One of the simplest models involves connections in series (Reuss prediction) or in parallel (Voigt prediction) of the components, and leads to the lowest lower bound and highest upper bound for the moduli, respectively. It corresponds to a mixing rule on the compliance and modulus of the components, respectively. Regardless of the morphology of the binary system, the experimental modulus value lies between these two limits that can be very different.

9.2.1 Mean field approach

In the pioneering work on tunicin nanocrystal reinforced poly(S-co-BuA), an attempt was made to understand the unusual reinforcing effect provided at high temperatures by the nanoparticles from their high modulus. Modeling of the mechanical behavior of these nanocomposites was performed using the theoretical model of Halpin–Kardos (Halpin and Kardos, 1972) which has been extensively used to predict the elastic shear modulus of short-fiber composites and semicrystalline polymers. This model was chosen for its simplicity and ability to account for the viscoelastic behavior. The

Halpin–Kardos model is known as the "upper bond" of the mean field approaches. It is worth noting that for temperatures much larger or well below Tg, the mechanical behavior can be considered in a first estimation as elastic. In fact the loss component of the modulus is more than 10 times lower than the elastic one. For this reason, only the elastic behavior was considered. In this mean field approach, fibers are assumed to be dispersed in the matrix to form a homogeneous continuum. The composite is assimilated to a "quasi isotropic" material framed by four layers of oriented plies (at 0°, 45°, 90°, and −45°) (Figure 9.2).

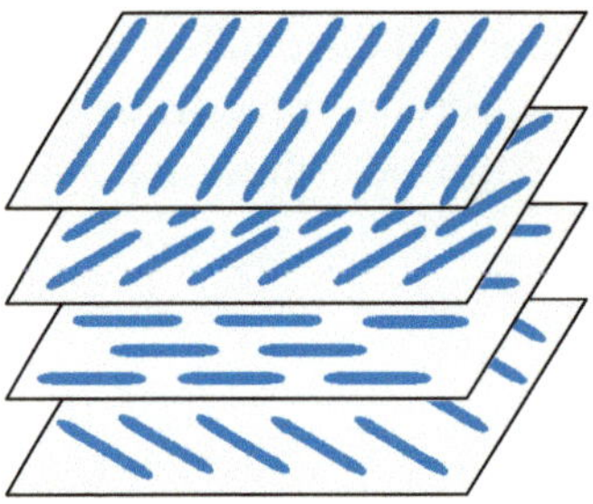

Fig. 9.2: Schematic representation of a composite with four layers of oriented fillers (0°, 45°, 90°, −45°) in the Halpin–Kardos model.

The mechanical properties become isotropic as the number of distinct equal-angular ply orientations increases. The mechanical properties of a single ply (unidirectional short fibers) are given by the Halpin–Tsaï micromechanics equations ("self-consistent" approach) (Tsaï et al., 1968):

$$\frac{E_{ii}}{E_m} = \frac{(1+\xi_{ii}\phi_f)\,E_{iif} + \xi_{ii}\,(1-\phi_f)\,E_m}{(1-\phi_f)\,E_{iif} + (\xi_{ii}+\phi_f)\,E_m} \qquad (i = 1, 2)$$

$$\frac{G_{12}}{G_m} = \frac{(1+\phi_f)\,G_f + (1-\phi_f)\,G_m}{(1-\phi_f)\,G_f + (1+\phi_f)\,G_m} \tag{9.1}$$

where E_{11} is the stiffness in the fiber direction, E_{22} the stiffness estimate perpendicular to the fiber direction, G_{12} the in-plane shear modulus estimate, and ϕ_f the fiber volume fraction. The subscripts m and f refer to the matrix and the filler, respectively. The geometry of the filler is involved through the ξ_{ii} parameters, where L, ℓ and e are the length, the width, and the thickness of the fibers, respectively. For cellulose nanocrystals assimilated to rods, ℓ = e = d, the diameter of the nanocrystal, and it follows (Halpin and Kardos, 1972):

$$\xi_{11} = 2 \cdot \frac{L}{e}$$

$$\xi_{22} = 2 \cdot \frac{\ell}{e} = 2 \tag{9.2}$$

The engineering constants characteristic of the unidirectional plies are then given as follows:

$$Q_{ii} = \frac{E_{ii}}{1 - \nu_{12} \cdot \nu_{21}} \qquad (i = 1, 2)$$

$$Q_{12} = \nu_{12} \cdot Q_{22} = \nu_{21} \cdot Q_{11}$$

$$Q_{66} = G_{12} \tag{9.3}$$

which leads to the following expressions for the invariant terms:

$$U_1 = \frac{1}{8}\left(3Q_{11} + 3Q_{22} + 2Q_{12} + 4Q_{66}\right)$$

$$U_2 = \frac{1}{8}\left(Q_{11} + Q_{22} - 2Q_{12} + 4Q_{66}\right) \tag{9.4}$$

The tensile modulus of the "quasi-isotropic" laminate, the behavior of which is assumed to be close to the one of the short fiber composite, is then given by:

$$E_c = \frac{4U_2\left(U_1 - U_2\right)}{U_1} \tag{9.5}$$

The Poisson's ratio ν_{12} is approximately given by a mixing rule:

$$\nu_{12} = \nu_m \phi_m + \nu_f \phi_f = \frac{Q_{11}}{Q_{22}} \cdot \nu_{21} \tag{9.6}$$

Therefore, in this approach the predicted properties of the composite depend on the size, shape, and volume fraction of the filler and on the mechanical properties of both the matrix and the fibers, including the mechanical anisotropy of the filler. Hence, the high modulus of cellulose nanocrystals is accounted for but it is worth noting that no interactions between fibers are taken into account as shown by the scheme in Figure 9.2.

This model has been applied to predict the mechanical behavior of poly(S-co-BuA) nanocomposites reinforced with tunicin (Favier et al., 1995a; Hajji et al., 1996) and wheat straw (Helbert et al., 1996) cellulose nanocrystals, as well as poly(vinyl

Cellulose Nanocrystals

Parameter	Source of CNC	Value	Reference
E_{11f}	Tunicin	150 GPa	(Favier et al., 1995a)
		130 GPa	(Hajji et al., 1996; Chazeau et al., 1999a)
	Wheat Straw	150 GPa	(Helbert et al., 1996; Dufresne et al., 1997)
E_{22f}	Tunicin	15 GPa	(Favier et al., 1995a; Chazeau et al., 1999a)
		5 GPa	(Hajji et al., 1996)
	Wheat Straw	15 GPa	(Helbert et al., 1996; Dufresne et al., 1997)
G_f	Tunicin	5 GPa	(Favier et al., 1995a; Chazeau et al., 1999a)
		1.77 GPa	(Hajji et al., 1996)
	Wheat Straw	5 GPa	(Helbert et al., 1996; Dufresne et al., 1997)
L /d	Tunicin	100	(Favier et al., 1995a)
		66.5	(Chazeau et al., 1999a)
	Wheat Straw	45	(Helbert et al., 1996)
ν_f	Tunicin	0.3	(Favier et al., 1995a; Helbert et al., 1996; Hajji et al., 1996; Chazeau et al., 1999a)
	Wheat Straw		

Polymeric Matrix

Parameter	Polymer	Temperature (K)	Value	Reference
G_m	Poly(S-co-BuA)	325	0.1 MPa	(Favier et al., 1995a)
	PVC/DOP	235	740 MPa	(Chazeau et al., 1999a)
		280	0.53 MPa	
E_m	Poly(S-co-BuA)	233	980 MPa	(Hajji et al., 1996)
		296	0.2 MPa	
		225	1.97 GPa	(Helbert et al., 1996)
		325	0.26 MPa	
ν_m	Poly(S-co-BuA)	325	0.5	(Favier et al., 1995a; Helbert et al., 1996)
		233	0.3	(Hajji et al. 1996)
		296	0.5	
		225	0.3	(Helbert et al., 1996)
	PVC/DOP	225	0.35	(Chazeau et al., 1999a)
		325	0.5	

Table 9.1: Mechanical parameters used for cellulose nanocrystals and polymeric matrix for Halpin–Kardos modeling.

chloride) (PVC) reinforced with tunicin nanocrystals (Chazeau et al., 1999a). The parameters used in the calculation are reported in Table 9.1.

The mechanical properties of cellulose nanocrystals were taken from the literature (Chapter 1, Table 1.4). The aspect ratio (L/d) can be averaged from microscopic observations and the Poisson's ratio of cellulose nanoparticles was taken as 0.3 for cellulose in the glassy crystalline state. The mechanical properties of the polymer matrix were experimentally determined and its Poisson's ratio was taken as 0.5 in the rubbery state and 0.3–0.35 in the glassy state. Therefore, no adjustable parameter was used except in (Hajji et al., 1996) for which the aspect ratio was adjusted to fit the experimental data.

It was found that the predicted rubbery modulus failed to describe the experimental data. The glassy modulus of the composite was overestimated whereas the rubbery modulus was underestimated (Helbert et al., 1996; Chazeau et al., 1999a). However, a good agreement was reported in the glassy state of the matrix for modulus values derived from tensile tests (Hajji et al., 1996). A numerical identification of the potential of whisker- and platelet-filled polymers was reported and it appeared that the widely used Halpin–Tsaï equation systematically and considerably underestimates the actual potential of these materials (Gusev, 2001). It was difficult to question the parameters used in the Halpin–Kardos prediction except possibly the aspect ratio of the nanoparticle. Indeed, it was suggested that cellulose nanocrystals can connect together to form longer entities or strings of nanoparticles. For nanocrystals prepared from wheat straw fibers, the aspect ratio was used as an adjustable parameter and experimental data were successfully fitted with an aspect ratio of 450, i.e. 10 times higher than the one derived from morphological observations (45) (Helbert et al., 1996). A mechanical percolation phenomenon was therefore suspected.

9.2.2 Percolation approach

The term percolation for the statistical-geometry model was first introduced in 1957 (Hammersley, 1957). It is a statistical theory, which can be applied to any system involving a high number of species likely to be connected. The aim of the statistical theory is to forecast the behavior of an incompletely connected set of objects. By varying the number of connections, this approach allows describing the transition from a local to an infinite "communication" state. The percolation threshold is defined as the critical volume fraction separating these two states. Various parameters, such as particle interactions (Balberg and Binenbaum, 1983), orientation (Balberg et al., 1984) or aspect ratio (de Gennes, 1976) can modify the value of the percolation threshold. The use of this approach to describe and predict the mechanical behavior of cellulosic nanocrystal-based composites suggests the formation of a rigid network of nanocrystals which should be responsible for the unusual reinforcing effect observed at high temperatures. The modeling consists of three important steps:

(i) The first step consists in determining the percolation threshold (v_{Rc}). The volume fraction of cellulose nanoparticles required to achieve geometrical percolation can be calculated using a statistical percolation theory for cylindrical shape particles according to their aspect ratio and the effective skeleton of nanocrystals (Favier et al., 1997a). A three-dimensional set of 5000 rods was generated at random, both in location and in orientation, and possible intersections and formation of connected nanocrystals corresponding to infinite length branch of nanoparticles connecting both sides of the specimen were investigated. The critical volume fraction needed to reach geometrical percolation was found to decrease sharply when increasing the aspect ratio of the rods and for aspect ratios significantly different from unity the following relation was determined:

$$v_{Rc} = \frac{0.7}{L/d} \tag{9.7}$$

For nanorods with an aspect ratio of 67, such as tunicin nanocrystals, a volume fraction around 1% is then required to achieve a 3D geometrical percolation. This calculation can help to identify the effective skeleton of nanocrystals, i.e. those that effectively belong to an infinite length branch, by eliminating finite length branches. For wheat straw cellulose nanocrystals with an aspect ratio of 45, the v_{Rc} value was found to be around 2 vol% (Dufresne et al., 1997). However, it can be shown that the percolation threshold calculated for a non-uniform length distribution is lower than the one calculated for a uniform length distribution with the same average length (Balberg and Binenbaum, 1983).

Nevertheless, the concept of percolation appears rather fuzzy since mechanical percolation is not identical to geometrical percolation, unlike electrical percolation. Formation of a continuous pathway of conducting particles in an insulating matrix is enough to make the composite system conducting. DC conductivity measurements were performed to characterize the connectivity of cellulose nanocrystals and evaluate their percolation threshold (Flandin et al., 2000). In this study, the surface of the tunicin nanocrystals was covered with conductive polypyrrole before incorporation in a poly(S-co-BuA) latex matrix. The coated nanocrystals were characterized by scanning electron microscopy (SEM) and it was found that almost the entire cellulose surface was covered with polypyrrole. The thickness of the polypyrrole layer was around 70 nm and the final nanoparticles had an average length of about 2 μm and a diameter of 160 nm, leading to an aspect ratio close to 15. The percolation threshold was determined to be about 3 vol% in agreement with the percolation theory and experimental data (Nan, 1993) for an aspect ratio close to 15.

However, the situation is somewhat different for mechanical stiffness because if a continuous pathway of rigid particles is built throughout a compliant matrix, it is not enough to ensure the rigidity of the whole system. Indeed, it largely depends on the bonds linking the particles.

(ii) The second step consists therefore in the estimation of the stiffness of the percolating filler network. It is obviously different from the one of individual nanoparticles and depends on the origin of the cellulose, preparation procedure of the nanocrystals and nature and strength of inter-particles interactions. Basically, this modulus can be assumed to be similar to that of a paper sheet for which the hydrogen bonding forces provide the basis of its stiffness. As reported in Chapter 3 (Section 3.8), experimental mechanical tests have been performed on films prepared by water evaporation of cellulose nanocrystal suspensions. Values ranging between 500 MPa and 15 GPa were reported (Chapter 3, Figure 3.12) depending on the source of cellulose. These values are close to those reported for sheets prepared from bacterial cellulose (17 GPa on average) (Yamanaka et al., 1989) or isotropic sheets of hydrolyzed crystallites of cotton cellulose (10 GPa) (Marchessault et al., 1978). However, the experimental conditions such as cross head speed used for tensile tests and moisture content of the sample can affect the stiffness of the specimen. Indeed, disengagement of the nanocrystal network due to competitive hydrogen bonding with water is expected to reduce the interactions between the nanocrystals and therefore the stiffness of the film (Dagnon et al., 2012) as shown for MFC in Figure 2.9 (Chapter 2). It is interesting to note that a correlation exists between the aspect ratio of the nanocrystals and mechanical stiffness of the mat (as shown in Chapter 3, Figure 3.12). The stiffness of the film was found to increase with the aspect ratio of the nanocrystals (Bras et al., 2011). It therefore means that the use of higher aspect ratio cellulose nanocrystals is more interesting from a mechanical point of view because it first induces a decrease of the critical percolation threshold and also stiffens the formed continuous network. Correlation between nanocrystal aspect ratio and stronger inter-nanoparticle interactions was also proposed to account for the high reinforcing capability of cellulose nanocrystals extracted from *Syngonanthus nitens* (Capim Dourado) (Siqueira et al., 2010a).

The mechanical percolation effect has been investigated using a two-dimensional finite element approach (Favier et al., 1997b). Two-dimensional handmade meshes were created from a random distribution of nanorods and the nanocrystal area fractions ranged from 0.9% to 9.5%, the 2D percolation threshold being close to 4.8%. The points where nanorods crossed each other were taken as nodes of the finite element model. The nodes were assumed to provide “moment-less” hinge or “rotation-less” rigid connections. In the former case, angle change was allowed between two fibers meeting at the nodes, whereas in the latter case any angle change was not allowed. These elements were called bar elements and beam elements, respectively. A bar network is suitable if the bond is considered to be weak, whereas a beam network is appropriate for strong bond situations. For a hard matrix (glassy state), the modulus was found to vary almost linearly with the nanocrystal content and no significant difference between bar and beam models was observed. Above Tg of the matrix, the results revealed that a cellulose nanocrystal percolating network acts more like a network of beams rather than bars, and that connections between nanorods are rigid and do not allow the relative rotation of connected nanocrystals. It was suggested that hydrogen bonds played an important role in the mechanical behavior of cellulose nanocrystal network. The number of hydrogen bonds between two tunicin nanocrystals in contact was calculated assuming parallelepipedic rods with a square section $d^2 = 15 \cdot 15\ nm^2$. For the smallest contact area, d^2, it was found that more than 1500 hydrogen bonds can be formed between two perpendicular nanocrystals (Favier et al., 1997b).

The apparent tensile modulus of a tunicin nanocrystal network was also calculated by a 3D finite elements simulation (Brechet et al., 2001). The linking elements were considered as beams with adjustable stiffness. All the calculated values were lower than 1 GPa. For link modulus values below 1 GPa, the network modulus was found to increase with increasing nanocrystal concentration and seemed to increase linearly with the link modulus. For higher linking modulus, the modulus of the percolating network tends toward the value for totally rigid links. It was shown that rather tight bonds between the nanocrystals are needed to reproduce a sufficient rigidity of the material, which can be achieved by the hydrogen bonds that can be created when nanocrystals come into contact. Based on an evaluation of the volume fraction of the effective skeleton, a simple Voigt model seemed enough to reproduce well the observed stiffness.

(iii) Finally, the third step for the prediction of the mechanical properties of cellulose nanocrystal reinforced nanocomposites is the description of the composite using a model involving three different phases, viz. the matrix, the filler percolating network and the non-percolating filler phase. The simplest model consists of two parallel phases, namely the effective nanocrystal skeleton and the rest of the sample. For studying the mechanical behavior of poly(methyl methacrylate) and poly(S-co-BuA) blends, the classic phenomenological series-parallel model (Takayanagi et al., 1964) has been extended to propose a model in which the percolating filler network is set in parallel with a series part composed of the matrix and the non-percolating filler phase (Ouali et al., 1991) (Figure 9.3).

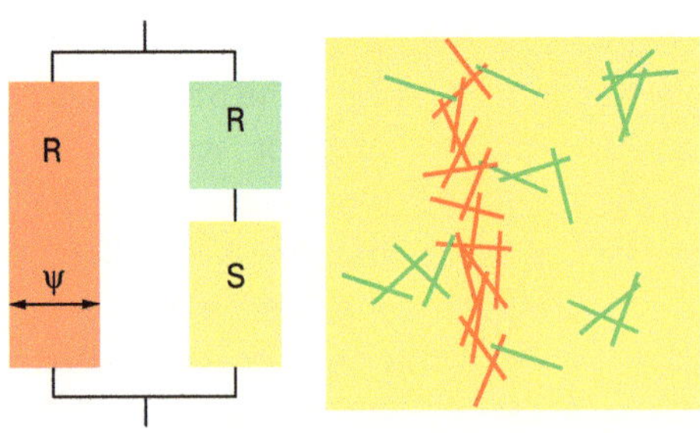

Fig. 9.3: Schematic representation of the series-parallel model. R and S refer to the rigid (cellulosic filler) and soft (polymeric matrix) phases, respectively, and ψ is the volume fraction of the percolating rigid phase. Red and green rods correspond to percolating and unpercolating nanoparticles, respectively.

In this approach, the elastic tensile modulus E_c of the composite is given by the following equation:

$$E_c = \frac{(1 - 2\psi + \psi v_R) E_S E_R + (1 - v_R) \psi E_R^2}{(1 - v_R) E_R + (v_R - \psi) E_S} \tag{9.8}$$

The subscripts S and R refer to the soft and rigid phase, respectively. The adjustable parameter, ψ, involved in the series-parallel model corresponds in the adapted prediction to the volume fraction of the percolating rigid phase. With b being the critical percolation exponent, ψ can be written as:

$$\psi = 0 \quad \text{for} \quad v_R < v_{Rc}$$

$$\psi = v_R \left(\frac{v_R - v_{Rc}}{1 - v_{Rc}} \right)^b \quad \text{for} \quad v_R > v_{Rc} \tag{9.9}$$

where b = 0.4 (de Gennes, 1979; Stauffer, 1985) for a 3D network.

At high temperatures when the polymeric matrix could be assumed to have zero stiffness ($E_S \sim 0$), the calculated stiffness of the composites is simply the product of the percolating filler network and the volume fraction of percolating filler phase:

$$E_c = \psi E_R \tag{9.10}$$

In the former study dealing with tunicin nanocrystals reinforced poly(S-co-BuA), a good agreement between experimental and predicted data was reported when using the Takayanagi's series-parallel model modified to include a percolation approach (Favier et al., 1995a; Favier et al., 1995b). It was therefore suspected that all the stiffness of the material was due to infinite aggregates of cellulose nanocrystals. Above the percolation threshold the cellulosic nanoparticles can connect and form a 3D continuous pathway through the nanocomposite film. The formation of this cellulose network was supposed to result from strong interactions between nanocrystals, like hydrogen bonds. This mechanical percolation effect allowed explaining both the high reinforcing effect and the thermal stabilization of the composite modulus up to 500 K for evaporated films (Figure 9.1).

A unified description of the moduli of nanocomposites containing elongated filler particles over a range of volume fractions spanning the filler percolation threshold has been provided (Chatterjee, 2006). A model has been developed for both the tensile and shear elastic moduli of three-dimensional fiber networks using the semi empirical Halpin–Tsai equations combined with results from percolation theory. Estimates have been developed for the strains and at the elastic limits under tensile and shear deformation, and model calculations have been presented for the dependencies of composite moduli and yield strains on particle aspects ratios and volume fractions.

The dependence of the percolation threshold in systems of cylindrical rod-like nanoparticles upon polydispersity and number-averaged aspect ratio was also examined (Chatterjee, 2008). In this study, a model based on excluded volume considerations was presented for the connectedness percolation thresholds. Calculations employing a model integrating results from percolation theory with micromechanical and effective medium approaches were compared to experimental measurements reported for composites reinforced with cellulosic nanoparticles (Prokhorova and Chatterjee, 2009). For the different systems, a semi-quantitative description of the observed mechanical behavior was provided. The impact of particle clustering upon the percolation threshold and percolation probability was also investigated (Chatterjee, 2011). Percolation was described by interpenetrable rod-like particles in terms of the Bethe lattice and the model was generalized to account for local clustering effects. The formation of local, physically connected cliques of particles was shown to raise the percolation threshold whilst reducing the percolation and backbone fractions for a fixed volume fraction of particles.

The existence of such a three-dimensional percolating nanoparticle network was evidenced by performing successive tensile tests on crab shell chitin nanocrystals reinforced natural rubber (NR) (Gopalan Nair and Dufresne, 2003). The experiment consisted in stretching the material up to a given elongation and then releasing the force, and stretching the material again up to a higher elongation, and so on (see Section 9.7.2 of the present chapter). From these experimental data, the true stress and the true strain were calculated. For each cycle, the tensile modulus was determined and the relative tensile modulus, that is the ratio of the modulus measured during a given cycle to the one measured during the first cycle, was plotted as a function of both the number of cycles and the nanocrystal content. For the unfilled NR matrix, an increase of the relative modulus during successive tensile tests was observed. It was ascribed to the well-known strain-induced crystallization of the NR matrix. For composites, the behavior was reported to be totally different. During the successive tensile tests, the relative tensile modulus first decreased and then slowly increased. The initial decrease of the modulus for composite materials was ascribed to the progressive damaging of the polysaccharide nanocrystal network. This is an indication that the tensile behavior of the composites is mainly governed by the percolating nanoparticle network. After the complete destruction of the network, the modulus started to slowly increase as a result of the strain-induced crystallization of the matrix already observed for the unfilled sample. In this strain range, the composite modulus becomes matrix dominated.

Therefore, any factor that affects the formation of the percolating nanocrystal network or interferes with it changes the mechanical performances of the composite (Dufresne, 2006). Three main parameters were reported to affect the mechanical properties of such materials, viz. the morphology and dimensions of the nanoparticles, the processing method, and the microstructure of the matrix and matrix/filler interactions. Obviously, the potential relative reinforcing effect of cellulosic nanoparticles increases as the stiffness of the neat matrix decreases. Indeed, for semicrystalline polymers the rubbery modulus is known to depend on the degree of crystallinity of the material. The crystalline regions act as physical cross-links for the elastomer and in this physically cross-linked system, the crystalline regions would also act as filler particles due to their finite size, which would increase the modulus substantially and therefore blur the reinforcing capacity of nanocellulose. A simple mixing rule allows accounting for this phenomenon.

The effect of these parameters on the mechanical performances of nanocellulose reinforced nanocomposite is reported and discussed below.

9.3 Influence of the morphology of the nanoparticles

Cellulose nanoparticles occur as rod-like nanoparticles when prepared by a selective acid hydrolysis treatment (nanocrystal) and long entangled filament network when prepared by mechanical shearing (microfibrillated cellulose – MFC).

For rod-like particles, the geometrical aspect ratio (L/d, L being the length and d the diameter of the particle) is of course an important factor since it determines the percolation threshold value. Moreover, a higher aspect ratio value results in higher stiffness of the percolating nanoparticle network (Bras et al., 2011). This factor is linked to the source of cellulose and acid hydrolysis conditions. Rods with a high aspect ratio give the highest reinforcing effect. For instance, the rubbery storage tensile modulus was systematically lower for wheat straw nanocrystal/poly(S-co-BuA) composites than for tunicin nanocrystal-based materials (Dufresne, 2006). In addition, for the former system, a gradient of nanocrystal concentration between the upper and lower faces of the composite film was reported and evidenced by scanning electron microscopy (SEM), wide angle X-ray scattering and DMA (Dufresne et al., 1997). It was ascribed to a translational motion of the nanocrystals leading to their sedimentation during the evaporation step. The mechanical behavior of wheat straw cellulose-based composites was well described by using a multilayered model consisting in layers parallel to the film surface. High cellulose nanocrystal content reinforced nanocomposites were rarely studied but the difference in mechanical behavior should most probably vanish in this composition range regardless of the aspect ratio of the nanoparticle. This is well predicted from the series-parallel model of Takayanagi modified to include the percolation approach that shows a sharp increase of the reinforcing capability of polysaccharide nanoparticles just above the percolation threshold and a leveling-off of this effect at higher filler contents. The nanocrystal network becomes sufficiently closed to smooth this effect.

Also, the flexibility and tangling possibility of the nanofibers play an important role. This was exemplified for poly(S-co-BuA) reinforced with cellulose nanoparticles extracted from sugar beet and submitted to different hydrolysis conditions (Azizi Samir et al., 2004a). Unhydrolyzed MFC and MFC submitted to a hydrolysis treatment using different sulfuric acid concentrations were used as the reinforcing phase. As the acid concentration increased, it was observed that the length of the microfibrils decreased because of the increased probability of removal of the amorphous material with the concentration of acid. For the strongly hydrolyzed material, most of the amorphous domains were hydrolyzed and rigid rod-like nanocrystals were obtained. DMA experiments performed on poly(S-co-BuA) reinforced with these nanoparticles did not show significant differences by varying the strength of the hydrolysis step and then the length and flexibility of the nanoparticles. However, from non-linear mechanical tensile tests it was observed that as the hydrolysis strength increased both the modulus and the strength of the composite decreased, whereas the elongation at break increased. This result showed the strong influence of flexible nanoparticle entanglements on the mechanical behavior of the nanocomposites. Conversely, increased reinforcement capacity was observed for enzyme treated bacterial cellulose (BC) reinforced glycerol plasticized starch films (Woehl et al., 2010). It was ascribed to two different mechanisms. The first one was the elimination of less organized regions between fibers causing entanglements in the original material, thus allowing better dispersion within the matrix. The second effect was the reduction of defects at the surface of fibers that could act as crack propagator. The crack-stopping capability of cellulose microfibrils and improvement of strength were also observed for BC reinforced phenol formaldehyde (Nakagaito and Yano, 2004) and MFC reinforced amylopectin films (Plackett et al., 2010).

Similar investigations were reported for cellulose nanoparticles extracted from sisal (Siqueira et al., 2009), date palm tree (Bendahou et al., 2010) and alfa fibers (Ben Mabrouk et al., 2012). For each system, the possibility of entanglement and higher aspect ratio of MFC has been emphasized to explain the higher stiffness of nanocomposites compared to nanocrystal reinforced materials. Moreover, the presence of residual lignin, extractive substances and fatty acids at the surface of MFC was also invoked as potential compatibilizer with the polymeric matrix (Bendahou et al., 2010). These entanglements limit the deformation of the matrix and a lower elongation at break was reported for MFC-based composites.

Cellulosic nanoparticles prepared from sisal fibers using different processing routes, viz. a combination of mechanical shearing, acid and enzymatic hydrolysis, were shown to display a broad range of morphologies (Siqueira et al., 2010b). The reinforcing capacity of these nanoparticles was investigated by preparing nanocomposites using NR as the matrix (Siqueira et al., 2011). A broad spectrum of mechanical behaviors was observed for similar composite compositions (the cellulose content was restricted to 6 wt%) with relative tensile modulus and elongation at break values ranging from 1.7 to 218 and from 0.01 to 1.47, respectively. The hydrolysis time and therefore dimensions of cellulose nanocrystals were shown to have a greater effect on the toughness than on the tensile strength of glycerol plasticized starch films (Chen et al., 2009).

A significantly higher reinforcing effect was observed when comparing cellulose nanocrystals and cellulose nanoballs prepared from fully bleached pine kraft pulp as nanoreinforcer in an acrylic latex matrix (Pu et al., 2007). The former had an average aspect ratio of 20–50 whereas the latter had an average diameter around 80 nm and an aspect ratio of 1. No reinforcing effect was observed for nanoballs.

9.4 Influence of the processing method

The processing method governs the homogeneous dispersion and possible formation of a continuous nanoparticle network and then the final properties of the nanocomposite material. As seen in Chapter 7, in most studies nanocellulose reinforced nanocomposites are prepared in liquid medium, using polymer solution or polymer dispersion (latex). Its main advantage is that it allows preserving the dispersion state of the nanoparticles in the liquid medium. This dispersion state is good in aqueous medium and can be acceptable in some organic solvents. The polymer melt approach is probably the most convenient processing technique as stated in Chapter 7, because it is applicable to all thermoplastic polymers and constitutes a greener approach, but it is challenging because of the inherent incompatibility of cellulose with most polymeric matrices and thermal stability issues.

Slow wet processes such as casting/evaporation were reported to give the highest mechanical performance materials compared to other processing techniques. Indeed, during liquid evaporation strong interactions between nanoparticles can settle and promote the formation of a strong percolating network. Comparisons between experimental data and predicted values calculated from the percolation approach can be used to ensure that good dispersion and effective percolation occur. Cast/evaporated samples correspond to the highest mechanical properties that can be reached for a given polymeric system.

The effect of processing technique on mechanical properties was reported for tunicin nanocrystal reinforced poly(S-co-BuA) (Hajji et al., 1996). Three types of composite films were prepared from aqueous mixture of the nanoparticle suspension and polymer latex by (i) casting/evaporation, (ii) freeze-drying and hot-pressing, and (iii) freeze-drying, extruding and then hot-pressing. The nanocrystal content was varied between 0 and 6 wt% and the mechanical behavior was investigated through tensile tests in both the glassy and rubbery state of the matrix. The processing methods have been classified in ascending order of their reinforcement efficiency as extrusion < hot-pressing < evaporation. It was ascribed to a breaking and/or orientation effect of the nanoparticles induced during shear stresses induced by freeze-drying/molding or freeze-drying/extruding/molding techniques. A similar observation was reported for MFC obtained from sugar beet pulp reinforced poly(S-co-BuA) nanocomposites using evaporation and freeze-drying/hot-pressing techniques (Dalmas et al., 2006).

During slow water (or organic liquid) evaporation, because of Brownian motion in the suspension or solution (whose viscosity remains low up to the end of the process when the latex particle or polymer concentration becomes very high), the rearrangement of the nanoparticles is possible. They have time enough to interact and connect to form a percolating network which is the basis of their reinforcing effect. The resulting structure (after the coalescence of latex particles or and/or interdiffusion of polymeric chains) is completely relaxed and direct contacts between nanocrystals or microfibrils are then created. Conversely, during the freeze-drying/hot-pressing process, the nanoparticle arrangement in the suspension is first frozen, and then, during the hot-pressing stage, because of the polymer melt viscosity, the particle rearrangements are strongly limited. Thus, in this case, contacts are made through a certain amount of polymer matrix. However, although the freeze-drying/hot-pressing process limits the possibility of creation of hydrogen bonds it is expected that for high nanoparticle content some bonds may evenly be created.

For wheat straw nanocrystal reinforced poly(S-co-BuA), the higher reinforcing effect observed for evaporated samples compared to freeze-dried and hot-pressed specimens was also explained in terms of sedimentation of the nanoparticles during the evaporation step (Dufresne et al., 1997). In this study, the samples were carefully cut with a razor blade in the film thickness and dynamic mechanical characterization was carried out on these two half-samples. The modulus of the upper face was found to be systematically lower than the one of the lower face and the behavior of the whole sample was closer to the one of the stiffer lower face rather than the one of the softer upper face as expected for a two-phase system in parallel arrangement (Figure 9.4).

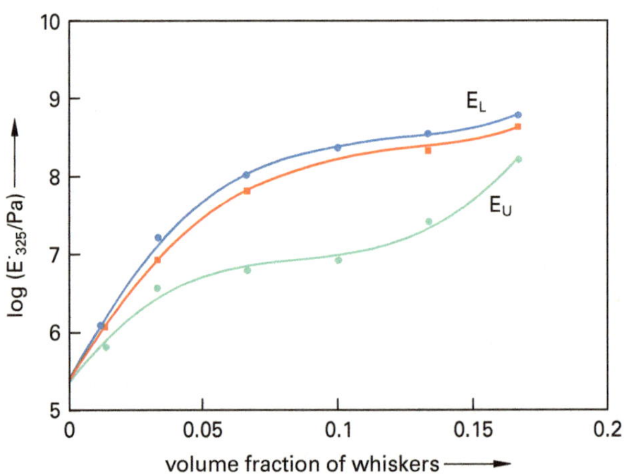

Fig. 9.4: Evolution of the logarithm of the relaxed tensile modulus at Tg + 50 K (E'325) of the poly(S-co-BuA) matrix as a function of the volume fraction of wheat straw nanocrystals. (■) refers to the whole sample, (●) to the lower half-sample and (●) to the upper half-sample (Dufresne et al., 1997).

Indeed, the system can be assimilated to the Voigt model for which the mechanical behavior is governed by the rigid phase. The sedimentation phenomenon and higher nanocrystal content in the lower part of the sample was also evidenced from SEM observation and X-ray diffraction experiments.

It was shown that nanocrystal dimensions affect their ability to align within a shear field (Orts et al., 1998). Neutron and X-ray scattering data were used to monitor shear alignment of softwood-derived nanocrystals with different sizes and it was shown that, as the aspect ratio falls below 20, shear alignment of nanocrystals, and the ability to maintain this alignment dropped significantly. In addition, this difference was suggested to be due to the predominance of the filler/filler interactions and their contribution to the overall reinforcing effect in the evaporated films in relation to homogeneous filler dispersion into the matrix.

The mechanical behavior of ramie nanocrystal reinforced polyethylene oxide (PEO) films prepared by casting/evaporation or freeze-drying/extrusion was also compared (Alloin et al., 2011). The rheological characterization of the nanocomposites showed that the film obtained by casting/evaporation has the typical behavior of a solid material with a stabilization of the strain during creep experiments whereas the extruded specimen behaves as a liquid with a continuous evolution of the strain as a function of time (Figure 9.5(a)). As a consequence, the reinforcing effect of nanocrystals in extruded sample was dramatically reduced, suggesting the absence of a strong mechanical network or at least, the presence of a weak percolating network (Figure 9.5(b)). Moreover, no significant orientation of the nanocrystals was observed during extrusion as shown by the overlapping of the DMA curves obtained for the sample cut in the extrusion and cross-sectional directions (Figure 9.5(b)). On the contrary,

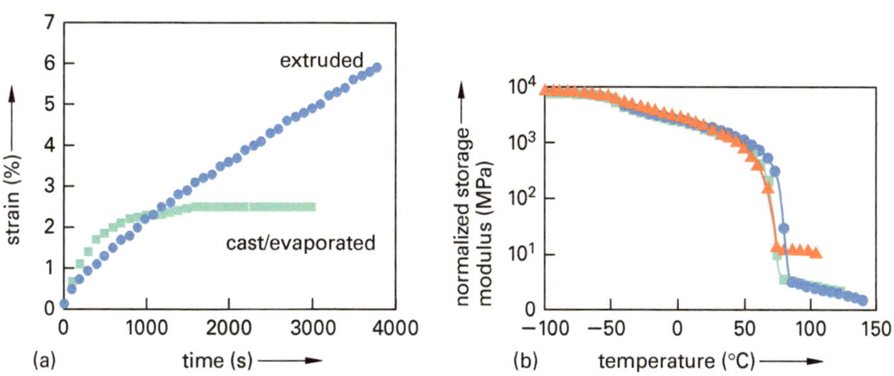

Fig. 9.5: (a) Creep measurements (torque = 5 μNm) at 90°C under inert atmosphere for cast-evaporated (■) and freeze-dried/extruded (●), and (b) normalized storage modulus as a function of temperature for cast-evaporated (▲) and freeze-dried/extruded in the extrusion direction (●) and cross-sectional direction (■) for PEO nanocomposite films reinforced with 6 wt% ramie nanocrystals (Alloin et al., 2011).

an increase of the tensile strength due to shear alignment of cellulose nanocrystals in extruded starch was reported (Orts et al., 2005). Similarly, cellulose nanocrystals orientation induced by a strong magnetic film was found to increase the modulus in the orientation direction (Kvien and Oksman, 2007).

A method for making nanocomposite materials that allows otherwise immiscible components to be combined and also permits high filler loading has been developed (Capadona et al., 2007). In this approach, rather than adding the nanocrystals to the polymer, a network of nanoparticles serves as a template that can be filled with a polymer (see Chapter 7, Section 7.3.2). Through the utilization of a template approach, nanocomposites based on an ethylene oxide/epichlorohydrin (EO-EPI) copolymer (Capadona et al., 2007; Capadona et al., 2008; Capadona et al., 2009), polyvinyl acetate (PVAc) (Shanmuganathan et al., 2010a; Shanmuganathan et al., 2010b), poly(butyl methacrylate) (PBMA) (Shanmuganathan et al., 2010c) or polyurethane (PU) (Mendez et al., 2011) and cellulose nanocrystals have been produced that display the maximum mechanical reinforcement predicted by the percolation model. It was also shown that the mechanical properties of these materials can be selectively and reversibly controlled through the formation and decoupling of the three-dimensional network of well-individualized nanocrystals in response to specific chemical or thermal stimuli (Figure 9.6). These stimuli-responsive materials behave as the sea cucumber dermis which has the ability to rapidly and reversibly alter its stiffness. The materials adapted their original stiffness upon drying (Figure 9.6).

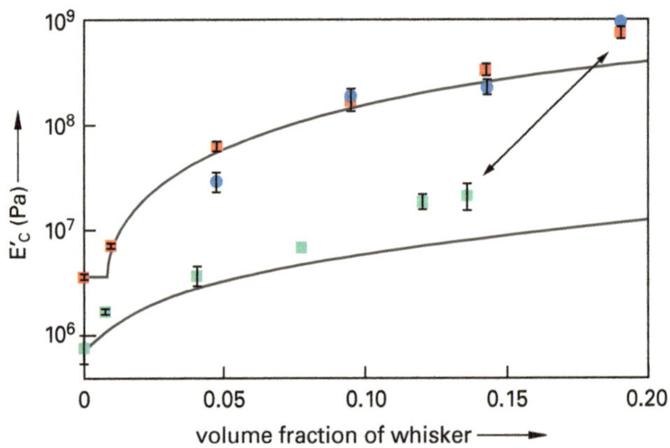

Fig. 9.6: Storage tensile modulus of neat EO-EPI and EO-EPI/cellulose nanocrystal nanocomposites in the dry state (■), swollen with deionized water (■), and re-dried after swelling with water (●). The arrow indicates changes in modulus and volume fraction of nanocrystals resulting from aqueous swelling of one selected sample (19 vol% nanocrystals). Upper and lower lines represent the values predicted by the percolation and Halpin–Kardos models, respectively (Capadona et al., 2008).

When using a polymeric matrix in latex form, the particle size also seems to play a predominant role on the mechanical behavior of cellulose nanocrystal reinforced nanocomposites (Dubief et al., 1999). Larger latex particle size resulted in higher mechanical properties. Indeed, the polymeric particles act as impenetrable domains to nanoparticles during the film formation due to their high viscosity. Increasing latex particle size leads to an increase of the excluded volume and these geometrical constraints seem to affect the nanocrystal network formation. Cellulose nanocrystal reinforced epoxy resin nanocomposites have been prepared (Tang and Weder, 2010). Suspensions of the nanoparticles in dimethylformamide (DMF) were combined with an oligomeric difunctional diglycidyl ether of bisphenol A and a diethyl toluendiamine-based curing agent. Nanocomposite films were obtained by casting and curing the mixture. No modification of the mechanical performance was observed regardless of the initial state of the nanocrystals, i.e. dispersed in DMF *via* a solvent exchange route or redispersed from a freeze-dried state.

9.5 Filler/matrix interfacial interactions

Hence, the ability in order to produce polymer nanocomposites, which comprise a percolating, three-dimensional network of well-individualized nanoparticles, is important to maximize the reinforcing effect of the nanoparticles. This means that filler/matrix interactions can interfere with filler/filler interactions and affect the mechanical behavior of the nanoparticle reinforced nanocomposites. Classic composite science tends to privilege the former as a fundamental condition for optimal performance. In nanocellulose-based composite materials the opposite trend is generally observed when the materials are processed via a casting/evaporation method. The higher the affinity between the polysaccharide filler and the host matrix, the lower the mechanical performances are. This unusual behavior is ascribed to the originality of the reinforcing phenomenon of nanocellulose resulting from the formation of a percolating network thanks to hydrogen bonding forces. When using other processing techniques, enhanced filler/matrix interactions generally improve the mechanical behavior of the material by promoting the dispersion of the nanoparticles within the polymeric matrix compared to the low-interaction system that can lead to aggregation.

The challenge consists therefore in promoting the homogeneous dispersion of the cellulosic nanoparticles and avoiding agglomeration during processing, thus requiring eventually favorable filler/matrix interactions, and at the same time promoting filler/filler interactions to allow the beneficial formation of a percolating network of nanoparticles. These two requirements are simply conflicting.

The filler/matrix interaction can be altered by different means. It obviously depends on the chemical structure of the matrix. Polar matrices tend to strongly interact with the cellulosic surface whereas apolar matrices weakly interact. The polarity of the nanoparticle can be tuned by a surface chemical modification treatment allowing favorable interactions with a non-polar matrix. The microstructure of the interfacial zone in the vicinity of the reinforcing phase can also be locally altered by the presence of the filler and affect the mechanical performance of the material.

The mechanical characterization techniques used for different nanocellulose reinforced polymer systems are summarized in Table 9.2. In view of the extensive literature on the mechanical properties of nanocellulose reinforced composites, this list is obviously not exhaustive.

Polymer Matrix	Nano-particle	Source of Cellulose	Mechanical Characterization	Reference
Acrylic	CNC	Tunicin	DMA	(Favier et al., 1995a; 1995b)
			Tensile Tests	(Hajji et al., 1996)
		Wheat Straw	DMA	(Helbert et al., 1996; Dufresne et al., 1997)
		Sugar Beet	DMA, Tensile Tests	(Azizi Samir et al., 2004a)
		Acacia Pulp	Tensile Tests	(Pu et al., 2007)
		Alfa	DMA, Tensile Tests	(Ben Elmabrouk et al., 2009)
	MFC	Sugar Beet	DMA, Tensile Tests	(Azizi Samir et al., 2004a)
		Opuntia ficus indica Cladodes	DMA, Tensile Tests	(Malainine et al., 2005)
		Swede Root Pulp	Tensile Tests	(Bruce et al., 2005)
		Sugar Beet	Tensile Tests	(Dalmas et al., 2006)
			DMA	(Dalmas et al., 2007)
		Eucalyptus Pulp	DMA	(Besbes et al., 2011a; 2011b)
	BC	–	DMA, Tensile Tests	(Trovatti et al., 2010)
Cellulose and Cellulose Derivatives	CNC	BC	DMA	(Grunert and Winter, 2002)
		Cotton	DMA, Tensile Tests	(Choi and Simonsen, 2006)
		Cotton Linter	Tensile Tests	(Qi et al., 2009)
		MCC	Tensile Tests	(Ma et al., 2011)
	MFC	Sulfite Pulp	Tensile Tests	(Zimmermann et al., 2004)
			Nanoindentation	(Zimmermann et al., 2005)
		Daicel	Tensile Tests	(Duchemin et al., 2009)
		Wheat Straw, Beech Wood Pulp	Tensile Tests	(Zimmermann et al., 2010)
		Sulfite Softwood Pulp	Tensile Tests	(Sehaqui et al., 2011)

Polymer Matrix	Nano-particle	Source of Cellulose	Mechanical Characterization	Reference
Chitosan	MFC	Bleached Sulfite Softwood	Tensile Tests	(Nordqvist et al., 2007)
		BC	Compression Tests	(Nge et al., 2010)
Epoxy	CNC	Tunicin	DMA	(Matos Ruiz et al., 2000; 2001)
	MFC	Swede Root Pulp	Tensile Tests	(Bruce et al., 2005)
		Daicel	DMA	(Lu et al., 2008)
			DMA, Tensile Tests, Fracture Toughness Tests	(Gabr et al., 2010)
			DMA, Tensile Tests	(Shibata and Nakai, 2010)
Glucomannan	CNC	MCC	DMA, Tensile Tests	(Mikkonen et al., 2010)
	MFC	Sulfite Softwood Pulp	DMA, Tensile Tests	(Mikkonen et al., 2011)
Hemicellulose	MFC	Swede Root Pulp	Tensile Tests	(Bruce et al., 2005)
Melamine Formaldehyde	MFC	Bleached Wood Sulfite Pulp	DMA, Tensile Tests	(Henriksson and Berglund, 2007)
NR	CNC	Date Palm Tree	DMA, Tensile Tests	(Bendahou et al., 2009; 2010; 2011)
		Capim Dourado	DMA	(Siqueira et al., 2010a)
		Cassava Bagasse	DMA	(Pasquini et al., 2010)
		Sugarcane Bagasse	DMA, Tensile Tests	(Bras et al., 2010)
		Sisal	DMA, Tensile Tests	(Siqueira et al., 2011)
		MCC	Tensile Tests, Hardness	(Xu et al., 2012)
	MFC	Date Palm Tree	DMA, Tensile Tests	(Bendahou et al., 2010)
		Sisal	DMA, Tensile Tests	(Siqueira et al., 2011)
PCL	CNC	Ramie	DMA, Tensile Tests	(Habibi and Dufresne, 2008; Habibi et al., 2008)
			DMA	(Zoppe et al., 2009; Goffin et al., 2011a)
		Sisal	DMA, Tensile Tests	(Siqueira et al., 2009)
	MFC	Sisal	DMA, Tensile Tests	(Siqueira et al., 2009)

Polymer Matrix	Nano-particle	Source of Cellulose	Mechanical Characterization	Reference
PE	CNC	Tunicin	DMA	(Chauve et al., 2005)
		Ramie	DMA, Tensile Tests	(de Menezes et al., 2009)
	MFC	Soybean Stock	Tensile Tests	(Wang and Sain, 2007a)
		Soybean Pods	DMA, Tensile Tests	(Wang and Sain, 2007b)
PEO	CNC	Tunicin	DMA	(Azizi Samir et al., 2004b; 2004c; 2004d; 2004e; 2006)
			Tensile Tests	(Schroers et al., 2004)
			DMA, Tensile Tests	(Azizi Samir et al., 2004f; 2005)
		BC	Tensile Tests	(Park et al., 2007)
		MCC	Tensile Tests	(Goetz et al., 2010)
		Ramie	DMA	(Alloin et al., 2011)
	BC	–	DMA	(Brown and Laborie, 2007)
PHA	CNC	Tunicin	DMA	(Dubief et al., 1999; Dufresne et al., 1999; Dufresne, 2000)
		MCC	DMA, Tensile Tests	(Jiang et al., 2008; Ten et al., 2010))
Phenolic Resin	CNC	MCC	DMA	(Liu and Laborie, 2011)
	MFC	Sugar Beet Pulp	Tensile Tests	(Leitner et al., 2007)
		Kraft Pulp	Three-Point Bending Tests	(Nakagaito and Yano, 2004)
		Daicel	Three-Point Bending Tests	(Nakagaito and Yano, 2005)
		Kraft Pulp	Tensile Tests	(Nakagaito and Yano, 2008a; 2008b)
	BC	–	Three-Point Bending Tests, Tensile Tests	(Nakagaito et al., 2005)

Polymer Matrix	Nano-particle	Source of Cellulose	Mechanical Characterization	Reference
PLA	CNC	MCC	Tensile Tests	(Oksman et al., 2006)
			DMA	(Petersson et al., 2007)
			DMA, Tensile Tests	(Bondeson and Oksman, 2007a; 2007b; Petersson and Oksman, 2007)
			Tensile Tests	(Sanchez-Garcia and Lagaron, 2010; Fortunati et al., 2012)
		Grass of Korea (*genus Zoysia*)	Tensile Tests	(Pandey et al., 2009)
		Cotton	Tensile Tests	(Pei et al., 2010)
		Ramie	DMA	(Goffin et al., 2011b)
		Cotton Linter	DMA, Tensile Tests	(Lin et al., 2009; 2011)
	MFC	Hemp	Tensile Tests	(Wang and Sain, 2007c)
		Daicel	DMA, Tensile Tests	(Iwatake et al., 2008)
			DMA, Tensile Tests	(Nakagaito et al., 2009)
			DMA, Tensile Tests	(Suryanegara et al., 2009)
			DMA, Tensile Tests, Cantilever Beam Test	(Suryanegara et al., 2011)
		Bleached Sulfite Wood Pulp	DMA	(Tingaut et al., 2010)
		Kenaf Pulp	DMA, Tensile Tests	(Jonoobi et al., 2010)
Polysulfone	CNC	Cotton	Tensile Tests	(Noorani et al., 2007)
PP	CNC	Tunicin	DMA, Tensile Tests	(Ljungberg et al., 2005; 2006)
		MCC	DMA, Tensile Tests	(Bahar et al., 2012)
	MFC	Soybean Pods	DMA, Tensile Tests	(Wang and Sain, 2007b)
PS	CNC	Cotton	DMA	(Rojas et al., 2009; Kim et al., 2009; Mendez et al., 2011)
	BC	–	Tensile Tests	(Peng et al., 2011)

Polymer Matrix	Nano-particle	Source of Cellulose	Mechanical Characterization	Reference
PU	CNC	MCC	Tensile Tests	(Marcovich et al., 2006; Auad et al., 2011)
			DMA, Tensile Tests	(Auad et al., 2008)
		Flax	Tensile Tests	(Cao et al., 2007)
		Cottonseed Linter	DMA, Tensile Tests	(Cao et al., 2009; Wang et al., 2010)
	MFC	Hard Wood	DMA, Tensile Tests	(Seydibeyoğlu and Oksman, 2008)
PVA	CNC	Cotton	DMA, Tensile Tests	(Roohani et al., 2008; Uddin et al., 2011)
			Tensile Tests	(Paralikar et al., 2008)
		MCC	Tensile Tests	(Lee et al., 2009; Frone et al., 2011)
			DMA, Tensile Tests	(Cho and Park, 2011)
		Ramie	DMA	(Peresin et al., 2010)
	MFC	Sulfite Pulp	Tensile Tests	(Zimmermann et al., 2004)
		Swede Root Pulp	Tensile Tests	(Bruce et al., 2005)
		Sugar Beet Pulp	Tensile Tests	(Leitner et al., 2007)
		Soybean Stock	Tensile Tests	(Wang and Sain, 2007a)
		Soybean Pods	DMA, Tensile Tests	(Wang and Sain, 2007b)
	BC	–	Tensile Tests	(Gea et al., 2010)
PVAc	CNC	Tunicin	DMA	(Chauve et al., 2005; Shanmuganathan et al., 2010b)
		Cotton	DMA	(Shanmuganathan et al., 2010c)
	MFC	Bleached Beech Pulp	DMA	(López-Suevos et al., 2010)
PVC	CNC	Tunicin	DMA	(Chazeau et al., 1999a; 1999b)
			Tensile Tests	(Chazeau et al., 1999c)
			Compression Tests	(Chazeau et al., 2000)
Silk Fibroin	CNC	Tunicin	Tensile Tests	(Noishiki et al., 2002)
Soy Protein	CNC	Cotton Linter	DMA	(Wang et al., 2006)

Polymer Matrix	Nano-particle	Source of Cellulose	Mechanical Characterization	Reference
Starch	CNC	Tunicin	DMA, Tensile Tests	(Anglès and Dufresne, 2001; Mathew et al., 2008)
		Cottonseed Linter	DMA, Tensile Tests	(Lu et al., 2005)
		Ramie	DMA, Tensile Tests	(Lu et al., 2006)
		MCC	DMA, Tensile Tests	(Kvien et al., 2007)
		Hemp	Tensile Tests	(Cao et al., 2008a)
		Cassava Bagasse	DMA, Tensile Tests	(Teixeira et al., 2009)
		Pea Hull Fiber	Tensile Tests	(Chen et al., 2009)
		Flax	Tensile Tests	(Cao et al., 2008b)
		Bamboo	Tensile Tests	(Liu et al., 2010)
	MFC	Potato Pulp	DMA	(Dufresne and Vignon, 1998)
			Tensile Tests	(Dufresne et al., 2000)
		Bleached Sulfite Softwood Pulp	Tensile Tests	(Orts et al., 2005; López-Rubio et al., 2007)
			DMA, Tensile Tests	(Svagan et al., 2007)
			Compression Tests	(Svagan et al. 2008; Svagan et al., 2011))
			Tensile Tests	(Plackett et al., 2010)
		Wheat Straw	DMA, Tensile Tests	(Kaushik et al., 2010)
	BC	–	Tensile Tests	(Grande et al., 2009; Woehl et al., 2010)
Xylan	CNC	Bleached Softwood	Tensile Tests	(Saxena et al.,2009)

Table 9.2: Mechanical characterization of polymer nanocomposites obtained from nanocellulose and polymeric matrix.

9.5.1 Polarity of the matrix

Polar matrices display favorable interactions with cellulose, hence improving the stress transfer at the interface. The most-used polar matrices for the processing of nanocellulose reinforced composites are PEO, polyvinyl alcohol (PVA) and starch (see Table 9.2). Regenerated cellulose was also used as matrix (Qi et al., 2009; Duchemin et al., 2009; Ma et al., 2011).

Interactions between cellulose nanocrystals and PEO have been quantified using heat flow microcalorimetry (Azizi Samir et al., 2004b). The heat of immersion was directly measured by mixing tunicin nanocrystals with water, dodecane and oligoethers. Dodecane was used to evaluate the contribution of the alkane part of PEO to the global interactions. Obviously, the highest heat of immersion was observed for water whereas dodecane displayed the weakest interactions with cellulose. For oligoethers, intermediate values were obtained and stronger affinity of cellulose nanocrystal surface with hydroxyl end groups of PEO rather than its ether oxygen groups was reported. Figure 9.7 shows the temperature dependence of the storage tensile modulus of unfilled PEO and composites reinforced with tunicin nanocrystals (Azizi Samir et al., 2004b). The main relaxation associated with the glass-rubber transition of PEO amorphous phase was observed around −60°C. Above this temperature, a higher rubbery modulus was observed for composites. It was ascribed to a reinforcing effect of the nanoparticles since no change of the degree of crystallinity of the matrix was reported in this filler content range. However, the main effect of tunicin nanocrystals on the mechanical behavior of PEO was observed at higher temperatures, above the melting point of the matrix. Whereas the modulus of the unfilled matrix dropped irremediably with the melting of the crystalline PEO domains, nanocrystals brought a thermal stabilization effect. This effect was ascribed to the formation of a percolating cellulose network and the high temperature modulus, whose value increased with increasing nanoparticle/nanoparticle interaction probability and density of the cellulose network, was well predicted from the percolation approach. The establishment of this rigid percolating nanocrystal network was not influenced when using PEO-lithium trifluoromethyl sulfonyl imide (LiTFSI) polymer electrolyte plasticized with tetra(ethylene) glycol dimethyl ether (TEGDME) (Azizi samir et al., 2004d) or cross-linked polyether-LiTFSI (Azizi samir et al., 2004e) as matrix. Nor did the processing medium (water or DMF) alter the percolation of the nanocrystals (Azizi Samir et al.,

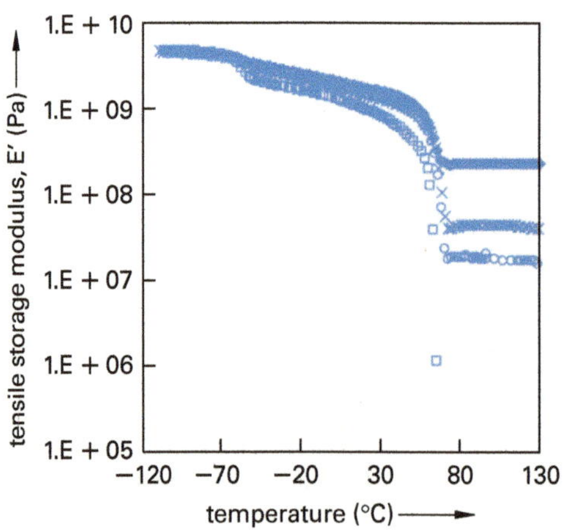

Fig. 9.7: Storage tensile modulus as a function of temperature at 1 Hz for PEO reinforced with 0 (□), 3 (○), 6 (×) and 10 wt% (◇) tunicin nanocrystals (Azizi Samir et al., 2004b).

2004f). Improved thermal stabilization of the modulus of the PEO matrix above its melting point was also reported when adding BC (Brown and Laborie, 2007).

Electrospun PEO fibers were reinforced with low cellulose nanocrystal contents (Park et al., 2007). Improved tensile modulus, tensile strength and elongation at break were reported when adding 0.2 or 0.4 wt% nanocrystals. In situ co-cross-linked nanocomposite hydrogels were prepared by dispersing and cross-linking cellulose nanocrystals with poly(methyl vinyl ether-co-maleic acid)-polyethylene glycol matrix (Goetz et al., 2010). Compared to percolating cellulose nanostructures, the in situ cross-linked materials displayed higher toughness, but lower stiffness and strength because of the absence of direct contact between nanoparticles.

Strong interactions between cellulose nanocrystals prepared from cottonseed linters (Lu et al., 2005), ramie fibers (Lu et al., 2006), hemp fibers (Cao et al., 2008a), flax fibers (Cao et al., 2008b) and pea hull fibers (Chen et al., 2009) and glycerol plasticized starch matrix were reported to play a key role in reinforcing properties. Similar behavior was observed for sorbitol plasticized starch (Mathew et al., 2008). However, the reported values remain lower than expected from the percolation concept. Reduced water uptake of glycerol plasticized starch upon cellulose nanocrystals addition was also suspected to contribute to the enhancement of the tensile strength and Young's modulus (Liu et al., 2010). For MFC reinforced glycerol plasticized starch, favorable filler-matrix interactions were also suspected to contribute to the improvement in mechanical properties (Svagan et al., 2007; Kaushik et al., 2010). Moreover, the formation of the MFC network during water evaporation in the film preparation stage was evoked. The nanostructured characteristics of MFC, in combination with favorable interfacial interactions, were found to delay material damage during deformation and explain the observed ductility and high toughness (Svagan et al., 2007). The crack-stopping capability of MFC at higher load was also suggested to explain the strength increase of amylopectin films (Plackett et al., 2010).

Similarly, strong interactions with cellulose have been suggested to contribute to the reinforcing effect of PVA upon adding nanocrystals (Lee et al., 2009; Peresin et al., 2010; Uddin et al., 2011; Frone et al., 2011) or MFC (Wang and Sain, 2007a; Wang et al., 2007b). Strong interactions between cellulose and soy protein isolate (SPI) plastics due to hydrogen bonding were also suggested to account for the enhanced mechanical properties of cellulose nanocrystal reinforced SPI (Wang et al., 2006).

In non-percolating systems, for instance for materials processed in toluene from freeze-dried cellulose nanocrystals, strong matrix/filler interactions enhance the reinforcing effect of the filler. This observation was reported using random copolymers of poly(ethylene-co-vinyl acetate) (EVA) matrices with different vinyl acetate contents and then different polarities (Chauve et al., 2005). The more polar matrices exhibited, the higher the storage modulus values. It also appeared that a minimum number of acetate groups was needed to observe a reinforcing effect above the melting point of the matrix and prevent the flow of the matrix. Moreover, a leveling-off of this effect appeared for copolymers with vinyl acetate content of 40 and 100 wt% for which the storage modulus was approximately the same.

Nanocomposite materials were also prepared from copolymers of PVA and polyvinyl acetate (PVAc) and cellulose nanocrystals prepared from cotton linter (Roohani et al., 2008). The degree of hydrolysis of the matrix was varied in order to vary the hydrophilic character of the polymer matrix and then the degree of interaction between the filler and the matrix. Nanocomposite films were conditioned at various moisture contents, and the mechanical properties were characterized in both the linear and non-linear range. The order of magnitude of experimental mechanical data agreed with the percolation approach. Figure 9.8 shows the evolution of the relative tensile modulus, defined as the ratio of the modulus of the composite to the one of the unfilled matrix, of these copolymers reinforced with 12 wt% nanocrystals and conditioned at 0% relative humidity (RH) as a function of the degree of hydrolysis of the matrix. A clear increase was observed that was assigned to increased filler-matrix interactions. However, as the water content in the film increased, a lower reinforcing effect was observed. It was ascribed to strong water/cellulose interactions that hinder the inter-nanocrystal connections.

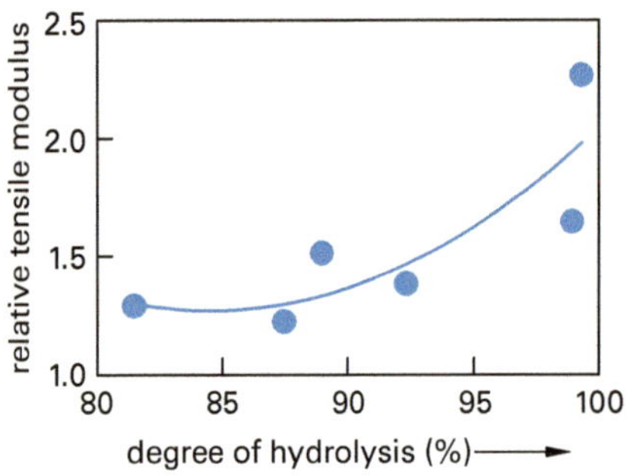

Fig. 9.8: Evolution of the relative tensile modulus of poly(vinyl alcohol-co-vinyl acetate) reinforced with 12 wt% cotton nanocrystals and conditioned at 0% RH as a function of the degree of hydrolysis of the matrix. The solid line serves to guide the eye (Roohani et al., 2008).

The extent of filler/matrix interaction can also be accessed from DMA experiments through the evolution of the loss angle. A shift of the main relaxation process associated with the glass-rubber transition of the matrix towards higher temperatures is an indication of an increase of Tg value resulting from strong interactions. However, this effect can be affected by the so-called mechanical coupling effect because the temperature position of the relaxation depends on both Tg and the associated magnitude of the modulus drop. A broadening of the main relaxation process can also be observed or even a splitting into two peaks. This effect is exemplified in Figure 9.9 for MFC reinforced NR films (Bendahou et al., 2010). The main relaxation was found to split into two well-defined peaks, and the relative magnitude of the high temperature peak was found to increase with increasing filler content. This splitting was not observed for cellulose nanocrystals extracted from the same source (date palm tree) and it was ascribed to strong interactions between MFC and the NR matrix. The presence of residual lignin, extractive substances and fatty acids at the surface of MFC was suggested to compatibilize it with the matrix. These favorable interactions could lead to the formation of an interfacial layer surrounding the filler whose mobility is restricted compared to the bulk matrix.

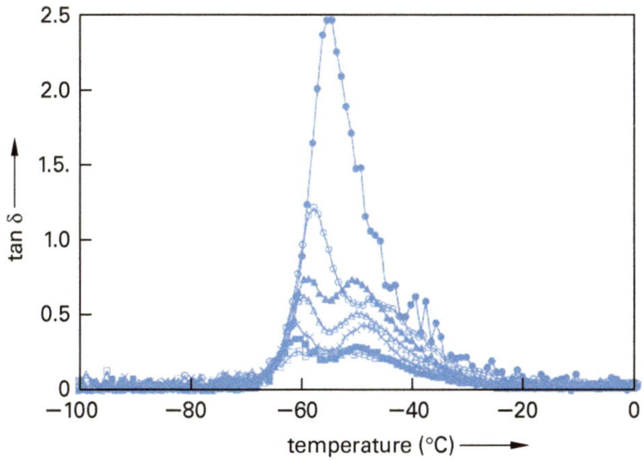

Fig. 9.9: Loss angle tangent tan δ as a function of temperature at 1 Hz for natural rubber reinforced with 0 (●), 1 (○), 2.5 (▲), 5 (△), 7.5 (×), 10 (■) and 15 wt% (□) date palm tree MFC (Bendahou et al., 2010).

The magnitude of the relaxation process is generally found to decrease when increasing the filler content. It is ascribed to a decrease of the matrix material amount, responsible for damping properties, i.e. a decrease in the number of mobile units participating to the relaxation phenomenon. However, new damping mechanisms can be introduced by the filler particles. Possible new damping mechanisms include: (i) particle-particle slippage or friction, (ii) particle-polymer motion at the filler interface, and (iii) change in the properties of the polymer by adsorption onto the filler particle. It is generally accepted that if significant interactions between the polymer and the filler occur, this tends to create a layer of polymer surrounding each filler particle. The resulting polymeric layer has different properties compared with those of the bulk polymer. Assuming the dispersed phase particles to be rigid, this leads to an immobilized polymer layer contributing to the effective filler volume fraction in the compound.

9.5.2 Chemical modification of the nanoparticles

One of the drawbacks of cellulose nanoparticles is their high tendency to agglomerate due to the large number of hydroxyl groups on their surface (highly polar and hydrophilic). This makes dispersion of nanocellulose very difficult in polymer matrices, especially those that are non-polar or hydrophobic.

The hydrophilic character of cellulose nanoparticles can be tuned by surface chemical grafting or coating with surfactant. The surface chemical modification of cellulose nanoparticles has a dual effect on the mechanical properties of ensuing polymer nanocomposites. The decoration of the nanoparticles with grafted moieties obviously restricts the possibility of inter-particle interactions and then the establishment of a percolating network within the continuous medium. Therefore, the outstanding properties resulting from this phenomenon are definitively lost. However, depending on the matrix and processing conditions it can allow a better dispersion of the filler within the matrix and then higher mechanical performances to be reached compared to the aggregated state. It can be used to process composites from an organic liquid medium by improving the dispersion state of the nanoparticles in this medium. However, for an accurate comparison the cellulose content should be similar, i.e. the grafted moieties content should be known and subtracted from the grafted nanoparticle content.

Nanocomposites based on cellulose acetate butyrate (CAB) and cellulose nanocrystals prepared from BC were obtained by casting/evaporation in acetone (Grunert and Winter, 2002). A higher reinforcing effect was reported for unmodified cellulose nanoparticles than for trimethylsilylated nanocrystals. Apart from the fact that 18% of the weight of the silylated crystals was due to the silyl groups, this difference was attributed to restricted filler/filler interactions. Similar results and loss of mechanical properties were reported for NR-based nanocomposites reinforced with both unmodified and surface chemically modified chitin (Gopalan Nair et al., 2003) and starch nanocrystals (Angellier et al., 2005).

Grafting organic acid chlorides presenting different lengths of the aliphatic chain by an esterification process on the surface of cellulose nanocrystals was found to have a beneficial effect on the dispersion of nanoparticles within a low density polyethylene matrix by extrusion (de Menezes et al., 2009). However, compared to unmodified nanoparticles, no improvement of the mechanical properties was observed upon chemical grafting. It was ascribed to two antagonistic effects, viz. improvement of the homogeneity of the nanocomposites and hindering of inter-nanoparticle interactions. However, improved elongation at break was reported for nanocomposites reinforced with long chain (C18)-grafted nanoparticles.

For poly(lactic acid) (PLA)-based nanocomposites, both improved dispersion and mechanical properties were observed by grafting PLA chains (Goffin et al., 2011b), acetylation (Tingaut et al., 2010; Lin et al., 2011) or silylation (Pei et al., 2010; Raquez et al., 2012) of the nanoparticles or modifying the surface of cellulose nanocrystals with a surfactant (Petersson et al., 2007; Fortunati et al., 2012). However, it is worth noting that enhanced crystallinity of the matrix is generally induced by grafted and surfactant-coated nanoparticles that contributes to the enhancement of the mechanical performances of the nanocomposite film. Moreover, for PLA-grafted nanocrystal reinforced PLA a shift towards lower temperatures of the main relaxation peak was observed and attributed to a plasticizing effect of short PLA chains (grafted or not) (Goffin et al., 2011b). When using a large amount of surfactant, a decrease of the stiffness of PLA was observed because of decreased crystallinity and increased porosity in the material (Petersson et al., 2007). Decreased mechanical performance was also reported for unmodified cellulose nanocrystal reinforced PLA compared to organically modified bentonite clay, and it was ascribed to lack of favorable interfacial interaction (Petersson and Oksman, 2006). Filler-induced plasticization by sorbed moisture was also evoked to explain the lower mechanical properties of composites compared to neat PLA (Sanchez-Garcia and Lagaron, 2010).

Enhancement of the dispersion and mechanical properties was reported for N-octadecyl isocyanate- (Siqueira et al., 2009) and poly(ε-caprolactone) (PCL)-grafted cellulose nanocrystals blended with PCL using a grafting onto (Habibi and Dufresne, 2008) or grafting from approach (Habibi et al., 2008; Goffin et al., 2011a). Compared to unmodified nanocrystal reinforced PCL, both a high modulus and high elongation at break were reported (Habibi and Dufresne, 2008; Habibi et al., 2008; Siqueira et al., 2009). In addition, rheological analyses were performed at 90°C, i.e. above the melting point of the PCL matrix (Goffin et al., 2011a). The frequency dependence of the storage shear modulus G' is shown in Figure 9.10 for unmodified (panel (a)) and PCL-grafted nanocrystal (panel (b)) reinforced nanocomposites. The addition of ungrafted nanocrystals did not provide any effect on the viscoelastic properties of the polyester matrix, regardless of the filler content, because of the weak dispersion state of the nanoparticles and lack of interactions between the matrix and the cellulose nanocrystals (Figure 9.10(a)). On the contrary, G' was found to strongly increase when adding PCL-grafted nanoparticles (Figure 9.10(b)) and a progressive cross-over of G' and G" curves was observed. A solid-like behavior was observed for the highly filled composites assumed to result from the formation of a percolating network. This network provided the material with a higher resistance to the applied deformation over the whole range of low frequency. Because of the grafting process, the formation of this network structure was likely to result from the formation of a polymer physical network based on the entanglement of the surface-grafted polymer chains with the free PCL chains of the matrix. To attest this hypothesis, shorter PCL chain-grafted nanocrystals were prepared and ensuing nanocomposites were tested (Figure 9.10(c)). Limitation of the ring-opening polymerization reaction time was imposed to reduce the length of grafted polymer chains. No observation of G' and G" cross-over was observed, confirming that it is necessary to get a sufficient PCL chains lengths to form the network.

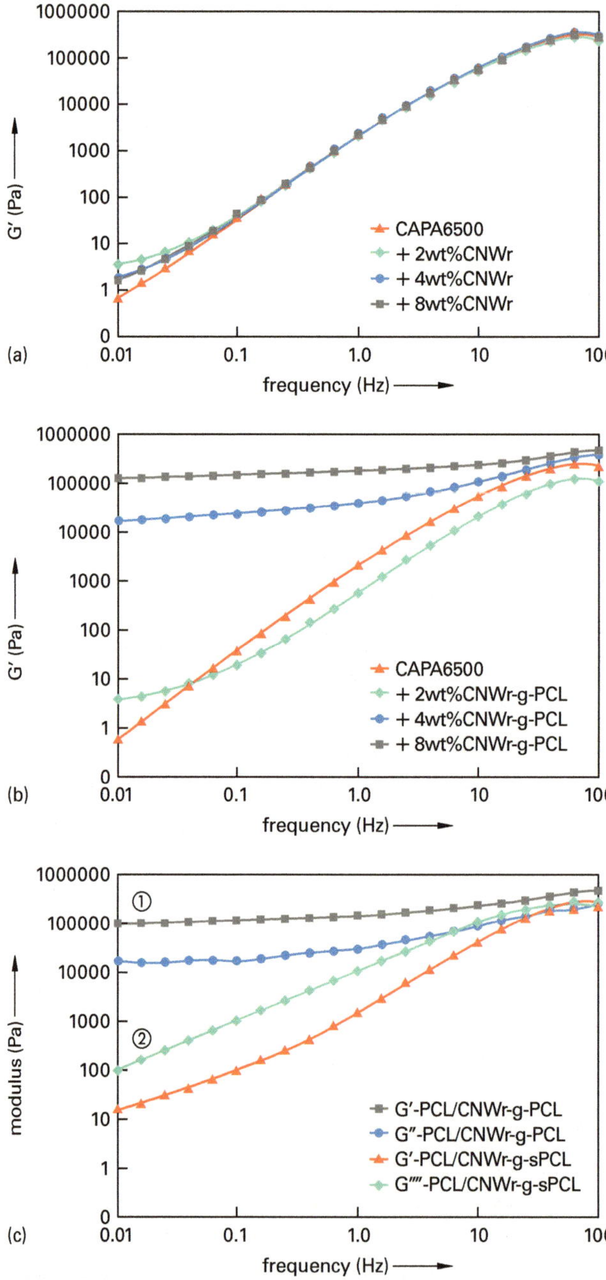

Fig. 9.10: Frequency dependence of the storage shear modulus G' at 90°C for the unfilled PCL matrix (CAPA6500) and related nanocomposites reinforced with 2, 4 and 8 wt% (a) ungrafted and (b) PCL-grafted cellulose nanocrystals. (c) Frequency dependence of the storage G' and loss G" shear moduli at 90°C for PCL reinforced with 8 wt% PCL-grafted cellulose nanocrystals with long (A) and short PCL chains (B) (Goffin et al., 2011a).

Nanocomposites were prepared from amorphous atactic polypropylene (aPP) and tunicin nanocrystals (Ljungberg et al., 2005). Three types of nanoparticles were used with various surface and dispersion characteristics, viz. aggregated without surface modification, aggregated grafted with maleated PP, and surfactant modified. A significant increase of the rubbery modulus was reported, ascribed to filler-filler interactions, and it was found to be independent of the dispersion quality. However, for modified nanoparticles a negative slope was observed with increasing temperature because of the weakening of these interactions. This phenomenon was also evidenced from tensile tests performed on cellulose nanocrystal films. In the non-linear range where the filler-filler interactions were at least partially destroyed, the filler-matrix interactions and dispersion quality grew in importance. Improvement of matrix-filler interactions by using cellulose nanocrystals coated with a surfactant was shown to play a major role, especially on the elongation at break. Similar results were reported for semicrystalline isotactic PP (iPP) (Ljungberg et al., 2006).

Improved dispersion and mechanical reinforcement was observed for silane-treated MFC dispersed in an epoxy resin system (Lu et al., 2008).

9.5.3 Local alteration of the matrix in the presence of the nanoparticles

Conventionally, the interface in composites is considered to have zero thickness and results from interactions between matrix and filler surface. However, changes in local morphology may lead to the formation of an interphase around the filler particles with properties different from those of the bulk matrix, i.e. the composite can be regarded as a three-phase material (Dufresne and Lacabanne, 1993). The properties of the interfacial zone or interphase can play a major role in overall properties of the nanocellulose reinforced composite materials. In many cases, the interphase can be the controlling element in composite performances. It was suggested that the interphase exists from some point in the filler where the local properties begin to change from the filler bulk properties through the interphase into the matrix, where the local properties again equal the bulk properties (Drzal, 1985).

Silk fibroin/tunicin nanocrystals composites were prepared by casting-evaporation (Noishiki et al., 2002). Tensile tests performed at room temperature showed a nearly linear and additive dependence on the mixing ratio for the Young's modulus, whereas the tensile strength and elongation at break showed a maximum around 70–80 wt% cellulose, reaching five times those of fibroin-alone or cellulose-alone films. This behavior was ascribed to the formation of the β structure (silk II) from the random coil state in solution that was evidenced from infrared spectroscopy. This structure of fibroin usually forms under shearing or elongating stress and it was supposed that a flat and ordered crystalline cellulosic surface served as a template through its specific interactions with the fibroin molecules.

The viscoelastic behavior of plasticized poly(vinyl chloride) (PVC) reinforced with cellulose nanocrystals has been reported (Chazeau et al., 1999a). To overcome the discrepancy observed between experimental and predicted data (Halpin–Kardos model) the existence of a hypothetical interphase of immobilized matrix in contact with the surface of the nanoparticles was assumed. However, the discrepancy still persisted. In further papers, an analysis based on the quasi-point defect theory was reported to describe the behavior of the composite in the linear (Chazeau et al., 1999b) and nonlinear range (Chazeau et al., 2000). The matrix was described as a parallel assembly of phases with different plasticizer concentration and satisfactory modeling of the composite was obtained. Small-angle neutron scattering (SANS) study of stretched cellulose nanocrystal filled PVC composites evidenced a dewetting of the nanocrystals by the matrix and microvoid formation during the tensile deformation process (Chazeau et al., 1999b).

For cellulose nanocrystal reinforced polyurethane a significant increase of the tensile modulus was observed at very low filler loadings (0.5–5 wt%) (Marcovich et al., 2006). However, no step increment was reported and it was ascribed to the absence of percolation in this system. Indeed, strong filler-matrix interaction developed during curing because of a chemical reaction occurring between the nanocrystals and the isocyanate component and inducing an increase in the cross-linking density of the matrix.

For tunicin nanocrystal reinforced semicrystalline poly(hydroxyoctanoate) (PHO), a transcrystallization phenomenon was suspected (Dufresne et al., 1999). The mechanical behavior of this system is reported in Figure 9.11 and compared to the one obtained for amorphous PHO-based nanocomposites. For low nanocrystal contents (1 wt%, below the percolation threshold), the rubbery storage tensile modulus of amorphous PHO-based system was much lower than the one of the semicrystalline matrix system. This indicates that the mechanical behavior of the composite was matrix-dominated. At higher nanocrystal contents (3 and 6 wt%, above the percolation threshold), the differences between both kinds of matrix faded out, indicating that the mechanical behavior became nanocrystal-dominated. In this filler content range, the mechanical stiffness of the material was ascribed to the percolating network of nanocrystals. However, as the melting point of the semicrystalline PHO (around 330 K) was reached, a sharp drop of the modulus was observed due to the breakup of crystalline domains whereas the one of the amorphous PHO-based nanocomposites remained constant. It was suggested that percolation of the nanocrystals occurred through crystalline PHO domains.

In these systems, the filler/matrix interactions and distance away from the surface at which the molecular mobility of the amorphous PHO phase is restricted were quantified using a physical model predicting the mechanical loss angle (Dufresne, 2000). The determination of the ratio of experimental to predicted magnitude of the main relaxation process allowed removal of the filler reinforcement effect keeping only the interfacial effect, and was used to calculate the thickness of the interphase.

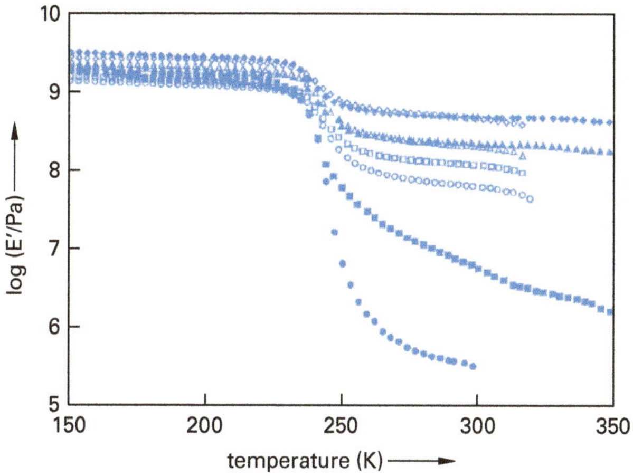

Fig. 9.11: Logarithm of the storage tensile modulus as a function of temperature at 1 Hz for amorphous (filled symbols) and semicrystalline (open symbols) PHO reinforced with 0 (●,○), 1 (■,□), 3 (▲,△) and 6 wt% (♦,◊) tunicin nanocrystals (Dufresne et al., 1999).

It was shown that when using semicrystalline PHO as matrix, the molecular mobility of amorphous PHO chains was only slightly affected by the presence of tunicin nanocrystals, owing to a possible transcrystallization phenomenon leading to the coating of the nanoparticles with the crystalline PHO phase. The thickness of the transcrystalline layer, around 2.7 nm, was found to be independent of the cellulose nanocrystal content. In contrast, when using amorphous PHO as matrix, the flexibility of polymeric chains in the surface layer was lowered by the conformational restrictions imposed by cellulose surface. This resulted in a broader interphase and in a broadening of the main relaxation process of the matrix.

The presence of MFC was found to accelerate the crystallization of PLA (Suryanegara et al., 2009). It was shown that the combination of MFC reinforcement and crystallization of PLA in the matrix both contributed to improve the heat resistance of the material. Phenylphosphonic acid zinc (PPA-Zn) was reported to be more effective in accelerating the crystallization of PLA than MFC (Suryanegara et al., 2011). The combination of faster crystallization induced by PPA-Zn and higher stiffness provided by MFC was found to give favorable conditions for accelerating the injection molding cycle of PLA.

For glycerol plasticized starch reinforced with cellulose nanocrystals a relatively very low reinforcing effect was reported (Anglès and Dufresne, 2001). This observation was explained by competitive interactions between the components and by a plasticizer accumulation in the interfacial zone. This plasticizer accumulation, enhanced in moist conditions was suggested to interfere with hydrogen-bonding forces that are likely to hold the percolating cellulose nanocrystal network within the matrix. In highly moist conditions, a possible transcrystalline zone around the nanocrystals was observed for amylopectin chains located in the glycerol-rich domains (Anglès et al., 2000). The coating of the nanoparticles by a soft plasticizer-rich interphase hindered the stress transfer at the filler/matrix interface when the material was submitted to a high strain tensile test, resulting in poor mechanical properties of the composites. A similar behavior was reported for cassava starch plasticized with glycerol or a mixture of glycerol and sorbitol reinforced with cellulose nanocrystals extracted from cassava bagasse (Teixeira et al., 2009), and glucomannan films plasticized with glycerol reinforced with cellulose nanocrystal extracted from MCC (Mikkonen et al., 2010). This strong loss of performance demonstrates the outstanding importance of the filler/filler interactions to ensure the mechanical stiffness and thermal stability of these composites. When using unhydrolyzed cellulose microfibrils extracted from potato pulp rather than cellulose nanocrystals to reinforce glycerol plasticized thermoplastic starch a completely different mechanical behavior was reported (Dufresne and Vignon, 1998; Dufresne et al., 2000). A significant reinforcing effect of unhydrolyzed microfibrils was observed. It was suspected that tangling effect contributed to this high reinforcing effect (Anglès and Dufresne, 2001).

However, for PEO-based nanocomposites it was found that the formation of the percolating cellulose network was not altered by the crystallization of the matrix and filler/PEO interactions (Azizi Samir et al., 2004b). Significant improvement of the mechanical properties was reported for cellulose nanocrystal reinforced waterborne polyurethane (WPU) composites (Cao et al., 2009). The composites were prepared by in situ polymerization and part of the pre-synthesized WPU chains was grafted on the surface of the nanoparticles. Co-crystallization between grafted and polymer matrix chains was suggested to induce good dispersion and strong interfacial adhesion between the filler and the matrix, inducing the improvement of the mechanical properties compared to the neat matrix.

9.6 Synergistic reinforcement

Few studies have reported the reinforcement of polymer matrices with nanocellulose in association with another filler. For instance, waterborne polyurethane (WPU) was reinforced with starch and cellulose nanocrystals obtained by acid hydrolysis of waxy maize starch granules and cotton linter pulp, respectively (Wang et al., 2010). A synergistic effect was observed when adding 1 wt% starch and 0.4 wt% cellulose nanocrystals with a significant improvement in tensile strength, Young's modulus and tensile energy at break without significant loss for the elongation at break. The mechanical performance was found to be higher than for individual filler, but it is worth noting that the total filler content was different. In the ternary system, the formation of a much jammed network consisting of nanoparticles with different geometrical characteristics was suggested to play an important role in the enhancement of the crosslinked network. Moreover, strong hydrogen bonding interactions between the nanoparticles and between the nanoparticles and the hard segments of WPU matrix was suspected to improve the mechanical properties.

Reinforcement of NR with cellulose nanocrystals prepared by acid hydrolysis of the rachis of date palm tree and natural montmorillonite was reported (Bendahou et al., 2011). The nanocomposite films were prepared by changing the weight content of each filler keeping the total filler content equal to 5 wt%. However, it is worth noting that because of the difference in density of both fillers, the filler volume content was not constant for all samples. The mechanical properties of NR were significantly improved upon filler addition but a synergism effect was observed for the film containing 1 wt% montmorillonite and 4 wt% cellulose nanocrystals. Microscopic observations showed that cellulose nanoparticles were connected within the polymer matrix and formed a continuous macroscale 3D network. Small stacks of intercalated montmorillonites were homogeneously dispersed within the NR matrix in the binary montmorillonite reinforced NR nanocomposites. However, the montmorillonite dispersion was totally modified in the presence of cellulose nanocrystals. The small intercalated montmorillonite stacks were found to be located in the vicinity of the cellulosic nanoparticles in the ternary systems. This specific morphology was explained by the polar character of both fillers compared to the less polar matrix.

The partial replacement of silica in NR by silane-modified cellulose nanocrystals was investigated (Xu et al., 2012). Accelerated curing rate, reduced Payne effect and better processing performances were reported. Moreover, improved modulus, tear strength and hardness, as well as heat built-up, compression set and dynamic mechanical performance were observed.

9.7 Specific mechanical characterization

Except tensile tests and DMA, other techniques have been more marginally used in the literature to characterize the mechanical properties of nanocellulose reinforced polymer nanocomposites. Some of them are described below.

9.7.1 Compression test

A compression test consists in determining the behavior of materials under crushing loads. The specimen is compressed and deformation at various loads is recorded. The apparatus used for this experiment is basically similar to the one used for tensile tests. However, instead of applying a uniaxial tensile load, a uniaxial compressive load is applied, resulting in a negative strain. For some materials, completely different behavior is obtained during tensile and compression tests, such as polystyrene at room temperature which displays a brittle behavior in tension and behaves as a ductile material in compression. The compression test is mainly used for brittle materials and expanded cellular materials.

Compression tests were performed for plasticized PVC reinforced with tunicin nanocrystals (Chazeau et al., 2000). Measurements were carried out in the glassy state of the matrix to investigate the plastic behavior of the materials. The compression modulus and yield strength were found to increase with the nanocrystal content and decrease with temperature. Moreover, a more rapid hardening and a whitening of the sample were observed in the presence of the filler. Successive compression tests were also performed and the residual strain was found to increase during the successive tests. The hardening also increased with increasing the total strain.

Compressive tests have also been performed on amylopectin foams reinforced with MFC (Svagan et al., 2008; Svagan et al., 2011). At low strains, the compressive stress-strain curves showed linear elasticity and a plateau-like region followed, resulting from the collapse of cells. When the cells had almost completely collapsed, the stress raised steeply with strain because opposing cell walls were in contact and further strain compressed the solid itself corresponding to the densification regime. Improved modulus, yield strength and level of the plateau zone were observed up to 40 wt% MFC whereas a decrease was reported at 70 wt% MFC. This decrease was ascribed to the ill-defined cell structure in the highly filled foam. The area under the curve can be used to determine the energy absorption. The plot of absorbed energy per unit volume versus stress is useful in the design of packaging materials. Cell wall properties, cell structure and density of the foam affect the compressive properties of the material.

Similar tests were carried out on BC/chitosan porous scaffolds of open pore microstructure with interconnecting pores (Nge et al., 2010). TEMPO-oxidized BC was used and scaffolds were prepared by the freezing and freeze-drying method. The compressive modulus and strength increased upon BC addition.

9.7.2 Successive tensile test

Successive tensile tests can be performed to characterize the damage process occurring during tensile tests which modifies the mechanical properties of the material. It consists in stretching the material up to a certain elongation or deformation, then releasing the force, and stretching the material again up to a higher elongation. This procedure is repeated with increasing deformation, until break (Gopalan Nair and Dufresne, 2003). It is also possible to maintain the sample at a higher temperature after each force release to allow relaxation of the material and partial recovery of the deformation (Chazeau et al., 1999c). The tensile modulus and shrinkage can also be determined for each successive cycle (Bendahou et al., 2010).

This procedure has been applied to cellulose nanocrystal reinforced plasticized PVC (Chazeau et al., 1999c) and NR reinforced with cellulose nanocrystals and MFC

(Bendahou et al., 2010). For composites, the observed behavior was close to that observed for filled rubber and known as the Mullins effect (Bueche, 1960; Bueche, 1961). The composite sample stretched to a given elongation and when released did not follow the same stress-strain curve when it was stretched once again. It appeared softer during the test when the deformation was below the maximum deformation reached during the previous cycle. For further elongation, the stress-strain curve followed the behavior of the stress-strain curve obtained from the monotone tensile test. Moreover, the residual strain measured after release increased with the strain imposed during the previous tensile cycle. The damage of the material can be characterized following an energetic approach (Heuillet et al., 1992).

For PVC reinforced with cellulose nanocrystals, the damage was ascribed to the debonding at the filler-matrix interface (Chazeau et al., 1999b). The ensuing voids, revealed by a whitening of the sample, were likely to grow from the interface. For nanocellulose reinforced NR, it was observed that the tensile modulus decreased continuously during successive tensile tests (Bendahou et al., 2010). For the neat NR matrix, and for a given cycle, the modulus was lower than for nanocomposites evidencing the reinforcing effect of nanocellulose. However, for higher elongations, no more difference was observed between the unfilled matrix and composites and then the relative decrease of the modulus was higher for composites than for the neat matrix. This indicates that stretched composites behaved as the neat matrix and that the reinforcing phases did not play any role in the stiffness of the material because of the breakage of the percolating cellulose network. The shrinkage was found to increase during successive tensile test and for a given cycle it was lower for composites than for the neat matrix, showing that cellulosic nanoparticles increased the elastic behavior of the material. For a given cycle, the shrinkage was systematically lower for the nanocomposite sample reinforced with nanocrystals compared to MFC-based materials.

9.7.3 Bulge test

The bulge test has been used to characterize poly(3-hydroxybutyrate-co-3-hydroxyvalerate) (PHBV) nanocomposites reinforced with cellulose nanocrystals extracted from MCC (Ten et al., 2010). The bulge test is a standard technique to characterize mechanical properties of thin films. It has the advantage of being able to characterize the residual stress, elastic modulus, and other important parameters such as yield strength and fracture toughness. In bulge testing the composite films are wax-mounted onto an aluminum substrate with a pre-drilled hole (Figure 9.12). A varying gas pressure is applied to the film through the hole using a pressure/vacuum variator and the resulting film deflection is monitored with a scanning laser vibrometer. By fitting the curve of pressure versus deflection to a cubic polynomial, Young's modulus of the film could be calculated from the slope of the polynomial using the model of circular membrane behavior under uniform pressure (Small and Nix, 1992):

$$P = \frac{8Et}{3a^4}h^3 + \frac{4t\sigma}{a^2}h \quad (9.11)$$

where P is the gas pressure, E the sample modulus, t the sample thickness, a the bulge radius, h the deflection and σ the residual stress in the film determined by $Pa^2/(4ht)$. A good agreement between Young's modulus values determined by tensile and bulge tests were reported (Ten et al., 2010). Compared to neat PHBV, the modulus of the nanocomposite reinforced with 5 wt% cellulose nanocrystals was increased by 77% (tensile test) and 91% (bugle test), respectively.

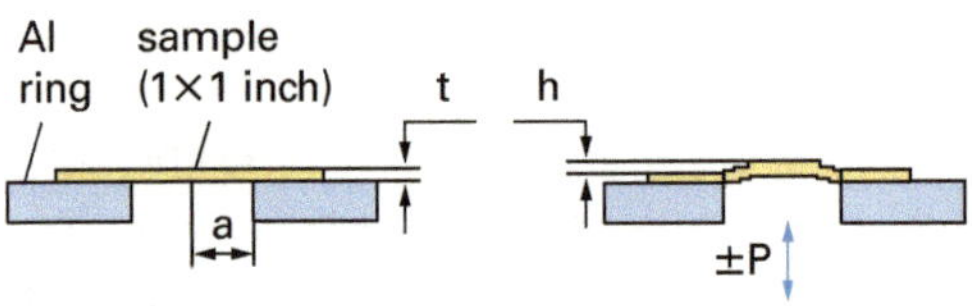

Fig. 9.12: Schematic representation of the bulge test setup (Ten et al., 2010).

9.7.4 Raman spectroscopy

The use of Raman spectroscopy for the determination of the mechanical properties of cellulose fibers and nanoparticles has been described in Chapter 1 (Sections 1.7.1, 1.7.2 and 1.7.3). It can also be used for the interfacial characterization of composites. Indeed, this technique allows to quantify the level of deformation of cellulose nanoparticles embedded in a polymer matrix and therewith indirectly also the stress transfer within the composite. This method has been applied to cellulose nanocrystals extracted from tunicate (Šturcová et al., 2005; Rusli et al., 2011) and cotton (Rusli and Eichhorn, 2008; Rusli et al., 2011). In this experiment, the nanocrystals are dispersed in the epoxy resin and placed on the surface of a beam of the neat epoxy which is deformed in tension and compression using a four-point bending device. A strain gauge is secured to the surface of the beam. No interference from the Raman spectrum for the epoxy resin with the main vibration located at approximately 1095 cm^{-1} from the cellulose nanocrystal occurs. This enables the monitoring of the position of this band with deformation. The Raman peak located at 1095 cm^{-1} corresponds to C–O stretching, both within the cellulose ring and along the glycosidic linkage.

As the composite beam deforms in tension, the position of the 1095 cm^{-1} Raman band shifts toward a lower wavenumber position whereas it shifted toward a higher wavenumber position when the sample was compressed. This shift is nearly linear as a function of the applied strain. The shift in the Raman band is indicative of molecular deformation and is a measure of the extent of stress transfer from the less-stiff matrix to the reinforcing cellulose nanocrystals. The value of the slope of this shift is an indication of the level of the interfacial stress transfer since experiments were performed for composites below the nanoparticle percolation threshold (Rusli et al., 2011). Therefore, the stress transfer is thought not to involve a nanocrystal network and relies primarily on nanocrystal-matrix interactions. Moreover, the absolute value of the band shift upon mechanical solicitation was found to be similar in tension and compression suggesting that the stress transfer mechanisms are the same.

A greater level of deformation was observed for high aspect ratio tunicin nanocrystals compared to low aspect ratio cotton nanocrystals (Rusli et al., 2011). Therefore, the aspect ratio influences stress transfer in nanocomposites. Interestingly, for nanocrystals isolated by hydrolysis with hydrochloric acid, no detectable shift in the position of the Raman peak located at 1095 cm^{-1} was observed nor in tension or compression contrarily to sulfuric acid hydrolyzed-nanocrystals (Rusli et al., 2011). It was suggested to result from aggregation of HCl-hydrolyzed nanocrystals that induces reduction of the aspect ratio and surface area, hence lowering the stress transfer efficiency. This indicates that the surface charge of the nanocrystals also plays a key role in the interfacial mechanics of nanocomposites.

A weakening or a breakdown of the nanocrystal-matrix interface is thought to occur where these loading data plateau. This breakdown can be due to either matrix yielding at the interface or debonding. The ability to follow the breakdown of the interface is of importance in understanding how these nanocomposites may respond in engineering applications.

9.7.5 Atomic force microscopy

Nanomechanical properties of the interphase were quantitively characterized in cellulose nanocrystal reinforced PVA-poly(acrylic acid) (PAA) nanocomposites using peak force tapping mode in atomic force microscopy (AFM) (Pakzad et al., 2012). Direct measurements of the gradient in adhesion and elastic modulus in nanocomposites at the nanoscale interphase were performed. The variation in matrix properties as a function of distance from the surface of nanocrystals was studied before and after the cross-linking process. The interphase in the composites containing PAA exhibited a higher gradient in adhesive and mechanical properties in comparison to the samples with no PAA. It was ascribed to the increased density of ester linkages from the polymer matrix to the nanocrystal interface. Moreover, the interphase thickness increased with the increase in nanoparticle diameter. Depending on the size of the nanocrystal, the thickness of the interphase was found to vary between 4 and 35 nm

9.8 Conclusions

Outstanding mechanical properties can be obtained by blending nanocellulose and polymer matrix. These originate from the high stiffness of crystalline cellulose that provides the strength to higher plants, nanoscale dimensions and high aspect ratio of the nanoparticles, and the high reactivity of cellulose. In suitable conditions, a mechanically percolating stiff network of nanoparticles can form within the polymer matrix that supports the mechanical solicitation. The formation of this network is conditioned by the homogeneous dispersion of the filler, the percolation threshold that depends on the aspect ratio of the nanoparticles and strength of the filler/filler interactions. In these conditions, the host polymeric matrix does not play any role on the mechanical stiffness of the material. It corresponds to the highest mechanical reinforcement effect that can be obtained from these nanoparticles. However, many parameters can affect this phenomenon. When the formation of this percolating nanoparticle network is inhibited, only the high stiffness of crystalline cellulose, nanoscale dimensions, high aspect ratio and dispersion of the nanoparticles, and filler/matrix interactions are involved in the reinforcing phenomenon.

9.9 References

Alloin, F., D'Aprea, A., Dufresne, A., El Kissi, N. and Bossard, F. (2011). Poly(oxyethylene) and ramie whiskers based nanocomposites: Influence of processing: Extrusion and casting/evaporation. Cellulose *18*, 957–973.

Angellier, H., Molina-Boisseau S. and Dufresne, A. (2005). Mechanical properties of waxy maize starch nanocrystal reinforced natural rubber. Macromolecules *38*, 9161–9170.

Anglès, M.N. and Dufresne, A. (2000). Plasticized starch/tunicin whiskers nanocomposites: 1. Structural analysis. Macromolecules *33*, 8344–8353.

Anglès, M.N. and Dufresne, A. (2001). Plasticized starch/tunicin whiskers nanocomposites: 2. Mechanical behavior. Macromolecules *34*, 2921–2931.

Auad, M.L., Contos, V.S., Nutt, S., Aranguren, M.I. and Marcovich, N.E. (2008). Characterization of nanocellulose-reinforced shape memory polyurethanes. Polym. Inter. *57*, 651–659.

Auad, M.L., Richardson, T., Orts, W.J., Medeiros, E.S., Mattoso, L.H.C., Mosiewicki, M.A., Marcovich, N.E. and Aranguren, M.I. (2011). Polyaniline-modified cellulose nanofibrils as reinforcement of a smart polyurethane. Polym. Inter. *60*, 743–750.

Azizi Samir, M.A.S., Alloin, F., Paillet, M. and Dufresne, A. (2004a). Tangling effect in fibrillated cellulose reinforced nanocomposites. Macromolecules *37*, 4313–4316.

Azizi Samir, M.A.S., Alloin, F., Sanchez, J.-Y. and Dufresne, A. (2004b). Cellulose nanocrystals reinforced poly(oxyethylene). Polymer *45*, 4149–4157.

Azizi Samir, M.A.S., Alloin, F., Gorecki, W., Sanchez, J.-Y. and Dufresne, A. (2004c). Nanocomposite polymer electrolytes based on poly(oxyethylene) and cellulose nanocrystals. J. Phys. Chem. B *108*, 10845–10852.

Azizi Samir, M.A.S., Montero Mateos, A., Alloin, F., Sanchez, J.-Y. and Dufresne, A. (2004d). Plasticized nanocomposite polymer electrolytes based on poly(oxyethylene) and cellulose whiskers. Electrochim. Acta *49*, 4667–4677.

Azizi Samir, M.A.S., Alloin, F., Sanchez, J.Y. and Dufresne, A. (2004e). Cross-linked nanocomposite polymer electrolytes reinforced with cellulose whiskers. Macromolecules *37*, 4839–4844

Azizi Samir, M.A.S., Alloin, F., Sanchez, J.Y., El Kissi, N. and Dufresne, A. (2004f). Preparation of cellulose whiskers reinforced nanocomposites from an organic medium suspension. Macromolecules *37*, 1386–1393.

Azizi Samir, M.A.S., Chazeau, L., Alloin, F., Cavaillé, J.-Y., Dufresne, A. and Sanchez, J.-Y. (2005). POE-based nanocomposite polymer electrolytes reinforced with cellulose whiskers. Electrochim. Acta *50*, 3897–3903.

Azizi Samir, M.A.S., Alloin, F. and Dufresne, A. (2006). High performance nanocomposite polymer electrolytes. Compos. Interfaces *13*, 545–559.

Bahar, E., Ucar, N., Onen, A., Wang, Y., Oksüz, M., Ayaz, O., Ucar, M. and Demir, A. (2012). Thermal and mechanical properties of polypropylene nanocomposite materials reinforced with cellulose nano whiskers. J. Appl. Polym. Sci. *125*, 2882–2889.

Balberg, I. and Binenbaum, M. (1983). Computer study of percolation threshold in a two-dimensional anisotropic system of conducting sticks. Phys. Rev. B *28*, 3799–3812.

Balberg, I., Binenbaum, M. and Wagner, N. (1984). Percolation thresholds in the three-dimensional sticks system. Phys. Rev. Lett. *52*, 1465–1468.

Ben Elmabrouk, A., Thielemans, W., Dufresne, A. and Boufi, S. (2009). Preparation of poly(styrene-co-hexylacrylate)/cellulose whiskers nanocomposites via miniemulsion polymerization. J. Appl. Polym. Sci. *114*, 2946–2955.

Ben Mabrouk, A., Kaddami, H., Boufi, S., Erchiqui, F. and Dufresne, A. (2012). Cellulosic nanoparticles from alfa fibers (*Stipa tenacissima*): Extraction procedures and reinforcement potential in polymer nanocomposites. Cellulose *19*, 843–853.

Bendahou, A., Habibi, Y., Kaddami, H. and Dufresne, A. (2009). Physico-chemical characterization of palm from *Phoenix dactylifera* – L, preparation of cellulose whiskers and natural rubber–based nanocomposites. J. Biobased Mat. Bioenergy *3*, 81–90.

Bendahou, A., Kaddami, H. and Dufresne, A. (2010). Investigation on the effect of cellulosic nanoparticles' morphology on the properties of natural rubber based nanocomposites. Eur. Polym. J. *46*, 609–620.

Bendahou, A., Kaddami, H., Espuche, E., Gouanvé, F. and Dufresne, A. (2011). Synergism effect of montmorillonite and cellulose whiskers on the mechanical and barrier properties of natural rubber. Macromol. Mat. Eng. *296*, 760–769.

Besbes, I., Alila, S. and Boufi, S. (2011a). Nanofibrillated cellulose from TEMPO-oxidized eucalyptus fibres: Effect of the carboxyl content. Carbohydr. Polym. *84*, 975–983.

Besbes, I., Rei Vilar, M. and Boufi, S. (2011b). Nanofibrillated cellulose from alfa, eucalyptus and pine fibres: preparation, characteristics and reinforcing potential. Carbohydr. Polym. *86*, 1198–1206.

Bondeson, D. and Oksman, K. (2007a). Polylactic acid/cellulose whisker nanocomposites modified by polyvinyl alcohol. Composites Part A *38*, 2486–2492.

Bondeson, D. and Oksman, K. (2007b). Dispersion and characteristics of surfactant modified cellulose whiskers nanocomposites. Compos. Interfaces *14*, 617–630.

Bras, J., Hassan, M.L., Bruzesse, C., Hassan, E.A., El-Wakil, N.A. and Dufresne, A. (2010). Mechanical, barrier, and biodegradability properties of bagasse cellulose whiskers reinforced natural rubber nanocomposites. Ind. Crops Prod. *32*, 627–633.

Bras, J., Viet, D., Bruzzese, C. and Dufresne, A. (2011). Correlation between stiffness of sheets prepared from cellulose whiskers and nanoparticles dimensions. Carbohydr. Polym. *84*, 211–215.

Brechet, Y., Cavaillé, J.Y., Chabert, E., Chazeau, L., Dendievel, R., Flandin, L. and Gauthier, C. (2001). Polymer based nanocomposites: Effect of filler-filler and filler-matrix interactions. Adv. Eng. Mater. *3*, 571–577.

Brown, E.E. and Laborie, M.P.G. (2007). Bioengineering bacterial cellulose/poly(ethylene oxide) nanocomposites. Biomacromolecules *8*, 3074–3081.

Bruce, D.M., Hobson, R.N., Farrent, J.W. and Hepworth, D.G. (2005). High-performance composites from low-cost plant primary cell walls. Compos. Part A: Appl. Sci. Manufact. *36*, 1486–1493.

Bueche, F. (1960). Molecular basis for the Mullins effect. J. Appl. Polym. Sci. *4*, 107–114.

Bueche, F. (1961). Mullins effect and rubber-filler interaction. J. Appl. Polym. Sci. *5*, 271–281.

Cao, X., Dong, H and Li, C.M. (2007). New nanocomposite materials reinforced with flax cellulose nanocrystals in waterborne polyurethane. Biomacromolecules *8*, 899–904.

Cao, X., Chen, Y., Chang, P.R., Stumborg, M. and Huneault, M.A. (2008a). Green composites reinforced with hemp nanocrystals in plasticized starch. J. Appl. Polym. Sci. *109*, 3804–3810.

Cao, X., Chen, Y., Chang, P.R., Muir, A.D. and Falk G. (2008b). Starch-based nanocomposites reinforced with flax cellulose nanocrystals. eXPRESS Polym. Let. *2*, 502–510.

Cao, X., Habibi, Y. and Lucia, L.A. (2009). One-pot polymerization, surface grafting, and processing of waterborne polyurethane-cellulose nanocrystal nanocomposites. J. Mater. Chem. *19*, 7137–7145.

Capadona, J.R., Van den Berg, O., Capadona, L., Schroeter, M., Tyler, D., Rowan, S.J. and Weder, C. (2007). A versatile approach for the processing of polymer nanocomposites with self assembled nanofibre templates. Nature Nanotechnology *2*, 765–769.

Capadona, J.R., Shanmuganathan, K., Tyler, D.J., Rowan, S.J. and Weder, C. (2008). Stimuli-responsive polymer nanocomposites inspired by the sea cucumber dermis. Science *319*, 1370–1374.

Capadona, J.R., Shanmuganathan, K., Trittschuh, S., Seidel, S., Rowan, S.J. and Weder, C. (2009). Polymer nanocomposites with nanowhiskers isolated from microcrystalline cellulose. Biomacromolecules *10*, 712–716.

Chatterjee, A.P. (2006). A model for the leastic moduli of three-dimensional fiber networks and nanocomposites. J. Appl. Phys. *100*, 054302-1-054302-8.

Chatterjee, A.P. (2008). Percolation threshold for rod-like particles: Polydispersity effects. J. Phys.: Condens. Matter *20*, 255250-1-255250-5.

Chatterjee, A.P. (2011). Connectedness percolation in monodisperse rod systems: Clustering effects. J. Phys.: Condens. Matter *23*, 375101-1-255250-7.

Chauve, G., Heux, L., Arouini, R. and Mazeau, K. (2005). Cellulose poly(ethylene-co-vinyl acetate) nanocomposites studied by molecular modeling and mechanical spectroscopy. Biomacromolecules *6*, 2025-2031.

Chazeau, L., Cavaillé, J.Y., Canova, G.R., Dendievel, R. and Boutherin, B. (1999a). Viscoelastic properties of plasticized PVC reinforced with cellulose whiskers. J. Appl. Polym. Sci. *71*, 1797–1808.

Chazeau, L., Cavaillé, J.Y. and Terech, P. (1999b). Mechanical behaviour above Tg of a plasticized PVC reinforced with cellulose whiskers, a SANS structural study. Polymer *40*, 5333–5344.

Chazeau, L., Paillet, M. and Cavaillé, J.Y. (1999c). Plasticized PVC reinforced with cellulose whiskers: 1-linear viscoelastic behavior analyzed through the quasi point defect theory. J. Polym. Sci. Polym. Phys. *37*, 2151–2164.

Chazeau, L., Cavaillé, J.Y. and Perez, J. (2000). Plasticized PVC reinforced with cellulose whiskers. II. Plastic behavior. J. Polym. Sci. B *38*, 383–392.

Chen, Y., Liu, C., Chang, P.R., Cao, X. and Anderson, D.P. (2009). Bionanocomposites based on pea starch and cellulose nanowhiskers hydrolyzed from pea hull fibre: Effect of hydrolysis time. Carbohydr. Polym. *76*, 607–615.

Cho, M.J. and Park, B.D. (2011). Tensile and thermal properties of nanocellulose-reinforced poly(vinyl alcohol) nanocomposites. J. Ind. Eng. Chem. *17*, 36–40.

Choi, Y.J. and Simonsen, J. (2006). Cellulose nanocrystal-filled carboxymethyl cellulose nanocomposites. J. Nanosci. Nanotechnol. *6*, 633–639.

Dagnon, K.L., Shanmuganathan, K., Weder, C. and Rowan, S.J. (2012). Water-triggered modulus changes of cellulose nanofiber nanocomposites with hydrophobic polymer matrices. Macromolecules *45*, 4707–4715.

Dalmas, F., Chazeau, L., Gauthier, C., Cavaillé, J.Y., and Dendievel, R. (2006). Large deformation mechanical behaviour of flexible nanofiber filled polymer nanocomposites. Polymer *47*, 2802–2812.

Dalmas, F., Cavaillé, J.Y., Gauthier, C., Chazeau, L. and Dendievel, R. (2007). Viscoelastic behaviour and electrical properties of flexible nanofiber filled polymer nanocomposites. Influence of processing conditions. Compos. Sci. Technol. *67*, 829–839.

de Gennes, P.G. (1976). On a relation between percolation theory and the elasticity of gels. J. Phys. Lett. *37*, l–2.

de Gennes, P.G. (1979). Scaling Concepts in Polymer Physics (Comell University Press, Ithaca, NY).

de Menezes, A.J., Siqueira, G., Curvelo, A.A.S. and Dufresne, A. (2009). Extrusion and characterization of functionalized cellulose whisker reinforced polyethylene nanocomposites. Polymer *50*, 4552–4563.

Drzal, L.T. (1985). The epoxy interphase in composites. Adv. Polym. Sci. *75*, 1–32.

Dubief, D., Samain, E. and Dufresne, A. (1999). Polysaccharide microcrystals reinforced amorphous poly(b-hydroxyoctanoate) nanocomposite materials. Macromolecules *32*, 5765–5771.

Duchemin, B.J.C., Mathew, A.P. and Oksman, K. (2009). All-cellulose composites by partial dissolution in the ionic liquid 1-butyl-3-methylimidazolium chloride. Composites A *40*, 2031–2037.

Dufresne, A. and Lacabanne, C. (1993). A study of interphase in model composites via low temperature thermo stimulated creep spectroscopy. Polymer *34*, 3173–3178.

Dufresne, A., Cavaillé, J.Y. and Helbert, W. (1997). Thermoplastic nanocomposites filled with wheat straw cellulose whiskers. Part II: Effect of processing and modeling, Polym. Compos. *18*, 198–210.

Dufresne, A. and Vignon, M.R. (1998). Improvement of starch films performances using cellulose microfibrils. Macromolecules *31*, 2693–2696.

Dufresne, A., Kellerhals, M.B. and Witholt, B. (1999). Transcrystallization in mcl-PHAs/ cellulose whiskers composites. Macromolecules *32*, 7396–7401.

Dufresne, A. (2000). Dynamic mechanical analysis of the interphase in bacterial polyester/ cellulose whiskers natural composites. Compos. Interfaces *7*, 53–67.

Dufresne, A., Dupeyre, D. and Vignon, M.R. (2000). Cellulose microfibrils from potato tuber cells: Processing and characterization of starch-cellulose microfibril composites. J. Appl. Polym. Sci. *76*, 2080–2092.

Dufresne. A. (2006). Comparing the mechanical properties of high performances polymer nanocomposites from biological sources. J. Nanosci. Nanotechnol. *6*, 322–330.

Favier, V., Canova, G.R., Cavaillé, J.Y., Chanzy, H., Dufresne, A. and Gauthier, C. (1995a). Nanocomposites materials from latex and cellulose whiskers. Polym. Adv. Technol. *6*, 351–355.

Favier, V., Chanzy, H. and Cavaillé, J.Y. (1995b). Polymer nanocomposites reinforced by cellulose whiskers. Macromolecules *28*, 6365–6367.

Favier, V., Dendievel, R., Canova, G., Cavaillé, J.Y. and Gilormini, P. (1997a). Simulation and modeling of three-dimensional percolating structures: Case of a latex matrix reinforced by a network of cellulose fibers. Acta Mater. *45*, 1557–1565.

Favier, V., Canova, G.R., Shrivastava, S.C. and Cavaillé, J.Y. (1997b). Mechanical percolation in cellulose whiskers nanocomposites. Polym. Eng. Sci. *37*, 1732–1739.

Flandin, L., Bidan, G., Brechet, Y. and Cavaillé, J.Y. (2000). New nanocomposite materials made of an insulating matrix and conducting fillers: Processing and properties. Polym. Compos. *21*, 165–174.

Fortunati, E., Armentano, I., Zhou, Q., Iannoni, A., Saino, E., Visai, L., Berglund, L.A. and Kenny, J.M. (2012). Multifunctional bionanocomposite films of poly(lactic acid), cellulose nanocrystals and silver nanoparticles. Carbohydr. Polym. *87*, 1596–1605.

Frone, A.N., Panaitescu, D.M., Donescu, D., Spataru, C.I., Radovici, C., Trusca, R. and Somoghi, R. (2011). Preparation and characterization of PVA composites with cellulose nanofibers obtained by ultrasonication. BioResources *6*, 487–512.

Gabr, M.H., Elrahman, M.A., Okubo, K. and Fujii, T. (2010). Effect of microfibrillated cellulose on mechanical properties of plain-woven CFRP reinforced epoxy. Compos. Struct. *92*, 1999–2006.

Gea, S., Bilotti, E., Reynolds, C.T., Soykeabkeaw, N. and Peijs, T. (2010). Bacterial cellulose-poly(vinyl alcohol) nanocomposites prepared by an *in-situ* process. Mater. Lett. *64*, 901–904.

Goetz, L., Foston, M., Mathew, A.P., Oksman, K. and Ragauskas, A.J. (2010). Poly(methyl vinyl ether-co-maleic acid)-polyethylene glycol nanocomposites cross-linked *in situ* with cellulose nanowhiskers. Biomacromolecules *11*, 2660–2666.

Goffin, A.-L., Raquez, J.-M., Duquesne, E., Siqueira, G., Habibi, Y., Dufresne, A. and Dubois, Ph. (2011a). Poly(e-caprolactone) based nanocomposites reinforced by surface-grafted cellulose nanowhiskers via extrusion processing: Morphology, rheology, and thermo-mechanical properties. Polymer *52*, 1532–1538.

Goffin, A.-L., Raquez, J.-M., Duquesne, E., Siqueira, G., Habibi, Y., Dufresne, A. and Dubois, Ph. (2011b). From interfacial ring-opening polymerization to melt processing of cellulose nanowhisker-filled polylactide-based nanocomposites. Biomacromolecules *12*, 2456–2465.

Gopalan Nair, K. and Dufresne, A. (2003). Crab shell chitin whisker reinforced natural rubber nanocomposites. 2. Mechanical behaviour. Biomacromolecules *4*, 666–674.

Gopalan Nair, K., Dufresne, A., Gandini, A. and Belgacem, M.N. (2003). Crab shell chitin whisker reinforced natural rubber nanocomposites. 3. Effect of chemical modification of chitin whiskers. Biomacromolecules *4*, 1835–1842.

Grande, C.J., Torres, F.G., Gomez, C.M., Troncoso, O.P., Canet-Ferrer, J. and Martinez-Pastor, J. (2009). Development of self-assembled bacterial cellulose-starch nanocomposites. Mat. Sci. Eng. C *29*, 1098–1104.

Grunert, M. and Winter, W. T. (2002). Nanocomposites of cellulose acetate butyrate reinforced with cellulose nanocrystals. J. Polym. Envir. *10*, 27–30.

Gusev, A.A. (2001). Numerical identification of the potential of whisker- and platelet-filled polymers. Macromolecules *34*, 3081–3093.

Habibi, Y. and Dufresne, A. (2008). Highly filled bionanocomposites from functionalized polysaccharide nanocrystals. Biomacromolecules *9*, 1974–1980.

Habibi, Y., Goffin, A.L., Schiltz, N., Duquesne, E., Dubois, P. and Dufresne, A. (2008). Bionanocomposites based on poly(ε-caprolactone)-grafted cellulose nanocrystals by ring opening polymerization. J. Mat. Chem. *18*, 5002–5010.

Hajji, P., Cavaillé, J.Y., Favier, V., Gauthier, C. and Vigier, G. (1996). Tensile behavior of nanocomposites from latex and cellulose whiskers. Polym. Compos. *17*, 612–619.

Halpin, J.C. and Kardos, J.L. (1972). Moduli of crystalline polymers employing composite theory. J. Appl. Phys. *43*, 2235–2241.

Hammersley, J.M. (1957). Percolation processes. The connective constant. Proc. Cambridge Phil. Soc. *53*, 642–645.

Helbert, W., Cavaillé, J.Y. and Dufresne, A. (1996). Thermoplastic nanocomposites filled with wheat straw cellulose whiskers. Part I: Processing and mechanical behavior. Polym. Comp. *17*, 604–611.

Henriksson, M. and Berglund, L.A. (2007). Structure and properties of cellulose nanocomposite films containing melamine formaldehyde. J. Appl. Polym. Sci. *106*, 2817–2824.

Heuillet, P., Neviere, R., Nottin, J.P. and Deneuville, P. (1992). Damage characterization for a solid propellant by an energetic approach. In: Damage detection in composite materials, J.E. Masters, ed. (American Society for testing and Materials, Philadelphia, PA), pp. 121–136.

Iwatake, A., Nogi, M. and Yano, H. (2008). Cellulose nanofiber-reinforced polylactic acid. Compos. Sci. Technol. *68*, 2103–2106.

Jiang, L., Morelius, E., Zhang, J., Wolcott, M. and Holbery, J. (2008). Study of the poly(3-hydroxybutyrate-co-3-hydroxyvalerate)/cellulose nanowhiskers composites prepared by solution casting and melt processing. J. Compos. Mat. *42*, 2629–2645.

Jonoobi, M., Harun, J., Mathew, A.P. and Oksman, K. (2010). Mechanical properties of cellulose nanofiber (CNF) reinforced polylactic acid (PLA) prepared by twin screw extrusion. Compos. Sci. Technol. *70*, 1742–1747.

Kaushik, A., Singh, M. and Verma, G. (2010). Green nanocomposites based on thermoplastic starch and steam exploded cellulose nanofibrils from wheat straw. Carbohydr. Polym. *82*, 337–345.

Kim, J., Montero, G. Habibi, Y., Hinestroza, J.P., Genzer, J., Argyropoulos, D.S. and Rojas, O.J. (2009). Dispersion of cellulose crystallites by nonionic surfactants in a hydrophobic polymer matrix. Polym. Eng. Sci. *49*, 2054–2061.

Kvien, S. and Oksman, K. (2007). Orientation of cellulose nanowhiskers in polyvinyl alcohol (PVA). Appl. Phys. A *87*, 641–643.

Kvien, I., Sugiyama, J., Votrubec, M. and Oksman, K. (2007). Characterization of starch based nanocomposites. J. Mater. Sci. *42*, 8163–8171.

Lee, S.Y., Mohan, D.J., Kang, I.A., Doh, G.H., Lee, S. and Han, N.O. (2009). Nanocellulose reinforced PVA composite films: Effects of acid treatment and filler loading. Fibers and Polymers *10*, 77–82.

Leitner, J., Hinterstoisser, B., Wastyn, M., Keckes, J. and Gindl, W. (2007). Sugar beet cellulose nanofibril-reinforced composites. Cellulose *14*, 419–425.

Lin, N., Huang, J., Chang, P.R., Feng, J. and Yu, J. (2011). Surface acetylation of cellulose nanocrystal and its reinforcing function in poly(lactic acid). Carbohydr. Polym. *83*, 1834–1842.

Liu, D., Zhong, T., Chang, P.R., Li, K. and Wu, Q. (2010). Starch composites reinforced by bamboo cellulosic crystals. Bioresource Technol. *101*, 2529–2536.

Liu H. and Laborie, M.-P. (2011). Bio-based nanocomposites by *in situ* cure of phenolic prepolymers with cellulose whiskers. Cellulose *18*, 619–630.

Ljungberg, N., Bonini, C., Bortolussi, F., Boisson, C., Heux, L. and Cavaillé, J.Y. (2005). New nanocomposite materials reinforced with cellulose whiskers in atactic polypropylene: Effect of surface and dispersion characteristics. Biomacromolecules *6*, 2732–2739.

Ljungberg, N., Cavaillé, J.Y. and Heux, L. (2006). Nanocomposites of isotactic polypropylene reinforced with rod-like cellulose whiskers. Polymer *47*, 6285–6292.

López-Rubio, A., Lagaron, J.M., Ankerfors, M., Lindström, T., Nordqvist, D., Mattozzi A. and Hedenqvist, M.S. (2007). Enhanced film forming and film properties of amylopectin using micro-fibrillated cellulose. Carbohydr. Polym. *68*, 718–727.

López-Suevos, F., Eyholzer, C., Bordeanu, N. and Richter K. (2010). DMA analysis and wood bonding of PVAc latex reinforced with cellulose nanofibrils. Cellulose *17*, 387–398.

Lu, J., Askeland, P. and Drzal, L.T. (2008). Surface modification of microfibrillated cellulose for epoxy composite applications. Polymer *49*, 1285–1296.

Lu, Y, Weng, L. and Cao X. (2005). Biocomposites of plasticized starch reinforced with cellulose crystallites from cottonseed linter. Macromol. Biosci. *5*, 1101–1107.

Lu, Y., Weng, L. and Cao, X. (2006). Morphological, thermal and mechanical properties of ramie crystallites – Reinforced plasticized starch biocomposites. Carbohydr. Polym. *63*, 198–204.

Ma, H., Zhou, B., Li, H.S., Li, Y.Q. and Ou, S.Y. (2011). Green composite films composed of nanocrystalline cellulose and a cellulose matrix regenerated from functionalized ionic liquid solution. Carbohydr. Polym. *84*, 383–389.

Malainine, M.E., Mahrouz, M. and Dufresne, A. (2005). Thermoplastic nanocomposites based on cellulose microfibrils from *Opuntia ficus indica* parenchyma cell. Compos. Sci. Technol. *65*, 1520–1526.

Marchessault, R.H., Dubé, M., St Pierre, J. and Revol, J.F. (1978). New insight into fibre swelling, interfibre bonding and wet strength. In: Fiber-water interactions in paper making (Oxford, Technical Division, The British Paper and Board Industry Federation, London, UK), pp. 795–818.

Marcovich, N.E., Auad, M.L., Belessi, N.E., Nutt, S.R. and Aranguren, M.I. (2006). Cellulose micro/nanocrystals reinforced polyurethane. J. Mater. Res. *21*, 870–881.

Mathew, A.P., Thielemans, W. and Dufresne, A. (2008). Mechanical properties of nanocomposites from sorbitol plasticized starch and tunicin whiskers. J. Appl. Polym. Sci. *109*, 4065–4074.

Matos Ruiz, M., Cavaillé, J.Y., Dufresne, A., Gérard, J.F. and Graillat, C. (2000). Processing and characterization of new thermoset nanocomposites based on cellulose whiskers. Compos. Interfaces *7*, 117–131.

Matos Ruiz, M., Cavaillé, J.Y., Dufresne, A., Graillat, C. and Gérard, J.F. (2001). New waterborne epoxy coatings based on cellulose nanofillers. Macromol. Symp. *169*, 211–222.

Mendez, J., Annamalai, P.K., Eichhorn, S.J., Rusli, R., Rowan, S.J., Foster, E.J. and Weder, C. (2011). Bioinspired mechanically adaptive polymer nanocomposites with water-activated shape-memory effect. Macromolecules *44*, 6827–6835.

Mikkonen, K.S., Mathew, A.P., Pirkkalainen, K., Serimaa, R., Xu, C., Willför, S., Oksman, K. and Tenkanen, M. (2010). Glucomannan composite films with cellulose nanowhiskers. Cellulose *17*, 69–81.

Mikkonen, K.S., Stevanic, J.S., Joly, C., Dole, P., Pirkkalainen, K., Serimaa, R., Salmén, L. and Tenkanen, M. (2011). Composite films from spruce galactoglucomannans with microfibrillated spruce wood cellulose. Cellulose *18*, 713–726.

Nakagaito, A.N. and Yano, H. (2004). The effect of morphological changes from pulp fiber towards nano-scale fibrillated cellulose on the mechanical properties of high-strength plant fiber based composites. Appl. Phys. A-Mater. Sci. Process *78*, 547–552.

Nakagaito, A.N. and Yano, H. (2005). Novel high-strength biocomposites based on microfibrillated cellulose having nano- orderunit web-like network structure. Appl Phys A-Mater. Sci. Process *80*, 155–159.

Nakagaito, A.N., Iwamoto, S. and Yano H. (2005). Bacterial cellulose: the ultimate nano-scalar cellulose morphology for the production of high-strength composites. Appl. Phys. A-Mater. Sci. Process *80*, 93–97.

Nakagaito, A.N. and Yano, H. (2008a). Toughness enhancement of cellulose nanocomposites by alkali treatment of the reinforcing cellulose nanofibres. Cellulose *15*, 323–331.

Nakagaito, A.N. and Yano, H. (2008b). The effect of fiber content on the mechanical and thermal expansion properties of biocomposites based on microfibrillated cellulose. Cellulose *15*, 555–559.

Nakagaito, A.N., Fujimura, A., Sakai, T., Hama, Y. and Yano, H. (2009). Production of microfibrillated cellulose (MFC)-reinforced polylactic acid (PLA) nanocomposites from sheets obtained by a papermaking-like process. Compos. Sci. Technol. *69*, 1293–1297.

Nan, C.W. (1993). Physics of inhomogeneous inorganic materials. Prog. Mater. Sci. *37*, 1–116.

Nge, T.T., Nogi, M., Yano, H. and Sugiyama, J. (2010). Microstructure and mechanical properties of bacterial cellulose/chitosan porous scaffold. Cellulose *17*, 349–363.

Noishiki, Y., Nishiyama, Y., Wada, M., Kuga, S. and Magoshi, J. (2002). Mechanical properties of silk fibroin-microcrystalline cellulose composite films. J. Appl. Polym. Sci. *86*, 3425–3429.

Noorani, S., Simonsen, J. and Atre, S. (2007). Nano-enabled microtechnology: Polysulfone nanocomposites incorporating cellulose nanocrystals. Cellulose *14*, 577–584.

Nordqvist, D., Idermark, J., Hedenqvist, M.S., Gällstedt, M., Ankerfors, M. and Lindström, T. (2007). Enhancement of the wet properties of transparent chitosan-acetic acid-salt films using microfibrillated cellulose. Biomacromolecules *8*, 2398–2403.

Oksman, K., Mathew, A. P., Bondeson, D. and Kvien, I. (2006). Manufacturing process of cellulose whiskers/polylactic acid nanocomposites. Compos. Sci. Technol. *66*, 2776–2784.

Orts, W.J., Godbout, L., Marchessault, R.H. and Revol, J.F. (1998). Enhanced ordering of liquid crystalline suspensions of cellulose microfibrils: A small angle neutron scattering study. Macromolecules *31*, 5717–5725.

Orts, W.J., Shey, J., Imam, S.H., Glenn, G.M., Guttman, M.E. and Revol, J.F. (2005). Application of cellulose microfibrils in polymer nanocomposites. J. Polym. Env. *13*, 301–306.

Ouali, N., Cavaillé, J.Y. and Perez, J. (1991). Elastic, viscoelastic and plastic behavior of multiphase polymer blends. Plast. Rubber Compos. Process Appl. *16*, 55–60.

Pakzad, A., Simonsen, J. and Yassar, R.S. (2012). Gradient of nanomaterial properties in the interphase of cellulose nanocrystal composites. Compos. Sci. Technol. *72*, 314–319.

Pandey, J.K., Chu, W.S., Kim, C.S., Lee, C.S. and Ahn, S.H. (2009). Bio-nano reinforcement of environmentally degradable polymer matrix by cellulose whiskers from grass. Composites: Part B *40*, 676–680.

Paralikar, S. A., Simonsen, J. and Lombardi, J. (2008). Poly (vinyl alcohol)/cellulose nanocrystals barrier membranes. J. Membr. Sci. *320*, 248–258.

Park, I., Kang, M., Kim, H.S. and Jin, H.J. (2007). Electrospinning of poly(ethylene oxide) with bacterial cellulose whiskers. Macromol. Symp. *249–250*, 289–294.

Pasquini, D., Teixeira, E.M., Curvelo, A.A.S., Belgacem, M.N. and Dufresne, A. (2010). Extraction of starch nanocrystals and cellulose whiskers from cassava bagasse and their applications as reinforcing agent in natural rubber. Ind. Crops Prod. *32*, 486–490.

Pei, A., Zhou, Q. and Berglund, L. (2010). Functionalized cellulose nanocrystals as biobased nucleation agents in poly(L-lactide) (PLLA) – Crystallization and mechanical property effects. Compos. Sci. Technol. *70*, 815–821.

Peng, K., Wang, B., Chen, S., Zhong, C. and Wang, H. (2011). Preparation and properties of polystyrene/bacteria cellulose nanocomposites by *in situ* polymerization. J. Macromol. Sci. B *50*, 1921–1927.

Peresin, M.S., Habibi, Y., Zoppe, J.O., Pawlak, J.J. and Rojas, O.J. (2010). Nanofiber composites of polyvinyl alcohol and cellulose nanocrystals: Manufacture and characterization, Biomacromolecules *11*, 674–681.

Petersson, L. and Oksman, K. (2006). Biopolymer based nanocomposites: comparing layered silicates and microcrystalline cellulose as nanoreinforcement. Compos. Sci. Technol. *66*, 2187–2196.

Petersson, L., Kvien, I. and Oksman, K. (2007). Structure and thermal properties of poly(lactic acid)/cellulose whiskers nanocomposites materials. Compos. Sci. Technol. *67*, 2535–2544

Plackett, D., Anturi, H., Hedenqvist, M., Ankerfors, M., Gällstedt, M., Lindström, T. and Siró, I. (2010). Physical properties and morphology of films prepared from microfibrillated cellulose and microfibrillated cellulose in combination with amylopectin. J. Appl. Polym. Sci. *117*, 3601–3609.

Prokhorova, D.A. and Chatterjee, A.P. (2009). Elastic moduli of cellulose nanoparticle-reinforced composites: A micromechanical model. Biomacromolecules *10*, 3259–3265.

Pu, Y., Zhang, J., Elder, T., Deng, Y., Gatenholm, P. and Ragauskas, A.J. (2007). Investigation into nanocellulosics versus acacia reinforced acrylic films. Composites: Part B *38*, 360–366.

Qi, H., Cai, J., Zhang, L. and Kuga, S. (2009). Properties of films composed of cellulose nanowhiskers and a cellulose matrix regenerated from alkali/urea solution. Biomacromolecules *10*, 1597–1602.

Raquez, J.M., Murena, Y., Goffin, A.L., Habibi, Y., Ruelle, B., DeBuyl, F. and Dubois, P. (2012). Surface-modification of cellulose nanowhiskers and their use as nanoreinforcers into polylactide: A sustainably-integrated approach. Compos. Sci. Technol. *72*, 544–549.

Roohani, M., Habibi, Y., Belgacem, N.M., Ebrahim, G., Karimi, A.N. and Dufresne, A. (2008). Cellulose whiskers reinforced polyvinyl alcohol copolymers nanocomposites. Eur. Polym. J. *44*, 2489–2498.

Rusli, R. and Eichhorn, S.J. (2008). Determination of the stiffness of cellulose nanowhiskers and the fiber-matrix interface in a nanocomposite using Raman spectroscopy. Appl. Phys. Lett. *93*, 033111.

Rusli, R. Shanmuganathan, K., Rowan, S.J., Weder, C. and Eichhorn, S.J. (2011). Stress transfer in cellulose nanowhisker composites – Influence of whisker aspect ratio and surface charge. Biomacromolecules *12*, 1363–1369.

Sanchez-Garcia, M.D. and Laragon, J.M. (2010). On the use of plant cellulose nanowhiskers to enhance the barrier properties of polylactic acid. Cellulose *17*, 987–1004.

Saxena, A., Elder, T.J., Pan, S. and Ragauskas, A.J. (2009). Novel nanocellulosic xylan composite film. Composites Part B *40*, 727–730.

Schroers, M., Kokil, A. and Weder, C. (2004). Solid polymer electrolytes based on nanocomposites of ethylene oxide-epichlorohydrin copolymers and cellulose whiskers. J. Appl. Polym. Scl. *93*, 2883–2888.

Sehaqui, H., Zhou, Q. and Berglund, L.A. (2011). Nanostructured biocomposites of high toughness – A wood cellulose nanofiber network in ductile hydroxyethylcellulose matrix. Soft Matter *7*, 7342–7350.

Seydibeyoğlu, M.Ö. and Oksman, K. (2008). Novel nanocomposites based on polyurethane and micro fibrillated cellulose. Compos. Sci. Technol. *68*, 908–914.

Shanmuganathan, K., Capadona, J.R., Rowan, S.J. and Weder, C. (2010a). Biomimetic mechanically adaptive nanocomposites. Prog. Polym. Sci. *35*, 212–222.

Shanmuganathan, K., Capadona, J.R., Rowan, S.J. and Weder, C. (2010b). Bio-inspired mechanically-adaptive nanocomposites derived from cotton cellulose whiskers. J. Mater. Chem. *20*, 180–186.

Shanmuganathan, K., Capadona, J.R., Rowan, S.J. and Weder, C. (2010c). Stimuli-responsive mechanically adaptive polymer nanocomposites. ACS Appl. Mater. Interfaces *2*, 165–174.

Shibata, M. and Nakai, K. (2010). Preparation and properties of biocomposites composed of bio-based epoxy resin, tannic acid, and microfibrillated cellulose. J. Polym. Sci. B *48*, 425–433.

Siqueira, G., Bras, J. and Dufresne, A. (2009). Cellulose whiskers vs. microfibrils: Influence of the nature of the nanoparticle and its surface functionalization on the thermal and mechanical properties of nanocomposites. Biomacromolecules *10*, 425–432.

Siqueira, G., Abdillahi, H., Bras, J. and Dufresne, A. (2010a). High reinforcing capability cellulose nanocrystals extracted from *Syngonanthus nitens* (Capim Dourado). Cellulose *17*, 289–298.

Siqueira, G., Tapin-Lingua, S., Bras, J., da Silva Perez, D. and Dufresne, A. (2010b). Morphological investigation of nanoparticles obtained from enzymatic and acid hydrolysis of sisal fibers. Cellulose *17*, 1147–1158.

Siqueira, G., Tapin-Lingua, S., Bras, J., Da Silva Perez, D. and Dufresne, A. (2011). Mechanical properties of natural rubber nanocomposites reinforced with cellulosic nanoparticles obtained from enzymatic and acid hydrolysis of sisal fibers. Cellulose *18*, 57–65.

Small, M.K. and Nix, W.D. (1992). Analysis of the accuracy of the bulge test in determining the mechanical properties of thin films. J. Mater. Res. *7*, 1553–1563.

Stauffer, D. (1985). Introduction to Percolation Theory (Taylor and Francis, London and Philadelphia).

Šturcova, A., Davies, G.R. and Eichhorn, S.J. (2005). Elastic modulus and stress-transfer properties of tunicate cellulose whiskers. Biomacromolecules *6*, 1055–1061.

Suryanegara, L., Nakagaito, A.N. and Yano, H. (2009). The effect of crystallization of PLA on the thermal and mechanical properties of microfibrillated cellulose-reinforced PLA composites. Compos. Sci. Technol. *69*, 1187–1192.
Suryanegara, L., Okumura, H., Nakagaito, A.N. and Yano, H. (2011). The synergetic effect of phenylphosphonic acid zinc and microfibrillated cellulose on the injection molding cycle time of PLA composites. Cellulose *18*, 689–698.
Svagan, A.J., Azizi Samir, M.A.S. and Berglund, L.A. (2007). Biomimetic polysaccharide nanocomposites of high cellulose content and high toughness. Biomacromolecules *8*, 2556–2563.
Svagan, A.J., Azizi Samir, M.A.S. and Berglund, L.A. (2008). Biomimetic foams of high mechanical performance based on nanostructured cell walls reinforced by native cellulose nanofibrils. Adv. Mat. *20*, 1263–1269.
Svagan, A.J., Berglund, L.A. and Jensen, P. (2011). A cellulose nanocomposite biopolymer foam – Hierarchical structure effects on energy absorption. ACS Appl. Mater. Interfaces *3*, 1411–1417.
Takayanagi, M., Uemura, S. and Minami, S. (1964). Application of equivalent model method to dynamic rheo-optical properties of a crystalline polymer. J. Polym. Sci. C *5*, 113–122.
Tang and Weder (2010). Cellulose whisker/epoxy resin nanocomposites. ACS Appl. Mater. Interfaces *2*, 1073–1080.
Teixeira, E.M., Pasquini, D., Curvelo, A.A.S., Corradini, E., Belgacem, M.N. and Dufresne, A. (2009). Cassava bagasse cellulose nanofibrils reinforced thermoplastic cassava starch. Carbohydr. Polym. *78*, 422–431.
Ten, E., Turtle, J., Bahr, D., Jiang, L. and Wolcott, M. (2010). Thermal and mechanical properties of poly(3-hydroxybutyrate-co-3-hydroxyvalerate)/cellulose nanowhiskers composites. Polymer *51*, 2652–2660.
Tingaut, P., Zimmermann, T. and Lopez-Suevos, F. (2010). Synthesis and characterization of bionanocomposites with tunable properties from poly(lactic acid) and acetylated microfibrillated cellulose. Biomacromolecules *11*, 454–464.
Trovatti, E., Oliveira, L., Freire, C.S.R., Silvestre, A.J.D., Neto, C.P., Cruz Pinto, J.J.C. and Gandini, A. (2010). Novel bacterial cellulose-acrylic resin nanocomposites. Compos. Sci. Technol. *70*, 1148–1153.
Tsaï, S.W., Halpin, J.C. and Pagano, N.J. (1968). Composite Materials Workshop (Technomic Publishing Co. Inc., Stamford, Conn.), pp. 233–253.
Uddin, A.J., Araki, J. and Gotoh, Y. (2011). Toward "strong" green nanocomposites: Polyvinyl alcohol reinforced with extremely oriented cellulose whiskers. Biomacromolecules *12*, 617–624.
Wang, B. and Sain, M. (2007a). Dispersion of soybean stock-based nanofiber in a plastic matrix. Polym. Inter. *56*, 538–546.
Wang, B. and Sain, M. (2007b). Isolation of nanofibers from soybean source and their reinforcing capability on synthetic polymers. Compos. Sci. Technol. *67*, 2521–2527.
Wang, B. and Sain, M. (2007c). The effect of chemically coated nanofiber reinforcement on biopolymer based nanocomposites. BioResources *2*, 371–388.
Wang, Y., Cao, X. and Zhang, L. (2006). Effects of cellulose whiskers on properties of soy protein thermoplastics. Macromol. Biosci. *6*, 524–531.
Wang, Y., Tian, H. and Zhang, L. (2010). Role of starch nanocrystals and cellulose whiskers in synergistic reinforcement of waterborne polyurethane. Carbohydr. Polym. *80*, 665–671.
Woehl, M.A., Canestraro, C.D., Mikowski, A., Sierakowski, M.R., Ramos, L.P. and Wypych, F. (2010). Bionanocomposites of thermoplastic starch reinforced with bacterial cellulose nanofibres: Effect of enzymatic treatment on mechanical properties. Carbohydr. Polym. *80*, 866–873.
Xu S.H., Gu J., Luo Y.F. and Jia D.M. (2012). Effects of partial replacement of silica with surface modified nanocrystalline cellulose on properties of natural rubber nanocomposites. eXPRESS Polym. Lett. *6*, 14–25.

Yamanaka, S., Watanabe, K. and Kitamura, N. (1989). The structure and mechanical properties of sheets prepared from bacterial cellulose. J. Mat. Sci. *24*, 3141–3145.

Zimmermann, T., Pohler, E. and Geiger, T. (2004). Cellulose fibrils for polymer reinforcement. Adv. Eng. Mat. *6*, 754–761.

Zimmermann, T., Pohler, E. and Geiger, T. (2005). Mechanical and morphological properties of cellulose fibril reinforced nanocomposites. Adv. Eng. Mat. *7*, 1156–1161.

Zimmermann, T., Bordeanu, N. and Strub E. (2010). Properties of nanofibrillated cellulose from different raw materials and its reinforcement potential. Carbohydr. Polym. *79*, 1086–1093.

Zoppe, J.O., Peresin, M.S., Habibi, Y., Venditti, R.A. and Rojas, O.J. (2009). Reinforcing poly(e-caprolactone) nanofibers with cellulose nanocrystals. ACS Appl. Mater. Interfaces *1*, 1996–2004.

10 Swelling and barrier properties

The largest interest in cellulose nanoparticles for composite applications originates from the outstanding mechanical properties imparted to the material. Improved properties can often be reached for low filler volume fraction without detrimental effect on other properties, such as impact resistance of plastic deformation capability. However, there is an increasing interest in the barrier properties of nanocellulose films or related nanocomposites due to increased tortuosity provided by nanoparticles. Indeed, because of their small size, the surface-to-volume ratio of the nanoparticles is significantly greater than for microparticles. Most materials used for food packaging are practically non-degradable petrochemical-based polymers, representing a serious environmental problem. The main reason for their use is due to their easiness of processability, low cost and excellent barrier properties. Moreover, the low permeability of cellulose can be enhanced by the highly crystalline nature of cellulose nanoparticles and ability to form a dense percolating network. Provided that strong particle-polymer molecular interactions exist, the smaller particles have a greater ability to bond to the surrounding polymer material, thereby reducing the chain segmental mobility and thus the penetrant diffusivity.

Barrier properties using bio-based materials are becoming increasingly advisable in our society to develop environmentally-friendly efficient material in different applications.

10.1 Swelling and sorption properties

The swelling process and its kinetics allow quantifying the capacity of a linear or branched polymer to dissolve or of a cross-linked polymer to swell in different liquid and vapor media. The interaction of polymeric materials with solvents is a huge problem from both an academic and a technological point of view. The mass and dimensions of polymer systems may be changed due to the penetration of solvents into swollen specimens. When a cross-linked polymer is brought into contact with a solvent, the network absorbs a certain amount of solvent, which strongly depends on temperature, molecular weight of this solvent, cross-linking density of the polymer and polymer/solvent interactions, besides the ingredients added.

For composite materials, swelling is a method currently used to determine if the filler creates supplementary cross-links due to specific interactions between the filler and the matrix. The choice of the solvent used for the experiment is of importance. It must be a good solvent of the matrix to allow its swelling or dissolution, but it must not be able to break the eventual links between the matrix and the filler. If this last condition is not respected, the experiments cannot be conclusive. For samples sensitive to the liquid medium, it is preferable to condition the material in vapor medium. For instance, in the case of water the moisture content can be achieved by conditioning the samples in desiccators at controlled humidity conditions containing saturated salt solutions. Conditioning needs to be achieved for a period of time ensuring equilibration of the solvent content in the films with that of the atmosphere (stabilization of the sample weight). Examples of saturated salt solutions and corresponding relative humidity (RH) conditions at room temperature are reported in Table 10.1.

Saturated Salt Solution		**RH (%)**
Cesium Fluoride	CsF	3.39 ± 0.94
Lithium Bromide	LiBr	6.37 ± 0.52
Zinc Bromide	$ZnBr_2$	7.75 ± 0.39
Potassium Hydroxide	KOH	8.23 ± 0.72
Sodium Hydroxide	NaOH	8.24 ± 2.1
Lithium Chloride	LiCl	11.30 ± 0.27
Calcium Bromide	$CaBr_2$	16.50 ± 0.20
Lithium Iodide	LiI	17.56 ± 0.13
Potassium Acetate	CH_3COOK	22.51 ± 0.32
Potassium Fluoride	KF	30.85 ± 1.3
Magnesium Chloride	$MgCl_2$	32.78 ± 0.16
Sodium Iodide	NaI	38.17 ± 0.50
Potassium Carbonate	K_2CO_3	43.16 ± 0.39
Magnesium Nitrate	$MG(NO_3)_2$	52.89 ± 0.22
Sodium Bromide	NaBr	57.57 ± 0.40
Cobalt Chloride	$CoCl_2$	64.92 ± 3.50
Potassium Iodide	KI	68.86 ± 0.24
Strontium Chloride	$SrCl_2$	70.85 ± 0.04
Sodium Nitrate	$NaNO_3$	74.25 ± 0.32
Sodium Chloride	NaCl	75.29 ± 0.12
Ammonium Chloride	NH_4Cl	78.57 ± 0.40
Potassium Bromide	KBr	80.89 ± 0.21
Ammonium Sulfate	$(NH_4)_2SO_4$	80.99 ± 0.28
Potassium Chloride	KCl	85.06 ± 0.38
Strontium Nitrate	$Sr(NO_3)_2$	85.06 ± 0.38
Potassium Nitrate	KNO_3	93.58 ± 0.55
Potassium Sulfate	K_2SO_4	97.30 ± 045
Potassium Chromate	K_2CrO_4	97.88 ± 0.49

Table 10.1: Equilibrium relative humidity (RH) of some saturated salt solutions at 25°C.

The kinetics of solvent absorption consists generally in first drying and weighing the sample, and then immersing in the liquid solvent or exposing to the vapor medium. The specimen generally consists of a rectangular strip whose thickness is sufficiently thin compared to lateral dimensions (typically ten times lower) so that the molecular diffusion can be considered one dimensional. The sample is then removed at specific intervals, blotted with a filter paper, and weighed until an equilibrium value is reached. The swelling rate or solvent content or solvent uptake (SU), of the sample can be calculated by dividing the gain in weight by the initial weight:

$$SU(\%) = \frac{M_t - M_o}{M_o} \cdot 100 \tag{10.1}$$

where M_t and M_o correspond to the weight of the sample at time t and before exposure to the liquid or vapor medium, respectively. Generally, the short-time behavior displays a fast absorption phenomenon whereas at longer times, the kinetics of absorption is slowed down and leads to a plateau corresponding to the solvent uptake at equilibrium (M_∞). The mass of solvent sorbed at time t (M_t–M_o) can be expressed as (Comyn, 1985):

$$\frac{M_t - M_o}{M_\infty} = 1 - \sum_{n=0}^{\infty} \frac{8}{(2n+1)^2\pi^2} \cdot \exp\left[\frac{-D(2n+1)^2\pi^2 t}{4L^2}\right] \tag{10.2}$$

where, M_∞ is the mass of liquid sorbed at equilibrium, D the diffusivity or diffusion coefficient, and 2L the thickness of the sample film. At short times, this equation can be written as:

$$\frac{M_t - M_o}{M_\infty} = \frac{2}{L}\left(\frac{D}{\pi}\right)^{1/2} t^{1/2} \tag{10.3}$$

The diffusion coefficient of the liquid in the material can therefore be estimated from the slope of the plot of $(M_t–M_o)/M_\infty$ as a function of $(t/L^2)^{1/2}$. For $(M_t–M_o)/M_\infty \leq 0.5$ the error made in using this last equation instead of the previous one to determine the diffusion coefficient is of the order of 0.1% (Vergnaud, 1991).

The water sorption isotherms can also be measured using a dynamic vapor sorption (DVS) apparatus. The sample is first dried and weighted, and placed in a quartz sample pan under controlled constant temperature. In this gravimetric technique the vapor concentration surrounding the sample is varied and the induced weight change is measured. It is an alternative to the desiccator/saturated salt solution method. The values of solvent content at equilibrium are measured for each vapor concentration and used to build the sorption isotherm. When using water vapor, it consists of the plot of the equilibrium water content against water activity a_w (with $a_w = P/P_{sat}$, the ratio of pressure P to saturation vapor pressure P_{sat}).

The swelling of films can be studied using quartz crystal microbalance with dissipation (QCM-D) measurements. QCM-D is a technique used for in situ adsorption studies at the solid-liquid interface that gives a mass per unit area by measuring the change in frequency of a quartz crystal resonator. It also enables studies of the kinetics of swelling and associated viscoelastic changes. Indeed, the dissipation is a parameter quantifying the damping in the system, and is related to the sample's viscoelastic properties. The dissipation is equal to the ratio of bandwidth and frequency. The QCM-D measures simultaneously changes in frequency and dissipation at the fundamental resonance frequency, 5 MHz, and its overtones 15, 25, 35, 45, 55 and 75 MHz. Without adsorbate, the crystal oscillates at a resonant frequency f_o, and after adsorption, the resonant frequency decreases to f. For uniform, rigidly adsorbed films, the change in frequency, Δf, is proportional to the adsorbed mass per unit surface, Δm, according to:

$$\Delta m = -\frac{C \cdot \Delta f}{n} \tag{10.4}$$

where n is the overtone number (1, 3, 5, 7, 9, 11, or 13) and C is a device-sensitive constant.

Several mathematical models can be used to correlate the water content of polymer films at equilibrium with the surrounding relative humidity. The GAB (Guggenheim–Anderson–De Boer) model is the most widely used and is applicable for water activities (a_w) ranging between 0.05 and 0.95. This model not only allows calculating the water content of the monolayer, but also determining the heat of sorption of monolayer and multilayer. The GAB equation can be written in the following form:

$$X_w = \frac{C \cdot k \cdot X_m \cdot a_w}{(1 - k \cdot a_w) \cdot (1 - k \cdot a_w + C \cdot k \cdot a_w)} \tag{10.5}$$

where X_w is the equilibrium moisture content (g of water/g of dry mass), X_m is the monolayer water content, C is the Guggenheim constant related to temperature, and k is a constant linked to the adsorption enthalpy difference between the first layer and the following. The determination of C and k coefficients is as follows:

$$\begin{aligned} C &= C_o \cdot \exp\left(\frac{\Delta H_c}{R \cdot T}\right) \\ k &= k_o \cdot \exp\left(\frac{\Delta H_k}{R \cdot T}\right) \end{aligned} \tag{10.6}$$

with:

$$\begin{aligned} \Delta H_c &= H_m - H_n \\ \Delta H_k &= H_1 - H_n \end{aligned} \tag{10.7}$$

where T is the temperature (K), R the ideal gas constant, H_m the heat of sorption of the monolayer, H_n the heat of sorption of the multilayer, and H_1 the heat of condensation of water vapor.

10.2 Barrier properties

Barrier properties are commonly in connection with particular permeable objects, which include commonly gases, water vapor, liquid and organic substance, etc. In polymers they are necessarily associated with their inherent ability to permit the exchange, to a higher or lower extent, of low molecular weight substances through mass transport processes like permeation. The whole transmitting process includes adsorption, dissolution, diffusion, and desorption. Gases or water vapor enter the surface of materials from higher density side. After diffusing inside the material, it desorbs on the low-density side. The permeation of low-molecular weight chemical species usually takes place through the polymer amorphous phase and the crystalline phase is considered impenetrable.

10.2.1 Water vapor transfer rate and water vapor permeability

The permeability to water vapor is expressed as water vapor permeability (WVP) or water vapor transmission rate (WVTR). The latter is most commonly used. The WVP is the volume of water vapor which perspires per thickness unit and surface unit from a specimen during a given period of time, under specified temperature, controlled RH and under a vapor pressure difference. The unit of WVP is $g \cdot m^{-1} \cdot s^{-1} \cdot Pa^{-1}$ and it can be determined as:

$$WVP = \frac{\Delta m \cdot e}{A \cdot \Delta t \cdot \Delta P} \tag{10.8}$$

The WVTR corresponds to the quantity of water which penetrates a surface unit sample of defined thickness during 24 hours under specified temperature, controlled RH and under a vapor pressure difference. The unit is $g \cdot m^2 \cdot day^{-1}$. The WVTR corresponds to the steady state rate at which water vapor permeates through a film under specified conditions of temperature and relative humidity. It can be determined by sealing the film specimen (thickness e and area A) on a test cup containing a desiccant. The tests are then performed in a controlled temperature and humidity atmosphere by placing in the bottom of the test cup a saturated salt solution or deionized water (wet side) to create a partial water vapor pressure difference, ΔP, across the specimen. The cup is weighted for different exposure times, Δt, to determine the mass increase of the desiccant Δm. WVTR is the standard measurement by which films are compared for their ability to resist moisture transmission. Lower values indicate better moisture protection. Only values reported at the same temperature and humidity can be compared, because transmission rates are affected by both of these parameters.

10.2.2 Gas permeability

Gas permeation experiments are carried out using a permeation cell consisting of two compartments separated by the studied membrane. Before any measurement, the permeation cell is completely evacuated by applying vacuum on both sides of the film. The gas under test is introduced at a given pressure P_1 in the upstream compartment of the cell and the pressure variations P_2 in the downstream compartment are measured with a pressure gauge as a function of time. The permeability coefficient P expressed in barrer unit (1 barrer = 10^{-10} cm^3 STP cm s^{-1} cm^{-2} $cmHg^{-1}$), can be calculated using the variable pressure method from the slope of the straight line obtained in the steady state assuming $P_1 >> P_2$:

$$P = \frac{J_{st} \cdot L}{P_1} \tag{10.9}$$

where J_{st} is the steady-state gas flux obtained from the slope of the steady-state part of the curve P_2 vs. time and L is the sample thickness. The diffusion coefficient D can be deduced from the time lag t_L provided by the extrapolation of this straight line on the time axis:

$$D = \frac{L^2}{6t_L} \tag{10.10}$$

The solubility coefficient S is given by the ratio of the permeability to diffusion coefficients:

$$S = \frac{P}{D} \tag{10.11}$$

This equation has also often been considered to describe the gas transport properties of composites composed of impermeable fillers dispersed in a polymer matrix. Assuming that the local characteristics of the matrix are not affected by the presence of particles and that the polymer-filler interactions are strong enough to avoid void creation at the interface, the gas solubility in the heterogeneous system can be expressed as:

$$S = (1 - \phi)\, S_m \tag{10.12}$$

where S_m and S are the solubility coefficients in the neat matrix and composite, respectively, and ϕ is the filler volume fraction.

As far as the particles act as impenetrable obstacles, the diffusing molecules have to follow a more tortuous pathway to proceed through the composite film (Figure 10.1). The diffusion rate is slowed down and can be expressed as (Nielsen, 1967):

$$D = \frac{D_m}{\tau} \tag{10.13}$$

where D_m and D correspond to the diffusion coefficients in the neat polymer and composite, respectively, and τ to the tortuosity.

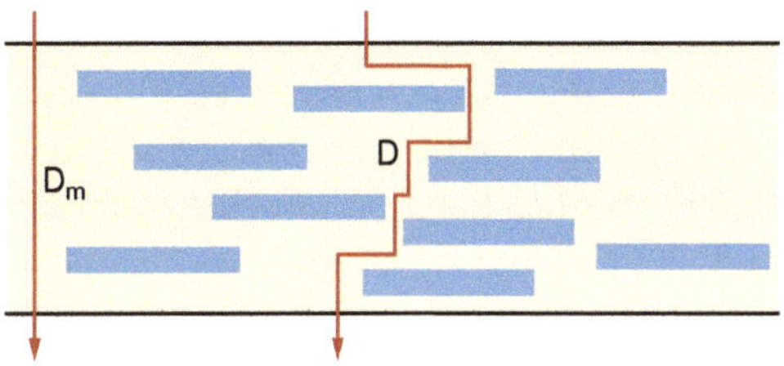

Fig. 10.1: Schematic representation of the tortuosity induced by the nanoparticles within a polymer matrix.

Combining these equations leads to the following expression of the relative permeability, i.e. the permeability of composite P divided by the one of the neat matrix, P_m, for completely aligned particles (all fillers have their larger surface parallel to the film surface):

$$\frac{P}{P_m} = \frac{(1 - \phi)}{\tau} = \frac{(1 - \phi)}{(1 + a\phi)} \tag{10.14}$$

with a being the filler aspect ratio (for square fillers of length/width L and thickness W, a = L/2W).

10.3 Water sorption and swelling properties of microfibrillated cellulose films

The water absorption of MFC films was determined by immersing the films in deionized water for 10 min (Spence et al., 2010). Different wood pulps with various chemical compositions were used. Compared to original wood pulps, either bleached or unbleached, refining pretreatment and further homogenization treatments were both found to decrease the equilibrium water sorption. It was concluded that the MFC film structure was significantly more compact, hence less water can penetrate into the film compared to those generated from the original pulp. The lignin-containing samples adsorbed more water in the original form, but were similar to the bleached samples after treatment.

The DVS method was used to characterize the water sorption of MFC films prepared from softwood dissolving pulp (Henriksson and Berglund, 2007; Svagan et al., 2009) and films obtained from MFC and cellulose nanocrystals prepared from sisal fibers (Belbekhouche et al., 2011). For softwood MFC, the moisture content at 30°C and 90% RH was around 15% (Henriksson and Berglund, 2007). At 33°C and 30% RH, it was around 3% (Svagan et al., 2009). It was also reported that the equilibrium water content for a given vapor pressure depended on the direction from which equilibrium was approached (Henriksson and Berglund, 2007). This was observed as a sorption hysteresis between the adsorption-desorption curves when moisture content was plotted as a function of relative vapor pressure. Indeed, a larger number of hydroxyl groups are accessible during desorption as compared with adsorption from the dry state.

The water sorption behavior of films prepared from MFC and cellulose nanocrystals extracted from the same source (sisal) has been investigated (Belbekhouche et al., 2011). Similar sigmoid profile water sorption isotherms generally reported for hydrophilic materials were observed for both films despite different structural differences (morphology, degree of crystallinity, and polarity) between both kinds of nanoparticles as shown in Figure 10.2. In addition, completely reversible and repeatable systems were observed as shown by similar data obtained during sorption-desorption-sorption cycles. The water sorption behavior of these films was described using the GAB and Park models. The experimental data were well fitted with the Park model. The water diffusion coefficient was higher for nanocrystal films compared to MFC film. It was explained by the presence of residual lignin, extractive substances and fatty acids at the surface of MFC. The difference in surface chemistry of both sisal nanoparticles was well evidenced from contact angle measurements, nanocrystal and MFC films displaying water contact angles of 44.6° and 59.4°, respectively (Siqueira et al., 2009). Surprisingly, the first half-sorption diffusion coefficient corresponding to the diffusion of water through the surface of the nanoparticles was found to be higher than the second half-sorption diffusion coefficient more representative of the diffusion in the core. This means that the diffusion of water is rather controlled by the surface than by the core, probably because of a barrier effect related to the presence of water at the surface during the sorption kinetics. The parallel exponential kinetics (PEK) model successfully predicted the water sorption kinetics.

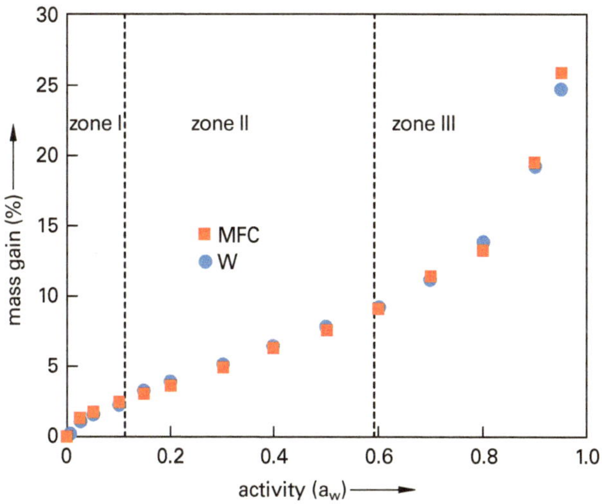

Fig. 10.2: Water sorption isotherms for sisal nanocrystal (●) and sisal MFC (■) films at 25°C (Belbekhouche et al., 2011).

Native cellulose model films were prepared by spin-coating aqueous MFC dispersions onto silica substrates (Ahola et al., 2008; Aulin et al., 2009). The swelling of the films was studied using QCM-D measurements.

The interaction with water of cellulose model films with different degrees of crystalline ordering was evaluated using QCM-D (Aulin et al., 2009). Indeed, cellulose does not naturally dissolve in water, but water can penetrate inside the amorphous part causing swelling of the entire material. Conversely, crystalline cellulose does not swell because water molecules cannot penetrate into the crystalline structure (Müller et al., 2000). Low-charged MFC and cellulose nanocrystal films were prepared by spin-coating and compared to crystalline cellulose II and amorphous cellulose films. As expected, the amorphous cellulose film prepared from cellulose dissolved in a solution of lithium chloride (LiCl) in N,N-dimethylacetamide (DMAc) displayed the highest degree of swelling (48%). However, surprisingly the swelling of cellulose nanocrystal film was reported to be higher (26%) than that of MFC film (7%). Water penetration between the nanocrystals that caused their separation and allowed for a further imbibition into the film was suspected. The sulfuric acid-prepared nanocrystals contain sulfate groups aiding in creating an osmotic pressure separating the crystals in the film and exposure to liquid water. It was concluded that the difference in both crystalline ordering and the mesostructure of the films affected the swelling of the cellulose films. Consequently, both properties affect the adsorption and interaction behavior of the substrates.

10.3.1 Influence of pretreatment

The moisture sorption behavior of films prepared from two types of MFC obtained from the same source (bleached sulfite softwood pulp) but differing by the nature of the pretreatment was reported (Minelli et al., 2010). The material labeled MFC G1 was processed using a combined refining and enzymatic pretreatment, whereas the material labeled MFC G2 was carboxymethylated. Both pretreated materials were then subjected to high-pressure homogenization. As a result of the different processing methods, the two MFC types had different fibril diameters (17–30 nm for MFC G1 and 5–15 nm for MFC G2) and charge densities (40 $\mu equiv \cdot g^{-1}$ for MFC G1 and 586 $\mu equiv \cdot g^{-1}$ for MFC G2). Films prepared from MFC G2 exhibited higher water sorption and lower diffusivity than those prepared from MFC G1. A good fit of experimental data was provided by the dual-mode (or Park) model except for MFC G2 at higher water activities. It was suggested that the higher charge density of carboxymethylated MFC enhanced the tendency to absorb water between the fibril microstructure whereas the higher compaction during film forming resulted in lowest diffusivity. The addition of glycerol increased the diffusion coefficient and decreased/increased the water uptake at low/high water activity.

The effect of fibril charge density, electrolyte concentration, and pH on swelling was investigated using QCM-D (Ahola et al., 2008). MFC films with low and high charge density were prepared by using enzymatic and carboxymethylation pretreatment, respectively, before homogenization. A denser film with lower roughness was obtained for high charge density films because of smaller and more uniform fibril dimensions induced by the carboxymethylation pretreatment. Both films were stable in aqueous solutions but desorption of some loosely bound fibrils from the low charge density film was observed that did not affect the coverage. A small stiffening of the high charge density film was reported after the addition of water. Electrolyte addition caused water uptake in both films but it was found to be much higher for the high charge density film. The increase in electrolyte concentration caused an increase in pH within the film (Donnan effect) and therefore more carboxyl groups can be dissociated increasing the swelling forces. The swelling observed was found to be reversible and correlated with AFM force measurements.

10.3.2 Influence of post-treatment

Post-treatments applied to MFC can also influence the moisture sorption of MFC films. For instance, impregnation of an MFC film with melamine formaldehyde (MF) followed by hot-pressing allows reducing the porosity of the film and hence moisture uptake (Henriksson and Berglund, 2007). This reduction was higher than the one predicted from a rule of mixtures. Interaction between the hydroxyl groups at the cellulose surface and MF, leaving fewer hydroxyl groups accessible for the water molecules, was also suggested to explain the low moisture adsorption in the nanocomposite.

Adsorption of cationic surfactants was proposed to control the degree of water wettability of MFC (Xhanari et al., 2011). Wetting and adhesion of water onto MFC films were determined by contact angle measurements. The results showed that the adhesion of water can be lowered up to 25% when surfactants were adsorbed. However, after reaching a certain level of surfactant adsorption (approximately 0.57 $mmol \cdot g^{-1}$ in this case), independently of the MFC charge, additional surfactant adsorption provided a more hydrophilic surface presumably due to the formation of a double layer structure in which hydrophilic head groups of the surfactant were oriented towards the aqueous phase.

10.4 Water vapor transfer rate and water vapor permeability of microfibrillated cellulose films

MFC films have been prepared from kraft wood pulps from different sources (softwood and hardwood) and different lignin contents (Spence et al., 2010). The WVTR values of the MFC films were compared to those obtained from the initial pulps (wet cup method). A clear decrease (–20 to –30%) of WVTR was reported upon homogenization. Among the different wood sources, MFC from bleached hardwood presented the highest water vapor barrier properties. It was also shown that samples with increased lignin content had higher WVTR, an unexpected result, but hypothesized to be due to large hydrophobic pores in the film.

10.4.1 Influence of pretreatment

The pretreatment applied to the fibers before high-pressure homogenization can impact the permeability of MFC films to water vapor. WVP was measured for films prepared from MFC either enzymatically pretreated or carboxymethylated before homogenization (Minelli et al., 2010). WVP was found to increase as the water activity increased for both samples. For a given water activity, a lower WVP was observed for the carboxymethylated sample due to a more homogeneous and less porous structure. Refining pretreatment was also found to decrease the WVTR of MFC films by blocking pores with smaller diameter fibrils and fragments which otherwise promote mass transport (Spence et al., 2011a).

Incorporation of low amounts of water soluble polymers, viz. carboxymethylcellulose (CMC), methylcellulose (MC), and polyvinyl alcohol (PVA), in the culture medium of bacterial cellulose (BC) was found to allow controlling the water content of the material (Seifert et al., 2003). The described control of the water content offered the possibility of preparing materials with increased (0.5% MC and 2.0% CMC) or decreased (0.5% PVA) water absorption capacities. Moreover, MC and CMC biomaterial composites could be stored in the dried state and re-swollen before use, reaching higher water absorption than pure, never-dried BC.

10.4.2 Influence of post-treatment

A small amount of paraffin wax was found to reduce the WVTR of MFC films to a value similar to low density polyethylene (LDPE) (Spence et al., 2010). It was also found that after coating the MFC films with cooked starch, beeswax and paraffin using a dipping method, the obtained WVTRs were approximately half that of low density polyethylene (Spence et al., 2011b). It was ascribed to surface pore closure and filling of the pore network. A multilayer structure was used to model the structure. It was shown that the coating was the most resistant to water vapor transport of the composite.

Acetylation was expected to reduce the WVTR of MFC films (Rodionova et al., 2011). When increasing the acetylation reaction time and then the degree of substitution (DS), an initial reduction of WVTR was observed. Compared to unmodified MFC films (234 $g \cdot m^{-2} \cdot day^{-1}$), lower values of WVTR were obtained after 0.5–1 hour

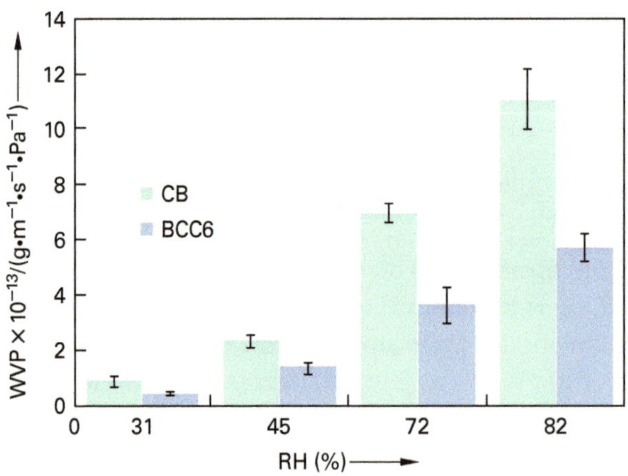

Fig. 10.3: Permeability towards water vapor (WVP) in bacterial cellulose membranes before (CB) and after esterification with hexanoyl chloride (BCC6) at different relative humidities (RH) (Tomé et al., 2010).

reaction time (167 $g \cdot m^{-2} \cdot day^{-1}$). This decrease was explained by the increasing fraction of acetylated hydroxyl groups which gradually decreased the solubility of water in the amorphous zones of the MFC structure. For higher reaction times, the WVTR increased again. It was suggested that as the DS continued to increase and less hydrogen bonding between the fibrils occurred, an increased pore volume fraction and more open network structure resulted. In parallel to these results, another method decreasing the WVTR was underlined (Rodionova et al., 2011). After successive solvent exchange using water, acetone and toluene, unmodified MFC films showed a drastically lower WVTR value of 91 $g \cdot m^{-2} \cdot day^{-1}$. WVP of BC membranes was decreased by conducting controlled heterogeneous chemical modification with hexanoyl chloride as shown in Figure 10.3 (Tomé et al., 2010). Measurements were performed at 26°C over a broad RH range from 31% to 82%.

10.5 Gas permeability of microfibrillated cellulose films

As with water vapor permeability, gas permeability and in particular the oxygen and air barrier properties play a key role in some applications such as food packaging.

The barrier properties of MFC films 20–33 μm thick (corresponding to basis weights between 15 and 30 $g \cdot m^{-2}$) obtained from bleached spruce sulfite pulp have been investigated (Syverud and Stenius, 2009). The air permeability and oxygen transmission rate (OTR) measured at 23°C and 0% RH were around 10 $nm \cdot Pa^{-1} \cdot s^{-1}$ and 17–18 $mL \cdot m^{-2} \cdot day^{-1}$, respectively. MFC films fulfill the requirements for oxygen transmission rates in modified atmosphere packaging and display similar performances to conventional synthetic polymer films of similar thickness. These good gas barrier properties were ascribed to the crystalline and dense structure of fibrils.

The carbon dioxide, nitrogen and oxygen permeability of films prepared from MFC and cellulose nanocrystals extracted from sisal was reported (Belbekhouche et al., 2011). Nanocrystal film showed significantly higher permeability to gases and diffusion coefficients than MFC film. It was suspected that gas molecules penetrated much more slowly in the MFC film because of a longer diffusion path. In addition to a different surface chemistry, it was supposed that entanglements of the fibrils and lower porosity of the MFC film represented barrier domains increasing the tortuosity of the diffusion pathway.

10.5.1 Effect of relative humidity

The oxygen permeability of carboxymethylated MFC films were studied at different RH level (Aulin et al., 2010; Minelli et al., 2010). The cast films (8 $g \cdot m^{-2}$, thickness 5.1 μm) displayed very low oxygen permeability value of 6 · 10^{-4} $cm^3 \cdot \mu m \cdot m^{-2} \cdot day^{-1} \cdot kPa^{-1}$ at 23°C/0% RH and 0.85 $cm^3 \cdot \mu m \cdot m^{-2} \cdot day^{-1} \cdot kPa^{-1}$ at 23°C/50% RH (Aulin et al.,

2010). Compared to non-pretreated MFC, the carboxylation pretreatment makes the fibrils highly charged and easier to liberate giving smaller and more uniform fibril dimensions. Additionally, the impact of different parameters was highlighted. The OTR decreased with increasing the film grammage (or thickness) as shown in Figure 10.4(a). It was suggested that most pores were located at the surface of the MFC films and that these pores were not connected as reported elsewhere (Syverud and Stenius, 2009; Fukuzumi et al., 2009; Minelli et al., 2010), thus contributing to the impermeable nature of the films. A slight increase in permeability and decrease in diffusivity were observed when MFC was combined with glycerol (Minelli et al., 2010). Contrarily to water vapor, the MFC film structure was not affected by the penetrant oxygen molecules and did not swell significantly. The only contribution of the glycerol was therefore the reduction in void volume inducing a decrease of the diffusivity.

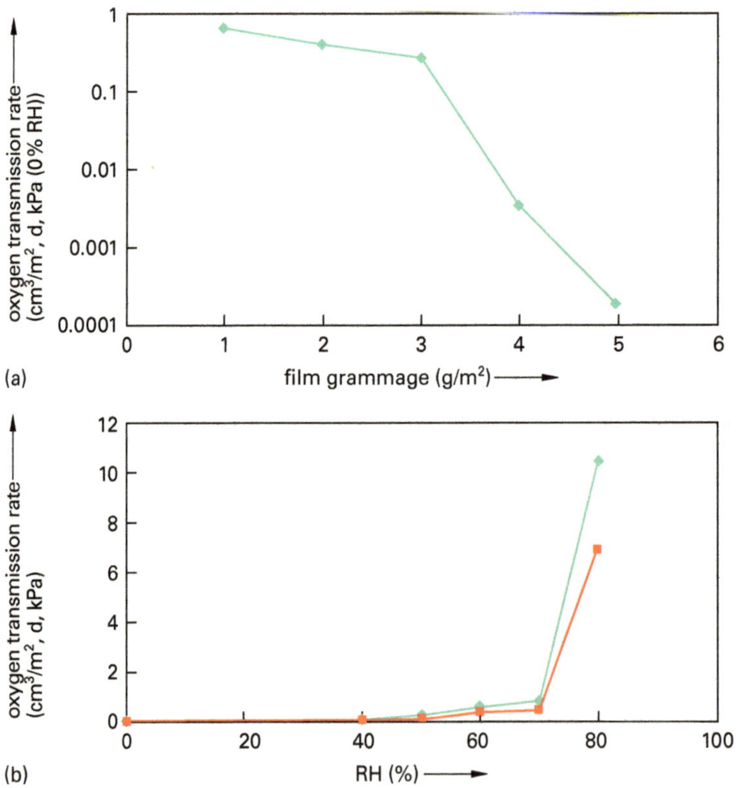

Fig. 10.4: (a) Effect of MFC film grammage on oxygen transmission rate (OTR) measured at 23°C and 0% RH, and (b) relative humidity (RH) on oxygen transmission rate (OTR) for carboxymethylated MFC films with grammage of 5 (♦) and 8 $g \cdot m^{-2}$ (■) (partial oxygen pressure 1 atm) (Aulin et al., 2010).

When the RH was increased from 0 to 80%, the OTR increased significantly with a sharp increase above 70% RH, which was related to the plasticization effect of the amorphous carboxymethylated MFC domains by sorbed water molecules (Figure 10.4(b)) (Aulin et al., 2010). For enzymatically pretreated and carboxymethylated MFC, a sudden change in oxygen permeability about two order of magnitude was also observed when changing the moisture conditions from dry conditions to moderate water activities (Minelli et al., 2010). When the structure was sufficiently swollen a plateau region was reached and a further increase in permeability was observed when the membrane was almost saturated. Very similar behavior was reported over the whole range of water activities when adding glycerol. A strong degradation of the oxygen barrier properties of TEMPO-oxidized MFC films was also reported when increasing the RH level from 0% to 50% (Fukuzumi et al., 2011). Finally, it was observed that an increase in the number of homogenization passes during MFC production did not induce any significant evolution of the OTR as confirmed in another investigation (Siró et al., 2011).

10.5.2 Improvement of gas barrier properties

Hydrophilic membranes are poor barrier under wet conditions. Indeed, when a hydrophilic membrane is exposed to water vapor, the water molecules adsorb in the material and increase the fractional free volume of the membrane. Water is a good swelling agent and a moving carrier for gases.

The oxygen barrier properties at high RH level were improved by preparing hybrid clay-MFC films (Liu et al., 2011; Liu and Berglund, 2012). A water-based paper-making procedure was used to prepare the film consisting of a cellulose fibril network with random-in-the-plane orientation distribution as the continuous matrix phase and montmorillonite platelets as the inorganic dispersed phase. An inorganic content as high as 89 wt% was reached. Although the oxygen barrier properties of pure MFC film were very good in the dry state (0–50% RH), the addition of montmorillonite significantly improved the barrier properties at higher relative humidity. Improvement of oxygen barrier properties of MFC film was also observed at 0 and 50% RH by adding montmorillonite (Wu et al., 2012). Moreover, the composite film showed self-extinguishing characteristics when subjected to open flames and much delayed thermal degradation of cellulose. The process has been patented (Berglund and Liu, 2011). A further decrease of OTR was observed by adding 10 wt% chitosan (Liu and Berglund, 2012).

OTR measurements were also reported for films around 42–47 μm thick made of MFC modified through surface acetylation with acetic anhydride (Rodionova et al., 2011). Microscopic observation of the films revealed that the pore dimensions were unchanged regardless of the duration of the acetylation reaction whereas the pore area fraction increased as the reaction time increased from 0.5 to 3 hours due to a lower amount of hydrogen bonds. Unmodified and acetylated (3 hours reaction time) MFC films had an OTR value of 4.2 and 11.1 $mL \cdot m^{-2} \cdot day^{-1}$, respectively.

Gas permeation of carbon dioxide, oxygen and nitrogen in unmodified and esterified BC membranes were also evaluated at 25°C and 100% RH (Tomé et al., 2010). The esterification of BC membranes with hexanoyl chloride caused a significant decrease in the permeability towards CO_2, O_2 and N_2 because of the insertion of fatty aliphatic chains on the membrane surface and ensuing increase of its hydrophobic character.

10.5.3 Polymer coating

Coating of polymer films with MFC layers is a new way to produce good barrier materials and a possible solution to only keep the advantages of both MFC and polymers. Films were prepared by casting TEMPO-oxidized MFC dispersions obtained from softwood and hardwood pulps on a plasma-treated polylactic acid (PLA) film (Fukuzumi et al., 2009). The plasma treatment was carried out to improve the adhesion of the MFC layer. With this method, an MFC coated layer 0.4 μm thick was formed on the PLA film with 25 μm thickness. Despite this ultrathin MFC layer, a significant decrease in the oxygen permeability determined under dry conditions was observed from 746 $mL \cdot m^{-2} \cdot day^{-1} \cdot Pa^{-1}$ for the neat PLA film to 1 $mL \cdot m^{-2} \cdot day^{-1} \cdot Pa^{-1}$ for the coated film. This value was close to that of typical polymer films such as poly(vinylidene chloride) and polyethylene-poly(vinyl alcohol) copolymers that have high oxygen barrier functionality.

A similar coating process was conducted with poly(ethylene terephthalate) (PET) films and two kinds of TEMPO-oxidized MFC (Fujisawa et al., 2011). TEMPO-mediated oxidation was conducted using the TEMPO/NaBr/NaClO system at pH 10 to prepare MFC with a sodium carboxylate (COONa) content of 1.74 $mmol \cdot g^{-1}$. Additionally, TEMPO-oxidized MFC with free carboxylic groups (COOH) was obtained by adding HCl to the suspension. PET films 50 μm thick were coated with 1 μm thick MFC layer of similar density and oxygen permeability measurements were performed at 23°C and 0% RH. The oxygen permeability of PET film decreased from 0.31 $mL \cdot m^{-2} \cdot day^{-1} \cdot kPa^{-1}$ for the neat film to 0.049 and 0.0017 $mL \cdot m^{-2} \cdot day^{-1} \cdot kPa^{-1}$ for the films coated with COOH and COONa-functionalized MFC, respectively. The sodium carboxylate groups present on the surface of the fibrils thus effectively improved the oxygen barrier properties of the films but the detailed mechanisms remained unknown. Coating of 50 μm thick PET films with MFC layers 1–6 μm thick was also conducted with TEMPO-oxidized MFC prepared from Norway spruce and mixed eucalyptus pulps (Rodionova et al., 2012). Different reaction times were used to vary the carboxylate content of the MFC layers. The OTR of the coated PET films was found to decrease by increasing the carboxylate content of cellulose.

10.5.4 Paper coating

Combination of MFC with paper has been recently investigated, even if it was already suggested in 1983 (Turbak et al., 1983). MFC could improve the mechanical and barrier properties of papers and the first studies showed the interest of this combination in applications such as food packaging or printing.

MFC was deposited on the top of a wet base paper with a dynamic sheet former (Syverud and Stenius, 2009). The top layer and the base paper were thus combined wet in wet. The basis weights of the top layers ranged from 2 to 8 $g \cdot m^{-2}$ for a total basis weight of the sheet former of 90 $g \cdot m^{-2}$. The air permeability decreased significantly with the increase of the thickness of the MFC layer as shown in Figure 10.5. For 2 $g \cdot m^{-2}$ MFC, the air permeability was about 3.10^4 $nm \cdot Pa^{-1} \cdot s^{-1}$ whereas for an 8 $g \cdot m^{-2}$ MFC layer, the permeability dropped to about 360 $nm \cdot Pa^{-1} \cdot s^{-1}$, i.e. was 100 times lower. The improved barrier properties were correlated with the reduction of the surface porosity which was confirmed elsewhere (Aulin et al., 2010). Using another coating process (Print Coat instruments with a rod coater for sheet), two different papers, kraft and greaseproof papers, were coated with a 0.85 wt% MFC dispersion (Aulin et al., 2010). The air permeability decreased considerably for both papers, from 69,000 $nm \cdot Pa^{-1} \cdot s^{-1}$ to less than 1 $nm \cdot Pa^{-1} \cdot s^{-1}$ for kraft paper, and from 660 $nm \cdot Pa^{-1} \cdot s^{-1}$ to 1 $nm \cdot Pa^{-1} \cdot s^{-1}$ for the greaseproof paper (coat weight between 1 and 2 $\cdot m^{-2}$). The dense structure formed by the MFC layer and their ability to form hydrogen bonds were believed to contribute to the enhancement of the barrier properties of the films (Figure 10.6). The reduction of the surface porosity decreased the air permeability but also improved the oil resistance.

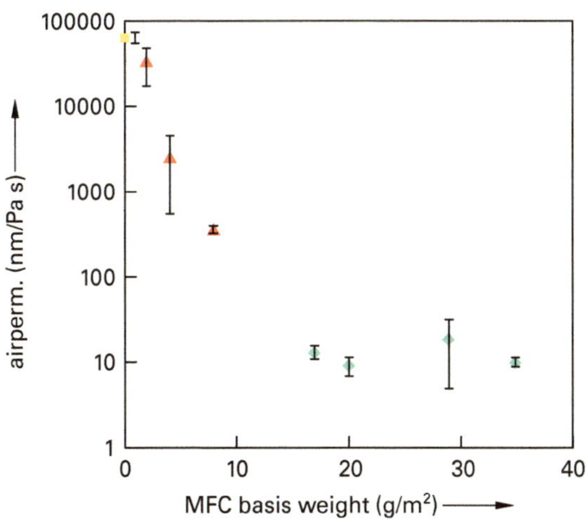

Fig. 10.5: Air permeability of the base paper (■), MFC-coated base paper (▲, total grammage 90 ± 1 $g \cdot m^{-2}$), and MFC films (◆) (Syverud and Stenius, 2009).

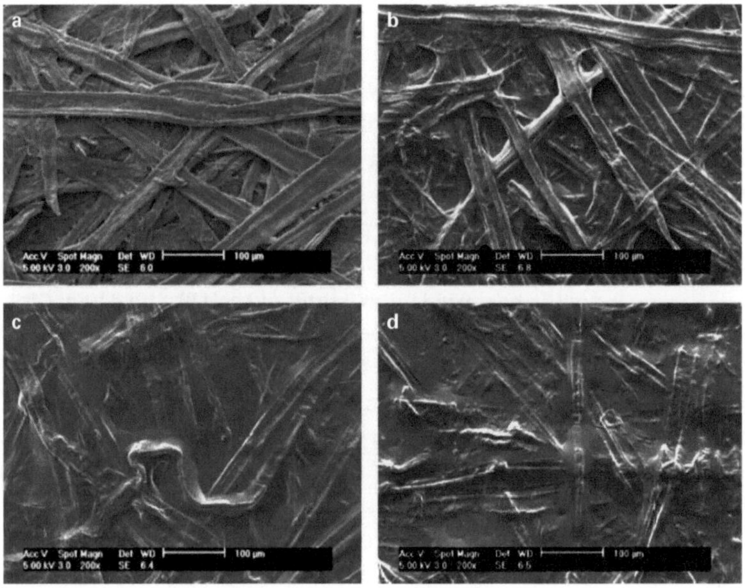

Fig. 10.6: Environmental scanning electron micrographs of (a) uncoated, and MFC-coated unbleached papers with coat weights of ca. (b) 0.9, (c) 1.3., and (d) 1.8 $g \cdot m^{-2}$, respectively. The scale bar is 100 μm (Aulin et al., 2010).

The oil barrier property has been rarely measured, even if it is often essential for food and packaging industries. Various standards are currently known to measure the oil resistance. Tappi T-454 test method was used to characterize the oil resistance of unbleached and greaseproof paper coated with MFC (Aulin et al., 2010). The penetration time of turpentine oil and castor oil was determined. When the air permeability decreased, i.e. increased coat weight, the oil resistance increased. It was ascribed to the reduced surface porosity induced by the MFC coating (Figure 10.6). The MFC-coated greaseproof paper showed better oil resistance.

MFC was combined with shellac and deposited on fiber-based paper or paperboard substrates and the air, oxygen and water vapor permeability properties were characterized to quantify the barrier effect of the applied coatings (Hult et al., 2010). Shellac is a natural resin which presents properties such as oil resistance, good moisture barrier, hydrophobic character and biodegradability. Two different MFC/shellac combinations were tested, i.e. as a one-layer coating using a MFC/shellac blend, and as a multi-layer system with MFC as a first layer and shellac as the top layer. Two coating methods were also compared, a bar coater and a dynamic sheet former (spray coating technique).

The air permeability of the paperboard and paper was significantly decreased by the MFC coating. However, it was found that the MFC coating did not completely cover the surface, whereas an additional shellac layer covers the holes leading to a considerably better barrier. The MFC/shellac combination with the bar coated process as either blend or multilayer system gave lower air permeability than paperboard and paper coated with shellac alone. MFC coating with dynamic sheet former obviously did not cover the entire surface of the substrate. The additional shellac layer coated secondly thus brought a supplementary reduction of the air permeability, already low with the first MFC coating (additional decrease from 80 to 98%). Regarding the oxygen permeability, the MFC layer decreased first the OTR values and the second shellac layer closed the surface nanopores on MFC layer and reduced these values further. However, the obtained OTR values remained too high (around 5,000 $cm^3 \cdot m^{-2} \cdot day^{-1}$) to consider these materials as high oxygen barriers (characterized by OTR values lower than 3 $cm^3 \cdot m^{-2} \cdot day^{-1}$ at 25°C and 50% RH). These results were attributed to the non-homogeneous coating of MFC as confirmed by SEM observations. The ability of the coated papers to act as a moisture barrier was also investigated by measuring the WVTR values. The MFC/shellac-coated substrates showed very low WVTR (7–8 $g \cdot m^{-2} \cdot day^{-1}$), close to the value required for a high moisture barrier material, i.e. 5 $g \cdot m^{-2} \cdot day^{-1}$ (for a 25 μm thick film). The shellac coating played a major role in the decrease of WVTR. Indeed, MFC displays a high WVTR, due to its hydrophilic nature but the MFC layer created a dense substrate that permitted a more homogeneous shellac coating.

10.6 Cellulose nanocrystal films

The swelling and barrier properties of neat cellulose nanocrystal films have been much less investigated than for MFC. It shows the greater potential envisioned in the packaging sector for MFC, whereas cellulose nanocrystals are probably more nanocomposite oriented.

The effect of the crystallinity of cellulose model films on the interaction with water was evaluated as reported in Section 10.3 of the present chapter (Aulin et al., 2009). Low-charged MFC and cellulose nanocrystal films were prepared by spin-coating and compared to crystalline cellulose II and amorphous cellulose films. The swelling of cellulose nanocrystal films was reported to be higher (26%) than the one of MFC films (7%). It was hypothesized that water penetration between the nanocrystals caused their separation and allowed for a further imbibition into the film. Moreover, the sulfuric acid-prepared nanocrystals contain sulfate groups aiding in creating an osmotic pressure separating the crystals in the film and exposure to liquid water.

The water sorption and gas barrier properties of films prepared from cellulose nanocrystals extracted from sisal have been reported (Belbekhouche et al., 2011). The results have been already discussed in Sections 10.3 and 10.5 since the properties of these films were compared to those obtained for films from sisal MFC. The water diffusion coefficient was higher for nanocrystal films compared to MFC films. It was explained by the presence of residual lignin, extractive substances and fatty acids at the surface of MFC. Nanocrystal films also showed significantly higher permeability to gases and diffusion coefficients than MFC films. It was suspected that gas molecules penetrated much more slowly in the MFC films because of a longer diffusion path. In addition to a different surface chemistry, it was supposed that entanglements of the fibrils and lower porosity of the MFC films represented barrier domains increasing the tortuosity of the diffusion pathway.

The permselective properties of cellulose nanocrystal membranes have been investigated using a cotton nanocrystal film (Thielemans et al., 2009). Indeed, the conventional method for preparing cellulose nanocrystals involves a sulfuric acid hydrolysis step that induces the formation of sulfate groups on the surface of the nanoparticles imparting thereby a negative surface charge. These negatively-charged surface groups should inhibit the transfer of negatively-charged species through the nanocrystal membrane, while the diffusion of neutral species should be only slightly hindered. Using rotating-disk electrode measurements, the diffusion of various species within the film was studied. The positively-charged species were adsorbed by the film, whereas the negatively-charges species were excluded from the film. Opportunities for the development of new selective membranes for separation technologies can be opened. Moreover, modification of the surface chemistry of cellulose nanocrystals to specific applications can be envisaged.

Cellulose nanocrystals were also used to prepare ultrafine membranes for water purification (Ma et al., 2011). Other polysaccharide nanofibers were also used in this study. A barrier layer of cellulose nanocrystals was obtained by casting on an electrospun polyacrylonitrile (PAN) nanofibrous scaffold. Very high permeation flux and high rejection rate were obtained compared to commercial ultrafine membranes for separation of oil/water emulsions. It was ascribed to the small pore size of the cellulose nanocrystal layer. Moreover, the negatively charged surface of cellulose nanocrystals was found to serve as a filter for virus adsorption.

10.7 Microfibrillated cellulose-based films

10.7.1 Swelling and sorption properties

The kinetics of water absorption of unplasticized and glycerol plasticized starch reinforced with MFC prepared from potato pulp was reported (Dufresne and Vignon, 1998; Dufresne et al., 2000). Samples were conditioned at room temperature and 95% RH instead of being immersed in water because starch is very sensitive to liquid water and can partially dissolve after longtime exposure to liquid water. The water uptake was found to be much higher for plasticized samples than for unplasticized. On the contrary, the water diffusion coefficient was systematically lower for plasticized specimens (Dufresne et al., 2000). Moreover, it was observed that both the water uptake at equilibrium and diffusion coefficient decreased with the MFC content.

Therefore, the presence of MFC conferred a water resistance to the starch-based films. It was ascribed to the presence of a three-dimensional intertwined cellulose microfibril network within the matrix preventing the swelling of the starch material when exposed to water or moist atmosphere.

DVS measurements were performed at 30% RH for glycerol plasticized and pure high-amylopectin starch films reinforced with MFC from wood (Svagan et al., 2009). The moisture uptake decreased with increasing cellulose content. The experimental data were compared to the theoretical moisture uptakes predicted from a rule of mixtures. Both values were similar at low MFC content but the predicted data were significantly higher than experimental at 70% MFC content. This discrepancy was ascribed to the cellulose nanofiber network reducing the swelling and thereby the moisture uptake. However, for the 40 wt% MFC reinforced glycerol-free composite the opposite was observed, i.e. the theoretical value was lower than the experimental value. The swelling yielded a moisture concentration-dependent diffusivity. The average and initial diffusion coefficients decreased with increasing MFC content and increased with increasing glycerol content. The observed reduction in moisture diffusivity was attributed to cellulose characteristics and geometrical impedance, swelling constraints due to a high-modulus/hydrogen bonded fiber network, and strong molecular interactions between cellulose nanofibers and with the amylopectin matrix. Marginal decrease of water uptake and diffusion coefficient was reported for glycerol plasticized maize starch when adding up to 10 wt% MFC prepared from wheat straw (Kaushik et al., 2010). Favorable filler-matrix interactions and reduced segmental mobility of the matrix were suggested to explain this behavior. At higher filler loading an increase of these parameters was observed and ascribed to the creation of diffusion pathways for water.

A similar decrease of the water uptake was observed for spruce galactoglucomannan films reinforced with MFC and attributed to the formation of a continuous hydrogen bonded network (Mikkonen et al., 2011). For all RH values ranging from 0 to 80–90% the experimental moisture uptake was systematically lower than theoretical values determined from a rule of mixtures. It was suspected that some of the water sorption sites available on galactoglucomannan were unavailable when MFC was added, indicating interaction establishment between the components of such ternary mixtures. Reduced moisture sorption was also reported for glucuronoxylan films reinforced with BC over a broad RH range (0–90% RH) (Dammström et al., 2005).

Diffusion coefficients of water were determined for ethylene vinyl alcohol copolymers and related composites reinforced with 2 wt% MFC (Fernândez et al., 2008). They were estimated from the desorption curves for samples conditioned at 100% RH (water sorption) and subsequently desorbed by conditioning at 0% RH. The permeability was also determined from the solubility estimated from the water uptake at equilibrium, density of the material and water vapor partial pressure. Higher diffusion rates and permeabilities were observed for the composites compared to the neat matrices. It was ascribed to changes in crystallinity and phase morphology promoted by MFC rather than the hydrophilic character of cellulose since no clear change in the solubility was detected. Diffusion and permeability coefficients were found to be positively decreased by ionizing gamma irradiation at low dose (30 kGy). It was associated with irreversible morphology change of the copolymer matrices such as cross-linking.

Moisture sorption of hydroxyethylcellulose (HEC) films reinforced with MFC was measured at 23°C and 50% RH (Sehaqui et al., 2011). Neat HEC was found to be only slightly more hygroscopic than the MFC film (7.6% vs. 6.1% moisture content). Moisture adsorption values for the MFC/HEC biocomposites were between those of HEC and MFC. The water adsorption/desorption isotherms were determined for hydroxypropyl methylcellulose (HPMC) films reinforced with three different types of cellulose nanoparticles, viz. MFC, TEMPO-oxidized MFC and cellulose nanocrystals (Bilbao-Sainz et al., 2011). The addition of the filler decreased the moisture uptake in the high water activity range ($a_w > 0.5$). Similar equilibrium moisture contents were obtained for the different cellulose nanoparticles at low loading level. For higher filler concentration, lower water affinity was reported for cellulose nanocrystal-based films which was ascribed to their higher degree of crystallinity compared to MFC. The experimental data were successfully fitted with the GAB model and a slight decrease of the monolayer water content from $0.054\ g \cdot g^{-1}$ to $0.050\ g \cdot g^{-1}$ was observed. This lower water-binding capacity was attributed to interactions between nanocrystals and the hydrophilic sites of HPMC, which substitute the HPMC-water interactions. A decrease of the water vapor diffusivity was reported and explained by a clustering phenomenon, i.e. once a monolayer of water molecules moistened the film, a further increase in moisture resulted in “free water” that did not interact with the polymer. The diffusivity was not altered by adding cellulose nanocrystals, whereas MFC increased the water diffusion coefficient because of amorphous zones in the nanoparticles that create a preferential pathway for the water vapor to diffuse.

Natural rubber (NR) films reinforced with MFC prepared from the rachis of the palm of date palm tree were characterized regarding their water uptake when immersed in distilled water (Bendahou et al., 2010). A continuous increase of the amount of water absorbed was observed when increasing the comparatively higher hydrophilic filler content within the NR matrix. Addition of MFC to PLA was also found to increase the moisture uptake of the material when conditioned at 25°C and 97% RH (Tingaut et al., 2010). However, for given filler content, the moisture sorption decreased as the acetyl content of MFC increased because of the reduction in the hydrophilic character of MFC upon acetylation. This means that the water sorption of PLA/MFC composites can be adjusted with the acetyl content and should allow the control of the biodegradation or dimensional stability of the composites.

The swelling behavior in toluene of poly(S-co-BuA) reinforced with MFC from *Opuntia ficus-indica* cladodes was also investigated (Malainine et al., 2005). A strong toluene resistance even at very low filler loading was observed. While the unfilled matrix completely dissolved in toluene and no residual gel was recovered at the end of the swelling experiment, only 27 wt% of the polymer was able to dissolve when filled with only 1 wt% MFC. This phenomenon was ascribed to the presence of a three-dimensional entangled cellulosic network which strongly restricted the swelling capability and dissolution of the matrix. For higher MFC content, no significant evolution was observed because of both the overlapping of the fibrils restricting the filler-matrix interfacial area and the decrease of the entrapping matrix fraction due to the densification of the MFC network. Similar experiments were conducted with MFC obtained from sugar beet pulp reinforced poly(S-co-BuA) (Dalmas et al., 2006). Toluene resistance of nanocomposites was found to be less significant than for MFC from *Opuntia ficus-indica* cladodes and to evolve with filler content. It was observed that the cohesion of composites prepared by evaporation was higher than that of freeze-dried/hot-pressed materials. This difference was ascribed to the presence of a hydrogen-bonded network in the former samples. It was concluded that the solvent did not have any effect on the hydrogen bonds of the cellulose network present in evaporated composites. On the contrary, for freeze-dried/hot-pressed materials, weaker interactions were supposed to form between fibrils and polymer chains were able to be more easily disentangled and dissolved by the solvent. Toluene uptake was also investigated for NR films reinforced with MFC (Bendahou et al., 2010). The relative toluene uptake at equilibrium, defined as the ratio of the weight gain to the initial weight, was measured after 4 h of immersion in toluene. It was shown that only 1 wt% of the cellulosic nanoparticles allowed preventing the disruption of the NR matrix and strongly restricted the swelling of the material. This behavior could not only be associated with the decrease of the fraction of material able to swell in toluene, i.e. NR, in the nanocomposites and it was suggested that adsorption of macromolecular chains at the filler/matrix interface through interactions between MFC and NR, and filler/filler interactions could also reduce swelling.

10.7.2 Water vapor transfer rate and water vapor permeability

Decreased WVTR was reported by adding low content (up to 1 wt%) MFC to glycerol plasticized starch (Savadekar and Mhaske, 2012). It was ascribed to nanometer size effect and high crystallinity of MFC, homogeneous dispersion within the polymeric matrix and strong filler/matrix interaction.

10.7.3 Oxygen permeability

A decrease in the oxygen permeability of amylopectin films was reported when adding MFC (Plackett et al., 2010). Both enzymatically pretreated and carboxymethylated MFC and glycerol-free and plasticized samples were used. Depending on the type of MFC and composition of the specimen, the average oxygen permeability (at 23°C, 50% RH) varied between 0.013 and 1.400 $mL \cdot mm \cdot m^{-2} \cdot day^{-1} \cdot atm^{-1}$. No significant difference was observed for glycerol-free samples composed of 50 or 100 wt% MFC. A similar decrease in oxygen permeability was reported for MFC reinforced glycerol plasticized starch (Savadekar and Mhaske, 2012).

10.8 Cellulose nanocrystal-based films

10.8.1 Swelling and sorption properties

Higher resistance to water of glycerol plasticized thermoplastic starch conditioned at 98% RH with increasing the cellulose nanocrystal content was reported (Anglès and Dufresne, 2000; Lu et al., 2005). These experiments were conducted using cellulose nanocrystals extracted from tunicate and cottonseed linter, respectively. Both the water uptake and the diffusion coefficient of water were found to decrease upon nanoparticle addition. The water uptake at equilibrium decreased non-linearly from about 60% for the unfilled material down to 40% and 37% for composites reinforced with 25 wt% tunicin (Figure 10.7) and 30 wt% cottonseed linter nanocrystals, respectively. These phenomena were ascribed to the presence of strong hydrogen bonding interactions between nanoparticles and between the starch matrix and cellulose nanoparticles. The hydrogen bonding interactions in the composites tend to stabilize the starch matrix when it is submitted to highly moist atmosphere. Moreover, the high crystallinity of cellulose might also be responsible for the decreased water uptake at equilibrium and diffusion coefficient of the materials. Lower water uptake and dependence on cellulose nanocrystal content were reported when using sorbitol rather than glycerol as plasticizer for the starch matrix (Mathew and Dufresne, 2002). An explanation was proposed based on the chemical structure of both plasticizers,

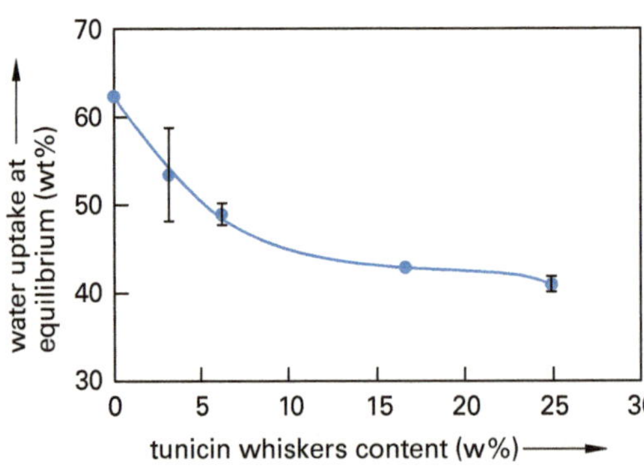

Fig. 10.7: Water uptake at equilibrium measured for glycerol plasticized waxy maize starch films conditioned at 25°C and 98% RH as a function of tunicin nanocrystal content (Anglès and Dufresne, 2000).

glycerol displaying about twice as many accessible end hydroxyl groups compared to sorbitol. A similar decrease of the water uptake of plasticized starch when conditioned at 98% RH (Lu et al., 2006; Cao et al., 2008a; Cao et al., 2008b; Chen et al., 2009), 75% RH (Liu et al., 2010) or 53% RH (Teixeira et al., 2009) upon adding cellulose nanocrystals prepared from different sources has been reported. The diffusion coefficient of water was also found to decrease when adding cellulose nanocrystals (Lu et al., 2006).

A higher water resistance of soy protein isolate (SPI) films conditioned at 98% RH when reinforced with cotton cellulose was also reported (Wang et al., 2006). Intermolecular hydrogen bonding between cellulose nanocrystals and the SPI matrix was proposed. Similar results were obtained for cotton cellulose nanocrystal reinforced glycerin plasticized carboxymethyl cellulose (CMC) (Choi and Simonsen, 2006). In addition, it was found that heat-treatment imparted water resistance to the glycerin-free composites. The degree of resistance to dissolution increased with increasing temperature of the treatment. It was ascribed to ester bond formation between carboxylic groups from CMC and hydroxyl groups from cellulose. Barrier membranes were prepared from poly(vinyl alcohol) (PVA) and cellulose nanocrystals (Paralikar et al., 2008). Poly(acrylic acid) (PAA) was used as a cross-linking agent to provide water resistance to PVA. Upon heat-treatment the carboxyl groups from PAA were able to form ester linkages with hydroxyl groups from PVA and cellulose. Optimal heat treatment at 170°C for 45 min was reported. The solubility of cross-linked PVA was significantly decreased because of the reduction of unreacted PVA molecules but no further decrease was found when adding cellulose nanocrystals. Carrageenan films reinforced with cellulose nanocrystals have been prepared and the water uptake was accessed by conditioning the samples at 11, 54 and 75% RH (Sanchez-Garcia et al., 2010). A strong reduction of the water uptake was observed when adding up to 3 wt% nanocrystals. However, for higher nanoparticle contents, agglomeration led to an increase of the water uptake.

For hydrophobic polymer matrices, the water uptake obviously increases when adding cellulose nanoparticles. For instance, the water sorption behavior of polyvinyl acetate (PVAc) films reinforced with cellulose nanocrystals was reported (Garcia de Rodriguez et al., 2006). A very small effect of adding nanocrystals was observed at low and intermediate water activities ($0 < a_w < 0.75$). For higher water activities between 0.75 and 0.98 a more pronounced water uptake was reported when increasing the nanocrystal content. However, the water content surprisingly stabilized around 12% (for samples conditioned at 98% RH) between 2.5 and 12 wt% nanocrystals as shown in Figure 10.8(a). Absorbed water molecules can accumulate in three regions within the composite, i.e. the PVAc matrix, the cellulose nanocrystals and the filler/matrix interface. The latter could be potentially predominant since neither crystalline cellulose nor PVAc are expected to significantly sorb water, but the hydrophilic nature of cellulose should induce water accumulation at the interface. The effect of interface can be quite substantial because of the nanoscale dimensions of the filler and since a water monolayer (3.1 Å) along the nanocrystal surface should increase the water uptake by 0.25% for 1 wt% nanocrystals in the composite. However, following this assumption, increasing nanocrystal content should cause a linear increase in equilibrium water uptake as the nanocrystal/polymer interfacial area increases. It was hypothesized that the formation of a three-dimensional network through hydrogen bonding above the percolation threshold of the filler (estimated around 1.40 wt%) could significantly inhibit swelling of the nanocomposites, similarly to the formation of chemical cross-links within a polymer phase.

For these cellulose nanocrystal reinforced PVAc systems, the kinetics of water diffusion was characterized by conditioning the samples at 98% RH (Garcia de Rodriguez et al., 2006). Figure 10.8(b) shows the evolution of the water diffusion coefficient determined from the initial slope of the water absorption data. Below the percolation threshold, the addition of cellulose nanocrystals had a very limited effect on water

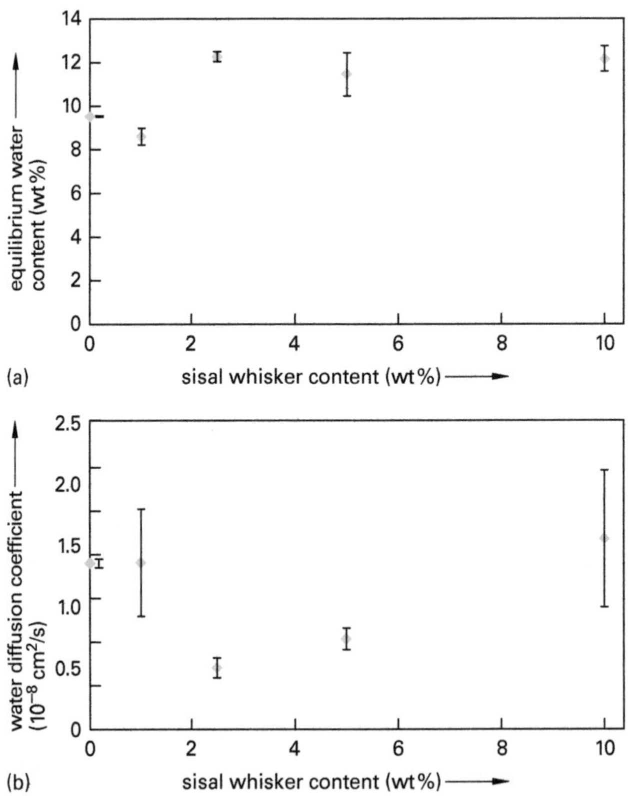

Fig. 10.8: (a) Water uptake at equilibrium and (b) water diffusion coefficient for polyvinyl acetate (PVAc) reinforced with sisal cellulose nanocrystals conditioned at 98% RH as a function of nanocrystal content (Garcia de Rodriguez et al., 2006).

diffusion which largely took place in the polymeric matrix. When the percolation threshold (around 1.40 wt%) was reached, the formation of a continuous pathway inhibited swelling of the polymeric matrix, thereby hindering and slowing down the water diffusion, and therefore reduced the water diffusion coefficient. With further increases in nanocrystal content, the diffusivity coefficient surprisingly increased. In fact, while the nanocrystal network effectively reduced the diffusivity through the polymeric matrix, the water molecules did not stay contained within this phase, and might migrate to and accumulate at the filler/surface interface. This water accumulation during transient water sorption depleted the polymer matrix phase, therefore speeding up diffusion, until equilibration saturation was reached at the interface and in the PVAc phase.

The swelling properties of cotton nanocrystal reinforced PVAc were characterized upon exposure to physiological conditions (immersion in artificial cerebrospinal fluid, ACSF, at 37°C) (Shanmuganathan et al., 2010a). Swelling increased with the nanocrystal content due to increased hydrophilicity of the nanocomposites (Figure 10.9). A similar observation was reported for the swelling behavior of tunicin nanocrystal reinforced PVAc and poly(butyl methacrylate) (PBMA) in the same liquid medium, but swelling of PVAc-based nanocomposites was lower for a given cellulose nanocrystal content (Shanmuganathan et al., 2010b). Moreover, it was also shown that the degree of swelling increased with temperature. Comparison of the swelling behavior obtained in identical conditions for similar PVAc nanocomposite films reinforced with tunicin nanocrystals was conducted (Shanmuganathan et al., 2010a). It was found that cotton nanocrystal containing materials swelled much less than the analogous PVAc/tunicin nanocrystal nanocomposites (Figure 10.9). Apart from being derived from a different source and having different dimensions, the major difference

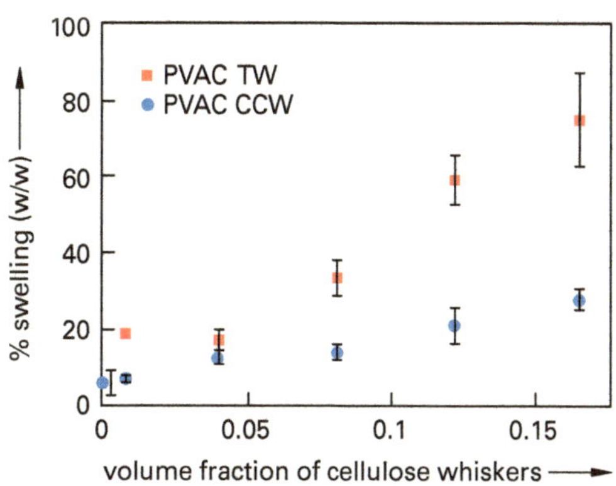

Fig. 10.9: Swelling of polyvinyl acetate (PVAc) reinforced with cotton (PVAC CCW) and tunicin nanocrystals (PVAC TW) conditioned for 1 week in physiological conditions (immersion in artificial cerebrospinal fluid, ACSF, at 37°C) as a function of nanocrystal content (Shanmuganathan et al., 2010a).

between tunicin and cotton cellulose nanocrystals used in this study was the density of charged sulfate groups. It was suggested that the low degree of swelling displayed by cotton nanocrystal nanocomposites was related to the lower sulfate charge density of the nanocrystals employed.

Cellulose nanocrystal reinforced NR nanocomposites were stored at 25°C and 75% RH and the moisture sorption was determined (Bras et al., 2010). A sharp increase was observed between 2.5 and 5 wt% nanocrystal whereas a slight decrease was noticed for higher filler contents (up to 12.5 wt%). The water uptake of natural rubber films filled with cellulose nanocrystal was also evaluated by immersing the samples in distilled water (Bendahou et al., 2010). It obviously increased when increasing the filler content because of its more hydrophilic nature compared to the NR matrix. Similar experiments were carried out on nanocomposite NR films reinforced with MFC prepared from the same source (rachis of the palm of date palm tree). For a given nanoparticle content the water uptake of nanocrystal reinforced NR films was found to be much higher than for MFC-based nanocomposites. It was ascribed to the structural and surface chemical differences between both cellulosic nanoparticles. Indeed, the presence of residual lignin and fatty acids at the surface of MFC could comparatively limit the hydrophilic character of the filler. Moreover, assuming that the filler/matrix compatibility was consequently lower for nanocrystal reinforced materials, water infiltration could be facilitated at the filler/matrix interface. A strong reduction of the water uptake of poly(D, L-lactide) (PDLLA) was observed when adding only 1 wt% cellulose nanocrystals, modifying the kinetics of the hydrolytic process in PDLLA (de Paula et al., 2011).

Polyvinylchloride (PVC) films reinforced with tunicin nanocrystals were submitted to swelling experiments in methyl ethyl ketone (MEK) (Chazeau et al., 1999). Among the good solvents for PVC, this solvent was chosen rather than tetrahydrofuran (THF) because the latter caused fractionation of samples in small aggregates that were impossible to weigh. THF might be strong enough to break the links between the matrix and the nanocrystals. A large decrease in swelling of PVC-based samples when immersed in MEK was observed when increasing the cellulose nanocrystal content. It was assumed to be due to the existence of an interphase making a link between nanoparticles, thus allowing the formation of a flexible network. The swelling behavior of NR films reinforced with cellulose nanocrystals was also characterized by immersing the samples in toluene (Bendahou et al., 2010). The swelling of the polymeric matrix by toluene was found to strongly decrease even when adding only 1 wt% of cellulose nanoparticles.

It was shown that only 1 wt% of the cellulosic nanoparticles allowed preventing the disruption of the NR matrix and strongly restricted the swelling of the material. This behavior could not only be associated with the decrease of the fraction of material able to swell in toluene, i.e. NR, in the nanocomposites and it was suggested that adsorption of macromolecular chains at the filler/matrix interface through interactions between MFC and NR, and filler/filler interactions could also reduce swelling.

10.8.2 Water vapor transfer rate and water vapor permeability

Reduction of water vapor transmission rate (WVTR) of PVA films was observed when adding PAA or cellulose nanocrystals (Paralikar et al., 2008). For PAA the decrease of WVTR was explained by the reduced number of hydroxyl groups (hence hydrophilicity) induced by the cross-linking reaction. For cellulose nanoparticles it was ascribed to the physical barrier role provided by nanocrystals creating a tortuous path for the permeating moisture. The lowest WVTR (more than 50% lower than that of the neat PVA matrix) was observed for films containing 10 wt% cellulose nanocrystals and 10 or 20 wt% PAA. For higher filler contents, agglomeration of the nanoparticles provided channels in the film that allowed for more rapid permeation.

Cellulose nanoparticles 50–100 nm in diameter coagulated from a NaOH/urea/H_2O solution of MCC using ethanol/HCl aqueous solution as the precipitant were used to reinforce glycerol plasticized starch (Chang et al., 2010). Decreased permeability was reported when adding up to 2 wt% nanoparticles but the value was found to plateau for higher loading levels because of agglomeration. WVTR experiments were also conducted for alginate-acerola edible films reinforced with cellulose nanocrystals prepared from cotton and coconut husk fibers (Azeredo et al., 2012). The filler content was varied from 0 to 15 wt%. A continuous reduction of WVTR was observed when increasing the nanocrystal content. A similar decrease of WVTR upon cellulose nanoparticle addition was reported for edible films from mango puree (Azeredo et al., 2009) or from glycerol plasticized chitosan (Azeredo et al., 2010). However, in these two studies the exact nature of the cellulose nanoparticle was ambiguous because it was supposed to be MCC but displayed nanoscale dimensions. Reinforcement of methylcellulose (MC)-based films with low cellulose nanocrystal contents (0.1–1 wt%) was also shown to decrease the WVP of the polymer matrix when conditioned at 25°C and 60% RH for 24 h (Khan et al., 2010). It was ascribed to increased tortuosity, interactions of cellulose nanocrystals with MC-based films components (mainly cellulose) and hypothetical interactions between nanofibers given that the filler content was much lower than the percolation threshold. Barrier properties of the films were further improved by exposing them to γ radiation (0.5–50 kGy).

Nanocomposite films were obtained by mixing cellulose nanocrystals with polysulfone using a solvent exchange process (Noorani et al., 2007). A continuous increase of the WVTR was observed when increasing the nanoparticle content. It was associated with the formation of a percolating cellulose network within the polymeric matrix acting as a conduit for the diffusion of water through the film. However, as cellulose crystals are rather impermeable to water, the formation of a weak interphase in which diffusion is facilitated compared to the bulk matrix was suggested. This was supported by the low compatibility between cellulose and the polysulfone matrix.

The effect on WVP of adding cellulose nanocrystals to another hydrophobic polymer, i.e. PLA, was studied (Sanchez-Garcia and Lagaron, 2010). Two routes of processing were used, viz. a solvent exchange procedure from water to chloroform and freeze-drying of the nanoparticle dispersion followed by redispersion in chloroform. Permeability of the films to water vapor was measured at 24°C and 75% RH. Reductions of the WVP of ca. 64, 78, 82 and 81% were obtained when adding 1, 2, 3 and 5 wt%, respectively, of nanocrystals to the PLA matrix using the freeze-drying method. Lower reductions of ca. 44, 49 and 21% were achieved for the films containing 1, 3 and 5 wt% nanocrystals, respectively, using the solvent exchange procedure. The difference was mainly ascribed to a higher dispersion level and higher nanocrystal-induced crystallinity of the PLA matrix for the films containing freeze-dried nanoparticles.

Similar results were obtained for unplasticized and glycerol plasticized carrageenan reinforced with cellulose nanocrystals (Sanchez-Garcia et al., 2010). Moreover, the nanocrystals were found to be more efficient in reducing WVP than microfibers chiefly due to nanodispersion of the filler. The best water barrier performance was observed for a nanocrystal content of ca. 3 wt%. Higher nanoparticle contents led to poor filler dispersion and agglomeration which was detrimental in terms of barrier enhancement. Similarly, an optimal concentration around 10 wt% was reported for xylan films reinforced with cellulose nanocrystals (Saxena and Ragauskas, 2009). Moreover, the reduction in water transmission was found to be more significant with sulfuric acid-hydrolyzed than with hydrochloric acid-hydrolyzed cellulose nanocrystals (Saxena et al., 2011).

The WVP of NR films reinforced with cellulose nanocrystals prepared from sugarcane bagasse was investigated (Bras et al., 2010). An increase of WVP was observed by adding the cellulosic nanoparticles up to 7.5 wt%. It was ascribed to the high hydrophilic nature of cellulose compared to the neat matrix. For higher filler loading a decrease was reported, ascribed to the formation of a percolation nanocrystal network.

10.8.3 Gas permeability

The oxygen permeability measured at 24°C and 80% RH was characterized for cellulose nanocrystal reinforced PLA (Sanchez-Garcia and Lagaron, 2010). Reductions of 83, 90, 90 and 88% were obtained when adding 1, 2, 3 and 5 wt%, respectively, of nanocrystals to the PLA matrix, showing therefore the highest barrier effect for 2–3 wt% filler loading. As for WVP experiments, this effect was attributed to the nanocrystal-induced crystallinity of the PLA matrix. This experimental permeability drop was higher than the one predicted by Nielsen and Fricke models even if experimental data were corrected by crystallinity alterations. Because experiments were conducted at 80% RH, it was hypothesized that sorbed moisture was thought to fill in the existing free volume therefore reducing the gas permeability.

The O_2 and CO_2 barrier properties of ternary systems consisting of NR reinforced with cellulose nanocrystals and natural montmorillonite were investigated (Bendahou et al., 2011). Nanocomposite films were prepared by changing the weight content of each filler keeping the total filler content equal to 5 wt%. However, it is worth noting that because of the difference in density of both fillers, the filler volume content was not constant for all samples. Improved gas barrier properties were reported and

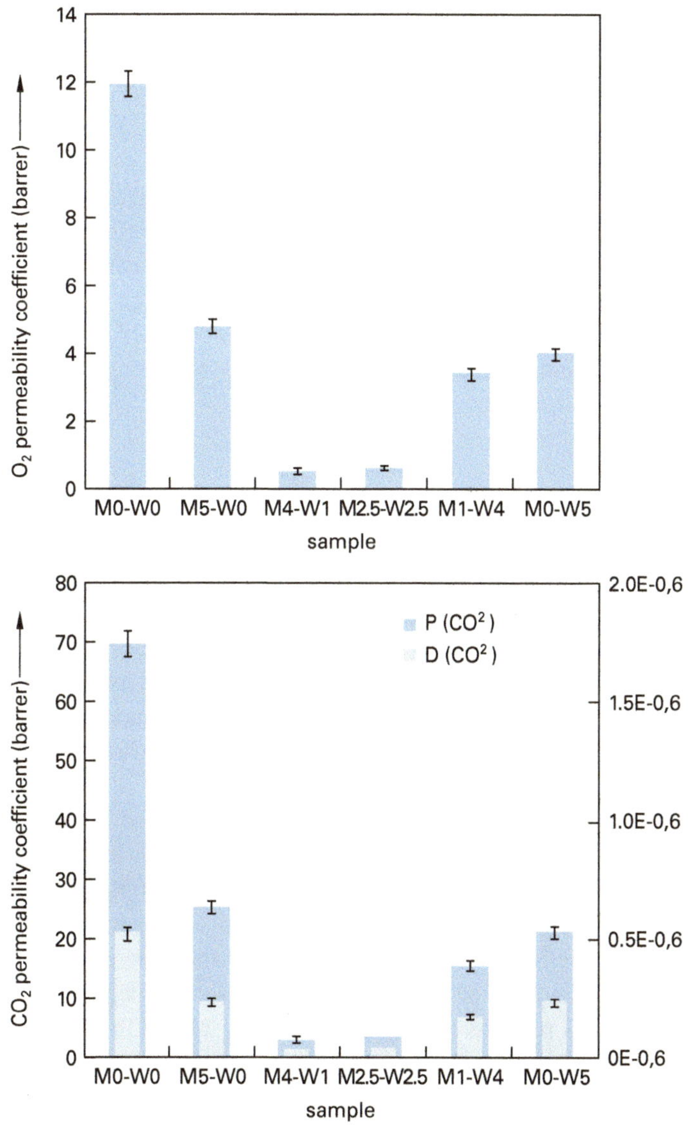

Fig. 10.10: Permeability coefficient and diffusion coefficient measured for natural rubber films reinforced with date palm tree cellulose nanocrystals (W) and montmorillonite (M) (the total filler content was fixed to 5 wt%): (a) for oxygen and (b) for carbon dioxide (Bendahou et al., 2011).

explained by a tortuosity effect. For binary systems, a more efficient barrier effect was observed for montmorillonite compared to cellulose nanocrystals. The tortuosity values calculated from permeability and diffusion coefficients indicated that the simultaneous use of both fillers greatly slowed down the gas diffusion leading to a synergistic effect. This phenomenon was ascribed to the formation of montmorillonite-cellulose nanocrystal subassembly, the most efficient effect being observed for montmorillonite contents higher than 2.5 wt% (see Figure 10.10)

10.8.4 Other substances permeability

Trichloroethylene (TCE) was chosen as a representative toxic industrial material to evaluate the barrier properties of PVA films cross-linked with PAA and reinforced with cellulose nanocrystals (Paralikar et al., 2008). It was found that increasing PAA or cellulose nanoparticle content increased the lag time and decreased the flux compared to the neat PVA membrane. A synergistic composition was observed at 10 wt% nanocrystals and 10 wt% PAA and minimal effect was reported when using carboxylated cellulose nanocrystals.

10.9 Conclusions

Cellulose is a hydrophilic polymer and it obviously absorbs water when immersed in liquid water or conditioned in moist atmosphere. However, the water vapor permeability is decreased when the cellulose fibers are disintegrated to the nanoscale level. Moreover, the sensitivity to moisture of the nanoparticles can be tuned by performing pretreatment prior to homogenization or post-treatment (polymer impregnation, surfactant adsorption, chemical grafting, ...). When nanocellulose is used as filler in a polymeric matrix, the moisture sorption behavior depends on the nature of the matrix and filler/matrix interactions. The moisture sorption is decreased for hydrophilic matrices, whereas it is increased for hydrophobic matrices. This is ascribed to strong bonding between nanofibers leading to the formation of a percolating network that can either restrict the swelling of the matrix or facilitate the diffusion of moisture.

The gas permeability is also reduced in a dry atmosphere when decreasing the size of the cellulosic particles because of the crystalline and dense structure of the nanoparticle film. However, this property is lost in a moist atmosphere. To offset the affinity of cellulose nanoparticles but still preserve their good gas barrier properties, they can be embedded in a polymeric matrix or coated on a substrate (polymer or paper). Whatever the treatment or the experimental conditions used to produce nanocellulose, it is seen as a new biomaterial for the conception of good barrier food packaging. Nanocomposite films extend food shelf-life, and also improve food quality as they can serve as carriers of some active substances such as antioxidants and antimicrobials.

10.10 References

Ahola, S., Salmi, J., Johansson, Laine, J. and Österberg, M. (2008). Model films from native cellulose nanofibrils. Preparation, swelling and surface interactions. Biomacromolecules *9*, 1273–1282.

Anglès, M.N. and Dufresne, A. (2000). Plasticized starch/tunicin whiskers nanocomposites: 1. Structural analysis. Macromolecules *33*, 8344–8353.

Aulin, C. Ahola, S., Josefsson, P., Nishino, T., Hirose, Y., Österberg, M. and Wågberg, L. (2009). Nanoscale cellulose films with different crystallinities and mesostructures – Their surface properties and interaction with water. Langmuir *25*, 7675–7685.

Aulin, C., Gällstedt, M. and Lindström, T. (2010). Oxygen and oil barrier properties of microfibrillated cellulose films and coatings. Cellulose *17*, 559–574.

Azeredo, H.M.C., Mattoso, L.H.C., Wood, D., Williams, T.G., Avena-Bustillos, R.J. and McHugh, T.H. (2009). Nanocomposite edible films from mango puree reinforced with cellulose nanofibers. J. Food Sci. *74*, N31–N35.

Azeredo, H.M.C., Mattoso, L.H.C., Avena-Bustillos, R.J., Filho, G.C., Munford, M.L., Wood, D. and McHugh, T.H. (2010). Nanocellulose reinforced chitosan composite films as affected by nanofiller loading and plasticizer content. J. Food Sci. *75*, N1–N7.

Azeredo, H.M.C., Miranda, K.W.E., Rosa, M.F., Nascimento, D.M. and de Moura, M.R. (2012). Edible films from alginate-acerola puree reinforced with cellulose whiskers. Food Sci. Technol. *46*, 294–297.

Belbekhouche, S., Bras, J., Siqueira, G., Chappey, C., Lebrun, L., Khelifi, B., Marais, S. and Dufresne, A. (2011). Water sorption behavior and gas barrier properties of cellulose whiskers and microfibrils films. Carbohydr. Polym. *83*, 1740–1748.

Bendahou, A., Kaddami, H. and Dufresne, A. (2010). Investigation on the effect of cellulosic nanoparticles' morphology on the properties of natural rubber based nanocomposites. Eur. Polym. J. *46*, 609–620.

Bendahou, A., Kaddami, H., Espuche, E., Gouanvé, F. and Dufresne, A. (2011). Synergism effect of montmorillonite and cellulose whiskers on the mechanical and barrier properties of natural rubber. Macromol. Mat. Eng. *296*, 760–769.

Berglund, L. and Liu, A. (2011). Strong nanopaper. Patent WO 2011/059398 A1.

Bilbao-Sainz, C., Bras, J., Williams, T., Sénechal, T. and Orts, W. (2011). HPMC reinforced with different cellulose nano-particles. Carbohydr. Polym. *86*, 1549–1557.

Bras, J., Hassan, M.L., Bruzesse, C., Hassan, E.A., El-Wakil, N.A. and Dufresne, A. (2010). Mechanical, barrier, and biodegradability properties of bagasse cellulose whiskers reinforced natural rubber nanocomposites. Ind. Crops Prod. *32*, 627–633.

Cao, X., Chen, Y., Chang, P.R., Stumborg, M. and Huneault, M.A. (2008a). Green composites reinforced with hemp nanocrystals in plasticized starch. J. Appl. Polym. Sci. *109*, 3804–3810.

Cao, X., Chen, Y., Chang, P.R., Muir, A.D. and Falk G. (2008b). Starch-based nanocomposites reinforced with flax cellulose nanocrystals. eXPRESS Polym. Let. *2*, 502–510.

Chang, P.R., Jian, R., Zheng, P., Yu, J. and Ma, X. (2010). Preparation and properties of glycerol plasticized-starch (GPS)/cellulose nanoparticle (CN) composites. Carbohydr. Polym. *79*, 301–305.

Chazeau, L., Cavaillé, J.Y., Canova, G.R., Dendievel, R. and Boutherin, B. (1999). Viscoelastic properties of plasticized PVC reinforced with cellulose whiskers. J. Appl. Polym. Sci. *71*, 1797–1808.

Chen, Y., Liu, C., Chang, P.R., Cao, X. and Anderson, D.P. (2009). Bionanocomposites based on pea starch and cellulose nanowhiskers hydrolyzed from pea hull fibre: Effect of hydrolysis time. Carbohydr. Polym. *76*, 607–615.

Choi, Y.J. and Simonsen, J. (2006). Cellulose nanocrystal-filled carboxymethyl cellulose nanocomposites. J. Nanosci. Nanotechnol. *6*, 633–639.

Comyn, J. (1985). Introduction to polymer permeability and the mathematics of diffusion. In: Polymer permeability, J. Comyn, ed. (Chapman & Hall, London, UK), pp. 1–10.

Dalmas, F., Chazeau, L., Gauthier, C., Cavaillé, J.Y., and Dendievel, R. (2006). Large deformation mechanical behaviour of flexible nanofiber filled polymer nanocomposites. Polymer *47*, 2802–2812.

Dammström, S., Salmén, L. and Gatenholm, P. (2005). The effect of moisture on the dynamic mechanical properties of bacterial cellulose/glucuronoxylan nanocomposites. Polymer *46*, 10364–10371.

de Paula, E.L., Mano, V. and Pereira, F.V. (2011). Influence of cellulose nanowhiskers on the hydrolytic degradation behavior of poly(D,L-lactide). Polym. Degrad. Stab. *96*, 1631–1638.

Dufresne, A. and Vignon, M.R. (1998). Improvement of starch film performances using cellulose microfibrils. Macromolecules *31*, 2693–2696.

Dufresne, A., Dupeyre, D. and Vignon, M.R. (2000). Cellulose microfibrils from potato tuber cells: Processing and characterization of starch-cellulose microfibril composites. J. Appl. Polym. Sci. *76*, 2080–2092.

Fernández, A., Sánchez, M.D., Ankerfors, M. and Lagaron, J.M. (2008). Effects of ionizing radiation in ethylene-vinyl alcohol copolymers and in composites containing microfibrillated cellulose. J. Appl. Polym. Sci. *109*, 126–134.

Fujisawa, S., Okita, Y., Fukuzumi, H., Saito, T. and Isogai, A. (2011). Preparation and characterization of TEMPO-oxidized cellulose nanofibril films with free carboxyl groups. Carbohydr. Polym. *84*, 579–583.

Fukuzumi, H., Saito, T., Iwata, T., Kumamoto, Y. and Isogai, A. (2009). Transparent and high gas barrier films of cellulose nanofibers prepared by TEMPO-mediated oxidation. Biomacromolecules *10*, 162–165.

Fukuzumi, H., Saito, T., Iwamoto, S., Kumamoto, Y., Ohdaira, T., Suzuki, R. and Isogai, A. (2011). Pore size determination of TEMPO-oxidized cellulose nanofibril films by positron annihilation lifetime spectroscopy. Biomacromolecules *12*, 4057–4062.

Garcia de Rodriguez, N.L., Thielemans, W. and Dufresne, A. (2006). Sisal cellulose whiskers reinforced polyvinyl acetate nanocomposites. Cellulose *13*, 261–270.

Henriksson, M. and Berglund, L.A. (2007). Structure and properties of cellulose nanocomposite films containing melamine formaldehyde. J. Appl. Polym. Sci. *106*, 2817–2824.

Hult, E.L., Iotti, M. and Lenes, M. (2010). Efficient approach to high barrier packaging using microfibrillar cellulose and shellac. Cellulose *17*, 575–586.

Kaushik, A., Singh, M. and Verma, G. (2010). Green nanocomposites based on thermoplastic starch and steam exploded cellulose nanofibrils from wheat straw. Carbohydr. Polym. *82*, 337–345.

Khan, R.A., Salmieri, S., Dussault, D., Uribe-Calderon, J., Kamal, M.R., Safrany, A. and Lacroix, M. (2010). Production and properties of nanocellulose-reinforced methylcellulose-based biodegradable films. J. Agric. Food Chem. *58*, 7878–7885.

Liu, A., Walther, A., Ikkala, O., Belova, L., and Berglund, L. A. (2011). Clay Nanopaper with Tough Cellulose Nanofiber Matrix for Fire Retardancy and Gas Barrier Functions. Biomacromolecules *12*, 633–641.

Liu, A. and Berglund, L.A. (2012). Clay nanopaper composites of nacre-like structure based on montmorillonite and cellulose nanofibers – Improvements due to chitosan addition. Carbohydr. Polym. *87*, 53–60.
Liu, D., Zhong, T., Chang, P.R., Li, K. and Wu, Q. (2010). Starch composites reinforced by bamboo cellulosic crystals. Bioresource Technol. *101*, 2529–2536.
Lu, Y, Weng, L. and Cao X. (2005). Biocomposites of plasticized starch reinforced with cellulose crystallites from cottonseed linter. Macromol. Biosci. *5*, 1101–1107.
Lu, Y., Weng, L. and Cao, X. (2006). Morphological, thermal and mechanical properties of ramie crystallites – Reinforced plasticized starch biocomposites. Carbohydr. Polym. *63*, 198–204.
Ma, H., Burger, C., Hsiao, B.S. and Chu, B. (2011). Ultrafine polysaccharide nanofibrous membranes for water purification. Biomacromolecules *12*, 970–976.
Malainine, M.E., Mahrouz, M. and Dufresne, A. (2005). Thermoplastic nanocomposites based on cellulose microfibrils from *Opuntia ficus indica* parenchyma cell. Compos. Sci. Technol. *65*, 1520–1526.
Mathew, A.P. and Dufresne, A. (2002). Morphological investigation of nanocomposites from sorbitol plasticized starch and tunicin whiskers. Biomacromolecules *3*, 609–617.
Mikkonen, K.S., Stevanic, J.S., Joly, C., Dole, P., Pirkkalainen, K., Serimaa, R., Salmén, L. and Tenkanen, M. (2011). Composite films from spruce galactoglucomannans with microfibrillated spruce wood cellulose. Cellulose *18*, 713–726.
Minelli, M., Baschetti, M.G., Doghieri, F., Ankerfors, M., Lindström, T., Siró, I. and Plackett, D. (2010). Investigation of mass transport properties of microfibrillated cellulose (MFC) films. J. Membrane Sci. *358*, 67–75.
Müller, M., Czihak, C., Schober, H., Nishiyama, Y. and Vogl, G. (2000). All disordered regions of native cellulose show common low-frequency dynamics. Macromolecules *33*, 1834–1840.
Nielsen, L.E. (1967). Models for the permeability of filled polymer systems. J. Macromol. Sci. *A1*, 929–942.
Noorani, S., Simonsen, J. and Atre, S. (2007). Nano-enabled microtechnology: Polysulfone nanocomposites incorporating cellulose nanocrystals. Cellulose *14*, 577–584.
Paralikar, S.A., Simonsen, J. and Lombardi, J. (2008). Poly(vinyl alcohol)/cellulose nanocrystal barrier membranes. J. Membrane Sci. *320*, 248–258.
Plackett, D., Anturi, H., Hedenqvist, M., Ankerfors, M., Gällstedt, M., Lindström, T. and Siró, I. (2010). Physical properties and morphology of films prepared from microfibrillated cellulose and microfibrillated cellulose in combination with amylopectin. J. Appl. Polym. Sci. *117*, 3601–3609.
Rodionova, G., Lenes, M., Eriksen, Ø. and Gregersen, Ø. (2011). Surface chemical modification of microfibrillated cellulose: improvement of barrier properties for packaging applications. Cellulose *18*, 127–134.
Rodionova, G., Saito, T., Lenes, M., Eriksen, Ø, Gregersen, Ø, Fukuzumi, H. and Isogai, A. (2012). Mechanical and oxygen barrier properties of films prepared from fibrillated dispersions of TEMPO-oxidized Norway spruce and Eucalyptus pulp. Cellulose *19*, 705–711.
Sanchez-Garcia, M.D. and Laragon, J.M. (2010). On the use of plant cellulose nanowhiskers to enhance the barrier properties of polylactic acid. Cellulose *17*, 987–1004.
Sanchez-Garcia, M.D., Hilliou, L. and Laragon, J.M. (2010). Morphology and water barrier properties of nanobiocomposites of k/i-hybrid carrageenan and cellulose nanowhiskers. J. Agric. Food Chem. *58*, 12847–12857.
Savadekar, N.R. and Mhaske, S.T. (2012). Synthesis of nano cellulose fibers and effect on thermoplastics starch based films. Carbohydr. Polym. *89*, 146–151.
Saxena, A. and Ragauskas, A.J. (2009). Water transmission barrier properties of biodegradable films based on cellulosic whiskers and xylan. Carbohydr. Polym. *78*, 357–360.

Saxena, A., Elder, T.J. and Ragauskas, A.J. (2011). Moisture barrier properties of xylan composite films. Carbohydr. Polym. *84*, 1371–1377.
Sehaqui, H., Zhou, Q. and Berglund, L.A. (2011). Nanostructured biocomposites of high toughness – A wood cellulose nanofiber network in ductile hydroxyethylcellulose matrix. Soft Matter *7*, 7342–7350.
Seifert, M., Hesse, S., Kabrelian, V. and Klemm, D. (2003). Controlling the water content of never dried and reswollen bacterial cellulose by the addition of water-soluble polymers to the culture medium. J. Polym. Sci. A *42*, 463–470.
Shanmuganathan, K., Capadona, J.R., Rowan, S.J. and Weder, C. (2010a). Bio-inspired mechanically-adaptive nanocomposites derived from cotton cellulose whiskers. J. Mater. Chem. *20*, 180–186.
Shanmuganathan, K., Capadona, J.R., Rowan, S.J. and Weder, C. (2010b). Stimuli-responsive mechanically adaptive polymer nanocomposites. ACS Appl. Mater. Interfaces *2*, 165–174.
Siqueira, G., Bras, J. and Dufresne, A. (2009). New process of chemical grafting of cellulose nanoparticles with a long chain isocyanate. Langmuir *26*, 402–411.
Siró, I., Plackett, D., Hedenqvist, M., Ankerfors, M. and Lindström, T. (2011). Highly transparent films from carboxymethylated microfibrillated cellulose: The effect of multiple homogenization steps on key properties. J. Appl. Polym. Sci. *119*, 2652–2660.
Spence, K., Venditti, R., Rojas, O., Habibi, Y. and Pawlak, J. (2010). The effect of chemical composition on microfibrillar cellulose films from wood pulps: Water interactions and physical properties for packaging applications. Cellulose *17*, 835–848.
Spence, K.L., Venditti, R.A., Rojas, O.J., Habibi, Y. and Pawlak, J.J. (2011a). A comparative study of energy consumption and physical properties of microfibrillated cellulose produced by different processing methods. Cellulose *18*, 1097–1111.
Spence, K.L., Venditti, R.A., Rojas, O.J., Pawlak, J.J. and Hubbe, M.A. (2011b). Water vapor barrier properties of coated and filled microfibrillated cellulose composite films. BioResources *6*, 4370–4388.
Svagan, A.J., Hedenqvist, M.S. and Berglund, L. (2009). Reduced water vapour sorption in cellulose nanocomposites with starch matrix. Compos. Sci. Technol. *69*, 500–506.
Syverud, K. and Stenius, P. (2009). Strength and barrier properties of MFC films. Cellulose *16*, 75–85.
Teixeira, E.M., Pasquini, D., Curvelo, A.A.S., Corradini, E., Belgacem, M.N. and Dufresne, A. (2009). Cassava bagasse cellulose nanofibrils reinforced thermoplastic cassava starch. Carbohydr. Polym. *78*, 422–431.
Thielemans, W., Warbey, C.R. and Walsh, D.A. 2009. Permselective nanostructured membranes based on cellulose nanowhiskers. Green Chem. *11*, 531–537.
Tingaut, P., Zimmermann, T. and Lopez-Suevos, F. (2010). Synthesis and characterization of bionanocomposites with tunable properties from poly(lactic acid) and acetylated microfibrillated cellulose. Biomacromolecules *11*, 454–464.
Tomé, L.C., Brandão, L., Mendes, A.M., Silvestre, A.J.D., Neto, C.P., Gandini, A., Freire, C.S.R. and Marrucho, I.M. (2010). Preparation and characterization of bacterial cellulose membranes with tailored surface and barrier properties. Cellulose *17*, 1203–1211.
Turbak, A.F., Snyder, F.W. and Sandberg, K.R. (1983). Microfibrillated cellulose: A new cellulose product: Properties, uses, and commercial potential. J. Appl. Polym. Sci. Polym. Symp. *37*, 815–827.
Vergnaud, J.M. (1991). Liquid transport processes in polymeric materials: Modeling and industrial applications, (Prentice-Hall, Englewood Cliffs), 362 p.
Wang, Y., Cao, X. and Zhang, L. (2006). Effects of cellulose whiskers on properties of soy protein thermoplastics. Macromol. Biosci. *6*, 524–531.
Wu, C.N., Saito, T., Fujisawa, S., Fukuzumi, H. and Isogai, A. (2012). Ultrastrong and high gas-barrier nanocellulose/clay layered composites. Biomacromolecules *13*, 1927–1932.

Xhanari, K., Syverud, K., Chinga-Carrasco, G., Paso, K. and Stenius, P. (2011). Reduction of water wettability of nanofibrillated cellulose by adsorption of cationic surfactants. Cellulose *18*, 257–270.

11 Other polysaccharide nanocrystals

Apart from cellulose, nanoparticles can be extracted from other semicrystalline polysaccharides, such as starch and chitin. A process similar to the one used to prepare cellulose nanocrystals involving acid hydrolysis of amorphous domains of the polysaccharide followed by a mechanical treatment can be applied to these polymers to release the crystalline domains. Recent review papers have been published on starch nanoparticles (Le Corre et al., 2010) and chitin nanocrystals (Zeng et al., 2012). Although less studied than cellulosic nanoparticles, they can offer specific properties because of their different morphology and chemical composition.

11.1 Starch

Starch is a natural, renewable, and biodegradable polysaccharide produced by many plants as a storage polymer. It is the major carbohydrate reserve in plant tubers and seed endosperm and it is found in plant roots, stalks, crop seeds, and staple crops such as rice, corn, wheat, tapioca and potato (Buléon et al., 1998). The starch industry extracts and refines starches by wet grinding, sieving and drying. After its extraction from plants, starch occurs as a flour-like white powder insoluble in cold water. This powder (called native starch) consists of microscopic granules with diameters ranging from 2 to 100 μm depending on the botanic origin, and with a density of 1.5.

11.1.1 Composition

Chemically, starches are polysaccharides, composed of a number of glucose monomers called α-D-glycopyranose (or α-D-glycose) when in cycle and linked together with α-D-(1→4) and/or α-D-(1→6) linkages. Starch composition was first determined by studying the residue of its total acid hydrolysis. It is a combination of two glucosidic macromolecules called amylose and amylopectin (Figure 11.1). In most common types of starch, the amylopectin content ranges between 72 and 82 wt%, and the amylose content ranges between 18 and 28 wt%. However, some mutant types of starch have very high amylose content (up to 70 wt% and more for amylomaize) and some very high amylopectin content (99 wt% for waxy maize starch). Starch is composed of 98–99% of these two molecules. Other trace elements are lipids, proteins, minerals, phosphorous, enzymes, amino acids, and nucleic acids. Although present in low content, these components can affect the physico-chemical properties of starch.

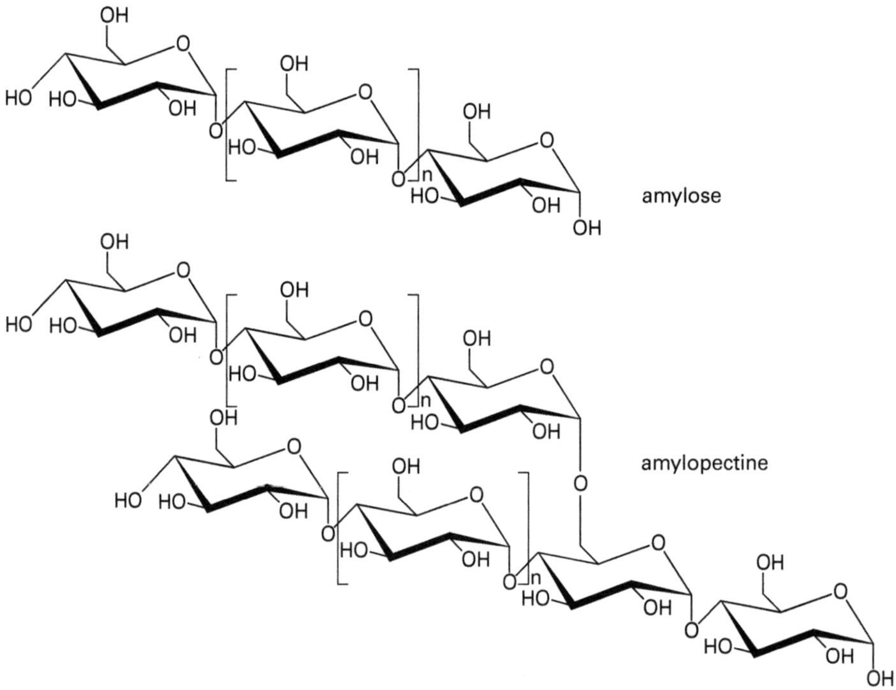

Fig. 11.1: Chemical structure of amylose and amylopectin.

Amylose is essentially a linear polymer constituted by glucose monomer units joined to one another head-to-tail by α-(1→4) glycosidic bonds, slightly branched by α-(1→6) linkages. Its degree of polymerization (DP) is up to 6,000. Amylose can form an extended shape (hydrodynamic radius 7–22 nm) (Parker and Ring, 2001) but generally tends to wind up into a rather stiff left-handed single helix or form even stiffer parallel left-handed double helical junction zones. Single helical amylose has hydrogen-bonding O2 and O6 atoms on the outside surface of the helix with only the ring oxygen pointing inwards. Hydrogen bonding between aligned chains causes retrogradation and releases some of the bound water. Retrogradation takes place in gelatinized starch upon cooling and corresponds to a rearrangement of the linear molecules, amylose and linear parts of amylopectin molecules, to a more crystalline structure. The aligned chains may then form double stranded crystallites that are resistant to amylases. These possess extensive inter- and intra-strand hydrogen bonding, resulting in a fairly hydrophobic structure of low solubility. The amylose content of starches is thus the major cause of resistant starch formation. Single helix amylose behaves similarly to the cyclodextrins by possessing a relatively hydrophobic inner surface that holds a spiral of water molecules, which are relatively easily lost to be replaced by hydrophobic lipid or aroma molecules. It is also responsible for the characteristic binding of amylose to chains of charged iodine molecules where each turn of the helix holds about two iodine atoms and a blue color is produced due to donor-acceptor interaction between water and the electron deficient polyiodides.

Amylopectin is a highly branched polymer consisting of relatively short branches of α-D-(1→4) glycopyranose that are interlinked by α-D-(1→6) glycosidic linkages. This branching is determined by branching enzymes that leave each chain with up to 30 glucose residues. Each amylopectin molecule contains a million of such residues, about 5 of which form the branch points. Each amylopectin molecule contains up to two million glucose residues in a compact structure with hydrodynamic radius 21–75 nm (Parker and Ring, 2001). The molecules are oriented radially in the starch granule and as the radius increases so does the number of branches required to fill up the space, with the consequent formation of concentric regions of alternating amorphous and crystalline structure. Some amylopectin (e.g. from potato) has phosphate groups attached to some hydroxyl groups, which increase its hydrophilicity and swelling power.

To describe the multiplicity in branching, the basic organization of the chains has been described in terms of A, B and C chains (Peat et al., 1952). The single C chain, with an average DP above 60, carries other chains as branches and contains the terminal reducing end of the amylopectin macromolecule. The A chains are glycosidically linked to the rest of the molecule by their reducing group through C6 of a glucose residue. The B chains are defined as bearing other chains as branches and are linked to the rest of the molecule by their reducing group on one side and by α-(1→6) linkage

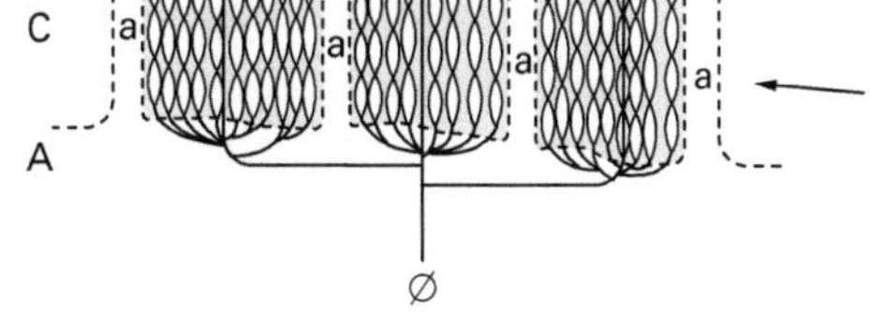

Fig. 11.2: Amylopectin cluster model: Each cluster contains between nine and 17 side chains, and the double helical structure of the polymer is represented. Amorphous zones are shown between both the crystalline lamellae and between the individual side chain clusters. The lamellae are not completely straight, parallel or of uniform thickness and, consequently, the starch polymers are not always aligned at right angles to the direction of the lamellae. The general direction of the lamellae is shown by the large black arrow. C: Crystalline lamellae (amylopectin side chain clusters, on average 6 nm length), A: amorphous lamellae (branching zone) on average 4 nm length, a: amorphous regions between crystalline clusters (Gallant et al., 1997).

on the other, thus being the backbone of the grape-like macromolecule. This organization led to several models referring to the cluster model presented in Figure 11.2.

Amylopectin (without amylose) can be isolated from waxy maize starch whereas amylose (without amylopectin) is best isolated after specifically hydrolyzing the amylopectin with pullulanase (Vorwerg et al., 2002). Genetic modification of starch crops has led to the development of starches with improved and targeted functionality (Jobling, 2004). Amylose and amylopectin are inherently incompatible molecules.

The primary function of starch in plants is to act as an energy storage molecule for the organism. In plants, simple sugars are linked into starch molecules by specialized cellular organs called amyloplasts. Starch molecules can be digested by hydrolysis, catalyzed by enzymes called amylases, which can break the glycosidic bonds between the α-glucose components of starch. Humans and other animals have amylases, so they can digest starch. Digestion of starches consists of the process of the cleavage of the starch molecules back into their constituent simple sugar units by the action of the amylases. The resulting sugars are then processed by further enzymes (such as maltase) in the body, in the same manner as other sugars in the diet.

11.1.2 Multi-scale structure of the granule

Starch structure has been under research for years and because of its complexity, a universally accepted model is still lacking (Buléon et al., 1998). However, one model seems predominant. It is a multi-scale structure shown in Figure 11.3 consisting in the (a) granule (2–100 μm), into which we find (b) growth rings (120–500 nm) composed of (d) blocklets (20–50 nm) made of (c) amorphous and crystalline lamellae (9 nm) (Gallant et al., 1997) containing (g) amylopectin and (h) amylose chains (0.1–1 nm).

The shape and particle size of granules depends strongly on their botanical origin. On the surface, pores can be observed as can be seen in Figure 11.3(a). They are thought to be going through the growth rings to the hilum (center of the granule). Observed under a microscope and polarized light, starch shows birefringence. The refracted characteristic "Maltese cross" corresponding to the crystalline region is characteristic of a radial orientation of the macromolecules. X-ray diffraction studies showed that starch is a semi-crystalline polymer (Katz, 1930). The semi crystalline nature of starch granules can be also visualized from transmission electron microscopy (TEM) observation of a partially hydrolyzed granule. Starch granules display a so-called onion-like structure with more or less concentric growth rings composed of alternating hard crystalline and soft semicrystalline shells. Starch granules consist therefore of concentric alternating amorphous and semicrystalline growth rings. They grow by apposition from the hilum of the granule. The number and thickness of these layers depend on the botanical origin of starch. They are thought to be 120–400 nm thick (French, 1984). Details on the structure of amorphous growth ring are not found in the literature.

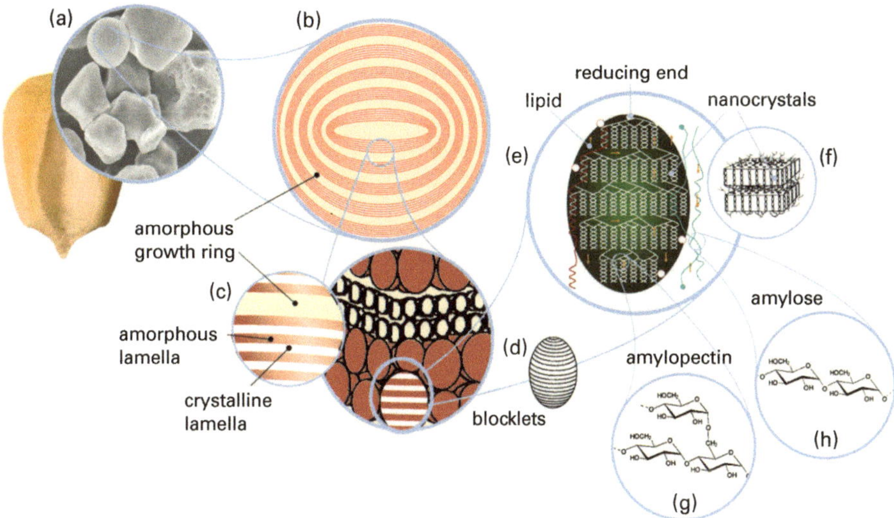

Fig. 11.3: Starch multi-scale structure: (a) starch granules from normal maize (30 μm), (b) amorphous and semicrystalline growth rings (120–500 nm), (c) amorphous and crystalline lamellae (9 nm) (magnified details of the semi crystalline growth ring), (d) blocklets (20–50 nm) constituting unit of the growth rings, (e) amylopectin double helixes forming the crystalline lamellae of the blocklets, (f) nanocrystals (other representation of the crystalline lamellae called starch nanocrystals when separated by acid hydrolysis), (g) amylopectin's molecular structure, (h) amylose's molecular structure (0.1–1 nm) (Le Corre et al., 2010).

The blocklets have an average size of 100 nm in diameter and are proposed to contain 280 amylopectin side chain clusters (Vandeputte and Delcour, 2004). Schematically, the semicrystalline growth rings consist of a stack of repeated crystalline and amorphous lamellae (Figure 11.3). The thickness of the combined layers is 9 nm regardless of the botanical origin. In reality, it is believed that the crystalline region is created by the intertwining of chains with a linear length above 10 glucose units to form double helixes which are packed and form the crystallites, and the amorphous region corresponds to branching points (Oates, 1997).

Crystallization or double helixes formation can occur either in the same amylopectin branch cluster or between adjacent clusters in three dimensions and is called the superhelical structure. Observation of amylopectin lamella within blocklets (about 10) supported the helical lamellar model (Gallant et al., 1997). Assuming that an amylopectin side chain cluster is 10 nm, a small blocklet (20–50 nm) is composed of about 2 to 5 side chain clusters. This model was illustrated as shown in Figure 11.3(d), making amylopectin the backbone of the blocklet structure (Tang et al., 2006).

Amylose molecules are thought to occur in the granule as individual molecules, randomly interspersed among amylopectin molecules and in close proximity with one another, in both the crystalline and amorphous regions (Oates, 1997). Depending on the botanical origin of starch, amylose is preferably found in the amorphous region (e.g. wheat starch), interspersed among amylopectin clusters in both the amorphous and crystalline regions (e.g. normal maize starch), in bundles between amylopectin clusters, or co-crystallized with amylopectin (e.g. potato starch) (Blanshard, 1987).

11.1.3 Polymorphism

X-ray diffraction analysis shows that starch is a semicrystalline polymer (Katz, 1930). Native starches contain between 15% and 45% of crystalline material (Zobel, 1988). Depending on their X-ray diffraction pattern, starches are categorized in three crystalline types referred to as A, B and C. Amylopectin chain length was found to be a determining factor for crystalline polymorphism (Hizukuri et al., 1983; Hizukuri, 1986). A-type is characteristic of cereal starches (wheat and maize starch). B-type is typical of tuber and amylose-rich cereal starches. C-type is characteristic of leguminous starches and corresponds to a mixture of A and B crystalline types. V-type, from German "Verkleisterung" (gelatinization), is observed during the formation of complexes between amylose and a complexing molecule (iodine, alcohols, cyclohexane, fatty acids, ...).

The appearance of starch X-ray diffraction patterns depends on the water content of granules during the measurement. The more starch is hydrated, the thinner the diffraction pattern rings become, up to a given limit. Water is therefore one of the component of the crystalline organization of starch. Determination of starch crystallinity is tricky because of both the influence of water content and absence of 100% crystalline standard. The crystalline to amorphous transition occurs at 60–70°C in water and this process is called gelatinization. In this amorphous state, hydrolysis is faster and this is why cooking food makes starchy food better digestible.

A model has been proposed for the double helixes packing configuration to explain differences between A- and B-type starches (Imberty et al., 1987; Imberty and Perez, 1988). A-type structures are closely packed with water molecules between each double helical structure, whereas B-types are more open and water molecules are located in the central cavity formed by 6 double helixes. It was envisaged that branching patterns of the different types of starch may also differ (Jane et al., 1997). It was also suggested that the B-type amylopectin branching points are clustered, forming a smaller amorphous lamella whereas A-type amylopectin branching points are scattered in both the amorphous and the crystalline regions, giving more flexibility to double helixes to pack closely. The distance between two $\alpha(1\rightarrow6)$ linkages and the branching density inside each cluster are determining factors for the development of crystallinity in starch granules (Gérard et al., 2000). Clusters with numerous short chains and short linkage distance produce densely packed structures which crystallize into the A allomorphic type. Longer chains and distances lead to a B-type.

The C-type starch pattern has been considered to be a mixture of both A- and B-types since its X-ray diffraction pattern can be resolved as a combination of the previous two. It has been suggested that C-type starch granules contain both types of polymorph: the B-type at the center of the granule and the A-type at the surrounding (Bogracheva et al., 1998). Several attempts of structural characterization of C-type starch were conducted using acid hydrolysis (Wang et al., 2008a; Wang et al., 2008b). It was shown that the core part of C-type starch was preferably hydrolyzed and that hydrolyzed starch showed the A-type diffraction pattern, suggesting that B-type polymorphs constitute mainly the amorphous regions and are more readily hydrolyzed than A-types constituting mainly the crystalline region. This shows that B-type starches are more acid-resistant than A-types. This conclusion is of importance for the preparation of starch nanocrystals. Another V-type was also identified as the result of amylose being complexed with other substances such as iodine, fatty acid, emulsifiers or butanol. This crystalline form is characterized by a simple left helix with six glucose units per turn (Averous and Halley, 2009).

11.2 Acid hydrolysis of starch

Aqueous suspensions of starch nanocrystals can be prepared according to the "lintnerization" procedure described in the literature (Robin et al., 1974; Battista, 1975). Acid hydrolysis is a chemical treatment largely used in industry to prepare glucose syrups from starch. Classically, the acid hydrolysis of starch is performed in aqueous medium with hydrochloric acid (Lintner, 1886) or sulfuric acid (Nägeli, 1874) at 35°C. Residues from hydrolysis are called "lintners" and "nägeli" or amylodextrin, respectively.

Degradation of native starch granules by acid hydrolysis depends on several parameters. It includes the botanical origin of starch, namely crystalline type, granule morphology (shape, size, surface state) and relative proportion of amylose and amylopectin. It also depends on the acid hydrolysis conditions, namely acid type, acid concentration, starch concentration, temperature, hydrolysis duration and stirring. The degradation of starch from different origins by hydrochloric acid has been studied in detail (Robin et al., 1975). The kinetics of lintnerization shows two main steps. For lower times (typically $t < 8$–15 days), the hydrolysis kinetics is fast and corresponds to the hydrolysis of amorphous domains. For higher times ($\sim t > 8$–15 days), the hydrolysis kinetics is slow and corresponds to the hydrolysis of crystalline domains. The critical time corresponding to fast/slow hydrolysis conditions depends on the botanical origin of starch (Singh and Ali, 2000; Jayakody and Hoover, 2002). It has also been reported that hydrolysis is faster when using hydrochloric acid rather than sulfuric acid (Muhr et al., 1984). Temperature favors the hydrolysis reaction but it is restricted to the gelatinization temperature of starch in acidic medium. Gelatinization corresponds to an irreversible swelling and solubilization phenomenon when native granules are heated above 60°C in excess water. As for temperature, the acid concentration favors the hydrolysis kinetics. However, above a given acid concentration, granule gelatinization occurs, around 2.5–3 N for hydrochloric acid (Robin, 1976).

The main drawbacks for the use of such hydrolysis residues in composite applications are the duration (40 days of treatment) and the yield (0.5 wt%) of the hydrochloric acid hydrolysis step (Battista, 1975). The process has been improved by performing periodic stirring of the suspension and the duration of the acid hydrolysis treatment has been reduced to 15 days using a 5 wt% starch suspension and 2.2 M HCl (Dufresne et al., 1996). Response surface methodology was used to investigate the effect of five selected factors on the selective sulfuric acid hydrolysis of waxy maize starch granules in order to optimize the preparation of aqueous suspensions of starch nanocrystals (Angellier et al., 2004). These predictors were temperature, acid concentration, starch concentration, hydrolysis duration and stirring speed. The preparation of aqueous suspensions of starch nanocrystals was achieved after 5 days of 3.16 M H_2SO_4 hydrolysis at 40°C, 100 rpm and with a starch concentration of 14.69 wt% with a yield of 15.7 wt%. This procedure has served for many studies as the standard recipe for the preparation of starch nanocrystals.

However, it was recently shown that starch nanocrystals are produced from a very early stage of the acid hydrolysis treatment (LeCorre et al., 2011a). It was observed that starch nanocrystals were formed, at least, after 24 h of sulfuric acid hydrolysis and that consequently, at any time including final suspension, both microscaled and nanoscaled particles can be found and coexist. The earlier formed nanocrystals might turn to sugar by the end of the batch production process explaining the low yields. This study clearly showed the need for a continuous production and extraction process of SNC. Differential centrifugation has been tested as an isolation process for separating these two kinds of particles, but did not seem fitted for fractionation due to hydrogen bonding and different densities within starch granules.

Filtration of the hydrolyzed residues using a microfiltration unit equipped with ceramic membranes to assess the cross-flow membrane filtration potential of starch nanocrystal suspensions was conducted (LeCorre et al., 2011b). Process parameters were monitored and the properties of feed, permeate and retentate were investigated. Cross-flow filtration was proved to be an efficient continuous operation for separating starch nanocrystals from the bulk suspension and non-fully hydrolyzed particles whatever the ceramic membrane pore size (0.2–0.8 μm). Analysis on permeate showed not only that collected nanoparticles were more crystalline than feed, but also that mostly B-type particles were produced during the first day of hydrolysis.

Based on this observation and as an attempt to establish a predictive model for the optimal parameter setting for preparing starch nanocrystals in one day, a statistical experimental design and a multi-linear regression method analysis were performed (Le Corre et al., 2012a).

The possibility of developing an enzymatic pretreatment of starch to reduce the acid hydrolysis duration was also investigated (LeCorre et al., 2012c). A screening of three types of enzymes, namely α-amylase, β-amylase, and glucoamylase, was proposed. The latter was the most efficient for producing microporous starch while keeping intact the semicrystalline structure of starch. With a 2 h pretreatment of waxy maize starch granules, the extent of acid hydrolysis currently reached in 24 and 120 h (5 days) were reached in only 6 and 45 h, respectively as shown in Figure 11.4.

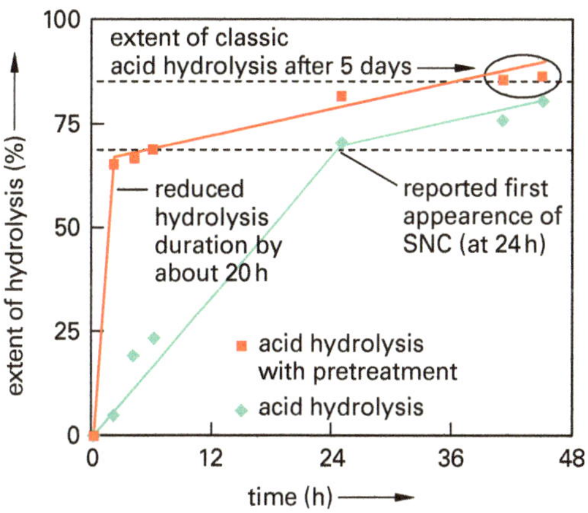

Fig. 11.4: Kinetics of sulfuric acid hydrolysis of non-pretreated (filled lozenges) and pretreated (grey squares) waxy maize starch (LeCorre et al., 2012c).

11.3 Starch nanocrystals

The first interest in starch nanocrystals for nanocomposite applications was reported by analogy with cellulose nanocrystals in 1996 (Dufresne et al., 1996). The so-called "microcrystalline starch" was reported to consist of agglomerated particles of a few tens of nanometers in diameter. The procedure consisted of hydrolyzing starch (5 wt%) in a 2.2 N HCl suspension for 15 days. The different starch sources used in the literature to prepare starch nanocrystals and extraction conditions are reported in Table 11.1.

Nanoparticles can also be prepared from starch following different strategies involving regeneration and precipitation and leading to particles with different properties, crystallinity, and shape (Le Corre et al., 2010). Moreover, the method for producing microfibrillated cellulose (MFC) has been transferred for producing starch colloids (Liu et al., 2009). A 5% slurry of high amylose corn starch was run through a Microfluidizer for several passes (up to 30). The particle size of the sample obtained from more than 10 passes was below 100 nm with a yield close to 100% and the gel-like suspension remained stable for more than one month. However, the ensuing

Source	Acid	Time (days)	Temperature (°C)	Acid-to-Starch Ratio (mL · g^{-1})	Reference
Amylo-maize	H_2SO_4 3.16 M	5	40	6.8	(LeCorre et al., 2011c)
	H_2SO_4 3 M	5	40	6.8	(LeCorre et al., 2012b)
Corn	H_2SO_4 2.87 mol.L^{-1}	7	45	6.8	(Song et al., 2008)
Normal Maize	H_2SO_4 3.16 M	5	40	6.8	(LeCorre et al., 2011c)
	H_2SO_4 3 M	5	40	6.8	(LeCorre et al., 2012b)
Pea	HCl 2.2 N	15	35	20	(Dubief et al., 1999)
	H_2SO_4 3.16 M	5	40	6.8	(Yu et al., 2008; Chang et al., 2009; Zheng et al., 2009)
Potato	HCl 2.2 N	15	35	20	(Dufresne et al., 1996)
			30	20	(Dufresne and Cavaillé, 1998)
	H_2SO_4 3.16 M	5	40	6.8	(Chen et al., 2008a; Chen et al., 2008b; Namazi and Dadkhah, 2008; LeCorre et al., 2011c)
	H_2SO_4 3 M	5	40	6.8	(LeCorre et al., 2012b)
Waxy Maize	HCl 2.2 N	15–42	36	20	(Putaux et al., 2003)
	H_2SO_4 3.16 M	1–9	35–40	6.8–20	(Angellier et al., 2004)
		5	40	6.8	(Angellier et al., 2005b; 2005c; 2005d; 2006; Thielemans et al., 2006; Labet et al., 2007; Viguié et al., 2007; Habibi and Dufresne, 2008; Garcia et al., 2009; 2011; LeCorre et al., 2011c; LeCorre et al., 2012c)
	HCl 2.2 N	2–40	35–40	20	(Angellier et al., 2005a)
	H_2SO_4 1.5–4 M	2–40	35–40	20	(Angellier et al., 2005a)
	HCl 2.2 N	30	35	20	(Kristo and Biliaderis, 2007)
	H_2SO_4 3 M	1–5	40	6.8	(LeCorre et al., 2011a)
		5	40	6.8	(LeCorre et al., 2012b)
	H_2SO_4 3–4.5 M	1–15 h	25–40	2.5–6.7	(LeCorre et al., 2012a)
Wheat	H_2SO_4 3 M	1	40	6.8	(LeCorre et al., 2011b)
	H_2SO_4 3.16 M	1	40	6.8	(LeCorre et al., 2011c)

Table 11.1: Hydrolysis conditions for the preparation of starch nanocrystals from different sources.

starch colloids were obtained from breaking down both amorphous and crystalline domains, rendering amorphous nanoparticles after 10 passes.

11.3.1 Aqueous suspensions

As for cellulose nanoparticles, starch nanocrystals are obtained as aqueous suspension. The stability of starch nanocrystal suspensions depends, as for their cellulose counterparts, on the dimensions of the dispersed particles, their size polydispersity and surface charge. The use of sulfuric acid for polysaccharide nanocrystals preparation leads to more stable aqueous suspension than that prepared using hydrochloric acid as shown in Figure 11.5 (Angellier et al., 2005a). Indeed, the H_2SO_4-prepared nanoparticles present a negatively charged surface while the HCl-prepared nanoparticles are not charged. A comparison between the effects of the two acids was performed with waxy maize starch (Angellier et al., 2005a). It was found that the use of sulfuric acid rather than hydrochloric acid allows reducing the possibility of agglomeration of starch nanoparticles and limits their flocculation in aqueous medium. Small angle light scattering experiments were performed on 3.4 wt% H_2SO_4-prepared starch nanocrystal aqueous suspensions in order to evaluate the kinetic of sedimentation of the nanoparticles (Angellier et al., 2005b). It was shown that there was no sedimentation of the nanocrystals for a period of at least 12 hours. However, the intensity of scattered light slightly increased, revealing that starch nanocrystals tend to aggregate in aqueous medium but not sufficiently to induce a sedimentation phenomenon.

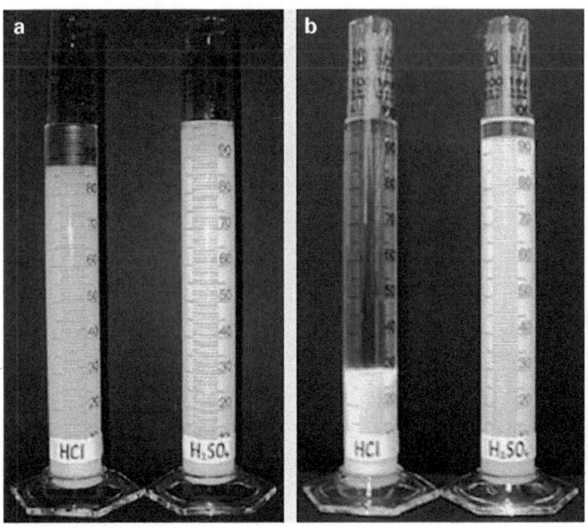

Fig. 11.5: Comparison of the sedimentation properties of HCl- (left tube) and H_2SO_4- (right tube) hydrolyzed starch nanocrystals suspended in water after (a) 5 min, and (b) 60 min (Angellier et al., 2005a).

11.3.2 Morphology

Compared to cellulose, the morphology of constitutive nanocrystals obtained from starch is completely different. Figures 11.6 and 11.7 show transmission electron micrographs (TEM) obtained from dilute suspensions of waxy maize starch nanocrystals

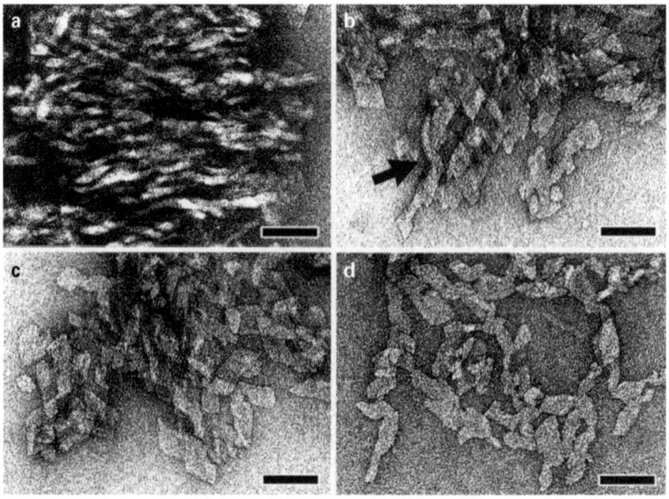

Fig. 11.6: TEM micrographs of negatively stained waxy maize starch samples: (a)-(c) fragments of waxy maize starch granules after 2 weeks of 2.2 N HCl hydrolysis at 36°C. In (a), a lamellar organization is clearly revealed with the platelets lying parallel to the incident electron beam. In (b) and (c), parallelepipedal platelets are seen lying flat on the carbon film. The arrow in (b) indicates a pyramidal stack of crystals. (d) Nearly individual waxy maize starch nanocrystals obtained after 6 weeks of hydrolysis (scale bars: 50 nm) (Putaux et al., 2003).

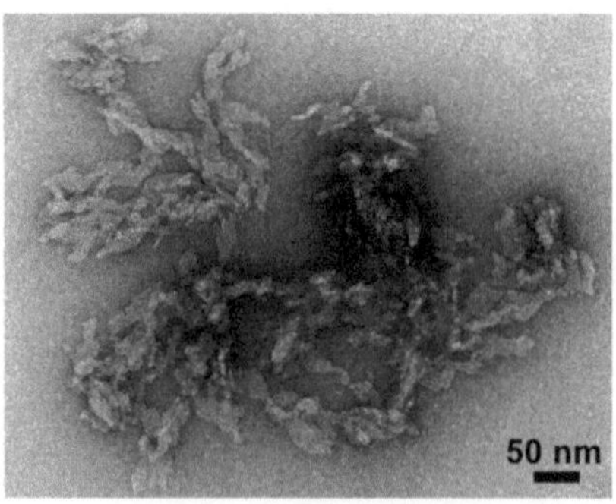

Fig. 11.7: TEM micrographs of negatively stained starch nanocrystals obtained by 3.16 M H_2SO_4 hydrolysis of waxy maize starch granules during 5 days, at 40°C, 100 rpm and with a starch concentration of 14.69 wt% (optimized conditions) (Angellier et al., 2004).

prepared by hydrochloric acid and sulfuric acid hydrolysis, respectively. They consist of 5–7 nm thick platelet-like particles with a length ranging from 20 to 40 nm and a width in the range 15–30 nm. The detailed investigation on the structure of these platelet-like nanoparticles was reported (Putaux et al., 2003; Putaux, 2005). Marked 60–65° acute angles were observed. TEM observations show that during acid hydrolysis, branching points are first hydrolyzed in amorphous domains, starch nanocrystals lying parallel to the incident electron beam (Figure 11.6(a)). As the acid hydrolysis progresses, the amorphous regions between crystalline lamellae become completely hydrolyzed and nanocrystals are seen lying flat on the carbon film (Figure 11.6(b)-(d)). Such nanocrystals are generally observed in the form of aggregates having an average size around 4.4 μm, as measured by laser granulometry (Angellier et al., 2005a).

The influence of the botanical origin and amylose content on the morphology of starch nanocrystals has been investigated (LeCorre et al., 2011c). Nanocrystals were prepared from five different starches, viz. normal maize, high amylose maize, waxy maize, potato and wheat, covering three botanical origins, two crystalline types, and three ranges of amylose content (0, 25, and 70%) for maize starch. Only a moderate influence of the botanical origin of starch was reported on properties such as size, size distribution, and thickness of the nanoparticles, as well as viscosity of the suspension. Differences were more pronounced when comparing shapes and crystallinity. Nanocrystals produced from A-type starches rendered square-like particles, whereas nanocrystals produced from B-type starches rendered round-like particles. This was explained by the different packing configurations of amylopectin chains for A- and B-type starches.

A detailed characterization of the molecular content of A-type nanocrystals prepared by acid hydrolysis of waxy maize starch granules was reported (Angellier-Coussy et al., 2009). Several populations of dextrins were found corresponding to different structural motifs. One of these had a DP of 14.2, which in the double-helical structure corresponds to a length of 5 nm and to the thickness of the crystalline lamellae within the starch granule. This clearly indicated that the nanocrystals correspond to the crystalline lamellae present in native starch granules. As the nanocrystals were described by parallelepipedal blocks with a length of 20–40 nm and a width of 15–30 nm (Putaux et al., 2003), this would indicate that between 150 and 300 double-helical components make up these crystalline domains. Further analysis indicated that roughly half of the dextrins in the nanocrystals were branched molecules, which was far more than previous investigations suggested. It was also concluded that they were equally distributed between populations A and B of high and low molecular weights, respectively. Taking into account the length of these branches and the thickness of the platelets, it was likely that the majority of the branching points were found at the reducing-end surface of the nanocrystals, whereas the rest were located at the non-reducing side.

11.3.3 Thermal properties

The thermal properties of five types of starch (waxy maize, normal maize, high amylose maize, potato and wheat) and their corresponding starch nanocrystals were characterized by differential scanning calorimetry (DSC) and thermogravimetric analysis (TGA) (LeCorre et al., 2012b) Native starches showed only one thermal transition, whereas nanocrystals showed two transitions. In excess water, the first peak was attributed to the first stage of crystallites melting (unpacking of the double helixes) and the second transition to the second stage of crystallites melting (unwinding of the helixes). B-type crystallinity starch nanocrystals gained more stability than A-type nanocrystals as they consist of more rigid crystallites. In the dry state, the peaks were attributed to crystallites melting, with a direct transition from packed helixes to unwound helixes. The presence of two peaks was attributed to the heterogeneity in crystallites quality. Limited influence of the amylose content of starch was observed. This study gave important information for the processing conditions of starch nanocrystal-based nanocomposites. It showed that starch nanocrystals can be used in wet processes, such as coating, if the temperature remains lower than 80–100°C, and in dry processes at temperatures below 150–200°C.

11.3.4 Surface chemical modification

As for cellulose nanoparticles, the chemical modification of starch nanocrystals involves the hydroxyl groups from the surface. Different grafting strategies have been investigated as shown in Table 11.2. The first report was conducted with alkenyl succinic anhydre (ASA) and phenyl isocyanate (Angellier et al., 2005d). Reaction was conducted in toluene/ (dimethylammo)pyridine medium to avoid hydrolysis of ASA. Modified nanoparticles were characterized by Fourier transform infrared (FTIR) and X-ray photoelectron (XPS) spectroscopies, contact angle measurements, TEM and X-ray diffraction analysis. The lower polarity of the modified nanocrystals was also demonstrated by a simple experiment. The pristine and the modified nanoparticles were mixed with two immiscible solvents with different polarities and densities and it was visually observed with which solvent they are best wetted. Distilled water and methylene chloride were chosen for the test and it was observed that unmodified starch nanocrystals remained in the water medium whereas modified nanoparticles migrated towards the methylene chloride phase.

A crystalline brush-like structure from the starch surface outward was observed from the stearate modification performed by the reaction of starch nanocrystals with stearic acid chloride in methyl ethyl ketone (MEK) (Thielemans et al., 2006). It was evidenced from X-ray diffraction experiments (Figure 11.8). The hydrolyzed unmodified waxy maize starch nanocrystals (Figure 11.8(a)) showed the expected scattering pattern for the A allomorph. Poly(ethylene glycol) methyl ether (PEGME) modification

Source of Starch	Reagent	Objective of the Modification	Reference
Corn	Polystyrene	Amphiphilic	(Song et al., 2008)
Pea	Microwave-assisted ROP of PCL	Blending with PLA	(Yu et al., 2008)
		Blending with PCL	(Chang et al., 2009)
Potato	$Sn(Oct)_2$-catalyzed ROP of PCL	Medical Applications	(Namazi and Dadkhah, 2008)
Waxy Maize	Alkenyl Succinic Anhydre	Dispersion in Dichloromethane	(Angellier et al., 2005d)
	Phenyl Isocyanate		
	Stearic Acid Chloride	Hydrophobization	(Thielemans et al., 2006)
	Poly(ethylene glycol) Methyl Ether		
	Poly(tetrahydrofuran)	Compatibilization with Polymer Matrices	(Labet et al., 2007)
	Poly(ethylene glycol) monobutyl Ether		
	Polycaprolactone		(Labet et al., 2007; Habibi and Dufresne, 2008)

Table 11.2: Surface chemical modification of starch nanocrystals.

did not have a pronounced effect on the diffraction pattern (Figure 11.8(b)). Hydrogen bonding between PEGME ether groups and unreacted starch hydroxyl groups were supposed to provide ample interactions to bend the surface-grafted chains onto the surface of the nanoparticles. On the contrary, significant crystallization of the stearate surface modification was evidenced from its diffraction pattern (Figure 11.8(c)). The stearate diffractions signals were clearly superimposed over the starch pattern. This was confirmed by DSC analysis, where a distinct melting endotherm appeared between 35 and 110°C.

Polycaprolactone-grafted starch nanocrystals were also obtained using "grafting onto" (Labet et al., 2007; Habibi and Dufresne, 2008) and "grafting from" (Yu et al., 2008; Namazi and Dadkhah, 2008; Chang et al., 2009) approaches. Polystyrene was also grafted on starch nanocrystals using a "grafting from" strategy (Song et al., 2008). It was systematically verified that the crystalline structure of the nanoparticles was not changed after grafting and that it only occurred on the surface. The surface coating of the nanoparticles allowed dispersion in organic solvents and compatibilization with apolar polymeric matrices. Amphiphilic starch nanocrystals prepared by the graft copolymerization of starch nanocrystals with styrene were well dispersed both in polar and nonpolar solvents (Song et al., 2008). Moreover, microscopic observations of modified starch nanocrystals showed the individualization of nanopar-

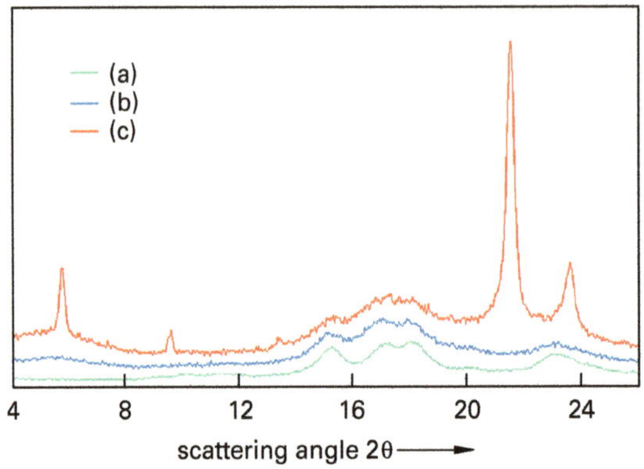

Fig. 11.8: X-ray diffraction patterns of (a) unmodified, (b) poly(ethylene glycol) methyl ether-, and (c) stearate-modified waxy maize starch nanocrystals (Thielemans et al., 2006).

ticles. The grafting efficiency of PCL chains onto the surface of starch nanocrystals decreased with the length of the polymeric chains, as expected (Labet et al., 2007).

11.4 Starch nanocrystal reinforced polymer nanocomposites

Nanocomposite materials have been prepared from starch nanocrystals as a filler using different polymer matrices as shown in Table 11.3.

The first investigation was performed using a copolymer of styrene and butyl acrylate (poly(S-co-BuA)) matrix in latex form (Dufresne et al., 1996). This aqueous dispersion was mixed with the aqueous suspension of starch nanocrystals and the mixture was freeze-dried and hot-pressed. The following works generally involved simpler casting/evaporation methods. However, care must be taken regarding the processing temperature to avoid gelatinization of starch nanocrystals. For instance, for the processing of thermoplastic starch reinforced with starch nanocrystals, the temperature of gelatinized starch was decreased to 40°C before adding starch nanocrystals (Angellier et al., 2006; Viguié et al., 2007).

11.4.1 Mechanical properties

The potential use of starch nanocrystals as a mechanically reinforcing phase in a polymeric matrix has been evaluated in both the linear (DMA) and non-linear range (tensile tests). In the pioneering work on potato starch nanocrystal reinforced poly(S-co-BuA), a high reinforcing effect of the filler was observed, especially above the glass transition temperature (Tg) of the matrix (Dufresne et al., 1996; Dufresne and Cavaillé, 1998). This reinforcing effect was later confirmed by most authors for dif-

Polymer	Source of Starch	Processing Technique	Reference
NR	Waxy Maize	Casting/Evaporation	(Angellier et al., 2005b; 2005c; Le Corre et al., 2012d)
	Amylomaize		(Le Corre et al., 2012d)
	Normal Maize		
	Potato		
	Wheat		
PCL	Waxy Maize	Casting/Evaporation	(Habibi and Dufresne, 2008)
PHO	Pea	Casting/Evaporation	(Dubief et al., 1999)
PLA	Pea	Casting/Evaporation	(Yu et al., 2008)
Poly(S-co-BuA)	Potato	Freeze-drying/Hot-pressing	(Dufresne et al., 1996; Dufresne and Cavaillé, 1998)
	Waxy Maize		(Angellier et al., 2005a; Garcia et al., 2009; 2011)
Pullulan/sorbito	Waxy Maize	Casting/Evaporation	(Kristo and Biliaderis, 2007)
PVA/glycerol	Pea	Casting/Evaporation	(Chen et al., 2008b)
Starch/glycerol	Waxy Maize	Casting/Evaporation	(Angellier et al., 2006)
Starch/sorbitol	Waxy Maize	Casting/Evaporation	(Viguié et al., 2007)
WPU	Potato	Casting/Evaporation	(Chen et al., 2008a)
	Pea		(Chang et al., 2009)
SPI/glycerol	Pea	Freeze-drying/Hot-pressing	(Zheng et al., 2009)

Table 11.3: Polymer nanocomposites obtained from starch nanocrystals and polymeric matrix.

ferent polymeric systems. In non-linear testing conditions, the introduction of starch nanocrystals induced an increase of both the tensile modulus and strength whereas the strain at break decreased. However, a decrease of the reinforcing capability of starch nanocrystals has been reported in some studies for higher filler contents because of self-aggregation within the polymeric matrix (Chen et al., 2008b; Zheng et al., 2009).

The effect of moisture content was also investigated for natural rubber (NR) based materials and the stiffness of the material was found to decrease when increasing the water content (Angellier et al., 2005c). For some systems, an increase of Tg of the matrix when increasing the starch nanocrystal content was observed, attributed to the existence of an interphase of immobilized matrix material in contact with particle surface (Angellier et al., 2006; Viguié et al., 2007; Kristo and Biliaderis, 2007). The formation of this interphase resulted from favorable filler/matrix interactions. Interestingly, a considerable slowing down of the recrystallization (retrogradation) of the thermoplastic starch matrix upon storage in humid atmosphere was observed when adding starch nanocrystals (Angellier et al., 2006; Viguié et al., 2007). For this system, the reinforcing effect was more significant than in NR because of strong interactions between the filler and amylopectin chains from the matrix and possible crystallization at the filler/matrix interface.

Waterborne polyurethane (WPU) was reinforced with starch and cellulose nanocrystals obtained by acid hydrolysis of waxy maize starch granules and cotton linter pulp, respectively (Wang et al., 2010). A synergistic effect was observed when adding 1 wt% starch and 0.4 wt% cellulose nanocrystals with a significant improvement in tensile strength, Young's modulus and tensile energy at break, without significant loss for the elongation at break. The mechanical performance was found to be higher than for individual filler, but it is worth noting that the total filler content was different. In the ternary system, the formation of much jammed network consisting of nanoparticles with different geometrical characteristics was suggested to play an important role in the enhancement of the cross-linked network. Moreover, strong hydrogen bonding interactions between the nanoparticles and between the nanoparticles and the hard segments of WPU matrix was suspected to improve the mechanical properties.

Lowering of the reinforcing effect was observed when using chemically modified starch nanocrystals in a NR (Angellier et al., 2005c), polylactic acid (PLA) (Yu et al., 2008) or waterborne polyurethane (WPU) (Chang et al., 2009) matrix compared to unmodified nanoparticles. This phenomenon was ascribed to improved filler/matrix interactions that develop at the expense of filler/filler interactions and then formation of a percolating starch network. However, enhancement of the elongation at break was reported for PCL-grafted starch nanocrystals dispersed in a PLA matrix (Yu et al., 2008).

The reinforcing effect of starch nanocrystals is generally ascribed to the formation of a hydrogen bonded percolating filler network above a certain starch content corresponding to the percolation threshold. However, contrarily to cellulose nanocrystals, this assumption is difficult to prove because the connecting particles are starch clusters of aggregates with ill-defined size and geometry. Attempts to model the mechanical behavior of starch nanocrystal reinforced poly(S-co-BuA) (Dufresne and Cavaillé, 1998) and NR (Angellier et al., 2005c) were conducted using simple models, such as the generalized Kerner of Guth equations modified for non-spherical particles. The main drawback of these phenomenological approaches lies in the use of an adjustable parameter, which has a value that was found to vary at the percolation threshold of starch aggregates.

A phenomenological modeling approach was developed to try to understand the reinforcing mechanism of starch nanocrystals in a non-vulcanized NR matrix (Mélé et al., 2011). Non-linear dynamic mechanical experiments highlighted the significant reinforcing effect of starch nanocrystals and the occurrence of the Mullins and Payne effects. Two models were used to predict the Payne effect considering that either filler-filler (Kraus model) or matrix-filler (Maier and Göritz model) interactions are preponderant. The use of the Maier and Göritz model demonstrated that phenomena of adsorption and desorption of NR chains on the filler surface governed non-linear viscoelastic properties, even if the formation of a percolating network for filler contents higher than 6.7 vol% (i.e. around 10 wt%) was evidenced by the Kraus model.

11.4.2 Swelling properties

By adding starch nanocrystals to poly(S-co-Bu-A) (Dufresne and Cavaillé, 1998) or NR (Angellier et al., 2005b; LeCorre et al., 2012d), the swelling by water increased because of the hydrophilic nature of starch. The coefficient diffusion of water increased as well but showed two well-defined regions. Below a critical starch nanocrystal concentration, the evolution of the diffusion coefficient was relatively low, whereas it was more significant above. It was assumed to be due to the formation of a starch nanocrystal network through hydrogen linkages between nanoparticle clusters and also to favorable interactions between the NR matrix and the filler. This critical concentration was around 20 wt% and 10 wt%, for poly(S-co-BuA) and NR, respectively. A slight increase of the water uptake of soy protein films was reported for increasing starch nanocrystal contents (Zheng et al., 2009). For sorbitol-plasticized pullulan (a hydrophilic system) a decrease of the water uptake was observed when adding starch nanocrystals particularly at high filler loading level (Kristo and Biliaderis, 2007). Again, it was ascribed to the formation of a three-dimensional network of nanoparticles. Conversely, for glycerol-plasticized thermoplastic starch the composites reinforced with starch nanocrystals were found to absorb more water than the unfilled matrix (Garcia et al., 2009). It was ascribed to a relocalization of glycerol around the nanoparticles leading to more hydroxyl groups in the matrix able to interact with water molecules.

Swelling by toluene of the NR matrix decreased when adding starch nanocrystals and the toluene diffusion coefficient decreased strongly for low filler contents and more progressively above 10 wt% (Angellier et al., 2005b). For low nanocrystal contents, a correlation between the toluene sorption behavior and calculated specific surface area of nanocrystals obtained from different botanical origin starches was observed (LeCorre et al., 2012d). The higher the theoretical specific surface area, the higher the toluene uptake and the lower the diffusivity were. For higher starch nanocrystal contents, no such observation was made. It was supposed to be due to aggregation phenomena at higher filler contents.

11.4.3 Barrier properties

Given their platelet-like morphology, starch nanocrystals were suspected, as nanoclays do, to create a tortuous diffusion pathway for penetrant molecules. However, few reports investigated the barrier properties of starch nanocrystal reinforced nanocomposites.

Continuous and significant reduction of the permeability to water vapor and oxygen was reported for NR films when adding starch nanocrystals up to 30 wt% (Angellier et al., 2005b). A substantial 40% decrease of the water vapor permeability (WVP) of cassava starch films plasticized with glycerol was also observed when adding only 2.5 wt% of waxy maize starch nanocrystals (Garcia et al., 2009). However, when using a glycerol-plasticized waxy maize starch matrix, a close association between starch nanocrystals and glycerol-rich domains was supposed to explain the unexpected increase of the WVP value when adding starch nanocrystals (Garcia et al., 2011). For sorbitol-plasticized pullulan films, no significant differences were observed in WVP when adding up to 20 wt% starch nanocrystals (Kristo and Biliaderis, 2007). Nevertheless, above this critical value, a significant decrease of WVP was reported. A detrimental effect of starch nanocrystals on the WVP of NR films was reported (LeCorre et al., 2012d). However, in this study WVP was measured under tropical condition (38°C, 90%RH), and it was suggested that the hydrophilic nature of starch nanocrystals was predominant

11.5 Chitin

Chitin is one of the main components in the cell walls of fungi, the exoskeleton of shellfish, insects and other arthropods, and in some other animals. It is believed to be the second most important natural polymer in the world and was first identified in 1884. Zooplankton cuticles (in particular small shrimps constituting krill) are the most important source of chitin. However, fishing of these tiny organisms (a few millimeters in length) is too difficult to consider for any industrial use. Despite the widespread occurrence of chitin, shellfish canning industry waste (shrimp or crab shells) in which the chitin content ranges between 8 and 33% constitutes the main source of this biopolymer. In industrial processing, chitin is extracted from crustaceans by acid treatment to dissolve calcium carbonate followed by alkaline extraction to solubilize proteins. In addition, a decolorization step is often carried out to remove leftover pigments and obtain a colorless product. These treatments must be adapted to each chitin source, owing to differences in the ultrastructure of the initial materials. The resulting extracted material needs to be graded in terms of purity and color since residual protein and pigment can cause problems for further utilization, especially for biomedical applications.

11.5.1 Chemical structure

Chitin is a polysaccharide, made out of units of acetylglucosamine (more completely, N-acetyl-D-glucose-2-amine) (Figure 11.9). These are linked together in β-1,4 fashion, the same as the glucose units that make up cellulose. So chitin may be thought of as cellulose, with one hydroxyl group on each monomer replaced by an acetylamino group. This allows for increased hydrogen bonding between adjacent polymer chains, giving the material increased strength.

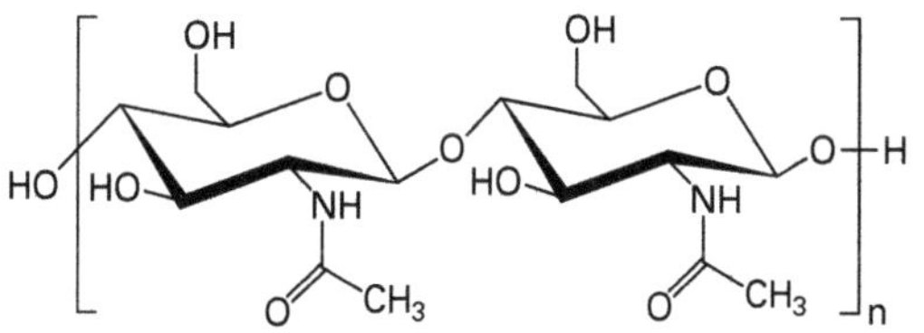

Fig. 11.9: Chemical structure of chitin.

Chitosan is a chitin derivative commercially produced by deacetylation of chitin using sodium hydroxide in excess as a reagent and water as a solvent. It is a linear polysaccharide composed of randomly distributed β-(1,4)-linked D-glucosamine (deacetylated unit) and N-acetyl-D-glucosamine (acetylated unit).

11.5.2 Polymorphism and structure

Native chitin is highly crystalline and depending on its origin it occurs in three forms identified as α-, β- and γ-chitin, which can be differentiated by infrared and solid-state nuclear magnetic resonance (NMR) spectroscopy together with X-ray diffraction. From a detailed analysis, it seems that the latter is just a variant of the α form (Atkins, 1985). In both α and β forms, the chitin chains are organized in sheets where they are tightly held by a number of intra-sheet hydrogen bonds. In α-chitin, all chains are arranged in an antiparallel fashion whereas the β form consists of a parallel arrangement. α-chitin is the most abundant and stable form since it constitutes arthropod cuticles and mushroom cellular walls. It occurs in fungal and yeast cell walls, krill, lobster and crab tendons and shells, shrimp shells, and insect cuticles. In addition to the native chitin, α form systematically results from recrystallization from solution (Persson et al., 1992; Helbert and Sugiyama, 1998), in vitro biosynthesis (Bartnicki-Garcia et al., 1994), or enzymatic polymerization (Sakamoto et al., 2000). The rarer β-chitin is found in association with proteins in squid pens (Rudall and Kenchington, 1973), tubes synthesized by pogonophoran and vestimetiferan worms (Blackwell et al., 1965; Gaill et al., 1992), aphrodite chaetae (Lotmar and Picken, 1950) and lorica built by some seaweeds or protozoa (Herth et al., 1997).

Chitin has been known to form microfibrillar arrangements embedded in a protein matrix, and these microfibrils have diameters ranging from 2.5 to 2.8 nm (Revol and Marchessault, 1993). Crustacean cuticles possess chitin microfibrils with diameters as large as 25 nm (Brine and Austin, 1975). Although it has never been specifically measured, the stiffness of chitin nanocrystals is at least 150 GPa, based on the observation that cellulose is about 130 GPa and the extra bonding in the chitin crystallite is going to stiffen it further (Vincent and Wegst, 2004).

11.6 Chitin nanocrystals

11.6.1 Acid hydrolysis

The structure of chitin is very similar to cellulose. They are both structural materials for living bodies and occur as highly crystalline nanoscale fibrils. Chitin microfibrils are embedded in a protein matrix. The method for preparing chitin nanocrystals is similar to the one used for cellulose nanocrystals and involves hydrolysis in strong acid aqueous medium. As for cellulose, before preparing chitin nanocrystals, chitin has to be purified from the living organism, in which it does not occur in pure form. Because the chitin content is generally low, intensive purification steps need to be performed. The first investigation on the preparation of chitin nanocrystals was reported in 1959 (Marchessault et al., 1959). In this study, purified chitin was first treated with 2.5 N hydrochloric acid solutions under reflux for 1 h, the excess acid was decanted, and then distilled water was added to obtain the suspension. During acid hydrolysis, disordered and low lateral ordered regions of chitin are preferentially hydrolyzed and dissolved in the acidic solution, whereas water-insoluble, highly crystalline residues that have higher resistance to acid attack remain intact. Acid-hydrolyzed chitin spontaneously dispersed into rod-like particles that could be concentrated to a liquid crystalline phase above a given concentration (Revol and Marchessault, 1993). On the basis of this procedure, chitin nanocrystals have been prepared from chitin of different origins as shown in Table 11.4.

11.6.2 Other treatments

Chitin nanocrystal suspensions were also prepared by 2,2,6,6-tetramethylpiperidine-1-oxyl radical (TEMPO)-mediated oxidation of α-chitin in water at pH 10 (Fan et al., 2008a). The formation of C6 carboxylate groups in chitin was monitored by controlling the amount of NaClO added in the TEMPO-mediated oxidation of chitin. When 5.0 mmol of NaClO per gram of chitin was used, the water-insoluble fraction in the

Source	Acid	Time (min)	Temperature (°C)	Acid-to-Chitin Ratio (mL · g^{-1})	Reference
Crab Shell	HCl 3 N	3 · 90	104	30	(Gopalan Nair and Dufresne 2003a; 2003b; Gopalan Nair et al., 2003; Lu et al., 2004; Feng et al., 2009; Tzoumaki et al., 2010)
		90	104	10	(Nge et al., 2003)
		360	120	30	(Hariraksapitak and Supaphol, 2010)
Riftia Tubes	HCl 3 N	3 · 90	104	–	(Morin and Dufresne, 2002)
Shrimp Shell	HCl 3 N	3 · 90	104	30	(Sriupayo et al., 2005a; 2005b)
			–	–	(Goodrich and Winter, 2007)
		180	105	100	(Phongying et al., 2007)
		3 · 360	104	30	(Wongpanit et al., 2007)
		360	120	30	(Junkasem et al., 2010)
			104	30	(Watthanaphanit et al., 2008)
		3 · 360	104	30	(Watthanaphanit et al., 2010)
Squid Pen	HCl 3 N	3 · 90	104	30	(Paillet and Dufresne, 2001)

Table 11.4: Hydrolysis conditions for the preparation of chitin nanocrystals from different sources.

TEMPO-oxidized chitin was maintained as high as 90 wt%, and the carboxylate content reached 0.48 mmol · g^{-1}. No N-deacetylation of the TEMPO-oxidized chitin was observed, irrespective of the amount of NaClO added in the oxidation, and the crystal structure was maintained showing that the C6 carboxylate groups formed only on nanoparticle surface. After ultrasonic treatment in water, mostly individualized chitin nanocrystals 340 nm long and 8 nm in diameter were obtained.

A procedure for preparing individualized chitin nanofibers 3–4 nm in cross-sectional width and few microns in length, that seem like microfibrillated cellulose, was also reported (Fan et al., 2008b). It consisted in a simple mechanical treatment in water at pH 3–4 without any chemical modification. Protonation or cationization of the C2 amino groups present on the crystallite surface of chitin under acid conditions was likely to be one of the most significant and necessary conditions for the nanofibers conversion. The original crystal structure of squid pen β-chitin used in this study was maintained, but the crystallinity index decreased from 0.51 to 0.37 as a result of the nanofiber conversion. However, this method of individualization of fibrils was only applicable to squid pen β-chitin, probably because of its low crystallinity and some structural defects compared to α-chitin. Partial deacetylation was applied to α-chitin to selectively increase the C2-primary amino groups on the crystalline fibril surface (Fan et al., 2010). By raising the cationic charge density on the crystalline fibril surface, individualization was achieved by enhanced electrostatic repulsion between cationically-charged fibrils through disintegration in water under acidic conditions. Mostly individualized α-chitin nanoparticles were obtained in yields of 85–90% by partial deacetylation with 33% NaOH at 90°C for 2–4 h and subsequent disintegration in water at pH 3–4. The obtained α-chitin nanoparticles had average width and length of 6.2±1.1 and 250±140 nm, respectively. Individual nanofibrils of more than 500 nm in length were also observed.

Chitin nanofibers were prepared from crab shell by a grinding treatment in a never-dried state (Ifuku et al., 2009). Nanofibers with a uniform width around 10–20 nm and high aspect ratio were obtained.

Chitosan nanocrystals have been prepared by deacetylation of chitin nanocrystals obtained by acid hydrolysis of chitin flakes from shrimp shells (Watthanaphanit et al., 2010). The average length and width of the nanocrystals were 309 and 64 nm, respectively, with an average aspect ratio around 4.8. Incorporation of these chitosan nanocrystals within alginate yarns imparted antibacterial activity against Gram-positive *Staphylococcus aureus* and Gram-negative *Escherichia coli*, which rendered the nanocomposite yarns as effectual dressing materials.

11.6.3 Morphology

Like cellulose nanocrystals, chitin nanocrystals also occur as rod-like nanoparticles. Figure 11.10 shows transmission electron micrographs (TEM) obtained from dilute suspensions of chitin fragments from different origins. The typical geometrical characteristics for crystallites derived from different species are reported in Table 11.5. Dimensions of chitin nanocrystals extracted from squid pen (Paillet and Dufresne, 2001), crab shell (Gopalan Nair and Dufresne, 2003a), and shrimp shell (Sriupayo et al., 2005b) were found to be close to those reported for cotton nanocrystals. For *Riftia* tubes, a quite exotic chitin source, the average length of nanocrystals was around 2.2 μm and the aspect ratio was 120 (Morin and Dufresne, 2002). *Riftia* tubes are secreted by a vestimetiferan worm called *Riftia* and were collected at a depth of 2500 m on the East-Pacific ridge. Dye absorption with Congo red was used to measure the specific surface area of α-chitin nanocrystals prepared from shrimp shells, indicating values near 350 $m^2 \cdot g^{-1}$ (Goodrich and Winter, 2007). This value was supported with calculations derived from X-ray crystallite size measurements.

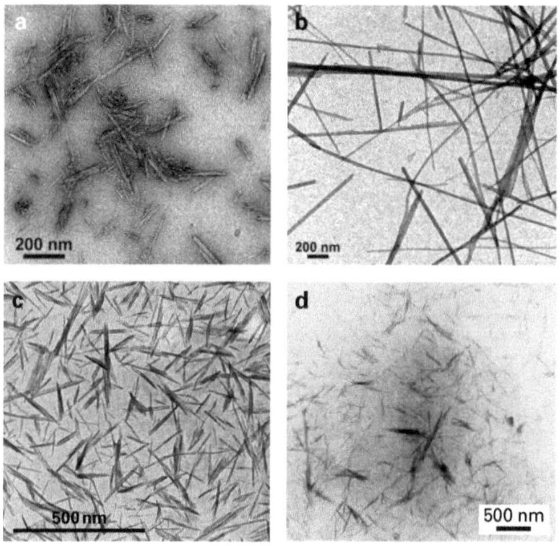

Fig. 11.10: Transmission electron micrographs from a dilute suspension of (a) squid pen (Paillet and Dufresne, 2001), (b) *Riftia* tubes (Morin and Dufresne, 2002), (c) crab shell (Gopalan Nair and Dufresne, 2003a), and (d) shrimp shell (Sriupayo et al., 2005b) nanocrystals.

Source	L (nm)	D (nm)	L/D	Reference
Crab Shell	50–300	6–8	–	(Marchessault et al., 1959)
	100–600	4–40	16	(Gopalan Nair and Dufresne, 2003a)
	80–350	8–12	10–30	(Nge et al., 2003)
	500±50	50±10	10±5	(Lu et al., 2004)
	250±140	6.2±1.1	–	(Fan et al., 2010)
Riftia Tubes	500–10,000	18	120	(Morin and Dufresne, 2002)
Shrimp Shell	150–800	5–70	17	(Sriupayo et al., 2005a; 2005b)
Squid Pen	50–300	10	15	(Paillet and Dufresne, 2001)

Table 11.5: Geometrical characteristics of chitin nanocrystals from various sources: length (L), cross section (D) and aspect ratio (L/d).

11.6.4 Surface chemical modification

As for cellulose and starch nanoparticles, the chemical modification of chitin nanocrystals involves the hydroxyl groups from the surface. Different grafting strategies have been investigated as shown in Table 11.6.

Source of Chitin	Reagent	Objective of the Modification	Reference
Crab Shell	Alkenyl Succinic Anhydre	Blending with NR	(Gopalan Nair et al., 2003)
	Phenyl Isocyanate		
	Isopropenyl-α, α'-dimethylbenzyl isocyanate		
	Acetic Anhydride	Compatibilization with Acrylic Resin	(Ifuku et al., 2010)
	$Sn(Oct)_2$-catalyzed ROP of PCL	Thermoformable Composites	(Feng et al., 2009)

Table 11.6. Surface chemical modification of chitin nanoparticles.

Acylation of chitin nanocrystals was achieved in a dioxane system with 4-(dimethyl-amino) pyridine as the catalyst (Gopalan Nair et al., 2003). The reaction was carried out for 1 week at 70°C. Fourier transform infrared (FTIR) spectroscopy, TEM and contact angle measurements were performed to prove the occurrence of the surface modification without any major morphological changes of the nanoparticles associated with the treatment applied. Stable toluene suspensions of the modified nanoparticles showing colloidal behavior when observed between crossed polars were obtained. Isocyanates, such as phenyl isocyanate (PI) and 3-isopropenyl-α-α'-dimethylbenzyl isocyanate (TMI), were also used to chemically modify the surface of chitin nanocrystals. Figure 11.11 shows a TEM micrograph of crab shell chitin nanocrystals before and after modification. After surface chemical modification with ASA and PI, the appearance of the chitin fragments changed. They seem to be entangled, and individual nanocrystals were difficult to observe. This binding agent was most probably the chemical coupling agent used for chemical modification of the chitin fragments. However, it seemed that these TEM micrographs reveal the absence of major morphological changes associated with the various treatments applied.

The surface of chitin nanocrystals was also functionalized by grafting PCL chains involving the "grafting from" approach (Feng et al., 2009). The ensuing grafted nanoparticles were directly shaped by thermoforming and injection-molding of this co-continuous material. When increasing the PCL content in grafted nanocrystals, the strength and elongation at break, as well as hydrophobicity increased. Chitin nanofibers were also acetylated to modify the fiber surface and were characterized in detail (Ifuku et al., 2010). The acetyl DS was found to be controlled from 0.99 to 2.96 by changing the reaction time. It was shown that the acetylation of chitin nanofibers changed their crystal structure, fiber thickness, and thermal degradation temperature.

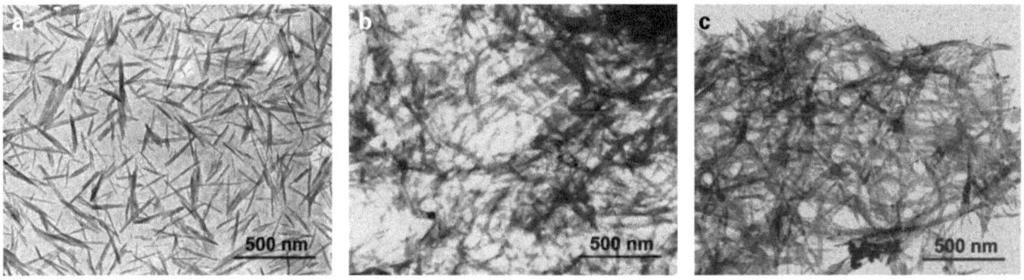

Fig. 11.11: Transmission electron micrographs from a dilute suspension of (a) unmodified, (b) ASA-modified, and (c) PI-modified crab shell chitin nanocrystals (Gopalan Nair et al., 2003).

11.7 Chitin nanocrystal reinforced polymer nanocomposites

Like other polysaccharide nanoparticles, chitin nanocrystals are usually incorporated into a polymer matrix to prepare polymer nanocomposite materials with expected improved properties. Because of the stability of aqueous dispersions of chitin nanocrystals, water is the preferred processing medium. Nanocomposite materials have been prepared from chitin nanocrystals as a filler using different polymer matrices as shown in Table 11.7.

Therefore, the pioneering works were conducted with poly(S-co-BuA) in the form of latex using chitin nanocrystals extracted from squid pen (Paillet and Dufresne, 2001) and *Riftia* tubes (Morin and Dufresne, 2002). The preparation of a latex of PCL using a copolymer of poly(ethylene oxide) and poly(propylene oxide) as the surfactant to process *Riftia* tubes chitin nanocrystal reinforced nanocomposites was also reported (Morin and Dufresne, 2002). In this study, two different techniques were used to prepare composites films: (i) casting and water evaporation, and (ii) freeze-drying followed by hot-pressing.

The use of water soluble polymers, such as soy protein isolate (SPI), the major component of soy bean (Lu et al., 2004), PVA (Sriupayo et al., 2005a; Junkasem et al., 2006; Junkasem et al., 2010), alginate (Watthanaphanit et al., 2008; Watthanaphanit et al., 2010), and hyaluronan (Hariraksapitak and Supaphol, 2010) was also reported. Reinforced chitosan films were also prepared by first dissolving chitosan in an aqueous solution of acid acetic and adding the chitin nanocrystal suspension (Sriupayo et al., 2005b).

11.7.1 Mechanical properties

Improvement in mechanical properties (stiffness) upon adding chitin nanocrystals was observed for almost all polymer matrices. For poly(S-co-BuA) films reinforced with squid pen chitin nanocrystals, the reinforcing effect was limited in the glassy

Polymer	Source of Chitin	Processing Technique	Reference
Acrylic Resin	Crab Shell	Impregnation	(Ifuku et al., 2010)
Alginate	Shrimp Shell	Wet Spinning	(Watthanaphanit et al., 2008; 2010)
Chitosan	Shrimp Shell	Casting/Evaporation	(Sriupayo et al., 2005b)
Hyaluronan	Crab Shell	Freeze-drying	(Hariraksapitak and Supaphol, 2010)
NR	Crab Shell	Casting/Evaporation	(Gopalan Nair and Dufresne, 2003a; 2003b; Gopalan Nair et al., 2003)
PCL	*Riftia* Tubes	Casting/Evaporation and Freeze-Drying/Hot-pressing	(Morin and Dufresne, 2002)
	Crab Shell	Melt Compounding	(Wu et al., 2007)
Poly(S-co-BuA)	Squid Pen	Casting/Evaporation	(Paillet and Dufresne, 2001)
PVA	Shrimp Shell	Casting/Evaporation	(Sriupayo et al., 2005a)
		Electrospinning and Casting/Evaporation	(Junkasem et al., 2006; 2010)
Silk Fibroin	Shrimp Shell	Freeze-drying	(Wongpanit et al., 2007)
SPI/glycerol	Crab Shell	Freeze-drying/Hot-pressing	(Lu et al., 2004)
Starch/glycerol	Crab Shell	Casting/Evaporation	(Chang et al., 2010)
WPU	Crab Shell	Casting/Evaporation	(Huang et al., 2011)

Table 11.7:. Polymer nanocomposites obtained from chitin nanocrystals and polymeric matrix.

state of the matrix but much more significant above Tg when adding at least 10 wt% nanocrystals (Paillet and Dufresne, 2001). Interactions between the nanoparticles to form a percolating network through the sample were again suggested to explain this behavior. Therefore, the limited reinforcing effect of squid pen chitin nanocrystals was ascribed to the low aspect ratio of the nanoparticles (15) and when *Riftia* tubes chitin nanocrystals with an aspect ratio of 120 were used to reinforce the same poly(S-co-BuA) matrix, a significant modulus increase was observed from 1 wt% loading (Morin and Dufresne, 2002).

However, the stiffness of this continuous network estimated from the tensile modulus of a film obtained by water evaporation of a dispersion of chitin nanocrystals was found to be lower than for cellulose nanocrystals. Values around 0.5 and 2 GPa were observed for squid pen (Paillet and Dufresne, 2001) and *Riftia* tubes (Morin and Dufresne, 2002), respectively. For the latter, both tensile test and DMA were used and the film was allowed to dry at 110°C for 20 min before measurements. Indeed, water was expected to reduce the interactions between the nanocrystals and therefore the stiffness of the film. In addition, it should be more representative of the behavior of the polysaccharide nanoparticles within a hydrophobic matrix.

The existence of such a three-dimensional percolating nanoparticle network was evidenced by performing successive tensile tests on crab shells chitin nanocrystal reinforced NR (Gopalan Nair and Dufresne, 2003b). The experiment consisted in stretching the material up to a given elongation and then releasing the force, and stretching the material again up to a higher elongation, and so on. From these experimental data, the true stress and the true strain were calculated. For each cycle, the tensile modulus was determined and the relative tensile modulus, i.e. the ratio of the modulus measured during a given cycle to the one measured during the first cycle, was plotted as a function of both the number of cycles and the nanocrystal content. For the unfilled natural rubber matrix, an increase of the relative modulus during successive tensile tests was observed. It was ascribed to the strain-induced crystallization of the matrix. For composites, the behavior was reported to be totally different. During the successive tensile tests, the relative tensile modulus first decreased and then slowly increased. The initial decrease of the modulus for composite materials was ascribed to the progressive damaging of the nanocrystal network. This was a good indication that the tensile behavior of the composites was mainly governed by the percolating nanoparticle network. After the complete destruction of the network, the modulus started to slowly increase as a result of the strain-induced crystallization of the matrix already observed for the unfilled sample.

Slow processes such as casting/evaporation were reported to give the highest mechanical performance materials compared to freeze-drying/hot-pressing techniques. This effect was observed for *Riftia* tubes chitin nanocrystals reinforced PCL (Morin and Dufresne, 2002), and crab shells chitin nanocrystals reinforced NR (Gopalan Nair and Dufresne, 2003b). It was ascribed to the probable orientation of these rod-like nanoparticles during film processing due to shear stresses induced by the hot-pressing step. Loss of mechanical properties was also reported for NR based nanocomposites reinforced with surface chemically modified chitin nanocrystals compared to unmodified (Gopalan Nair et al., 2003). This poorer reinforcing effect was ascribed to the limited hydrogen bonding interactions between nanocrystals after surface modification which reduce the driving force for network formation.

Optimum loading levels depending on the nature of the matrix and processing technique were also observed with chitin nanocrystals tending to aggregate for higher nanoparticle contents. For chitosan (Sriupayo et al., 2005a) and PVA (Sriupayo et al., 2005b) films reinforced with shrimp shell chitin nanocrystals, the tensile strength was found to increase with initial increase of filler loading to reach a maximum value around 3 wt% and decreased gradually for increased nanocrystal contents, whereas the elongation at break initially decreased and then leveled off for higher filler contents. For glycerol-plasticized potato starch (Chang et al., 2010), WPU (Huang et al., 2011), and hyaluronan-gelatin (Hariraksapitak and Supaphol, 2010) based nanocomposites, the maximum tensile strength was reported for 5, 3 and 2 wt% chitin nanocrystals, respectively.

Moreover, a decrease of the degree of crystallinity of PCL was reported when adding *Riftia* tubes chitin nanocrystals (Morin and Dufresne, 2002). It was suggested that during crystallization, the rod-like nanoparticles are most probably first ejected and then occluded in intercrystalline domains, hindering the crystallization of the polymer.

11.7.2 Swelling resistance

A higher water resistance of SPI films when reinforced with crab shells chitin nanocrystals was reported (Lu et al., 2004). Similar results were obtained for shrimp chitin nanocrystal reinforced chitosan (Sriupayo et al., 2005a) and PVA (Sriupayo et al., 2005b). In addition, it was found that heat-treatment decreased the affinity to water of shrimp chitin nanocrystal-based nanocomposite films.

The toluene uptake of crab shell chitin nanocrystal reinforced vulcanized and unvulcanized NR when immersed in toluene was investigated (Gopalan Nair and Dufresne, 2003a). For unvulcanized films, a comparison between samples prepared by freeze-drying/hot-pressing and casting/evaporation was reported. It was shown that cast/evaporated nanocomposites were more resistant to swelling than freeze-dried/hot-pressed ones because of the presence of a percolating chitin network in the former (Figure 11.12). The extent of filler-matrix interactions was also determined by calculating the relative weight loss (RWL) and fraction of bounded matrix (FBM) after 48 h immersion of samples in toluene and subsequent drying. For instance, the RWL and FBM values for a 20 wt% filled evaporated sample were 7.5% and 35.3% and those of the corresponding freeze-dried sample were 11.8% and 28.5%, respectively, giving quantitative indications of the effect of the three-dimensional chitin network in evaporated samples.

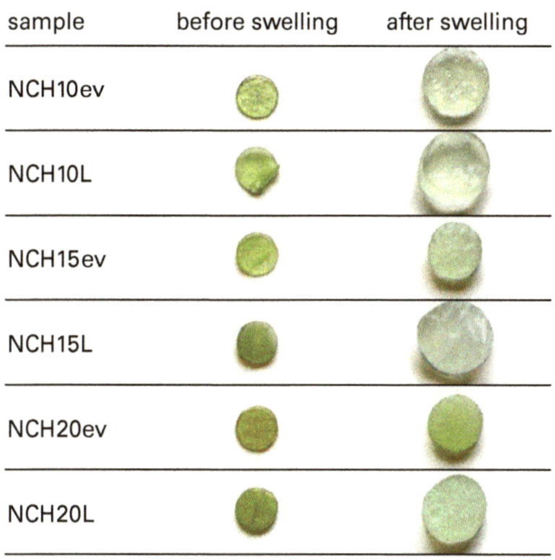

Fig. 11.12: Photographs of unvulcanized NR samples (prepared by casting/evaporation (NRev) and freeze-drying/hot-pressing (NRfdC) methods) reinforced with crab shell chitin nanocrystals before and just after swelling in toluene for 24 h at room temperature (25°C). The numbers correspond to the chitin nanocrystal contents and the diameter of all samples before swelling was d_0 = 7.5 mm. (Gopalan Nair and Dufresne, 2003a).

11.8 Conclusions

The controlled acid hydrolysis of starch granules and chitin fibers allows the release of highly crystalline nanoscale particles. The morphology of these nanocrystals is patterned after the microstructure of the crystalline domains in the substrate to be hydrolyzed. Elongated rod-like nanoparticles similar to cellulose nanocrystals are obtained from chitin. This is ascribed to the similar structural function of cellulose and chitin in living species. Similar applications can be envisaged for these two types of nanoparticles. Starch is a storage polymer and the constitutive nanocrystals occur as platelet-like nanoparticles. Compared to the two previous polysaccharides, this specific morphological feature limits their reinforcing capability when embedded within a polymeric matrix but may open the door to new applications. The acid hydrolysis procedure of starch is also slower because of the higher sensibility of starch towards temperature but can be optimized as recently shown in the literature.

11.9 References

Angellier, H., Choisnard, L., Molina-Boisseau, S., Ozil, P. and Dufresne, A. (2004). Optimization of the preparation of aqueous suspensions of waxy maize starch nanocrystals using a response surface methodology. Biomacromolecules *5*, 1545–1551.

Angellier, H., Putaux, J.L., Molina-Boisseau, S., Dupeyre, D. and Dufresne, A. (2005a). Starch nanocrystals fillers in an acrylic polymer matrix. Macromol. Symp. *221*, 95–104.

Angellier, H., Molina-Boisseau, S., Lebrun, L. and Dufresne, A. (2005b). Processing and structural properties of waxy maize starch nanocrystals reinforced natural rubber. Macromolecules *38*, 3783–3792.

Angellier, H., Molina-Boisseau, S. and Dufresne, A. (2005c). Mechanical properties of waxy maize starch nanocrystals reinforced natural rubber. Macromolecules *38*, 9161–9170.

Angellier, H., Molina-Boisseau, S., Belgacem, M.N. and Dufresne, A. (2005d). Surface chemical modification of waxy maize starch nanocrystals. Langmuir *21*, 2425–2433.

Angellier, H., Molina-Boisseau, S., Dole, P. and Dufresne, A. (2006). Thermoplastic starch-waxy maize starch nanocrystals nanocomposites. Biomacromolecules *7*, 531–539.

Angellier-Coussy, H., Putaux, J.L ., Molina-Boisseau, S., Dufresne, A., Bertoft, E. and Perez, S. (2009). The molecular structure of waxy maize starch nanocrystals. Carbohydr. Res. *344*, 1558–1566.

Atkins, E.D.T. (1985). Conformation in polysaccharides and complex carbohydrates. J. Biosci. *8*, 375–387.

Averous, L. and Halley, P.J. (2009). Biocomposites based on plasticized starch. Biofuels Bioprod. Biorefin. *3*, 329–343.

Bartnicki-Garcia, S., Persson, J. and Chanzy, H. (1994). An electron microscope and electron diffraction study of the effect of calcofluor and congo red on the biosynthesis of chitin in vitro. Arch. Biochem. Biophys. *310*, 6–15.

Battista, O.A. (1975). Microcrystal Polymer Science (McGraw-Hill Book Company, New York).

Blackwell, J., Parker, K.D. and Rudall, K.M. (1965). Chitin in pogonophore tubes. J. Mar. Biol. Assoc. UK *45*, 659–661.

Blanshard, J.M.V. (1987). Starch granule structure and function: A physicochemical approach. In: Starch: Properties and potentials, T. Galliard ed. (Society of Chemical Industry, London), vol. 13, pp. 16–54.

Bogracheva, T.Y., Morris, V.J., Ring, S.G. and Hedley, C.L. (1998). The granular structure of C-type pea starch and its role in gelatinization. Biopolymers *45*, 323–332.

Brine, C.J. and Austin, P.R. (1975). Renatured chitin fibrils, films and filaments. In: Marine chemistry in the coastal environment, T.D. Church ed. (ACS Symposium Series, American Chemical Society, Washington DC), pp. 505–518.

Buléon, A., Colonna, P., Planchot, V. and Ball, S. (1998). Starch granules: Structure and biosynthesis. Int. J. Biol. Macromol. *23*, 85–112.

Chang, P.R., Ai, F., Chen, Y., Dufresne, A. and Huang, J. (2009). Effects of starch nanocrystal-graft-polycaprolactone on mechanical properties of waterborne polyurethane-based nanocomposites. J. Appl. Polym. J. *111*, 619–627.

Chang, P.R., Jian, R., Yu, J. and Ma, X. (2010). Starch-based composites reinforced with novel chitin nanoparticles. Carbohydr. Polym. *80*, 420–425.

Chen, G., Wei, M., Chen, J., Huang, J., Dufresne, A. and Chang, P.R. (2008a). Simultaneous reinforcing and toughening: New nanocomposites of waterborne polyurethane filled with low loading level of starch nanocrystals. Polymer *49*, 1860–1870.

Chen, Y. Cao, X., Chang, P.R. and Huneault, M.A. (2008b). Comparative study on the films of poly(vinyl alcohol)/pea starch nanocrystals and poly(vinyl alcohol)/native pea starch. Carbohydr. Polym. *73*, 8–17.

Dubief, D., Samain, E. and Dufresne, A. (1999). Polysaccharide microcrystals reinforced amorphous poly(b-hydroxyoctanoate) nanocomposite materials. Macromolecules *32*, 5765–5771.

Dufresne, A., Cavaillé, J.Y. and Helbert, W. (1996). New nanocomposite materials: Microcrystalline starch reinforced thermoplastic. Macromolecules *29*, 7624–7626.

Dufresne, A. and Cavaillé, J.Y. (1998). Clustering and percolation effect in microcrystalline starch reinforced thermoplastic. J. Polym. Sci . B *36*, 2211–2224.

Fan, Y., Saito, T. and Isogai, A. (2008a). Chitin nanocrystals prepared by TEMPO-mediated oxidation of a-chitin. Biomacromolecules *9*, 192–198.

Fan, Y., Saito, T. and Isogai, A. (2008b). Preparation of chitin nanofibers from squid pen b-chitin by simple mechanical treatment under acid conditions. Biomacromolecules *9*, 1919–1923.

Fan, Y., Saito, T. and Isogai, A. (2010). Individual chitin nano-whiskers prepared from partially deacetylated a-chitin by fibril surface cationization. Carbohydr. Polym. *79*, 1046–1051.

Feng, L., Zhou, Z., Dufresne, A., Huang, J., Wei, M. and An, L. (2009). Structure and properties of new thermoforming bionanocomposites based on chitin whisker-graft-polycaprolactone. J. Appl. Polym. Sci. *112*, 2830–2837.

French, D. (1984). Organization of starch granules. In: Starch: Chemistry and technology, R.L. Whistler, J.N. BeMiller and E.F. Paschall eds. (Academic Press, New York), pp. 183–247.

Gaill, F., Persson, J., Sugiyama, J., Vuong, R. and Chanzy, H. (1992). The chitin system in the tubes of deep-sea hydrothermal vent worms. J. Struct. Biol. *109*, 116–128.

Gallant, D.J., Bouchet, B. and Baldwin, P.M. (1997). Microscopy of starch: Evidence of a new level of granule organization . Carbohydr. Polym. *32*, 177–191.

Garcia, N.L., Ribba, L., Dufresne, A., Aranguren, M.I. and Goyanes, S. (2009). Physico-mechanical properties of biodegradable starch nanocomposites. Macromol. Mater. Eng. *294*, 169–177.

Garcia, N.L., Ribba, L., Dufresne, A., Aranguren, M.I. and Goyanes, S. (2011). Effect of glycerol on the morphology of nanocomposites made from thermoplastic starch and starch nanocrystals. Carbohydr. Polym. *84*, 203–210.

Gérard, C., Planchot, V., Colonna, P. and Bertoft, E. (2000). Relationship between branching density and crystalline structure of A- and B-type maize mutant starches. Carbohydr. Res. *326*, 130–144.

Goodrich, J.D. and Winter, W.T. (2007). a-chitin nanocrystals prepared from shrimp shells and their specific surface area measurement. Biomacromolecules *8*, 252–257.

Gopalan Nair, K. and Dufresne, A. (2003a). Crab shell chitin whisker reinforced natural rubber nanocomposites. 1. Processing and swelling behaviour. Biomacromolecules *4*, 657–665.

Gopalan Nair, K. and Dufresne, A. (2003b). Crab shell chitin whisker reinforced natural rubber nanocomposites. 2. Mechanical behaviour. Biomacromolecules *4*, 666–674.

Gopalan Nair, K., Dufresne, A., Gandini, A. and Belgacem, M.N. (2003). Crab shell chitin whisker reinforced natural rubber nanocomposites. 3. Effect of chemical modification of chitin whiskers. Biomacromolecules *4*, 1835–1842.

Habibi, Y. and Dufresne, A. (2008). Highly filled bionanocomposites from functionalized polysaccharide nanocrystals. Biomacromolecules *9*, 1975–1980.

Hariraksapitak, P. and Supaphol, P. (2010). Preparation and properties of a-chitin-whisker-reinforced hyaluronangelatin nanocomposite scaffolds. J. Appl. Polym. Sci. *117*, 3406–3418.

Helbert, W. and Sugiyama, J. (1998). High resolution electron microscopy of cellulose II and a chitin single crystals. Cellulose *5*, 113–122.

Herth, W., Kuppel, A. and Schnepf, E. (1977). Chitinous fibrils in the lorica of the flagellate chrysophyte Poteriochromonas stipitata (syn. Ochromonas malhamensis). J. Cell. Biol. *73*, 311–321.

Hizukuri, S., Kaneko, T. and Takeda, Y. (1983). Measurement of the chain length of AMP and its relevance to the origin of crystalline polymorphism of starch granules. Biochim. Biophys. Acta *760*, 188–191.

Hizukuri, S. (1986). Polymodal distribution of the chain lengths of amylopectin and its significance. Carbohydr. Res. *147*, 342–347.

Huang, J., Zou, J.W., Chang, P.R., Yu, J.H. and Dufresne, A. (2011). New waterborne polyurethane-based nanocomposites reinforced with low loading levels of chitin whisker. eXPRESS Polym. Lett. *5*, 362–373.

Ifuku, S., Nogi, M., Abe, K., Yoshioka, M., Morimoto, M., Saimoto, H. and Yano, H. (2009). Preparation of chitin nanofibers with a uniform width as a-chitin from crab shells. Biomacromolecules *10*, 1584–1588.

Ifuku, S., Morooka, S., Morimoto, M., and Saimoto, H. (2010). Acetylation of chitin nanofibers and their transparent nanocomposite films. Biomacromolecules *11*, 1326–1330.

Imberty, A., Chanzy, H., Perez, S., Buléon, A. and Tran, V. (1987). New three-dimensional structure for A-type starch. Macromolecules *20*, 2634–2636.

Imberty, A. and Perez, S. (1988). A revisit to the three-dimensional structure of B-type starch. Biopolymers *27*, 1205–1221.

Jane, J.L., Wong, K.S. and McPherson, A.E. (1997). Branch structure differences in starches of A and B types X-ray patterns revealed by their Naegli dextrins. Carbohydr. Res. *300*, 219–227.

Jayakody, L. and Hoover, R. (2002). The effect of lintnerization on cereal starch granules. Food Res. Int. *35*, 665–680.

Jobling, S. (2004). Improving starch for food and industrial applications. Curr. Opin. Plant Biol. *7*, 210–218.

Junkasem, J., Rujiravanit, R. and Supaphol, P. (2006). Fabrication of a-chitin whisker-reinforced poly(vinyl alcohol) nanocomposite nanofibers by electrospinning. Nanotechnology *17*, 4519–4528.

Junkasem, J., Rujiravanit, R., Grady, B.P. and Supaphol, P. (2010). X-ray diffraction and dynamic mechanical analyses of α-chitin whisker-reinforced poly(vinyl alcohol) nanocomposite nanofibers. Polym. Int. *59*, 85–91.

Katz, J.R. (1930). Abhandlungen zur physikalischen Chemie der Stärke und der Brotbereitung. Z. Phys. Chem. *150*, 37–59.

Kristo, E. and Biliaderis, C.G. (2007). Physical properties of starch nanocrystal-reinforced pullulan films. Carbohydr. Polym. *68*, 146–158.

Labet, M., Thielemans, W. and Dufresne A. (2007). Polymer grafting onto starch nanocrystals. Biomacromolecules *8*, 2916–2927.

Le Corre, D., Bras, J. and Dufresne, A. (2010). Starch nanoparticles: A review. Biomacromolecules *11*, 1139–1153.

LeCorre, D., Bras, J. and Dufresne, A. (2011a). Evidence of micro and nano-scaled particles during starch nanocrystals preparation and their isolation. Biomacromolecules *12*, 3039–3046.

LeCorre, D., Bras, J. and Dufresne, A. (2011b). Ceramic membrane filtration for isolating starch nanocrystals. Carbohydr. Polym. *86*,1565–1572.

LeCorre, D., Bras, J. and Dufresne, A. (2011c). Influence of botanic orgin and amylose content on the morphology of starch nanocrystals. J. Nanopart. Res. *86*, 1565–1572.

Le Corre, D., Bras, J., Choisnard, L. and Dufresne, A. (2012a). Optimization of the batch preparation of starch nanocrystals to reach daily time-scale. Starch/Stärke *64*, 489–496.

LeCorre, D., Bras, J. and Dufresne, A. (2012b). Influence of native starch's properties on starch nanocrystals thermal properties. Carbohydr. Polym. *87*, 658–666.

LeCorre, D., Vahanian, E., Dufresne, A. and Bras, J. (2012c). Enzymatic pretreatment for preparing starch nanocrystals. Biomacromolecules *13*, 132–137.

Le Corre, D., Bras, J. and Dufresne, A. (2012d). Influence of starch nanocrystals' botanic origin on morphological and mechanical properties of natural rubber nanocomposites. Macromol. Mat. Eng. DOI:10.1002/mame.201100317.

Lintner, C.J. (1886). Studien über die Diastase. J. Prak. Chem. *34*, 378–386.

Lotmar, W. and Picken, L.E.R. (1950). A new crystallographic modification of chitin and its distribution. Experientia *6*, 58–59.

Liu, D., Wu, Q., Chen, H. and Chang, P.R. (2009). Transitional properties of micro- to nano-sized starch particles manufactured via a green process. J. Colloid Interface Sci. *339*, 117–124.

Lu, Y., Weng, L. and Zhang, L. (2004). Morphology and properties of soy protein isolate thermoplastics reinforced with chitin whiskers. Biomacromolecules *5*, 1046–1051.
Marchessault, R.H., Morehead, F.F. and Walter, N.M. (1959). Liquid crystal systems from fibrillar polysaccharides. Nature *184*, 632–633.
Mélé, P. Angellier-Coussy, H., Molina-Boisseau, S. and Dufresne, A. (2011). Reinforcing mechanisms of starch nanocrystals in a nonvulcanized natural rubber matrix. Biomacromolecules *12*, 1487–1493.
Morin, A. and Dufresne, A. (2002). Nanocomposites of chitin whiskers from *Riftia* tubes and poly(caprolactone). Macromolecules *35*, 2190–2199.
Muhr, A.H., Blanshard, J.M.V. and Bates, D.R. (1984). The effect of lintnerisation on wheat and potato starch granules. Carbohydr. Polym. *4*, 399–425
Nägeli, C.W. (1874). Beiträge zur näheren Kenntnis der Stärkegruppe. Annalen des Chemie *173*, 218–227.
Namazi, H. and Dadkhah, A. (2008). Surface modification of starch nanocrystals through ring-opening polymerization of ε-caprolactone and investigation of their microstructures. J. Appl. Polym. Sci. *110*, 2405–2412.
Nge, T.T., Hori, N., Takemura, A. and Ono, H. (2003). Phase behavior of liquid crystalline chitin/acrylic acid liquid mixture. Langmuir *19*, 1390-1395.
Oates, C.G. (1997). Towards an understanding of starch granule structure and hydrolysis. Trends Food Sci. Technol. *8*, 375–382.
Paillet, M. and Dufresne, A. (2001). Chitin whisker reinforced thermoplastic nanocomposites. Macromolecules *34*, 6527–6530.
Parker, R. and Ring, S.G. (2001). Aspects of the physical chemistry of starch. J. Cereal Sci. *34*, 1–17.
Peat, S., Whelan, W.J. and Thomas, G.J. (1952). Evidence of multiple branching in waxy maize starch. J. Chem. Soc. 4546–4548.
Persson, J.E., Domard, A. and Chanzy, H. (1992). Single crystals of a-chitin. Int. J. Biol. Macromol. *14*, 221–224.
Phongying, S., Aiba, S. and Chirachanchai, S. (2007). Direct chitosan nanoscaffold formation via chitin whiskers. Polymer *48*, 393–400.
Putaux, J.L., Molina-Boisseau, S., Momaur, T. and Dufresne, A. (2003). Platelet nanocrystals resulting from the disruption of waxy maize starch granules by acid hydrolysis. Biomacromolecules *4*, 1198–1202.
Putaux, J.L. (2005). Morphology and structure of crystalline polysaccharides: Some recent studies. Macromol. Symp. *229*, 66–71.
Revol, J.F. and Marchessault, R.H. (1993). In vitro chiral nematic ordering of chitin crystallites. Int. J. Biol. Macromol. *15*, 329–335.
Robin, J.P., Mercier, C., Charbonnière, R. and Guilbot, A. (1974). Lintnerized starches gel filtration and enzymatic studies of insoluble residues from prolonged acid treatment of potato starch. Cereal Chem. *51*, 389–406.
Robin, J.P., Mercier, C., Duprat, R.F., Charbonniere, R. and Guibot, A. (1975). Amidons lintnérisés. Etudes chromatographique et enzymatique des résidus insolubles provenant de l'hydrolyse chlorhydrique d'amidons de céréales, en particulier de maïs cireux. Die Stärke *27*, 36–45.
Robin, J.P. (1976). Comportement du grain d'amidon à l'hydrolyse acide ménagée. Etude physico-chimique et enzymatique de la fraction insoluble. Contribution à la connaissance de la structure de l'amylopectine. PhD Thesis, Université Pierre et Marie Curie de Paris.
Rudall, K.M. and Kenchington, W. (1973). The chitin system. Biol. Rev. *40*, 597–636.
Sakamoto, J., Sugiyama, J., Kimura, S., Imai, T., Itoh, T., Watanabe, T. and Kobayashi, S. (2000). Artificial chitin spherulites composed of single crystalline ribbons of alpha-chitin via enzymatic polymerization. Macromolecules *33*, 4155–4160.

Singh, V. and Ali, S.Z. (2000). Acid degradation of starch. The effect of acid and starch type. Carbohydr. Polym. *41*, 191–195.

Song, S., Wang, C., Pan, Z. and Wang, X. (2008). Preparation and characterization of amphiphilic starch nanocrystals. J. Appl. Polym. Sci. *107*, 418–422.

Sriupayo, J., Supaphol, P., Blackwell, J. and Rujiravanit, R. (2005a). Preparation and characterization of a-chitin whisker-reinforced chitosan nanocomposite films with or without heat treatment. Carbohydr. Polym. *62*, 130–136.

Sriupayo, J., Supaphol, P., Blackwell, J. and Rujiravanit, R. (2005b). Preparation and characterization of a-chitin whisker-reinforced poly(vinyl alcohol) nanocomposite films with or without heat treatment. Polymer *46*, 5637–5644.

Tang, H., Mitsunaga, T. and Kawamura, Y. (2006). Molecular arrangement in blocklets and starch granule architecture. Carbohydr. Polym. *63*, 555–560.

Thielemans, W., Belgacem, M.N. and Dufresne, A. (2006). Starch nanocrystals with large chain surface modifications. Langmuir *22*, 4804–4810.

Tzoumaki, M.V., Moschakis, T. and Biliaderis, C.G. (2010). Metastability of nematic gels made of aqueous chitin nanocrystal dispersions. Biomacromolecules *11*, 175–181.

Vandeputte, G.E. and Delcour, J.A. (2004). From sucrose to starch granule to starch physical behaviour: a focus on rice starch. Carbohydr. Polym. *58*, 245–266.

Viguié, J., Molina-Boisseau, S. and Dufresne, A. (2007). Processing and characterization of waxy maize starch films plasticized by sorbitol and reinforced with starch nanocrystals. Macromol. Biosci. *7*, 1206–1216.

Vincent, J.F.V. and Wegst, U.G.K. (2004). Design and mechanical properties of insect cuticle. Arthropod Structure and Development *33*, 187–199.

Vorwerg, W., Radosta, S. and Leibnitz, E. (2002). Study of a preparative-scale process for the production of amylose. Carbohydr. Polym. *47*, 181–189.

Wang, S., Yu, J., Jin, F. and Yu, J. (2008a). The new insight on ultrastructure of C-type starch granules revealed by acid hydrolysis. Int. J. Biol. Macromol. *43*, 216–220.

Wang, S., Yu, J., and Yu, J. (2008b). The semi-crystalline growth rings of C-type pea starch granule revealed by SEM and HR-TEM during acid hydrolysisThe semi-crystalline growth rings of C-type pea starch granule revealed by SEM and HR-TEM during acid hydrolysis. Carbohydr. Polym. *74*, 731–739.

Wang, Y., Tian, H. and Zhang, L. (2010). Role of starch nanocrystals and cellulose whiskers in synergistic reinforcement of waterborne polyurethane. Carbohydr. Polym. *80*, 665–671.

Watthanaphanit, A., Supaphol, P., Tamura, H., Tokura, S. and Rujiravanit, R. (2008). Fabrication, structure and properties of chitin whisker-reinforced alginate nanocomposite fibers. J. Appl. Polym. Sci. *110*, 890–899.

Watthanaphanit, A., Supaphol, P., Tamura, H., Tokura, S. and Rujiravanit, R. (2010). Wet-spun alginate/chitosan whiskers nanocomposite fibers: Preparation, characterization and release characteristic of the whiskers. Carbohydr. Polym. *79*, 738–746.

Wongpanit, P., Sanchavanakit, N., Pavasant, P., Bunaprasert, T., Tabata, Y. and Rujiravanit, R. (2007). Preparation and characterization of chitin whisker-reinforced silk fibroin nanocomposite sponges. Eur. Polym. J. *43*, 4123–4135.

Wu, X., Torres, F.G., Vilaseca, F. and Peijs, T. (2007). Influence of the processing conditions on the mechanical properties of chitin whisker reinforced poly(caprolactone) nanocomposites. J. Biobased Mat. Bioenergy *1*, 341–350.

Yu, J., Ai, F., Dufresne, A., Gao, S., Huang, J., Chang, P.R. (2008). Structure and mechanical properties of poly(lactic acid) filled with (starch nanocrystal)-graft-poly(e-caprolactone). Macromol. Mater. Eng. *293*, 763–770.

Zeng, J.B., He, Y.S., Li, S.L. and Wang, Y.Z. (2012). Chitin whiskers: An overview. Biomacromolecules *13*, 1–11.

Zheng, H., Ai, F., Chang, P.R., Huang, J. and Dufresne A. (2009). Structure and properties of starch nanocrystal-reinforced soy protein plastics. Polym. Compos. *30*, 474–480.
Zobel, H.F. (1988). Starch crystal transformation and their industrial importance. Starch/Sträke *40*, 1–7.

12 Conclusions, applications and likely future trends

Our industrialized world is driven in large part by petroleum. It is an important source of energy and many of the products that we depend upon for day-to-day living are derived from it. However, the cost of oil and gas has increased dramatically over the last decade and this upward spiral is expected to continue due to increased demand, finite quantities, unreliable supply chains, and other factors. In addition to the monetary cost, our petroleum-based economy comes at a significant environmental price that cannot be sustained. It is now widely recognized that more cost effective and environmentally benign alternatives to petroleum and the products derived from it will be required in order to realize a future with a sustainable economy and environment.

Abundant and renewable plant resources are obvious candidates. At a time when the world finally focuses on the question of the product's end-life and green design, we can hope that scientists and industries will exploit this new "green gold" by including also the values of sustainability, material life cycle, recyclability, and durability.

Despite being the most available natural polymer on earth, it is only quite recently that cellulose has gained prominence as a nanostructured material, in the form of cellulose nanocrystals and microfibrillated cellulose (MFC). However, the pioneering works were initiated a long time ago, 1947 for cellulose nanocrystals (Nickerson and Habrle, 1947) and 1983 (Herrick et al., 1983; Turbak et al., 1983) for microfibrillated cellulose. They have been academic curiosities for many years. Technological development depends on economic conditions. Companies primarily focus on their own interests and dig into the science where it seems financially attractive. Today there is a substantial amount of research on these nanocellulosic materials, and commercial development is now underway with some very promising applications.

Recently, a team of industry volunteers developed a four-minute video, available from TAPPI, to help raise awareness of the potential of trees in the nano world. They state that "We have been making useful products from trees for years. But now we are going even deeper into the tree for a new generation of products. Using the science of nanotechnology, we are unlocking the tiny secrets nature has held for centuries" This video is available for viewing and downloading (http://www.tappi.org/Groups/Divisions/Nanotechnology/nanovideo.aspx).

Now, after intensive research, several initiatives have emerged in the perspective of producing nanocellulose at large scale. A number of organizations have announced nanocellulose demonstration plants. Some of them are reported in Table 12.1. This development has been driven in most cases by economic reasons and to raise public awareness. Sustainability and industrial ecology are probably additional factors. The forestry industry in traditional locations such as Canada and Scandinavia is going through a major transition, and is strongly affected by the growing competition from emerging economies in Asia and South America. The economic struggle has simply become uneven. The utilization of new technologies is expected to provide a means for strengthening the competitiveness in the sector.

Organization	WEB Site	Comment
Bio Vision (Canada)	www.biovisiontechnology.com	Production of NANOCEL™ (carboxylated cellulose nanocrystals)
Borregaard Chemcell (Norway)	www.chemcell.com	Production of microfibrillated cellulose
CelluForce (Canada)	www.celluforce.com	Production of cellulose nanocrystals, joint venture between FPInnovations and Domtar, building a one ton/day demonstration plant
Centre Technique du Papier (France)	www.webctp.com	Production of microfibrillated cellulose
Engineered Fibers Technology (USA)	www.eftfibers.com	Production of nanofibrillated fibers from natural and synthetic materials
Jenpolymers (Germany)	www.jenpolymers.de	Production of bacterial cellulose for medical and high performance applications
Melodea (Israël)	www.melodea.eu	Production of cellulose nanocrystals from the sludge of paper production and bio-based resin composites
Rettenmaier (Germany)	www.jrs.de	
Stora Enso (Finland)	www.storaenso.com	Production of microfibrillated cellulose (pre-commercial plant at Imatra)
UPM Fibril cellulose (Finland)	www.upm.com/fibrilcellulose	Pre-commercial production of UPM Biofibrils in Otaniemi, Espoo (microfibrillated cellulose)

Table 12.1: Organizations that have announced nanocellulose demonstration plants.

Nanocellulose is being developed for use over a broad range of applications, even if a high number of unknowns remain at date. Tens of scientific publications and experts show its potential even if most of the studies focus on its mechanical properties as reinforcing phase and its liquid crystal self-ordering properties. The sound markets impacted by nanocellulose include composites, electronics (flexible circuits), energy (flexible batteries, such as Li-ion batteries, and solar panels), packaging, coatings (paints, and varnishes), detergents, adhesives, construction, pulp and paper, inks and printing, filtration, medicine and life science (scaffolds in tissue engineering, artificial skin and cartilage, wound healing, and vessel substitutes), optical devices (including reflective properties for security papers and UV or IR reflective barriers), rheological modifiers, cosmetics, and aerogels. Broadly, one can consider that the use of MFC is more related to the paper and coating applications, whereas cellulose nanocrystals are more composite oriented. The interest of paper and packaging industries for MFC is quite recent. In the field of polymer nanocomposites, nanocellulose offers the opportunity to process stiff thin films. However, the serious issues of dispersion in the polymer melt and development of suitable processing technologies for large scale have to be solved to broaden the applicability of these nanoparticles. The strong interactions between cellulose nanoparticles through hydrogen-bonding are beneficial to exploit their full potential and reach the highest mechanical reinforcement effect that can be obtained from these nanoparticles. At the same time, it limits their dispersion within a continuous medium.

The potential of nanotechnology and nanocomposites in various sectors of research and application is promising and attracting increasing investment. In addition, due to their abundance, renewability, high strength and stiffness, non-toxicity, low weight and biodegradability, nano-scale cellulose fiber materials serve as promising candidates for the preparation of bionanocomposites and other nano-devices. The mechanical modulus of crystalline cellulose is the basis of many potential applications. Moreover, the low thermal expansion coefficient caused by the high crystallinity of cellulose nanoparticles and high transparency without the presence of any existing polymers is highly advantageous for flexible display panels and electronic devices. For papermaking, in addition to improving the tensile strength, burst strength, tear, density, smoothness and also increasing the air permeability, the capacity of retaining the filler and the adsorption of a dye are also improved by the nanoparticles. Besides, the inherent high reactivity of cellulose and the pervasive surface

Reference	Focus/Content
(Azizi Samir et al., 2005)	First review on cellulose nanocrystals
(Kamel, 2007)	Nanocellulose
(Hubbe et al., 2008)	Cellulose nanocrystals and microfibrillated cellulose
(Ioelovich, 2008)	Cellulose nanocrystals and microfibrillated cellulose
(Eichhorn et al., 2010)	Current international research presented by a broad panel of experts
(Siró and Plackett, 2010)	Microfibrillated cellulose
(Dufresne, 2010)	Processing of nanocellulose-based nanocomposites
(Siqueira et al., 2010)	Nanocellulose
(Habibi et al., 2010)	Cellulose nanocrystals
(Ramires and Dufresne, 2011)	Cellulose nanocrystals
(Klemm et al., 2011)	Cellulose nanocrystals, microfibrillated cellulose and bacterial cellulose
(Moon et al., 2011)	Cellulose nanocrystals
(Eichhorn, 2011)	Cellulose nanocrystals
(Chinga-Carrasco, 2011)	Microfibrillated cellulose
(Lin et al., 2012)	Functional properties of nanocellulose
(Abdul Khalil et al., 2012)	Green cellulose nanocomposites
(Lavoine et al., 2012)	Microfibrillated cellulose for packaging and paper applications

Table 12.2: Reviews on nanocellulose and related materials.

hydroxyl groups associated with the nano-scale dimensions of cellulose nanocrystals and MFC open up new opportunities to develop new functional nanomaterials.

Properties of nanocellulose and related materials, and potential foreshadowed applications have been extensively reviewed during the last years. Some of them are reported in Table 12.2 and can be used to deepen some specific skills.

The future will tell if this sudden interest for nanocellulose was only a flash in the pan or a real corner in material science.

12.1 References

Abdul Khalil, H.P.S., Bhat, A.H. and Ireana Yusra, A.F. (2012). Green composites from sustainable cellulose nanofibrils: A review. Carbohydr. Polym. *87*, 963– 979.

Azizi Samir, M.A.S., Alloin, F. and Dufresne, A. (2005). A review of recent research into cellulosic whiskers, their properties and their application in nanocomposite field. Biomacromolecules, *6*, 612–626.

Chinga-Carrasco, G. (2011). Cellulose fibres, nanofibrils and microfibrils: The morphological sequence of MFC components from a plant physiology and fibre technology point of view. Nanoscale Res. Lett. *6*, 417.

Dufresne, A. (2010). Processing of polymer nanocomposites reinforced with polysaccharide nanocrystals. Molecules *15*, 4111–4128.

Eichhorn, S.J., Dufresne, A., Aranguren, M., Marcovich, N.E., Capadona, J.R., Rowan, S.R., Weder, C., Thielemans, W., Roman, M., Renneckar, S., Gindl, W., Veigel, H., Yano, K., Abe, M., Nogi, A.N., Nakagaito, A., Mangalam, J., Simonsen, A.S., Benight, S., Bismarck, A., Berglund, L.A. and Peijs, T. (2010). Review: Current international research into cellulose nanofibres and nanocomposites. J. Mat. Sci. *45*, 1–33.

Eichhorn, S.J. (2011). Cellulose nanowhiskers: Promising materials for advanced applications. Soft Matter *7*, 303–315.

Habibi, Y., Lucia, L.A. and Rojas, O.J. (2010). Cellulose nanocrystals: Chemistry, self-assembly, and applications. Chem. Rev. *110*, 3479–3500.

Herrick, F.W., Casebier, R.L., Hamilton, J.K. and Sandberg, K.R. (1983). Microfibrillated cellulose: Morphology and accessibility. J. Appl. Polym. Sci. Polym. Symp. *37*, 797–813.http://www.tappi.org/Groups/Divisions/Nanotechnology/nanovideo.aspx (consulted July 8, 2012).

Hubbe, M.A., Rojas, O.J., Lucia., L.A., and Sain, M. (2008). Cellulose nanocomposites: A review. BioResources *3*, 929–980.

Ioelovich, M. (2008). Cellulose as a nanostructured polymer: A short review. BioResources *3*, 1403–1418.

Kamel, S. (2007). Nanotechnology and its applications in lignocellulosic composites, A mini review. eXPRESS Polym. Lett. *1*, 546–575.

Klemm, D., Kramer, F., Moritz, S., Lindström, T., Ankerfors, M., Gray, D. and Dorris, A. (2011). Nanocelluloses: A new family of nature-based materials. Angew. Chem. Int. Ed. *50*, 5438–5466.

Lavoine, N., Desloges, I., Dufresne, A. and Bras, J. (2012). Microfibrillated cellulose - Its barrier properties and applications in cellulosic materials: A review. Carbohydr. Polym. *90*, 735–764.

Lin, N., Huang, J. and Dufresne A. (2012). Preparations, properties and applications of polysaccharide nanocrystals in advanced functional nanomaterials: A review. Nanoscale *4*, 3274–3294.

Moon, R.J., Martini, A., Nairn, J., Simonsen, J. and Youngblood, J. (2011). Cellulose nanomaterials review: Structure, properties and nanocomposites. Chem. Soc. Rev. *40*, 3941–3994.

Nickerson, R.F. and Habrle, J.A. (1947). Cellulose intercrystalline structure. Ind. Eng. Chem. *39*, 1507–1512.

Ramires, E.C. and Dufresne, A. (2011). A review of cellulose nanocrystals and nanocomposites. Tappi J. *10*, 9–16.

Siqueira, G., Bras, J. and Dufresne, A. (2010). Cellulosic bionanocomposites: Preparation, properties and applications. A review. Polymers *2*, 728–765.

Siró, I. and Plackett, D. (2010). Microfibrillated cellulose and new nanocomposite materials: A review. Cellulose *17*, 459–494.

Turbak, A.F., Snyder, F.W. and Sandberg, K.R. (1983). Microfibrillated cellulose: A new cellulose product: properties, uses, and commercial potential. J. Appl. Polym. Sci. Polym. Symp. *37*, 815–827.

13 Index

Bei Fragen zur Produktsicherheit wenden Sie sich bitte an:
If you have any questions regarding product safety,
please contact:

Walter de Gruyter GmbH
Genthiner Straße 13
10785 Berlin
productsafety@degruyterbrill.com